Methods in Enzymology

Volume 212
DNA STRUCTURES

Part B
Chemical and Electrophoretic Analysis of DNA

METHODS IN ENZYMOLOGY

EDITORS-IN-CHIEF

John N. Abelson   Melvin I. Simon

DIVISION OF BIOLOGY
CALIFORNIA INSTITUTE OF TECHNOLOGY
PASADENA, CALIFORNIA

FOUNDING EDITORS

Sidney P. Colowick and Nathan O. Kaplan

Methods in Enzymology

Volume 212

DNA Structures

Part B

Chemical and Electrophoretic Analysis of DNA

EDITED BY

David M. J. Lilley

DEPARTMENT OF BIOCHEMISTRY
THE UNIVERSITY
DUNDEE, SCOTLAND

James E. Dahlberg

DEPARTMENT OF BIOMOLECULAR CHEMISTRY
UNIVERSITY OF WISCONSIN-MADISON
MADISON, WISCONSIN

ACADEMIC PRESS, INC.
Harcourt Brace Jovanovich, Publishers
San Diego   New York   Boston
London   Sydney   Tokyo   Toronto

Academic Press, Inc.
1250 Sixth Avenue, San Diego, California 92101-4311

United Kingdom Edition published by
Academic Press Limited
24–28 Oval Road, London NW1 7DX

Library of Congress Catalog Number: 54-9110

International Standard Book Number: 0-12-182113-7

PRINTED IN THE UNITED STATES OF AMERICA
92 93 94 95 96 97   EB   9 8 7 6 5 4 3 2 1

Table of Contents

Section I. Gel Electrophoresis and Topological Methods for Analysis of DNA Structure

Section II. Probing DNA Structure *in Vitro*

Section III. Analysis of DNA Structure inside Cells

Section IV. Interaction of DNA and Proteins

Contributors to Volume 212

Article numbers are in parentheses following the names of contributors.
Affiliations listed are current.

FAREED ABOUL-ELA (5), *Department of Biochemistry, The University, Dundee DD1 4HN, Scotland*

EZRA S. ABRAMS (4), *Department of Biology, Massachusetts Institute of Technology, Cambridge, Massachusetts 02139*

JACQUELINE K. BARTON (12), *Department of Chemistry, California Institute of Technology, Pasadena, California 91125*

MICHAEL M. BECKER (14), *Institute of Biosciences and Technology, Texas A&M University, Houston, Texas 77030*

RICHARD BOWATER (5), *Department of Biochemistry, The University, Dundee DD1 4HN, Scotland*

CHRISTINE S. CHOW (12), *Department of Chemistry, California Institute of Technology, Pasadena, California 91125*

NICHOLAS R. COZZARELLI (6), *Department of Molecular and Cell Biology, University of California, Berkeley, Berkeley, California 94720*

DONALD M. CROTHERS (1, 3), *Department of Chemistry, Yale University, New Haven, Connecticut 06511*

STEPHAN DIEKMANN (2), *Abteilung Molekulare Biologie, Max-Planck-Institut für Biophysikalische Chemie am Fassberg, D-3400 Göttingen, Germany*

JACQUELINE DRAK (1, 3), *Department of Molecular Biology and Biochemistry, Harvard University, Cambridge, Massachusetts 02139*

PETER DRÖGE (6, 22), *Department of Biology, University of Konstanz, D-7750 Konstanz, Germany*

GREGORY GROSSMANN (14), *Department of Biochemistry, Case Western Reserve University, Cleveland, Ohio 44106*

ROBERT W. HOEPFNER (13), *Department of Molecular Genetics, Biochemistry and Microbiology, University of Cincinnati College of Medicine, Cincinnati, Ohio 45267*

HAN HTUN (15), *Division of Chemistry, California Institute of Technology, Pasadena, California 91125*

BRIAN H. JOHNSTON (10, 15), *Cell and Molecular Biology Laboratory, SRI International, Menlo Park, California 94025*

JASON D. KAHN (1), *Department of Chemistry, Yale University, New Haven, Connecticut 06511*

Y. KOHWI (9), *Cancer Research Center, La Jolla Cancer Research Foundation, La Jolla, California 92037*

T. KOHWI-SHIGEMATSU (9), *Cancer Research Center, La Jolla Cancer Research Foundation, La Jolla, California 92037*

STEPHEN D. LEVENE (1), *Department of Molecular and Cell Biology, University of California, Berkeley, Berkeley, California 94720*

DAVID M. J. LILLEY (5, 7), *Department of Biochemistry, The University, Dundee DD1 4HN, Scotland*

LEROY F. LIU (20), *Department of Biological Chemistry, The Johns Hopkins School of Medicine, Baltimore, Maryland 21205*

TIMOTHY M. LOHMAN (24, 25), *Department of Biochemistry and Molecular Biophysics, Washington University School of Medicine, St. Louis, Missouri 63110*

DAVID P. MASCOTTI (24, 25), *Department of Biochemistry and Molecular Biophysics, Washington University School of Medicine, St. Louis, Missouri 63110*

ALFRED NORDHEIM (22), *Institute for Molecular Biology, Hannover Medical School, D-3000 Hannover 61, Germany*

EMIL PALEČEK (8, 17), *Institute of Biophysics, Czechoslovak Academy of Sciences, 612 65 Brno, Czechoslovakia*

MARY ANN PRICE (11), *Department of Chemistry, The Johns Hopkins University, Baltimore, Maryland 21218*

DAVID G. SANFORD (21), *Department of Biochemistry, Tufts University School of Medicine, Boston, Massachusetts 02111*

RICHARD R. SINDEN (13, 18), *Institute of Biosciences and Technology, Texas A&M University, Houston, Texas 77030*

VINCENT P. STANTON, JR. (4), *Center for Cancer Research, Massachusetts Institute of Technology, Cambridge, Massachusetts 02139*

B. DAVID STOLLAR (21), *Department of Biochemistry, Tufts University School of Medicine, Boston, Massachusetts 02111*

ANDREW A. TRAVERS (23), *MRC Laboratory of Molecular Biology, Cambridge CB2 2QH, England*

THOMAS D. TULLIUS (11), *Department of Chemistry, The Johns Hopkins University, Baltimore, Maryland 21218*

WILLIAM G. TURNELL (23), *MRC Laboratory of Molecular Biology, Cambridge CB2 2QH, England*

DAVID W. USSERY (13, 18), *Institute of Biosciences and Technology, Texas A&M University, Houston, Texas 77030*

FRANZ WOHLRAB (16), *Department of Biochemistry, University of Alabama at Birmingham, Birmingham, Alabama 35294*

HAI-YOUNG WU (20), *Department of Pharmacology, Wayne State University School of Medicine, Detroit, Michigan 48201*

W. ZACHARIAS (19), *Department of Biochemistry, University of Alabama at Birmingham, Birmingham, Alabama 35294*

Preface

The structure of DNA has been known in general terms for about forty years now. Yet in the last decade we have become much more attuned to its subtleties and how they depend on the interplay of local base sequence, environmental conditions, covalent modification, and topology. For this reason the use of the plural form for structure in the titles of Volumes 211 and 212 is significant.

Structural alterations of DNA may be relatively subtle—essentially small (though very important, nevertheless) modifications to the standard B-DNA structure—or they may be very great. Until the late 1970s DNA was regarded as an invariant, straight, right-handed B-form double helix. In the intervening period DNA structures have been described that depart from virtually all the conventional patterns. DNA can be left-handed (Z-DNA), and base pairing can be rearranged (cruciform) or remodeled (H-triplex). The axis can be bent or kinked, and triple and quadruple helices are possible. This has changed our perspective of DNA structure to one of a highly polymorphic molecule. Not all of the structures that are possible may be biologically relevant, but some undoubtedly are. Some structures may even be conformational traps that must be avoided. But it is clear that if we are to understand the structures and dynamics of DNA and its interactions with ligands, a complete analysis of its conformational flexibility is required.

Progress in a number of areas had led to our current view of DNA as a more structurally adventurous molecule. Initially it was the advent of new methods for the synthesis of oligonucleotides of defined sequence and high purity and the consequent crystallographic analysis of a number of DNA structures at near atomic resolution that revealed the sequence dependence of DNA structure. This also gave us the first major surprise— left-handed Z-DNA. Short, well-defined DNA fragments may also be studied by many other physical techniques, and nuclear magnetic resonance spectroscopy has been particularly useful.

But in parallel with these general methods, a number of powerful techniques that are more specific to DNA have emerged, including the ability to synthesize any DNA sequence of interest, the facility to clone these sequences into plasmids, and the opportunity to manipulate DNA as a closed circle. Many of the perturbed structures that DNA can adopt are stabilized by negative supercoiling, and this can be exploited to analyze DNA structure and energetics. Thus band-shift methods and two-dimensional gel electrophoresis are now important methods for the study of

DNA structure. In addition, probing methods have become vital to the analysis of local structure in DNA, and an armory of chemical and enzyme probes are available for this task. Chemical probes are now being extended to the analysis of DNA inside the cell, an important aspect of their use if we are to assess the biological role of DNA structural polymorphism.

In view of the amount of activity in this area in the last decade and the new methods developed, we felt it was time to collect all the methods and approaches in one place, and these volumes are the result. Originally it was our intention to produce a single volume, which, thanks to the hard work of all our authors, eventually became two. We are most grateful to all the authors for their efforts, and we hope that their reward will be the existence of these books that should be a valuable resource for all of us in the field. We also thank the staff of Academic Press for their assistance in the preparation of these volumes.

DAVID M. J. LILLEY
JAMES E. DAHLBERG

METHODS IN ENZYMOLOGY

VOLUME XXVII. Enzyme Structure (Part D)
Edited by C. H. W. HIRS AND SERGE N. TIMASHEFF

VOLUME XXVIII. Complex Carbohydrates (Part B)
Edited by VICTOR GINSBURG

VOLUME XXIX. Nucleic Acids and Protein Synthesis (Part E)
Edited by LAWRENCE GROSSMAN AND KIVIE MOLDAVE

VOLUME XXX. Nucleic Acids and Protein Synthesis (Part F)
Edited by KIVIE MOLDAVE AND LAWRENCE GROSSMAN

VOLUME XXXI. Biomembranes (Part A)
Edited by SIDNEY FLEISCHER AND LESTER PACKER

VOLUME XXXII. Biomembranes (Part B)
Edited by SIDNEY FLEISCHER AND LESTER PACKER

VOLUME XXXIII. Cumulative Subject Index Volumes I–XXX
Edited by MARTHA G. DENNIS AND EDWARD A. DENNIS

VOLUME XXXIV. Affinity Techniques (Enzyme Purification: Part B)
Edited by WILLIAM B. JAKOBY AND MEIR WILCHEK

VOLUME XXXV. Lipids (Part B)
Edited by JOHN M. LOWENSTEIN

VOLUME XXXVI. Hormone Action (Part A: Steroid Hormones)
Edited by BERT W. O'MALLEY AND JOEL G. HARDMAN

VOLUME XXXVII. Hormone Action (Part B: Peptide Hormones)
Edited by BERT W. O'MALLEY AND JOEL G. HARDMAN

VOLUME XXXVIII. Hormone Action (Part C: Cyclic Nucleotides)
Edited by JOEL G. HARDMAN AND BERT W. O'MALLEY

VOLUME XXXIX. Hormone Action (Part D: Isolated Cells, Tissues, and
Organ Systems)
Edited by JOEL G. HARDMAN AND BERT W. O'MALLEY

VOLUME XL. Hormone Action (Part E: Nuclear Structure and Function)
Edited by BERT W. O'MALLEY AND JOEL G. HARDMAN

VOLUME XLI. Carbohydrate Metabolism (Part B)
Edited by W. A. WOOD

VOLUME XLII. Carbohydrate Metabolism (Part C)
Edited by W. A. WOOD

VOLUME XLIII. Antibiotics
Edited by JOHN H. HASH

VOLUME XLIV. Immobilized Enzymes
Edited by KLAUS MOSBACH

VOLUME XLV. Proteolytic Enzymes (Part B)
Edited by LASZLO LORAND

VOLUME XLVI. Affinity Labeling
Edited by WILLIAM B. JAKOBY AND MEIR WILCHEK

VOLUME XLVII. Enzyme Structure (Part E)
Edited by C. H. W. HIRS AND SERGE N. TIMASHEFF

VOLUME XLVIII. Enzyme Structure (Part F)
Edited by C. H. W. HIRS AND SERGE N. TIMASHEFF

VOLUME XLIX. Enzyme Structure (Part G)
Edited by C. H. W. HIRS AND SERGE N. TIMASHEFF

VOLUME L. Complex Carbohydrates (Part C)
Edited by VICTOR GINSBURG

VOLUME LI. Purine and Pyrimidine Nucleotide Metabolism
Edited by PATRICIA A. HOFFEE AND MARY ELLEN JONES

VOLUME LII. Biomembranes (Part C: Biological Oxidations)
Edited by SIDNEY FLEISCHER AND LESTER PACKER

VOLUME LIII. Biomembranes (Part D: Biological Oxidations)
Edited by SIDNEY FLEISCHER AND LESTER PACKER

VOLUME LIV. Biomembranes (Part E: Biological Oxidations)
Edited by SIDNEY FLEISCHER AND LESTER PACKER

VOLUME 81. Biomembranes (Part H: Visual Pigments and Purple Membranes, I)
Edited by LESTER PACKER

VOLUME 82. Structural and Contractile Proteins (Part A: Extracellular Matrix)
Edited by LEON W. CUNNINGHAM AND DIXIE W. FREDERIKSEN

VOLUME 83. Complex Carbohydrates (Part D)
Edited by VICTOR GINSBURG

VOLUME 84. Immunochemical Techniques (Part D: Selected Immunoassays)
Edited by JOHN J. LANGONE AND HELEN VAN VUNAKIS

VOLUME 85. Structural and Contractile Proteins (Part B: The Contractile Apparatus and the Cytoskeleton)
Edited by DIXIE W. FREDERIKSEN AND LEON W. CUNNINGHAM

VOLUME 86. Prostaglandins and Arachidonate Metabolites
Edited by WILLIAM E. M. LANDS AND WILLIAM L. SMITH

VOLUME 87. Enzyme Kinetics and Mechanism (Part C: Intermediates, Stereochemistry, and Rate Studies)
Edited by DANIEL L. PURICH

VOLUME 88. Biomembranes (Part I: Visual Pigments and Purple Membranes, II)
Edited by LESTER PACKER

VOLUME 89. Carbohydrate Metabolism (Part D)
Edited by WILLIS A. WOOD

VOLUME 90. Carbohydrate Metabolism (Part E)
Edited by WILLIS A. WOOD

VOLUME 91. Enzyme Structure (Part I)
Edited by C. H. W. HIRS AND SERGE N. TIMASHEFF

VOLUME 92. Immunochemical Techniques (Part E: Monoclonal Antibodies and General Immunoassay Methods)
Edited by JOHN J. LANGONE AND HELEN VAN VUNAKIS

VOLUME 128. Plasma Lipoproteins (Part A: Preparation, Structure, and Molecular Biology)
Edited by JERE P. SEGREST AND JOHN J. ALBERS

VOLUME 129. Plasma Lipoproteins (Part B: Characterization, Cell Biology, and Metabolism)
Edited by JOHN J. ALBERS AND JERE P. SEGREST

VOLUME 130. Enzyme Structure (Part K)
Edited by C. H. W. HIRS AND SERGE N. TIMASHEFF

VOLUME 131. Enzyme Structure (Part L)
Edited by C. H. W. HIRS AND SERGE N. TIMASHEFF

VOLUME 132. Immunochemical Techniques (Part J: Phagocytosis and Cell-Mediated Cytotoxicity)
Edited by GIOVANNI DI SABATO AND JOHANNES EVERSE

VOLUME 133. Bioluminescence and Chemiluminescence (Part B)
Edited by MARLENE DELUCA AND WILLIAM D. MCELROY

VOLUME 134. Structural and Contractile Proteins (Part C: The Contractile Apparatus and the Cytoskeleton)
Edited by RICHARD B. VALLEE

VOLUME 135. Immobilized Enzymes and Cells (Part B)
Edited by KLAUS MOSBACH

VOLUME 136. Immobilized Enzymes and Cells (Part C)
Edited by KLAUS MOSBACH

VOLUME 137. Immobilized Enzymes and Cells (Part D)
Edited by KLAUS MOSBACH

VOLUME 138. Complex Carbohydrates (Part E)
Edited by VICTOR GINSBURG

VOLUME 139. Cellular Regulators (Part A: Calcium- and Calmodulin-Binding Proteins)
Edited by ANTHONY R. MEANS AND P. MICHAEL CONN

VOLUME 140. Cumulative Subject Index Volumes 102–119, 121–134

VOLUME 204. Bacterial Genetic Systems
Edited by JEFFREY H. MILLER

VOLUME 205. Metallobiochemistry (Part B: Metallothionein and Related Molecules)
Edited by JAMES F. RIORDAN AND BERT L. VALLEE

VOLUME 206. Cytochrome P450
Edited by MICHAEL R. WATERMAN AND ERIC F. JOHNSON

VOLUME 207. Ion Channels
Edited by BERNARDO RUDY AND LINDA E. IVERSON

VOLUME 208. Protein–DNA Interactions
Edited by ROBERT T. SAUER

VOLUME 209. Phospholipid Biosynthesis
Edited by EDWARD A. DENNIS AND DENNIS E. VANCE

VOLUME 210. Numerical Computer Methods
Edited by LUDWIG BRAND AND MICHAEL L. JOHNSON

VOLUME 211. DNA Structures (Part A: Synthesis and Physical Analysis of DNA)
Edited by DAVID M. J. LILLEY AND JAMES E. DAHLBERG

VOLUME 212. DNA Structures (Part B: Chemical and Electrophoretic Analysis of DNA)
Edited by DAVID M. J. LILLEY AND JAMES E. DAHLBERG

VOLUME 213. Carotenoids (Part A: Chemistry, Separation, Quantitation, and Antioxidation) (in preparation)
Edited by LESTER PACKER

VOLUME 214. Carotenoids (Part B: Metabolism, Genetics, and Biosynthesis) (in preparation)
Edited by LESTER PACKER

VOLUME 215. Platelets: Receptors, Adhesion, Secretion (Part B) (in preparation)
Edited by JACEK HAWIGER

Section I

Gel Electrophoresis and Topological Methods for Analysis of DNA Structure

[1] DNA Bending, Flexibility, and Helical Repeat by Cyclization Kinetics

By DONALD M. CROTHERS, JACQUELINE DRAK, JASON D. KAHN, and STEPHEN D. LEVENE

Introduction

Visualizing DNA as a simple linear double helix is a serious oversimplification of its probable functional shape. DNA in a cell is packaged by histones and bent by regulatory proteins into three-dimensional shapes whose structures are largely unknown at this time. Crystallization of complexes containing multiple regulatory proteins to determine the path of the DNA helix in transcriptional or replicational complexes remains a distant vision. However, many aspects of the DNA helix axis trajectory can now be investigated quantitatively by measurement of the apparent concentration of one end of a DNA fragment about the other. For a small, stiff DNA molecule this concentration is very small, but it can be dramatically increased by introducing a bend into the helix at one or more points.

A simple way to measure the effective end-about-end concentration is to determine the probability of helix formation between complementary sequences at the two ends, whose hybridization produces a circular DNA molecule. Formation of such a helical segment is greatly dependent not only on the end-to-end distance distribution function, but also on colinearity of the two ends and on correct helical phasing of the two terminal segments. These factors in turn depend critically on the bending and torsional stiffness of DNA, on its length and helical repeat, and on any bends which may have been produced by sequence features or bound ligands.

Study of the cyclization of polymer molecules has a long history,[1] but it is the application of DNA ligase-catalyzed reactions to the problem that has elicited considerable current interest in the method.[2-5] This approach relies on detection of a small, steady-state concentration of molecules cyclized transiently by formation of a short double-helical segment between the ends. Such structures are substrates for DNA ligase, which

[1] H. Jacobson and W. H. Stockmayer, *J. Chem. Phys.* **18,** 1600 (1950).
[2] D. Shore, J. Langowski, and R. L. Baldwin, *Proc. Natl. Acad. Sci. U.S.A.* **78,** 4833 (1981).
[3] D. Shore and R. L. Baldwin, *J. Mol. Biol.* **170,** 957 (1983).
[4] W. H. Taylor and P. J. Hagerman, *J. Mol. Biol.* **212,** 363 (1990).
[5] H.-S. Koo, J. Drak, J. A. Rice, and D. M. Crothers, *Biochemistry* **29,** 4227 (1990).

converts them into covalently closed circles. Under appropriate conditions, as discussed below, the rate of the covalent cyclization reaction is proportional to the substrate concentration of noncovalent circles. Comparative measurements on the rate of bimolecular ligation allow conversion of the results to an effective end-about-end concentration.

The effective concentration has a straightforward interpretation in terms of polymer chain statistics. Consider a DNA molecule bearing cohesive (complementary) ends. The ring closure probability (J factor or Jacobson–Stockmayer factor) is defined [Eq. (1)] as the ratio

$$J = K_c/K_a \qquad (1)$$

of the equilibrium constant for cyclization (K_c) to that for bimolecular association (K_a)[1]; it can be interpreted as the effective concentration of one end of the molecule about the other, subject to the further restriction that the ends are in an orientation appropriate for the cyclization reaction. J is generally expressed in molar concentration units, and it can be thought of as the molar DNA concentration required to cause bimolecular joining to occur at the same rate as the corresponding cyclization reaction. Formation of a small DNA circle requires large fluctuations in the shape of the molecule since the ends must come together for the closure reaction to take place. The cyclization equilibria are therefore expected to be quite sensitive to the mechanical rigidity of DNA and to the conformation of the chain, thus providing an excellent system to study the structural and fluctuational features of such molecules in solution. Cyclization measurements have a much larger "dynamic range" than many other techniques, since J can be measured over a range of four or more orders of magnitude.

In this chapter, we discuss experimental approaches to the measurement of J and the theory used to analyze ring closure and thereby obtain fundamental physical information from the J factor. We illustrate these procedures mainly with examples from our own work on cyclization of bent DNA molecules (A tract multimers) and recent work on the application of cyclization methods to protein–DNA complexes. Cyclization kinetics is unusual among molecular biological techniques in its requirement for computation of the properties of models in order to interpret the experimental results in anything more than a highly qualitative way.

The theory of cyclization kinetics can also be applied to other systems where DNA assumes conformations where its ends are fixed in a defined spatial orientation. Protein–DNA loops are the best example: *in vivo* and *in vitro* results on *lac* repressor-mediated DNA loops have been interpreted using the cyclization paradigm,[6,7] and enhancement of cyclization has been used to provide evidence for looping.[8]

Finally, we note that the analysis of topoisomer distributions established by ligation under various conditions can provide similar or complementary information to that available from cyclization kinetics.[9,10] Further detailed discussion of looping or topoisomer studies is beyond the scope of this chapter.

Experimental Approaches

Pioneering Methodologies

The fraction of chain configurations leading to ring closure for a Gaussian chain was derived by Jacobson and Stockmayer,[1] providing the first connection between experimental measurements and chain mechanics. They were also able to derive the cyclization factor by considering the kinetics of the forward and reverse reactions.

Wang and Davidson[11,12] used the band sedimentation method to demonstrate experimentally that for a rather long DNA molecule (bacteriophage λ) the cyclization factor could also be obtained by taking the ratio of the forward rate constants for the cyclization and bimolecular association reactions, both mediated by base pairing of the 12-nucleotide complementary ends. Their results verified the assumption that the joining of the cohesive ends takes place via the same microscopic process for the cyclization and association reactions. The agreement between their experimental values and the theoretical J factor calculated assuming that a long DNA molecule can be modeled as a random coil supported the validity of such a model.

Current Technology

Use of T4 DNA Ligase in Circularization Experiments. Shore, Langowski, and Baldwin[2,3] followed the overall logic of Wang and Davidson, while introducing the use of DNA ligase as a method for assaying the concentration of a very small fraction of DNA molecules which are tran-

[6] G. R. Bellomy, M. C. Mossing, and M. T. Record, Jr., *Biochemistry* **27**, 3900 (1988).
[7] M. Brenowitz, A. Pickar, and E. Jamison, *Biochemistry* **30**, 5986 (1991).
[8] S. Mukherjee, H. Erickson, and D. Bastia, *Cell (Cambridge, Mass.)* **52**, 375 (1988).
[9] D. Shore and R. L. Baldwin, *J. Mol. Biol.* **170**, 983 (1983).
[10] D. S. Horowitz and J. C. Wang, *J. Mol. Biol.* **173**, 75 (1984).
[11] J. C. Wang and N. Davidson, *J. Mol. Biol.* **15**, 111 (1966).
[12] J. C. Wang and N. Davidson, *J. Mol. Biol.* **19**, 469 (1966).

A. Cyclization

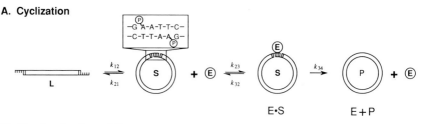

B. Bimolecular Association

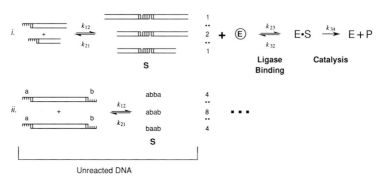

Unreacted DNA

FIG. 1. (A) Schematic representation of cyclization, hydrogen bond formation, ligase binding, and covalent joining. The *Eco*RI ends and adjacent base pairs are represented by hatch marks and shown above the drawing of the noncovalently closed circle. (B) Bimolecular association. Diagram (*i*) shows the association of DNA half-molecules of different lengths and the expected product distribution. Subsequent ligase binding and covalent joining are not shown. Diagram (*ii*) shows the association of molecules with two cohesive ends, giving the expected distribution of different dimers. The overall rate of reaction for the molecule with two cohesive ends is expected to be 4 times that for an equal total concentration of half-molecules. (Adapted from Shore *et al.*[2] Reprinted with permission.)

siently either dimerized or cyclized through a short double helix [2–4 base pairs (bp)] joining their ends. They showed that under certain conditions one can obtain the *J* factor from the relative rates of covalent closure into circles or dimeric molecules by T4 DNA ligase. This innovation enabled the Wang and Davidson experiment to be carried out on DNA restriction fragments, whose shorter complementary ends do not associate sufficiently strongly to allow detection of joined molecules by standard physical methods.

Cyclization occurs by a process (Fig. 1) in which a linear DNA molecule (L) is converted to a noncovalently joined circle (S) which is in turn a substrate for joining by T4 DNA ligase (E). The circular substrate binds to the enzyme and is converted to a covalently closed circle (P), with concomitant ATP hydrolysis. The kinetic equations are

$$L \underset{k_{21}}{\overset{k_{12}}{\rightleftharpoons}} S \tag{2}$$

$$E + S \underset{k_{32}}{\overset{k_{23}}{\rightleftharpoons}} ES \overset{k_{34}}{\longrightarrow} E + P \tag{3}$$

To produce a covalently closed circle, ligase must seal a break in each DNA strand. However, after the first strand has been closed the molecular ends are held together and the second strand is sealed rapidly.[3] Hence we can treat covalent closure as a single reaction.

Assuming that under the usual experimental conditions[2,3] dissociation of noncovalently joined cohesive ends is a fast reaction compared to ligase closure ($k_{21} \gg k_{23}[E]$), and applying the steady-state condition to both ES and S, one obtains expression (4) below for k_1, the first-order rate constant for covalent closure of circles[2,3]:

$$k_1 = \frac{k_{34}[E_o]K_c(1 - f_s)}{[(k_{32} + k_{34})/k_{23}] + [S]_{circular}} \tag{4}$$

$[E_o]$ is the total ligase concentration. The fraction of DNA molecules that are transiently cyclized and hence substrates for closure by ligase is f_s, which is equal to $[S]_{circular}/[\text{unreacted DNA}]$. K_c, the cyclization equilibrium constant, is the ratio k_{12}/k_{21}.

The corresponding equation for k_2, the second-order rate constant for joining half-molecules, is as follows:

$$k_2 = \frac{k_{34}[E_o]K_a(1 - f_s')^2}{[(k_{32} + k_{34})/k_{23}] + [S]_{bimolecular}} \tag{5}$$

Here f_s' is the fraction of DNA molecules that are substrates for dimerization by ligase, given by $2[S]_{bimolecular}/[\text{unreacted DNA}]$, and K_a is the association constant for noncovalent dimerization between distinguishable half-molecules.[3] Notice that for dimerization of molecules with two cohesive ends, the apparent rate constant for dimerization is given by $4k_2$, with k_2 as in expression (5).[4]

The fraction $(k_{32} + k_{34})/k_{23}$ is the K_m of ligase for the doubly nicked substrate S. Under most conditions $[S] \ll K_m$ for both cyclization and dimerization. This condition is necessary in order for cyclization kinetics to report faithfully the relative equilibrium populations of circular and bimolecular ligation substrates, as $[S]$ is not the same for the two types of intermediates (see below).

Taking the ratio of Eq. (4) to Eq. (5) and assuming $[S] \ll K_m$, one obtains for the relative rate constants

$$\frac{k_1}{k_2} = \frac{K_c(1 - f_s)}{K_a(1 - f_s')^2} \tag{6}$$

Provided that f_s and f_s' are negligible compared to 1, one can approximate this equation as[2,3]

$$J = K_c/K_a = k_1/k_2 \tag{7}$$

Negligibility of f_s. Equation (7) is the basic relationship needed to determine the J factor from the relative rate constants for cyclization and bimolecular joining, but in applying it one should keep in mind the constraints on the value of f_s, which are particularly important as J becomes large. (Bimolecular joining can always be carried out at sufficiently low concentration that f_s' is negligible.) With the relationship $K_c = f_s/(1 - f_s)$ (this corrects a sign error in this equation which crept into our earlier paper[5]), and the definition of J, one can readily show that

$$f_s = \frac{JK_a}{1 + JK_a} \tag{8}$$

from which one can see that f_s is no longer negligible when J approaches $1/K_a$. Failure of the assumption that f_s is negligible will cause Eq. (7) to be in error as shown below; hence, special caution is necessary for rapidly cyclizing systems, in which a bound protein or A tract bends act to position the molecular ends favorably for cyclization. This problem can be dealt with by reducing the value of K_a by shortening the complementary single-stranded segments at the ends of the molecule; when J values exceeding 10^{-6} M are anticipated, we recommend that experiments be designed to use staggered ends no more than 2 nucleotides in length.

Defining J_{app} as the apparent J factor given by the ratio of rate constants, it is simple to show[5] that

$$J_{app} = k_1/k_2 = f_s/K_a \tag{9}$$

When $f_s \ll 1$, $f_s = K_c$ and J_{app} is equal to J. When f_s approaches 1, J_{app} approaches an upper limit of $1/K_a$. This limit implies a real J value large enough to convert all of the DNA molecules to noncovalent circles ($f_s \to 1$). In such a case, further increase in K_c causes no accompanying increase in J_{app}.

Temperature-Jump Methods for Investigating K_a. DNA molecules having short complementary ends associate weakly, but when J is as large as, for example, 10^{-4} M, as in the case of A-tract-containing molecules,[5] an accurate experimental determination of J requires that K_a be less than about 10^3 M^{-1}. This is clearly a potential problem, exacerbated by longer overhanging ends which act to increase K_a. Koo *et al.*[5] used the tempera-ture-jump method[13] to estimate that for the CG overhanging sequence used in their studies, $K_a \leq 8 \times 10^2 \, M^{-1}$. This approach relies on detecting the absorbance increase that accompanies the melting of the duplex which joins two oligomers into a dimer. The concentration dependence of the relaxation rate can be used to extract the equilibrium constant K_a. This or some equivalent approach is an essential control in working with systems with large J values.

Experimental Determination of Rate Constants. In this section, we discuss the techniques and pitfalls of cyclization kinetics measurements. The examples are largely drawn from our work on cyclization of A tract multimers.[5] Additional considerations in studying protein–DNA com-plexes by cyclization are briefly treated.

Cyclization kinetics measurements can be done on end-labeled restric-tion fragments[2–4] or on body-labeled substrates, constructed by ligation[5] or via the polymerase chain reaction (PCR) (J. D. Kahn, unpublished, 1992) followed by restriction. Figure 2 shows the synthesis of a set of A tract multimer substrates used in our work on determining the A tract bend angle.[5]

The rate constants k_1 and k_2 are usually determined by quantifying conversion of linear monomer DNA to circular or dimeric/multimeric products, respectively, as a function of time. The conversion can be monitored by electrophoretic separation or by mung bean nuclease resis-tance of the ^{32}P label in the products.[2–5] We favor gel methods for several reasons: products are identified and quantified in the same experiment, many samples can be analyzed simultaneously, and gel methods are suit-able for analysis of bimolecular ligations of molecules with two cohesive ends. Gels can be quantified by autoradiography and densitometry, scintil-lation counting of excised bands, or direct detection of β particles [we use a Betascope blot analyzer, (Betagen, Waltham, MA)]. Agarose or polyacrylamide gels are used depending on fragment size. Ethidium bro-mide can be added to the gels to alter the mobility of the circular products, which is especially useful in identifying nicked circles or topoisomers.[3,9,10]

The identity of products is most simply determined by restriction or by resistance to exonucleases. Restriction at a unique site will give one product from a circle and up to five electrophoretically distinguishable products from the set of linear dimers (the number depends on whether one or both ends are cohesive). Circles are much more resistant than linear molecules to exonucleases such as *Bal*31 or exonuclease III.

Perhaps the most difficult aspects of the kinetic measurements are

[13] P. E. Cole and D. M. Crothers, *Biochemistry* **11**, 4368 (1972).

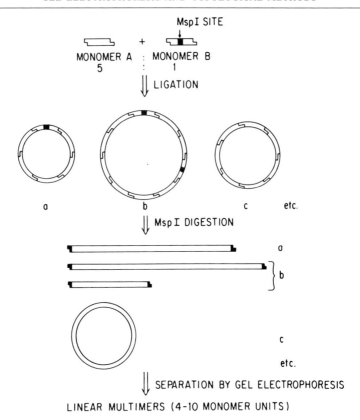

FIG. 2. Preparation scheme for linear multimers of a 21-bp bend DNA sequence. The monomers have 5-bp-long sticky ends and two A_6 tracts. Monomer B has a central *Msp*I site, and only monomer B was radioactively labeled; thus, the circular molecule C consisting only of A monomers was nonradioactive and did not interfere in the subsequent separation of linear multimers. A 5:1 excess of monomer A over B was used. The linear multimers were separated and purified by gel electrophoresis. The sequences of monomers A and B (one strand only) are as follows: A, 5'-GGGCAAAAAACGGCAAAAAAC; B, 5'-GGGCAAAAAACCGGAAAAAAC. (From Koo *et al.*[5] Reprinted with permission from *Biochemistry*.)

the standardization of cyclization versus bimolecular reactions and the standardization of ligase activity in reactions performed at different times on different molecules. The latter can be accomplished by ligation of a standard molecule in parallel with each experiment[2] or by the use of internal standards in the same reaction mixture.[13a] Both problems can be

[13a] J. D. Kahn and D. M. Crothers, *Proc. Natl. Acad. Sci. U.S.A.*, in press.

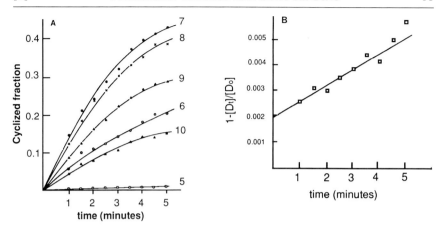

FIG. 3. (A) Cyclization kinetics of A tract multimers of different lengths. The cyclized fraction as a function of time is shown. The concentrations of reactants and ligase were 1×10^{-9} and 1.7×10^{-10} M [~10 New England Biolabs (Beverly, MA) cohesive end units/ml], respectively, except that a 10-fold higher ligase concentration was used for cyclization of the 105-bp molecule and the values shown were corrected accordingly. Products were analyzed by gel electrophoresis and scintillation counting of excised bands. (B) Bimolecular ligation kinetics of an 84-bp A tract multimer, plotted as the fraction of dimerized product versus time. $[D_t]$ is the concentration of remaining DNA reactant at time t. The concentrations of DNA and ligase were 1.05×10^{-8} and 8.5×10^{-9} M, respectively. The second-order rate constant k_2 was determined from the slope and the initial DNA concentration. (From Koo et al.[5] Reprinted with permission from Biochemistry.)

solved in principle by performing the measurements under conditions where circular and bimolecular products are formed in the same reaction[4]; however, this method has some drawbacks, which are discussed below.

Calibration of the cyclization and bimolecular reactions is difficult because in many cases different DNA and ligase concentrations are required for the two cases; much higher concentrations are often required to obtain a measurable bimolecular rate. Estimated corrections must then be applied to give the bimolecular rate constant under the conditions used for cyclization (discussed in detail by Koo et al.[5]). In general, because of these problems we believe that relative J factors among a set of molecules are more reliable than absolute numbers. Relative J factors can be obtained by comparing easily and consistently measured cyclization rate constants for a family of constructs.

Figure 3 shows examples of kinetic data from our A tract work.[5] The first-order rate constant k_1 is the slope obtained by replotting the data of Fig. 3A as $\ln([D_0]/[D_t])$ versus time, where $[D_t]$ is the concentration of unreacted monomer at time t. The data were obtained by scintillation counting of gel slices. The data shown in Fig. 3B on the bimolecular reaction were measured on a homologous A tract molecule (84 bp) which is too short to cyclize. We do not know the origin of the apparent burst phase in the bimolecular reaction. The bimolecular rate constant is assumed to be independent of DNA length, as found by Shore and Baldwin.[3]

In some cases the extent of reaction is large, but cyclization usually cannot be driven to completion (this appears to be a common problem in the literature). The fraction f_d of unreactive molecules can be corrected for by determining f_d for long times and fitting the kinetics according to the equation $[D_0]/[D_t] = \exp(k_1 t)/[1 - f_d + f_d \exp(k_1 t)]$. The unreactive DNA could be dephosphorylated, incorrectly restricted, or adsorbed to the reaction vessel.

The bimolecular calibration requires that the rate constants k_1 and k_2 depend linearly on ligase concentration. Linearity with ligase concentration must also be verified to ensure that the assumption of a rapid preequilibrium between monomer DNA and ligatable substrate is valid (see above). The proper dependence of reaction rates on DNA concentration must also be verified. We have found that 0.01% Nonidet P-40 (NP-40) in ligase dilutions and reaction mixtures greatly improves the linearity and reproducibility of ligase titrations. We also recommend the use of silanized microcentrifuge tubes and the standard precautions for accurate serial dilution.

Cyclization kinetics of protein–DNA complexes requires additional refinements.[14] Buffer conditions must be established which maintain ligase activity while allowing protein binding. The binding protein concentration must be optimized to ensure a homogeneous population of substrate protein–DNA complexes, that is, the protein binding site should be completely occupied but nonspecific binding should be minimized. For example, we have used gel retardation to define a concentration of catabolite activator protein (CAP) (6 nM for 2 nM DNA) which gives essentially 100% specific complex without substantial formation of nonspecific complexes. Titration of CAP in ligation reactions shows that at this concentration there is little inhibition of ligase activity, and there is a range of protein concentration over which the measured cyclization and bimolecular rate constants do not vary substantially. In these studies we prefer not to vary the DNA concentration over a wide range because the DNA concentration may affect the fraction of DNA bound in specific complexes. Finally, before analysis the binding protein should be removed to prevent the appearance of gel-shifted bands; we stop ligation reactions with 25 mM ethylenediaminetetraacetic acid (EDTA), 0.67 mg/ml proteinase K, 5%

---

[14] D. Kotlarz, A. Fritsch, and H. Buc, *EMBO J.* **5**, 799 (1986).

glycerol, and 0.025% dyes (final concentrations) and incubate for 15 min at 55° before loading the gel. This treatment also prevents the formation of DNA aggregates which otherwise remain in the well.

Use of Internal Controls. Hagerman and co-workers have recently presented a method for direct determination of the J factor in a single reaction.[4] Their experiments use a much higher DNA concentration ($\sim$100 nM) than those of Shore and Baldwin, so bimolecular ligation (second-order in DNA) can proceed at a rate comparable to unimolecular DNA cyclization. Taylor and Hagerman[4] show that at low conversion of linear DNA to circular or bimolecular products the J factor is obtained from the limit

$$J = 2M_0 \lim_{t \to 0} [C(t)/D(t)] \tag{10}$$

where M_0 is the input concentration of DNA monomer and $C(t)$ and $D(t)$ are the concentrations of circular and dimeric products, respectively, at time t. They used this approach to measure the J factor for a series of restriction fragments of lengths about 350 bp, showing that at added monovalent salt concentrations between 0 and 162 mM, in the presence of 1 mM Mg^{2+}, the persistence length, torsional modulus, and helical repeat of the molecules are essentially unchanged.

This method has the advantages that no standardization measurement of bimolecular kinetics is necessary and that variations in ligase concentration or activity do not affect the measured J values. Under appropriate conditions this makes the experimental determination of J simpler than previous methods, but certain cases of interest do not qualify. In particular, for molecules which cyclize efficiently ($J \gtrsim 10^{-6} M$) the measurements require high DNA concentrations (micromolar) to allow bimolecular joining to compete with cyclization. Under these conditions the ligatable substrate concentration can approach the K_m of DNA ligase (estimated at 0.6 μM[14a,b]), especially if long overhanging ends are used; the high ligase concentrations required to give a measurable extent of reaction may also perturb the cyclization and dimerization equilibria of the free DNA by sequestering ligatable substrate DNA. In this circumstance the relationship of kinetic measurements to the equilibrium J factor becomes problematic, as discussed above. Recently, we have used the cyclization of A tract multimers as internal standards for cyclization rate measurements on test samples, thereby avoiding any problems with variations in ligase concentration or activity.

[14a] A. J. Raae, R. K. Kleppe, and K. Kleppe, *Eur. J. Biochem.* **60,** 437 (1975).
[14b] A. Sugino, H. M. Goodman, H. L. Heynecker, J. Shrine, H. W. Boyer, and N. R. Cozarelli, *J. Biol. Chem.* **252,** 3987 (1977).

The bimolecular ligation reaction is measurable at low DNA concentrations for poorly cyclizing chains, and in these cases we use an approach similar to that of Taylor and Hagerman,[4] except that the kinetics of the complete time course for competing cyclization and dimerization are treated explicitly. The rate equations describing the competing reactions are as follows:

$$M \xrightarrow{k_1} C \tag{11}$$

$$2\,M \xrightarrow{k_2} D \tag{12}$$

where M is monomer; C, circular product; D, dimeric product; and k_1 and k_2 are the same as before. The dimeric product may be further reacted to give dimeric circles, trimeric products, and so forth; all of these can be summed in evaluating k_2 from experiment. The differential equations describing the system are as follows[4]:

$$d[C]/dt = k_1[M] \tag{13}$$
$$\tfrac{1}{2}(d[D]/dt) = 4k_2[M]^2 \tag{14}$$
$$d[M]/dt = -k_1[M] - 4k_2[M]^2 \tag{15}$$

Solution of the differential equations subject to the initial conditions $[C]|_{t=0} = [D]|_{t=0} = 0$ and $[M]|_{t=0} = M_0$, the initial molar concentration of ligatable DNA, gives the following expressions for the molar concentrations of reactant and products:

$$[M] = \frac{M_0\,e^{-k_1 t}}{1 + 4M_0(1 - e^{-k_1 t})k_2/k_1} \tag{16}$$

$$[C] = \frac{k_1}{4k_2}\{-k_1 t + \ln(M_0/[M])\} \tag{17}$$

$$[D] \equiv \tfrac{1}{2}(M_0 - [C] - [M]) \tag{18}$$

Initial estimates for the rate constants k_1 and k_2 are obtained from the time courses for the appearance of C and D considered separately. These estimates are refined iteratively using the above expressions: we fit k_1 using fixed k_2 in Eq. (17), fit k_2 with the new k_1 using Eq. (18), and so forth. The curve fitting program Kaleidagraph is used on an Apple Macintosh IIcx.

Theoretical Background of Chain Statistics Problem

The dependence of the J factor on helical repeat and bend phasing, DNA length, and protein binding can provide qualitative insight into the properties of the DNA, but the extraction of quantitative physical information requires some calculational effort. Here we give an introduction to DNA chain statistical mechanics, and in the following sections we discuss the use of Monte Carlo simulations to evaluate the J factor.

Consider a DNA molecule with complementary cohesive ends. For the case of a long and flexible DNA molecule, where the orientations of terminal chain elements can be taken as uncorrelated, Jacobson and Stockmayer[1] derived the following expression for J:

$$J = K_c/K_a = \exp\left[\frac{-(\Delta G_c^\circ - \Delta G_a^\circ)}{RT}\right] = W(0)/N_A \qquad (19)$$

where ΔG_c° and ΔG_a° refer to the standard free energy changes for the cyclization and dimerization reactions, respectively. $W(0)$ corresponds to the spatial probability density $W(r)$, where r is the end-to-end distance of the chain, evaluated at $r = 0$. N_A is Avogadro's number, and R is the gas constant. Assuming that the DNA can be modeled as a Gaussian chain composed of n elements of length l, $W(0)$ can be easily calculated and the J factor rewritten as:

$$J = \frac{1}{N_A}\left[\frac{3}{2\pi n l^2}\right]^{3/2} \qquad (20)$$

The Jacobson–Stockmayer theory agrees well with experiments performed on long DNA molecules, as demonstrated by the experiments of Wang and Davidson.[11,12] For short and rather stiff DNA fragments, however, the assumptions of Gaussian statistics and no correlation between the orientations of the ends of the chain do not hold. It is then useful to follow Kratky and Porod[15] and to model DNA as a wormlike chain.

Flory and co-workers[16] expressed the cyclization and dimerization probabilities in an elaborated form of the Jacobson–Stockmayer theory. They considered that the probability that a DNA chain will cyclize depends on the simultaneous fulfillment of three conditions: (1) the ends of the chain must be located at a distance such that the new bond can be formed, (2) the angle between the helix axes at the ends of the chain must be within the range allowed by the bendability of the DNA, and (3) the torsion angle between the ends must be such that the helix backbone is continuous at the newly created bond (Fig. 4). We define $\Gamma_R(\gamma)$ as the conditional probability density for the scalar product γ of the vectors parallel to the helix axes at the chain termini, at a specified value of the end-to-end distance R. Similarly, $\Phi_{R,\gamma}(\tau)$ is the conditional probability density for the torsion angle τ

[15] O. Kratky and G. Porod, *Rec. Trav. Chim.* **68,** 1106 (1949).
[16] P. J. Flory, U. W. Suter, and M. Mutter, *J. Am. Chem. Soc.* **98,** 5733 (1976).

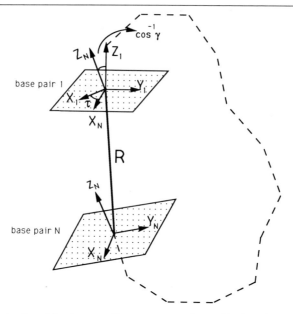

FIG. 4. Illustration of the three conditions necessary for cyclization of a short DNA chain, and the coordinate system used. The end-to-end distance R must be comparable to the length of the new bond to be formed; the helix axis at the end of the chain (Z_N) must be closely aligned to the axis direction at the beginning of the chain (Z_1), and the torsion angle τ must be close to the average rotation angle between DNA base pairs. The dashed line represents base pairs 2 to $N - 1$ of the DNA chain, the stippled planes represent base pairs, and $\cos^{-1} \gamma$ is the angle between initial and final helix axes.

between the vectors which specify the helical orientation of the chain ends, with R and γ specified. (Unless stated otherwise, all angular parameters are expressed in radians.) Then J can be written as follows:

$$J = 4\pi W(0)\Gamma_0(1)\Phi_{0,1}(\tau_0)/N_A \qquad (21)$$

where the spatial probability density is evaluated at $R = 0$, $\Gamma_R(\gamma)$ is evaluated when $R = 0$ at $\gamma = 1$ (parallel ends), and $\Phi_{R,\gamma}(\tau)$ is evaluated when $R = 0$ and $\gamma = 1$ at $\tau = \tau_0$, the angle corresponding to perfect torsional alignment of the chain. Equation (21) is the fundamental equation used for Monte Carlo computations of the J factor.

Monte Carlo Methods

For a wormlike chain, an exact closed form for the probability densities in Eq. (21) has not been derived. Some approximate distributions have been proposed,[17-19] but except for special cases[19] they are not generally applicable to short or intermediate length chains or to systematically bent molecules, leading us to use Monte Carlo simulations to calculate ring closure probabilities.[5,20-22] The simulations require specifying the persistence length, torsional modulus, and helical repeat of the DNA and some approximate geometrical description of any protein or sequence-directed bends.

Levene and Crothers[20,21] calculated ring closure probabilities for random sequences of different length by Monte Carlo simulation. Using a refined version of that program, we have subsequently simulated a set of systematically bend DNA molecules, multimers of a 21-bp-long deoxyoligonucleotide in which A_6 tracts were positioned 10 and 11 bp apart alternatively.[5] We calculated the ring closure probabilities (J factors) as a function of the bend angles at each junction, the helical repeat of these sequences, and the bending flexibility of DNA. Computations currently underway emphasize simulation of the effect of bound protein on DNA cyclization. Hagerman and co-workers[4,23] have also exploited the Monte Carlo approach for interpretation of experimental measurements.

In the following sections, we describe the Monte Carlo approach for the generation of the simulated DNA chain and the analysis of the J factor. All our recently published computer simulations[5] were carried out on a VAXserver 3500 or on a VAXstation 3200 using programs written in VMS-FORTRAN modified from the original code of Levene and Crothers.[20] More recently we have adapted the code to the Multiflow Trace computer and Silicon Graphics workstations. Software is available from the authors on request.

Generating the Ensemble of DNA Chains

We treat the DNA molecule as a discrete wormlike chain in which each base pair is represented by a vector normal to the mean plane of the base pair (the x–y plane) along the z axis (the helix axis) of a local coordinate frame. Propeller twisting or other deformations of the bases out of the plane perpendicular to the helix axis are ignored in this realization of

[17] H. E. Daniels, *Proc. R. Soc. Edinburgh* **63A**, 290 (1962).
[18] J. E. Hearst and W. H. Stockmayer, *J. Chem. Phys.* **37**, 1425 (1962).
[19] J. Shimada and H. Yamakawa, *Macromolecules* **17**, 689 (1984).
[20] S. D. Levene and D. M. Crothers, *J. Mol. Biol.* **189**, 73 (1986).
[21] S. D. Levene and D. M. Crothers, *J. Mol. Biol.* **189**, 61 (1986).
[22] N. Metropolis, A. W. Rosembluth, M. N. Rosembluth, A. H. Teller, and E. Teller, *J. Chem. Phys.* **21**, 1087 (1953).
[23] P. J. Hagerman and V. A. Ramadevi, *J. Mol. Biol.* **212**, 351 (1990).

the DNA molecule. Repeated application of standard rotation matrices generates a trajectory for the helix axis.[24]

Thermal motion is modeled by allowing the orientation of the helix axis to fluctuate through random small rotations about the x, y, and z axes at each base pair treated independently. The angles are chosen by assuming a Hooke's law restoring force for deviations from the mean angle, resulting in a Gaussian distribution of possible configurations. As shown by Levene and Crothers,[20] the probability that a base pair will be rotated with respect to its neighbor by tilt, roll, and twist angles θ, ϕ, and τ due to rotation about the x, y, and z axes, respectively, is $f(\theta, \phi)f(\tau)$, where

$$f(\theta, \phi) = C' \exp\left\{ - \left[\frac{(\theta - \theta_0)^2}{2\sigma_\theta^2} + \frac{(\phi - \phi_0)^2}{2\sigma_\phi^2} \right] \right\} \tag{22}$$

$$f(\tau) = C'' \exp\left\{ - \left[\frac{(\tau - \tau_0)^2}{2\sigma_\tau^2} \right] \right\} \tag{23}$$

The angles θ_0, ϕ_0, and τ_0 represent both the mean values for tilt, roll, and twist of the two adjacent base pairs with respect to each other and also the perfect helix parameters. C' and C'' are normalization constants, and σ_θ and σ_ϕ are chosen such that $\sigma_\theta^2 + \sigma_\phi^2 = 2l/P$, where l is the helix rise per base pair and P is the persistence length of the DNA.[25] Thermal bending is assumed to have no preferred direction, so the DNA is treated as an isotropic chain where σ_θ is equal to σ_ϕ. With a typical persistence length of 140 bp, the estimated root mean square (rms) fluctuations in θ and ϕ are about 0.084 radians or 4.8° each. Models incorporating anisotropic flexibility, with $\sigma_\theta \neq \sigma_\phi$, produce equivalent results as long as the persistence length is held constant,[20,21] for DNA without static bends. For the torsion angle distribution, elementary consideration of elasticity theory leads to $\sigma_\tau^2 = lk_B T/C$, where k_B is Boltzmann's constant, T is the absolute temperature, and C is the torsional modulus. For a C of 3.4×10^{-19} erg cm (see below) and a T of 300 K, this corresponds to an rms fluctuation of about 0.064 radians, or 3.7°, when l is 3.4 Å, corresponding to one base pair.

Value of Torsional Modulus. In our calculations[5,20,21] we have taken C to be 3.4×10^{-19} erg cm, a value which remains controversial. The primary basis for our value is the data of Shore and Baldwin[9] and Horowitz and Wang[10] on the linking number variance in DNA molecules in the size range up to about 600 bp. Our resulting number for C is larger than that of other

[24] S. D. Levene and D. M. Crothers, *J. Biomol. Struct. Dyn.* **1**, 429 (1983).
[25] J. A. Shellman, *Biopolymers* **13**, 217 (1974).

authors because our simulations[20] indicated that about 20% of the total torsional flexibility in such molecules is due to writhe, with the remaining 80% resulting from twist fluctuations as determined by C. Most other authors have not made such a correction for residual writhe, relying instead on the approximate calculation of LeBret,[26] according to which molecules below a critical size (roughly 1000 bp) have zero writhe. The result is to make our value of C about 25% larger than that found by other authors for analysis of the same data sets. Taylor and Hagerman[4] have misrepresented our calculation in stating that it is based on an extrapolation of the writhe variance to the long chain limit.

The value of C can also be determined from data such as those reported by Shore and Baldwin[3] on the dependence of cyclization rate on DNA chain length over an interval of a full helical turn. The rate is sensitive to torsional flexibility because cyclization of molecules whose number of helical turns is nonintegral requires torsional fluctuations in order to ensure that the ends join in proper helical register. These fluctuations are provided primarily by twist for molecules below the critical size, and primarily by writhe for larger molecules.

Taylor and Hagerman[4] reported a new data set of this kind for molecules of approximately 340 bp in size, and they deduced a value for C of 2.0×10^{-19} erg cm, substantially smaller than our value. They noted a small but systematic error in the plot used by Levene and Crothers[21] to fit the data of Shore and Baldwin[3] to a value of C and implied that this error caused our value of C to be too large. Figure 5 shows that replotting the data does not appreciably affect the value of C deduced, which is about 3 to 5×10^{-19} erg cm; this number is affected by the persistence length chosen, and is not highly accurate. Furthermore, the equations obtained from our simulations[20,21] when applied to the Taylor and Hagerman[4] data yield the same value as they obtained, namely, 2.0×10^{-19} erg cm, so there seems to be no underlying disagreement about the theory. However, the Taylor and Hagerman[4] value is clearly too small to be consistent with the data of Shore and Baldwin[9] and Horowitz and Wang[10] on the distribution of topoisomers for small circles, leading us to suspect that some unknown feature in the sequences studied by Taylor and Hagerman[4] confers an unusually large apparent torsional flexibility on DNA. For the present we consider C equal to 3.4×10^{-19} erg cm to be the best available general estimate. Future experiments will no doubt address this issue further.

[26] M. LeBret, *Biopolymers* **19**, 1709 (1979).

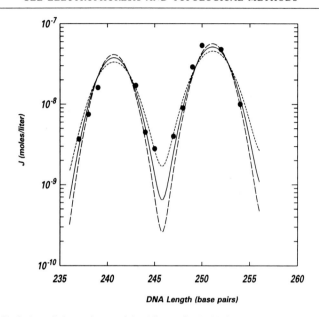

FIG. 5. Chain length dependence of the J factor for DNA fragments near 250 bp in length. The data points are the experimental data from Shore and Baldwin.[3] The smooth curves give the dependence of cyclization probability on DNA length calculated according to Levene and Crothers[20,21] for three values of the torsional rigidity, C. The parameters used were as follows: helical repeat, 10.46 bp/turn; persistence length P, 450 Å; and C, 3.0×10^{-19} erg cm (---), 4.0×10^{-19} erg cm (—), and 5.0×10^{-19} erg cm (- - -). The experimental data were also fit using a simplex method to determine the best-fit value of C for given values of P. With P equal to 450 Å the best fit was found to be $C = 4.8 \times 10^{-19}$ erg cm. It is clear that the quality of the fit to the data is not strongly affected by varying C over this range. (Reprinted with corrections from Levene and Crothers[20] with permission from the *Journal of Molecular Biology*.)

Computational Aspects

Our implementation of the theory uses one program to generate ensembles of chains, a second to apply the filtering conditions and select cyclizing chains, and finally a set of programs to convert the results from distribution functions to probability densities. In the generation of chains, a random number generator is used to sample the Gaussian distributions used for the angles θ, ϕ, and τ. Computation time is minimized through the use of two subchains, each composed of one-half the number of base pairs in the original DNA molecule.[21,27,28] Each subchain in one of the sets is paired

[27] Z. Alexandrowicz, *J. Chem. Phys.* **51**, 561 (1969).
[28] Y. Chen, *J. Chem. Phys.* **74**, 2034 (1981).

with all the subchains in the other set to generate an ensemble of full-length chains. To obtain a significant number of cyclized chains (especially with the shorter molecules) it is essential to start the simulations with at least 10^4 chains, so that after the "dimerization" procedure the ensemble contains 10^8 molecules. Otherwise the scatter in the angular distributions is so high that the final extrapolations required to calculate J are not reliable. We find that collecting adequate statistics for molecules with J on the order of $10^{-9} M$ requires an ensemble of about 10^{10} chains, requiring the equivalent of about 10 days on a VAXstation 3200. For simulations using more than 10^8 chains, at least 90% of the calculation time is spent in pairing subchains and applying filtering conditions as discussed below. (Calculation time for generating chains increases with $N^{1/2}$, where N is the number of chains in the final ensemble, whereas the time for joining and filtering rises with N.)

In simulating the cyclization of systematically bent A tract multimers,[5] the angular parameters for B-DNA were chosen as $\theta_0 = \phi_0 = 0$. The ratios of θ (tilt) to ϕ (roll) at the 5' and 3' junctions of the A_6 tract with B-DNA were chosen so that they correctly predict the observed anomalous gel mobilities found for bent DNA molecules of various sequences.[29] Unless stated otherwise a persistence length of 475 Å (140 bp) was used for both the A_6 tract and B forms of DNA[21] to calculate the value of $\sigma_\theta = \sigma_\phi$.

For simulation of a chain including a protein binding site, we replace the appropriate number of base pairs by a single "virtual segment" representing the protein-bound DNA. The rotation matrix for the virtual segment and the necessary displacement of the helix axis are calculated analytically given the assumed helical repeat and bend angle induced by the protein. The protein-bound DNA segment is assumed to be rigid. We also adjust the break point between subchains to retain an equal number of base pairs and hence "randomness" on each side of the new break point.

Statistical Mechanics of Ensemble and Application
of Ring Closure Conditions

To describe the closure reaction of the DNA chain, the ensemble of DNA molecules is filtered to meet the cyclic boundary conditions. The end-to-end distance is computed for each chain, and a radial distribution function for the ensemble is accumulated in a 250-point histogram. An end-to-end distance cutoff R_c is established, and chains whose end-to-end

[29] H.-S. Koo and D. M. Crothers, *Proc. Natl. Acad. Sci. U.S.A.* **85,** 1763 (1988).

distances fall outside R_c are discarded. For the remaining chains, the relative orientation of the initial and final vectors representing the helix axis is tested by obtaining the scalar product of the two vectors. The results are accumulated in a 100-point histogram corresponding to the angular distribution function. For chains whose ends are nearly parallel (those with scalar product greater than the angular cutoff γ_c), the angle between the initial and final x vectors is calculated (the x axis represents the torsional orientation of the base pair plane). Therefore, the final 72-point histogram representing the torsion angle distribution includes only those chains with $R \leq R_c$ and $\gamma \geq \gamma_c$.

It is quite important to choose the cutoff values R_c and γ_c judiciously. A very small radial cutoff (e.g., 3 bp) leads to sample attrition (a problem that can only be solved by using very large ensembles with the consequent increase in central processor unit hours and computer cost), whereas if R_c is too large (e.g., 10 bp), the $\Gamma_0(\gamma)$ distribution will be inaccurate. In the latter case the error can be significant when working with short DNA chains (<200–300 bp), where the $\Gamma_{R \leq R_c}(\gamma)$ distribution can depend strongly on R_c. Similarly, a stringent (large) cutoff for γ_c can generate too small an ensemble, and then the scattering of the data does not allow an accurate extrapolation when $\gamma_c \to 1$. Obviously, if γ_c is too small we would not be accurately representing chain cyclization. Taking into account the above considerations, in our simulations we have generally used compromise values of 6 bp for R_c and 0.9 for γ_c. Further refinements to this procedure are discussed below.

Next, we express the distribution functions as probability densities and extrapolate $W(R)$ to $R \to 0$ and $\Gamma_{R \leq R_c}(\gamma)$ to $\gamma_c \to 1$ using standard least-squares methodology to fit the experimental curves to even fourth-order polynomials.[21] Hagerman and Ramadevi[23] implied that our estimates[21] of $W(0)$ depend on R_c; however, in Ref. 21, $W(R)$ is extrapolated to $R \to 0$, with $W(0)$ obtained from the constant coefficient of the fitted polynomial. Finally, the values from the torsion angle distribution are normalized to give the corresponding probability density and fit to a Gaussian function $\Phi_{0,1}(\tau) = C \exp[-(\tau - A)^2/2\sigma^2]$. The torsional part of the ring closure probability is then given by $\Phi_{0,1}(\tau_0)$, where $\tau_0 = 360° - 360°/$(number of base pairs per helical turn at the cyclization junction). The J factor is then calculated according to Eq. (21).

In some cases, primarily when J is so small that it is difficult to accumulate adequate statistics, we collect several ensembles in the filtering process, representing chains meeting increasingly rigorous sets of cyclization conditions. The value for $\Gamma_0(1)$ and the parameters describing the torsional angular distribution are then obtained by simultaneously extrapolating the values obtained from the set of distributions to $R_c = 0$ and $\gamma_c = 1$. Details of the extrapolation procedure do not affect the final values substantially, because the parameters are slowly varying functions of the cutoffs. This procedure does not significantly increase calculation time for poorly cyclizing chains, because generating and filtering the great majority of the chains which do not meet any of the R_c conditions consumes most of the computation time. Hagerman and Ramadevi have also used extrapolation methods to solve the cutoff problem.[23]

Cyclization Kinetics and Monte Carlo Simulation: The Combined Approach

As an example of the application of the Monte Carlo methods to the simulation of experimental data, we discuss our work on determining the bend angle induced by A tracts in DNA. The magnitude of the overall bend was determined by fitting the calculated ring closure probabilities to the experimental values obtained from cyclization kinetics measurements on the same multimers[5]; the fitting process is described in detail below. The experimental results suggest that the optimum length for cyclization is approximately 155 bp (15 A tracts). This yields a preliminary estimate of $360°/15 = 24°$ bending per A_6 tract, which is refined to somewhat smaller values by the computer simulation experiments. As noted by Koo et al.,[5] there could be a significant error in the absolute magnitudes of the experimental J factors, but the relative values of J at different DNA lengths are thought to be quite reliable. In the fitting process both the absolute magnitudes of J values and their variation with DNA length were considered, but we focus mainly on relative values in assessing the quality of fits. The results of our best fits are illustrated in Fig. 6.[30] We can conclude from our studies that an A_6 tract bends the DNA helix by $17°$ to $21°$, an estimate that does not distinguish among the detailed proposed models to describe bending, whether junction, wedge, or other as preferred.

Our estimate for the A tract bend is significantly smaller than that of Ulanovsky et al.,[31] who, however, studied a different sequence. We believe that the discrepancy reflects mainly their approximations in neglecting the effect of DNA flexibility, the inexact phasing of multimers, and the complex kinetics observed in reactions where reactants are multimerized and cyclized in the same reaction. Under these conditions, smaller circles may be overrepresented relative to larger circles with more optimal geometry for cyclization, because as soon as the J factor of a growing multimer becomes larger than the DNA concentration, it can cyclize rapidly.

[30] J. Drak and D. M. Crothers, Proc. Natl. Acad. Sci. U.S.A. **88**, 3074 (1991).
[31] L. Ulanovsky, M. Bodner, E. N. Trifonov, and M. Choder, Proc. Natl. Acad. Sci. U.S.A. **83**, 862 (1986).

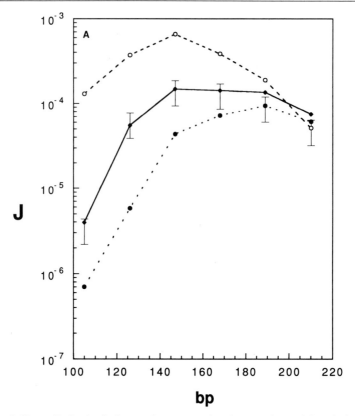

FIG. 6. Monte Carlo simulation results compared to the experimental data derived from Fig. 3. The experimental J factors are indicated with error bars, plotted against A tract multimer length. (A) Effect of varying the overall bend angle of an A tract on the theoretical J factor. The helical repeat used[30] was 10.35 bp/turn (see text). The bend angles shown correspond to 16° (filled circles), 18° (filled diamonds, solid line), and 22° (open circles) per A tract. (B) Effect of varying the helical repeat for an overall bend angle of 18°. The values of the helical repeat were 10.30 bp/turn (filled circles), 10.35 bp/turn (diamonds, solid line), and 10.40 bp/turn (open circles). (C) Effect of varying the thermal bending flexibility for $h = 10.35$ bp/turn and a bend angle of 18°. The values of σ_θ and σ_ϕ were increased (open circles) or decreased (filled circles) by 10%, corresponding to an increase or decrease in persistence length of 20% relative to the normal value of 475 Å (filled diamonds, solid line). (From Koo et al.[5] Reprinted with permission from Biochemistry.)

As our simulations involved several parameters, the theoretical cyclization probabilities will depend on the helical repeat and the thermal bending flexibility of the DNA sequences under study as well as on the values for the bend angles. To reduce the number of parameters we determined the average helical screw of A_6 tracts alternating with N_{4-5} stretches

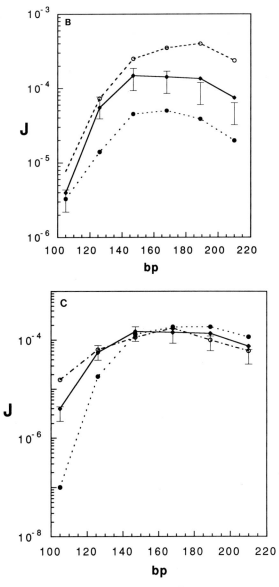

FIG. 6. (*continued*)

of B-DNA[30]; we found a value of 10.34 ± 0.04 bp/turn, which was used in most of the simulations. Taking the helical screw as 10.35 bp/turn we examined the effect of varying the overall bend angle, keeping the directions and relative magnitudes of the junction bends constant (5′ junction bend versus 3′ junction bend and tilt versus roll at each junction) as determined from the gel mobility data for bent DNA molecules of various sequences.[29] Typical results are illustrated in Fig. 6A.

We also wanted to study the effect of altering the helical repeat (h) while keeping the overall bend angle constant (18°). The best fit to the experimental data was obtained with h values ranging from 10.33 to 10.36 bp/turn (Fig. 6B). A relatively small increase in the helical repeat (from 10.35 to 10.40 bp/turn) shifts the maximum of the distribution of J values to longer molecules. On the other hand, reducing h from 10.35 to 10.30 bp/turn decreases the J factor for all the multimers. We found that the sensitivity of the simulations with respect to the helical repeat value is remarkable. The fact that with helical repeat values outside the range 10.35 ± 0.04 bp/turn we cannot obtain a good fit to the experimental results (by varying the junction bend angles) is consistent with our independent measurement of the helical screw of these sequences. The general effect of varying the chain flexibility in an isotropic manner is illustrated in Fig. 6C. We did not vary the torsional modulus in these calculations because the relative results for a series of molecules which all include approximately an integral number of helical turns are not very sensitive to the value of C.

With respect to the general methodology, it is important to emphasize the desirability of determining as many of the relevant parameters as is possible by independent experiments, since multiparameter fits to a limited data set are intrinsically inaccurate. For example, as mentioned above, we measured the helical repeat of A tract sequences by gel methods,[30] finding values consistent with best estimates from the multiparameter fit. Furthermore, even though we have not determined the persistence length of A tract-rich DNA, we have shown that, except for the smallest circles, the cyclization probabilities do not depend strongly on the persistence length within the range 390–590 Å, which covers the majority of the measured persistence length values. As shown in Fig. 6C, the best value of the persistence length of A tract-containing molecules is close to the standard value of 475 Å. A key element in our studies is the fact that we have simulated the ring closure probabilities of a *family* of constructs and we have compared the result of our calculations with experimental data for that same *series* of molecules; determination of the bend angle induced by an A tract would have been impossible from analysis of the same kind of data for only one molecule.

Cyclization of Protein–DNA Complexes

The importance of measuring the J factor for a family of molecules extends to the study of protein–DNA complexes as well. Experiments showing that binding of CAP protein,[14,32] HU protein,[33] or Cro protein[34] accelerates DNA cyclization have confirmed that these proteins bend DNA, but, in the absence of measurements at a variety of lengths or measurements of the J factor, quantitative information on the extent of protein-induced bending has not been obtained. Information potentially available from cyclization methods includes the extent of bending caused by the protein, any induced change in helical twist, and perhaps changes in the persistence length and torsional flexibility of the bound DNA.

In applying the combined cyclization and Monte Carlo methods to protein–DNA interactions we have been guided by three essential design principles. First, to maximize the sensitivity of the experiment to changes induced by the protein, the cyclization construct should be as small as possible, and a variety of lengths spanning at least one helical repeat should be used. This idea has also been discussed by others.[4,23,29] We note that the position of the protein binding site relative to the end of the DNA should not in principle affect the J factor, since the partition function of DNA conformations which close the chain is effectively independent of the position of the cyclization site, assuming the polymer to be a Markov chain (one for which fluctuations in the geometries of individual base-pair steps are not correlated with those located at any substantial distance away in the chain, thereby neglecting possible long-range polymer–polymer interactions). However, the protein binding site should not occlude the end and prevent ligase binding, and the sequence of the ends may affect the reaction.[14,32]

Second, the cyclization constructs should be large enough to cyclize with and without bound protein, both in the test tube and in the computer. In general, we believe that relative J factors are experimentally and computationally more reliable than absolute numbers, requiring that the molecule cyclize under both sets of conditions. With presently available computers and programs, molecules whose cyclization is slow but experimentally accessible ($J < 5 \times 10^{-10} M$) cannot be simulated easily.

Third, it is important to design the system such that K_a is as small as is practical; that is, a one- or two-base overhang at the cyclization site

[32] D. Dripps and R. M. Wartell, *J. Biomol. Struct. Dyn.* **5**, 1 (1987).

[33] Y. Hodges-Garcia, P. J. Hagerman, and D. E. Pettijohn, *J. Biol. Chem.* **264**, 14621 (1989).

[34] Y. Lyubchenko, L. Shlyakhtenko, B. Chernov, and R. E. Harrington, *Proc. Natl. Acad. Sci. U.S.A.* **88**, 5331 (1991).

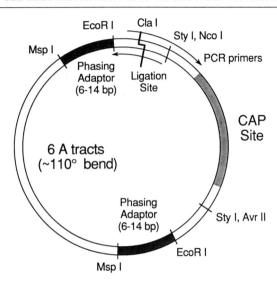

Fig. 7. DNA constructs for cyclization kinetics studies on CAP–DNA complexes (J. D. Kahn and D. M. Crothers, unpublished work). Restriction sites used in initial construction of circles are shown. The total length of the circles ranges from 150 to 166 bp depending on the phasing adaptors. Body-labeled substrates for ligation reactions are produced by PCR from cloned DNA using the primers shown, followed by ClaI restriction and gel purification.

should be used rather than a four-base overhang. A small molecule whose equilibrium conformation is appropriate for cyclization will cyclize extremely rapidly, since thermal fluctuations do not cause deviations from equilibrium as dramatic as those for larger molecules. In optimal cases, we expect that protein-bound DNA may have a very high J factor ($\sim 10^{-3}$ M). As discussed above, in these cases a substantial fraction of the population may be substrates for ligation [f_s in Eq. (6) may not be negligible], and the measured J_{app} will underestimate J. The cyclization rate may remain proportional to ligase concentration, so there may be no direct experimental indication that the measurement is in error.

These considerations have led us to the system shown in Fig. 7 for analysis of CAP-induced DNA bending. The inclusion of an A tract multimer sequence allows the size of the constructs to be reduced, and changing the phasing relationships between the sequence-directed and protein-induced bends amplifies the effect of the protein on cyclization. Our results show that in this context the binding of CAP can increase the J factor up to about 120-fold or reduce it by at least 10-fold depending on phasing. The second phasing adaptor, between the A tracts and the cyclization site, allows independent variation of the phasing between bends and the torsional alignment at the DNA ends. The *Cla*I 5'-CG overhang at the cyclization site allows efficient cyclization, and based on the temperature-jump experiments discussed above it should not give problems owing to nonnegligible values of f_s.

Conclusion

Cyclization kinetics is a valuable method for analysis of DNA in solution. The fundamental experimental observable is the *J* factor, the ratio of equilibrium constants for cyclization and bimolecular association; this ratio as opposed to the individual rates is obtained from theoretical treatments. The sensitivity of the *J* factor to helical repeat, flexibility, and conformation of DNA enables application of the method to the determination of both static and dynamic properties of DNA. The relative simplicity of the experiments and the large range of accessible *J* values give the method advantages over techniques such as fluorescence energy transfer for solution studies of long-range DNA properties. The method and the theory of cyclization kinetics also have applications in the study of DNA looping and topology, and in principle the experiments can be performed on complex nucleoprotein structures.

The sensitivity of these measurements to several different aspects of DNA structure and dynamics imposes several requirements for quantitative experiments. Throughout this chapter, we have emphasized the need to study families of related molecules and the advantages of checking or determining as many parameters as possible by independent methods. Our work has emphasized the use of comparative electrophoretic techniques in this regard (see the accompanying review[35]).

DNA length, helical repeat, flexibility, and conformation interact in complex ways to determine the *J* factor, and no single experiment can disentangle their effects. Therefore, whereas qualitative interpretation of cyclization results can indicate that a DNA sequence or a protein–DNA complex is bent or has an altered helical repeat, extracting quantitative information for most systems requires an accompanying computational effort. We have described a Monte Carlo methodology for the calculation of *J* factors from structural and dynamic DNA parameters. These methods give estimates which differ significantly from those of simpler theories for the torsional modulus of DNA and for the bend angle induced by an A tract segment. Cyclization kinetics is unusual among molecular biological techniques in requiring theoretical work for the interpretation of experimental results.

[35] D. M. Crothers and J. Drak, this volume [3].

[2] Analyzing DNA Curvature in Polyacrylamide Gels

By STEPHAN DIEKMANN

Introduction

This chapter focuses on sequence-induced DNA curvature or intrinsically bent DNA, terms which refer to net (time-averaged) deflections of the helix axis from straight linearity. The DNA structure is assumed to be in the equilibrium state with no external force applied; this is different from bent DNA structures which are formed due to external forces (e.g., by bound proteins[1]). Bending effects due to external forces result from different sequence-specific base pair properties than sequence-induced curved DNA. Because curved DNA migrates slower than straight DNA in polyacrylamide gels, this experimental effect has been used to analyze the properties of curved DNA. In this chapter the mobility analysis of DNA fragments in polyacrylamide gels is described in some detail, and the origin of DNA curvature is discussed.

Experimental Approach to Curved DNA

The different molecular properties of $dA \cdot dT$ and $dG \cdot dC$ base pairs result in a sequence-specific local modulation of the DNA structure.[2] For particular sequences or structural motifs this can lead to a deflection of the helix axis from straight linearity. When such sequences or motifs are repeated in phase with the helix repeat [~ 10.5 base pairs (bp)[3]], the local deflections point toward the same direction so that the DNA appears generally curved. Short tracts of the homopolymer $dA \cdot dT$ confer intrinsic curvature on the axis of the DNA double helix (for reviews, see Ref. 4–8). This ability is assumed to be a consequence of such tracts adopting a stable B′-DNA conformation which is distinct from that normally assumed by other DNA sequences. A major distinguishing feature of the B′ structure

[1] A. A. Travers, *Annu. Rev. Biochem.* **58**, 427 (1989).

[2] R. E. Dickerson and H. R. Drew, *J. Mol. Biol.* **149**, 761 (1981).

[3] S. Diekmann and J. C. Wang, *J. Mol. Biol.* **186**, 1 (1985).

[4] E. N. Trifonov, *Crit. Rev. Biochem.* **19**, 89 (1986).

[5] S. Diekmann, *in* "Nucleic Acids and Molecular Biology" (F. Eckstein and D. M. J. Lilley, eds.), Vol. 1, p. 138. Springer-Verlag, Berlin, 1987.

[6] A. A. Travers and A. Klug, *Philos. Trans. R. Soc. London, Ser. B* **317**, 537 (1987).

[7] P. J. Hagerman, *Annu. Rev. Biochem.* **59**, 755 (1990).

[8] D. M. Crothers, T. E. Haran, and J. G. Nadeau, *J. Biol. Chem.* **265**, 7093 (1990).

METHODS IN ENZYMOLOGY, VOL. 212

is a high propeller twist[9-14] that optimizes stacking interactions and narrows the minor groove. In principle such a propeller twist can only be readily adopted by base pairs with two, but not three, hydrogen bonds. The naturally occurring base pair with this property is dA · dT, but other base pairs, notably dI · dC, also share this characteristic. The B' structure might also have a large wedge roll angle between the dAA dinucleotides.[4,5,8]

DNA curvature can be detected using several experimental methods: circularization of DNA molecules,[15-18] electron microscopy,[19,20] and gel migration in gel electrophoresis.[3,21-27] Results obtained from these measurements can be compared with DNA structures obtained from nuclear magnetic resonance (NMR)[11,28] and X-ray crystallographic analysis.[12-14] Other techniques like Raman,[29] circular dichroism (CD),[29,30] and NMR[9,31,32] spectroscopy yield relevant information on the structural properties of curved DNA sequences.

[9] R. Behling and D. R. Kearns, *Biochemistry* **25**, 3335 (1986).

[10] D. G. Alexeev, A. A. Lipanov, and I. Y. Skuratovski, *Nature (London)* **325**, 821 (1987).

[11] A. A. Lipanov and V. P. Chuprina, *Nucleic Acids Res.* **15**, 5833 (1987).

[12] H. C. M. Nelson, J. T. Finch, B. F. Luisi, and A. Klug, *Nature (London)* **330**, 221 (1987).

[13] M. Coll, C. A. Frederick, A. H.-J. Wang, and A. Rich, *Proc. Natl. Acad. Sci. U.S.A.* **84**, 8385 (1987).

[14] A. D. DiGabriele, M. R. Sanderson, and T. A. Steitz, *Proc. Natl. Acad. Sci. U.S.A.* **86**, 1816 (1989).

[15] K. Zahn and F. Blattner, *Nature (London)* **317**, 451 (1985).

[16] L. Ulanovsky, M. Bodner, E. N. Trifonov, and M. Choder, *Proc. Natl. Acad. Sci. U.S.A.* **8**, 862 (1986).

[17] S. D. Levene and D. M. Crothers, *J. Mol. Biol.* **189**, 61 (1986).

[18] K. Zahn and F. Blattner, *Science* **236**, 416 (1987).

[19] J. Griffith, M. Bleyman, C. A. Rauch, P. A. Kitchin, and P. T. Englund, *Cell (Cambridge, Mass.)* **46**, 717 (1986).

[20] B. Theveny, D. Coulaud, M. LeBret, and B. Revet, *in* "Structure and Expression Volume 3: DNA Bending and Curvature" (W. K. Olson, M. H. Sarma, R. H. Sarma, and M. Sundaralingam, eds.), p. 39. Adenine Press, Schenectady, New York, 1988.

[21] J. C. Marini, S. D. Levene, D. M. Crothers, and P. T. Englund, *Proc. Natl. Acad. Sci. U.S.A.* **79**, 7664 (1982).

[22] J. C. Marini and P. T. Englund, *Proc. Natl. Acad. Sci. U.S.A.* **80**, 7678 (1983).

[23] H. M. Wu and D. M. Crothers, *Nature (London)* **308**, 509 (1984).

[24] P. J. Hagerman, *Biochemistry* **24**, 7033 (1985).

[25] S. Diekmann, *FEBS Lett.* **195**, 53 (1986).

[26] H. S. Koo, H. M. Wu, and D. M. Crothers, *Nature (London)* **320**, 501 (1986).

[27] P. J. Hagerman, *Nature (London)* **321**, 449 (1986).

[28] M. H. Sarma, G. Gupta, and R. H. Sarma, *Biochemistry* **27**, 3423 (1988).

[29] S. Brahms and J. G. Brahms, *Nucleic Acids Res.* **18**, 1559 (1990).

[30] S. R. Gudibande, S. D. Jayasena, and M. J. Behe, *Biopolymers* **27**, 1905 (1988).

[31] J.-L. Leroy, E. Charretier, M. Kochoyan, and M. Gueron, *Biochemistry* **27**, 8894 (1988).

[32] J. G. Nadeau and D. M. Crothers, *Proc. Natl. Acad. Sci. U.S.A.* **86**, 2622 (1989).

From the migration analysis in polyacrylamide gels, absolute values are not obtained for structural DNA parameters. This is a principal drawback of the gel assay. Therefore, conclusions obtained from this technique should be verified by other methods based on better understood physical techniques. However, rigorous techniques are often elaborate and do not detect the local structural variations which can be resolved in the gel (e.g., see Diekmann and Pörschke[33]). The gel assay is very convenient, and small amounts (a few nanograms) of DNA can be analyzed. Furthermore, the mobility of DNA molecules in gels is not affected by the presence of other DNA fragments or a variety of chemical compounds present in small quantities. The gel migration anomaly of considerably curved DNA fragments is a rather large effect since the anomaly increases with the square of the degree of the curvature.[34,35] The experimental results are easy to analyze, and the effects of temperature and salt on the sequence-specific structures can also be measured in the gel.[3,36]

Polyacrylamide Gel Assay

DNA molecules having a curved helix axis migrate slower through the acrylamide gel pores than do straight molecules, resulting in a migration anomaly. The pore sizes in polyacrylamide gels (1 to 8 nm) are in the range of the DNA diameter (2 nm). In gels with larger pores (like agarose gels) curved DNA fragments of up to a few hundred bp migrate normally. Attempts have been made to describe theoretically the migration of curved DNA fragments through gels,[37,38] yielding partial agreement with experimental data. However, a general quantitative theory for the migration of DNA molecules through gel pores that explains the migration anomaly of curved fragments is still not available. Therefore, the migration anomaly must be calibrated relative to the mobility of normal marker fragments[39] (see section on data analysis and Fig. 5).

A migration anomaly of DNA fragments is observed not only for curved DNA, other structural variations of DNA lead to a reduced mobility in gels as well. For example, a DNA four-way junction present in a DNA

---

[33] S. Diekmann and D. Pörschke, *Biophys. Chem.* **26**, 207 (1987).

[34] E. N. Trifonov and L. Ulanovsky, *in* "Unusual DNA Structures" (R. D. Wells and S. C. Harvey, eds.), p. 173. Springer-Verlag, New York, 1988.

[35] H. S. Koo and D. M. Crothers, *Proc. Natl. Acad. Sci. U.S.A.* **85**, 1763 (1988).

[36] S. Diekmann, *Nucleic Acids. Res.* **15**, 247 (1987).

[37] C. R. Calladine, H. R. Drew, and M. J. McCall, *J. Mol. Biol.* **201**, 127 (1988).

[38] S. D. Levene and B. H. Zimm, *Science* **245**, 396 (1989).

[39] N. C. Stellwagen, *Biochemistry* **22**, 6186 (1983).

fragment results in reduced mobility.[40,41] This structural origin of a migration anomaly in a polyacrylamide gel can be separated easily from curved DNA by running a gel in the presence of ethidium bromide: while the anomaly due to the four-way junction increases, the anomaly due to curved DNA is strongly reduced.[41] Also, protein–DNA complexes migrate slower in gels.[23,42–44]

Influence of Helical Repeats

Particular sequences or structural motifs (e.g., dA · dT tracts) lead to a deflection of the helix axis from straight linearity. When such sequences are repeated in phase with the helix repeat, the DNA axis is always deflected in the same direction, adding up to a planar curvature of the molecule. If the phase relation between sequence motif and helical repeat is not perfect, the overall curvature will not be planar; instead the DNA helix axis will adopt a superhelical form.[37] The migration anomaly in polyacrylamide gels is very sensitive to the phasing between sequence and helical repeat.[24,26] If a curved sequence motif is repeated every 9 or 12 bp, almost no migration anomaly is observed in polyacrylamide gels. Considerable effects are measured for repeat lengths 10 and 11 bp. The anomaly becomes large when two curved sequences are repeated, one 10 and one 11 bp, alternatively along the helix (repeated sequence length is 21 bp). Then the mean sequence repeat is 10.5 bp, which is very close to the helical repeat (10.5 bp/turn for B-form DNA,[45] 10.4 bp/turn for repeated dA · dT tracts[3]).

Many properties of curved DNA have been deduced from the electrophoretic behavior of sets of DNA molecules obtained from a partial ligation of double-stranded oligomers (often 10 bp in length[24,26,27,46,47]). These multimer distributions of oligomers show normal gel migration up to a length of about 50 bp. Beyond this length, a migration anomaly sets in that becomes large for longer molecules. If the sequence repeat matches the helical repeat (e.g., repeat length 21 bp), the anomaly continuously in-

[40] G. W. Gough and D. M. J. Lilley, *Nature (London)* **313**, 154 (1985).
[41] S. Diekmann and D. M. J. Lilley, *Nucleic Acids Res.* **15**, 5765 (1987).
[42] M. M. Garner and A. Revzin, *Nucleic Acids Res.* **9**, 3047 (1981).
[43] M. G. Fried and D. M. Crothers, *Nucleic Acids Res.* **9**, 6505 (1981).
[44] M. M. Garner and A. Revzin, *Trends Biochem. Sci.* **11**, 395 (1986).
[45] L. J. Peck and J. C. Wang, *Nature (London)* **292**, 375 (1981).
[46] H. S. Koo and D. M. Crothers, *Biochemistry* **26**, 3745 (1987).
[47] S. Diekmann, E. von Kitzing, L. W. McLaughlin, J. Ott, and F. Eckstein, *Proc. Natl. Acad. Sci. U.S.A.* **84**, 8257 (1987).

creases with the length of the molecules.[48] However, if the phase relation between sequence and helical repeat is not perfect (e.g., repeat length 10 bp), the anomaly levels off to reach a plateau at about 200 bp[48] (see Figs. 1 and 6). For both cases (10 and 21 bp repeat length), the anomaly values agree only for short molecular lengths, between 50 and 100 bp. Because the anomaly of longer molecules is influenced by the phase relation between sequence and helical repeat, the local curvature of a sequence motif should be deduced only from the anomaly of shorter molecules (≤ 100 bp).

General Sequence Design

If only the degree of the local curvature (of about one helix turn or less) is to be determined, the migration anomaly of shorter molecules (≤ 100 bp) should be analyzed. For longer molecules, the phase relation between sequence and helical repeat influences the measured anomaly value (see above). The shorter molecules have a smaller migration anomaly resulting in a larger relative error. The migration anomaly of a curved DNA fragment can be increased, however, by the addition of normally migrating DNA arms to both ends of the curved fragment, up to the persistence length (~ 150 bp[3]). To obtain molecules with the curved DNA sequence embedded between straight DNA sequences, the corresponding DNA sequence can be cloned.[25,36] Fragments convenient for experimental analysis containing the curved DNA can be cut out of the cloning vector by restriction endonucleases.[36] This approach yields DNA containing no nicks (single-strand breaks) within the fragment of interest. In vitro ligation techniques probably leave nicks; the structural influence of these single-strand breaks are unclear. It should be noted that (plasmid) DNA for sequence-specific structure analysis should not be prepared by a protocol in which drugs are added to the DNA. Some drugs bind very tightly to particular DNA sequences, and it is difficult to prove that all drugs are removed from the DNA before the experimental analysis. Therefore, for structural analysis plasmid DNA should not be purified in a CsCl gradient in the presence of ethidium bromide, but rather by high performance liquid chromatography (HPLC).[36]

Normally, the cloning approach cannot be used when the influence of chemically modified bases on curved DNA is analyzed. As an alternative

---

[48] S. Diekmann and E. von Kitzing, in "Structure and Expression, Volume 3: DNA Bending and Curvature" (W. K. Olson, M. H. Sarma, R. H. Sarma, and M. Sundaralingam, eds.), p. 57. Adenine Press, Schenectady, New York, 1988.

to cloning, multimer distributions of oligomers can be ligated between reporter arms.[49,50] This approach can be used for *in vitro* studies.

Method to Measure Small Sequence-Specific Structure Changes

Using a particular sequence design, the gel assay can be applied to measure very subtle sequence-specific structural influences on the DNA axis in solution. In particular, the method can be used to determine the relative order of the sequence-specific twist or wedge angles of DNA sequences.[51]

Wedge Angles. If only a single dinucleotide has a nonzero wedge angle within a longer sequence, the influence of this wedge angle on the DNA helix axis is in general too small to be detected by the gel assay. For example, multimers of 10-bp-long oligonucleotides containing a single dAA dinucleotide do not exhibit migration anomaly,[25,26] because the gel matrix is insensitive to such small degrees of curvature.[34,35]

Nevertheless, the effect of small wedge angles can be amplified using the following approach. A series of 21-bp-long double-stranded oligonucleotides is constructed[48] having the sequences

5'-dCAAAAACCGTGCCGCCCTCGA-3'
5'-dCAAAAACCGTGCCNNCCTCGA-3'

with dNN being the dinucleotide of interest. These 21-mers are ligated together to form multimer distributions and analyzed in polyacrylamide gels. The dA$_5$ tract at the 5' end creates a considerable migration anomaly. The rightmost 11 bp contain a sequence to be studied. The anomaly is unchanged when going from dCCGCC to, for example, dCCACC.[48] However, a linear increase in the migration anomaly is observed when going from dCCACC to dCCAAC and further to dCAAAC. A clear deviation from this linear increase is observed for dCAAAAC, suggesting a cooperative structural transition between dA$_3$ and dA$_4$.[48] In this way, the low sensitivity region of the gel for small degrees of curvature can be circumvented. The extreme sensitivity at higher migration anomalies is used to detect very small influences on the helix axis.[51] This experiment could clearly detect the migration anomaly difference between the sequences dCACC and dCAAC.

Twist Angles. For longer molecules the migration anomaly levels off

[49] P. J. Hagerman, *in* "Unusual DNA Structures" (R. D. Wells and S. C. Harvey, eds.), p. 224. Springer-Verlag, New York, 1988.
[50] P. J. Hagerman, *Biochemistry* **29**, 1980 (1990).
[51] S. Diekmann, *Electrophoresis* **10**, 354 (1988).

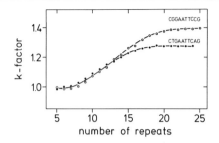

FIG. 1. Apparent length divided by sequence length (k factor) versus the number of repeats for multimers composed of decanucleotides of the sequence 5'-dCGGAATTCCG-3' and 5'-dCTGAATTCAG-3'. The k factor is determined in 10% polyacrylamide gels with *Hae*III and *Hin*fI digests of the plasmid pBR322 as length markers. The experimental error of the k factor is ±0.02. No migration anomaly is observed up to 70 bp. The anomalies of both sequences coincide up to a length of 130 bp, and anomaly differences become apparent beyond 150 bp. (From Ref. 52, with permission.)

owing to the nonperfect phase relation between sequence and helix repeat. The helix repeat reflects the sum of all dinucleotide twist angles of the repeated sequence. Thus, the observed value of the migration anomaly, particularly in the plateau region of the longer molecules (see section on helical repeats, above), depends not only on the degree of the local DNA axis deflection, but—via the phase relation—also on the sum of all dinucleotide twist angles involved. The value of the anomaly is modulated by the twist angles of those dinucleotides which do not contribute to the curvature (since they determine the helical repeat value). An example of the anomaly (displayed as the ratio apparent length in the gel divided by the sequence length, called the k factor, versus DNA sequence length, see section on data analysis below) of the two different sequences dCTGAATTCAG and dCGGAATTCCG is presented in Fig. 1.[52] Both molecules have the same curved sequence stretch (dAATT). Consequently, small molecular lengths (60 to 130 bp) have identical anomaly; anomaly differences are found only for higher molecular lengths (beyond 150 bp). The differences within the two sequences, dCAGCTG compared to dCCGCGG, are placed between the dAATT region. We interpret this result as being due to the different twist angles of the hexanucleotides. Thus, the gel assay can be used to measure the relative order of the twist angles of particular DNA sequences.[51]

[52] S. Diekmann and L. W. McLaughlin, *J. Mol. Biol.* **202**, 823 (1988).

DNA Preparation and Gel Run

Oligonucleotides

In general, DNA oligonucleotides are made on a DNA synthesizer. The oligonucleotides should be purified (if possible twice) by reversed-phase HPLC.[53,54] The oligonucleotides are purified with their chemical protecting group; then the protecting group is removed and the oligonucleotides are repurified. Oligonucleotides prepared in this way can be ligated together to produce high multimer numbers which give sharp bands in polyacrylamide gels. If this protocol is simplified and only the second (or none) HPLC purification is performed, additional bands might be observed in the ligation multimer distribution. Furthermore, the distribution might lack high multimer numbers, and the bands in the polyacrylamide gels sometimes become fuzzy.

The oligonucleotide sequence should be verified. After phosphorylation by polynucleotide kinase, annealing, and ligation, the multimer distribution can be radioactively labeled and chemically sequenced according to the protocol of Maxam and Gilbert.[55]

The DNA can be detected in the gel either by radioactivity or by ethidium fluorescence. The oligomers ligated to long ligation ladders for gel analysis can be radioactively labeled with $[\gamma\text{-}^{32}P]rATP$ by kinase. These longer multimers contain many radioactive labels which allow very small amounts of DNA to be detected. However, during the ligation process to form multimers, low DNA concentrations lead to an increased probability of circle formation; linear molecules with high multimer number might be obtained to a lesser extent. Alternatively, for fluorescence detection (the oligomers are kinased with "cold" rATP), larger amounts of molecules have to be prepared (1 to 3 μg for several gel runs), requiring higher amounts of enzymes. As a direct consequence, the higher DNA concentration (working also in small volumes) yields linear molecules with very high multimer numbers and small probabilities for circle formation (see Fig. 2). DNA which is not radioactively labeled can be detected by the fluorescence of intercalated ethidium (see below, gel run). Because binding has an influence on the DNA structure (DNA curvature is strongly reduced in the presence of ethidium bromide[3]), the reagent has to be added after the gel run.

[53] J. Ott and F. Eckstein, *Nucleic Acids Res.* **23,** 9137 (1984).
[54] L. W. McLaughlin and N. Piel, *in* "Oligonucleotide Synthesis: A Practical Approach" (M. J. Gait, ed.), p. 199. IRL Press, Oxford, 1984.
[55] A. M. Maxam and W. Gilbert, this series, Vol. 65, p. 499.

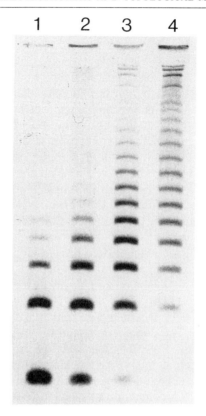

FIG. 2. Ligation ladders of the 21-bp oligonucleotide 5'-d(GTGCAAACCTCGA-CAAAAACC)-3' and its pairing strand 5'-d(CACGGTTTTTGTCGAGGTTTG)-3' run in a 6% polyacrylamide minigel. The kinase and ligation reaction was performed as described in the text. Aliquots of the ligation reaction were taken after 3 min (lane 1), 8 min (lane 2), 20 min (lane 3), and 45 min (lane 4), ethanol precipitated, and loaded on the gel (~0.5 μg DNA per lane). The DNA is detected by ethidium fluorescence. In lanes 3 and 4 DNA circles can be identified (gel bands close to the origin of the gel).

An example for the cold labeling of oligonucleotides is listed:

3 μg Single-stranded oligonucleotide	9 μl
10× Kinase buffer	2 μl
10 mM rATP freshly prepared from concentrated stock	2 μl
T4 polynucleotide kinase, 5 to 10 units	1 μl
Water	6 μl

The kinase reaction mixture is placed in a 37° water bath for 2 hr. Then, the single strand is mixed with its pairing strand which has been similarly treated; both 20-μl kinase reaction mixtures give a total of 40 μl. The 40-μl mixture is slowly cooled to 16°. Then 200 to 400 units (1 μl) of T4 DNA ligase (e.g., New England Biolabs, Beverly, MA) is added. Ligase works in the kinase buffer; addition of bovine serum albumin (BSA) or dithiothreitol is not necessary. (If the oligonucleotides are blunt-ended, good ligation yields are obtained with higher ligase concentration and at a higher temperature, e.g., 35°.) Ten-microliter aliquots are taken from the reaction mixture after 3, 8, 20, and 45 min. The aliquots are mixed immediately with 5 μl of 3 M sodium acetate, 35 μl water, and 100 μl ethanol for precipitation. The precipitation is immediate; neither low temperature ($-80°$) nor longer waiting times are required. However, it is essential to spin for at least 20 min at high speed in a bench-top centrifuge. Every aliquot yields a particular ligation multimer distribution which is visualized in a 6% polyacrylamide minigel (0.5 μg DNA per lane, see Fig. 2). Those multimer distributions are mixed which after mixing will give a uniform ligation ladder from short to very long molecules.

Polyacrylamide Gel Electrophoresis

Fresh acrylamide solution (e.g., 29 : 1, acrylamide to bisacrylamide) in 0.5× TBE [45 mM Tris (trishydroxymethylaminomethane), 45 mM boric acid, 1.25 mM EDTA (ethylenedinitrilotetraacetic acid disodium salt dihydrate)] and ammonium peroxodisulfate is filtered, mixed, and degassed. After addition of TMED (N,N,N',N'-tetramethylethylenediamine) the solution is poured between glass plates. The gel glass plates we use are 16 × 22 cm in size with 1.5 mm spacings. The comb and a platinum temperature detector are placed in the acrylamide solution between the glass plates. Polymerization is allowed to proceed for several hours. Finally, the comb is removed, and the wells are carefully washed with buffer. The glass plates including the gel and the temperature monitor are placed into a gel tank, the upper electrolyte compartment of which is normal size. The lower compartment is rather large (6 liters) and embeds the whole gel and the upper electrode compartment. Thermostatted water runs in glass helices through both compartments thermostatting the whole gel tank, which is thermally isolated.

The gel is prerun for several hours in 0.5× TBE running buffer at 4.5 V/cm until the temperature in the gel and the electrical current are constant. The DNA solutions are mixed with Ficoll and tracking dyes to a total volume of about 5 μl and loaded on the gel at the bottom of the wells. After the run (of several hours) the gel is stained for 1 hr in an ethidium bromide solution (~2 μg/ml) and washed 1 hr in water. The gel is then placed on a 254 nm UV transilluminator and photographed through a set

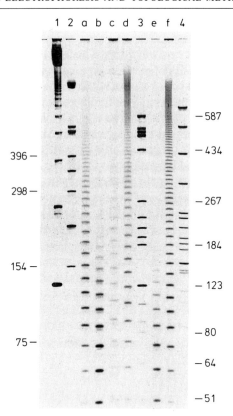

FIG. 3. Ligation ladders of the sequences 5'-d(GGGCRRRRRC) · 5'-d(CCCGYYYYYG). Lane 1, 123-bp ladder; lanes 2, 3, and 4, length markers of pBR322 origin (lane 2, *Hin*fI digest; lane 3, *Hae*III digest; lane 4, *Hpa*II digest); lanes a–d, dR_5 has the sequence dAIAIA and dY_5 has the sequence dUCUCU (lane a), dUMUMU (lane b), dTCTCT (lane c), or dTMTMT (lane d); lanes e and f, dR_5 has the sequence dAIIIA, and dY_5 has the sequence dTMMMT (lane e) or dTCCCT (lane f). M stands for methylated cytosine. The gel was 10% polyacrylamide in 0.5× TBE buffer run at room temperature and 4.5 V/cm. (From Ref. 65, with permission.)

of three optical filters (UV, yellow, orange) using a high-resolution black-and-white film [Kodak, Rochester, NY, AHU; this film should be processed with Neofin blue (Tetenal, Norderstedt, FRG) developer at 20° for 4 min]. The film negative is enlarged to at least the original gel size (16 × 22 cm) and analyzed. A typical gel is shown in Fig. 3.

The migration anomaly of curved DNA is temperature and salt dependent. For example, a change in migration anomaly was detected by the

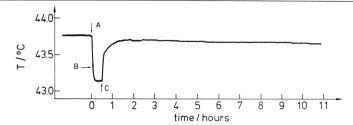

FIG. 4. Temperature measured during the course of an electrophoresis experiment. A 6% polyacrylamide gel was preelectrophoresed several hours at 150 V until temperature and current were stable. Then the voltage was disconnected (at point A) and the DNA samples loaded. The DNA was electrophoresed into the gel at 50 V (point B). After 25 min the voltage was increased back to 150 V (point C). During the gel run, the temperature in the gel was measured by a Pt element polymerized in the gel. The resistance readout was temperature calibrated and compared to a preset electrical value; only a difference value is indicated. This enables a very sensitive detection (±0.002°) independent of the temperature range analyzed. After loading, the temperature in the gel during the run was stable within ±0.03°. (From Ref. 41, with permission.)

removal of EDTA from the running buffer without addition of salt.[36] Therefore, the conditions in the gel should be monitored. When the influence of different salts is measured, the additional salt is added to the running buffer as well as to the acrylamide solution before polymerization. The conductivity of the buffer is measured by a Wayne–Kerr bridge before and after the gel run. During the gel electrophoresis, the temperature is measured by a Pt element polymerized in the gel. The temperature in a 6% gel during an electrophoretic run is plotted in Fig. 4.

If divalent ions are present in the buffer, the electrical heating during the gel run due to the current can increase the temperature in the gel over the buffer temperature in the electrolyte compartments by up to 10°. Under these conditions, the temperature has to be measured in the gel; temperature measurements in the electrolyte compartments would give values that are too low. In addition, the high current (up to 100 mA) produces electrode processes which change the buffer conditions, resulting in a reduction of the DNA migration in the gel. Higher concentrations of these contaminating compounds will modify the gel matrix. To avoid the production of these compounds and their influence on the gel, the buffer in the compartments should be pumped with 5 liters per hour each way (from the anode to the cathode compartment and vice versa). Alternatively, semipermeable ion-exchange membranes can be placed in the compartments between electrode and gel (particularly a cation-exchange membrane in the cathode compartment). Such membranes stop the penetration of electrode products into the gel.

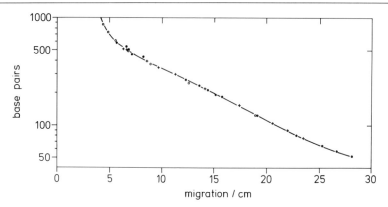

Fɪɢ. 5. Reference curve of the marker fragments in a 10% polyacrylamide gel run at 22°
in 0.5× TBE buffer and 4.5 V/cm. The graph shows pBR322 restriction digests with *Hin*fI
(+), *Hae*III (●), and a 123-bp ladder (○). (From Ref. 51, with permission.)

Migration anomaly of curved DNA in polyacrylamide gels depends on
many experimental parameters such as the gel concentration, the gel
thickness (defined by the spacers between the glass plates), the kind of
salt and salt concentration, the temperature, and the presence of drugs.
The anomaly also depends on the electrical field strength in the gel if the
value is larger than 5 V/cm. Good gels with sharp DNA bands are obtained
from freshly made acrylamide–bisacrylamide solutions. Thus, the migration anomaly of curved DNA should be measured under standardized
conditions. For quantitative comparison values of the migration anomaly
should be taken from neighboring lanes in the same gel.

Data Analysis: Gel Markers and k Factor

Various marker fragments should be loaded on the gel. A reasonable
DNA fragment distribution is obtained by using a *Hae*III and a *Hin*fI plus
(partial) *Eco*RI restriction digest of pBR322 together with a 123-bp DNA
fragment ladder (Gibco–BRL, Gaithersburg, MD). These markers are
loaded in different concentrations on both sides of the lanes to be analyzed.
Identical markers on both sides of the gel check for inhomogeneities in
the gel. The migration of the marker fragments is measured and plotted
[logarithm of the base pairs versus distance (in cm) migrated; see Fig. 5].
This graph is the calibration for the length determination of the multimer
distributions and is only roughly linear. The actual curve is taken for the
length determination, not the linear approximation.

Because short oligonucleotides (<50 bp) migrate according to their size
independent of any curvature, the short oligonucleotides in the multimer

distribution are used to identify the sequence length of the molecules. The apparent length of the DNA fragment divided by the actual sequence length defines a relative value (the k factor, R factor) which is plotted versus the actual fragment sequence length given in base pairs or the number of oligonucleotide repeats. For fragments with normal mobility in gels this relative factor is 1.0.

Models for DNA Curvature

When normal DNA sequences alternate with dA_n tracts in phase with the helix repeat, the DNA fragments have a reduced mobility in polyacrylamide gels; this is interpreted as DNA curvature.[21-27] It is common to all models and structures describing this effect that normal DNA sequences have a small propeller twist, whereas the dA tracts have a large propeller twist.[9,10,12-14,32,56] A large propeller twist can be formed easily only by base pairs with two hydrogen bonds, but not three. This effect is facilitated when the neighboring base pairs are of the same nature. Then, the large propeller twist is extended along the molecule. The gel migration data available can be interpreted in terms of a cooperative structure transition between dA_3 and dA_4 from a B-form to a B'-form DNA structure.[48,57] The cooperative structural parameter might be the extended interactions of neighboring base pairs with propeller twist.

Solution data indicate that the propeller twist is accompanied by a wedge roll angle.[9,10,11,58] From the experimental data available it remains unclear if the dAA dinucleotides in the dA tracts include a wedge roll angle.[59] Solution studies (reviewed by Crothers et al.[8]) and a molecular mechanics model[56] indicate the presence of a large wedge roll angle. However, structures obtained from X-ray crystallography display no angles[12,14] or very small[13] angles. However, in the crystals the DNA structure is deformed due to crystal packing forces.[14,32]

The above model for DNA curvature implies that any base pair which can adopt a high propeller twist could, in principle, confer intrinsic curvature on a DNA molecule. In addition to $dA \cdot dT$, another such base pair with this potential is the $dI \cdot dC$ pair. Indeed, the substitution of a $dI \cdot dC$ base pair for a $dA \cdot dT$ base pair in the center of a $dA_5 \cdot dT_5$ tract results in only a slight reduction in the apparent intrinsic curvature.[46,47] However, the sequence $dI_5 \cdot dC_5$, when repeated in phase with the helix repeat and

[56] E. von Kitzing and S. Diekmann, *Eur. Biophys. J.* **15**, 13 (1987).
[57] T. E. Haran and D. M. Crothers, *Biochemistry* **28**, 2763 (1989).
[58] S. P. Edmondson and W. C. Johnson, *Biopolymers* **24**, 825 (1985).
[59] L. Ulanovsky and E. N. Trifonov, *Nature (London)* **326**, 720 (1987).

not combined with any dA · dT base pairs, produces only a small gel migration anomaly (k factor of 1.10 in comparison with 2.10 for dA$_5$ · dT$_5$). Thus, the property of a base pair to have only two hydrogen bonds is not a sufficient prerequisite for a sequence block formed by these base pairs to generate DNA curvature. Additional sequence properties are important.

The relative stability of the B' structure of dA$_n$ tracts has been attributed to several possible factors: (1) optimal base stacking interactions consequent on the high propeller twist,[12,13,56] (2) bifurcated hydrogen bonds between adjacent dA · dT base pairs,[12–14] (3) interactions between the dT methyl groups,[50,60,61] and finally (4) a putative spine of water molecules in the minor groove.[62–64] These possibilities have been systematically investigated.

(1) In dA · dT tracts when dT is replaced by dU having no methyl group, the gel migration anomaly hardly changes,[46] suggesting that under the conditions used to determine curvature the methyl group of thymine does not contribute significantly to the stability of the structure. For particular sequences, position-dependent influences of the pyrimidine methyl groups on curvature were detected.[50]

(2) A second factor which has been suggested as a stabilizing influence on the B' structure is the presence of bifurcated hydrogen bonds.[12–14] However, the sequence dAIAIA · dTCTCT, in which the ability to form such bifurcated hydrogen bonds is lost, confers significant curvature.[47] Nevertheless, it can be speculated that although bifurcation and pyrimidine methyl group contributions alone may not be effective, combinations of these effects might be essential elements for B' structure formation and DNA curvature. To test these two factors in combination, oligonucleotides which were unable to form bifurcated hydrogen bridges but had different combinations of methylated pyrimidines were analyzed. Four different DNA oligonucleotides of 10 bp repeat lengths were constructed[65] with the homopurine block dAIAIA in one strand and different sequences in the pyrimidine strand: dTCTCT, dTMTMT, dUCUCU, and dUMUMU (M being methylated C). Considerable anomalous mobility was observed in all four cases (see Fig. 6). In particular, the pyrimidine sequence pairing with dAIAIA might be completely (dTMTMT) or not at all (dUCUCU)

[60] R. L. Jernigan, A. Sarai, K. L. Ting, and R. Nussinov, *J. Biol. Struct. Dyn.* **4,** 41 (1986).
[61] R. L. Jernigan, A. Sarai, B. Shapiro, and R. Nussinov, *J. Biol. Struct. Dyn.* **4,** 561 (1987).
[62] V. P. Chuprina, *FEBS Lett.* **186,** 98 (1985).
[63] V. P. Chuprina, *FEBS Lett.* **195,** 363 (1986).
[64] V. P. Chuprina, *Nucleic Acids Res.* **15,** 293 (1987).
[65] S. Diekmann, J. Mazzarelli, L. W. McLaughlin, E. von Kitzing, and A. A. Travers, in press.

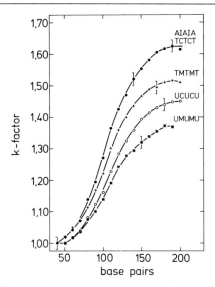

FIG. 6. Graph of k factor (i.e., apparent length relative to sequence length) versus the sequence length in base pairs of the oligonucleotides 5'-d(GGGCAIAIAC)·5'-d(CCCGY-YYYYG) with dY_5 being either dTCTCT (●), dTMTMT (+), dUCUCU (○), or dUMUMU (*). Data are taken from the polyacrylamide gel displayed in Fig. 3. These sequences cannot form bifurcated hydrogen bridges. The pyrimidines are completely, alternatingly, or not at all methylated. Nevertheless, they show considerable migration anomaly. (From Ref. 65, with permission.)

methylated with little influence on the anomaly. Thus, molecules neither forming bifurcated hydrogen bridges nor having methyl groups at the pyrimidines still show significant DNA curvature (k factor of 1.45 compared to 2.10 for $dA_5 \cdot dT_5$).[47] Thus, under the conditions of experimental analysis, neither bifurcation nor methyl groups nor both influences together are essential for DNA curvature.[65]

(3) It has been postulated that a spine of water molecules in the minor groove stabilizes the formation of a B' structure. The resolution of available crystal structures is still too low to observe this in $dA \cdot dT$ tracts, although its occurrence elsewhere has been noted.[66,67] The minor groove of $dI \cdot dC$ tracts is similar if not identical to that of $dA \cdot dT$ tracts, allowing for the same water spine formation. Thus, if the spine of hydration is the dominant stabilizing contribution to the structure of $dA \cdot dT$ stretches, $dI \cdot dC$ tracts should also be curved by a similar amount,[62] in contrast

[66] R. E. Dickerson, H. R. Drew, B. N. Conner, R. M. Wing, A. V. Fratini, and M. L. Kopka, *Science* **216,** 475 (1982).

[67] M. L. Kopka, A. V. Fratini, H. R. Drew, and R. E. Dickerson, *J. Mol. Biol.* **163,** 129 (1983).

to experimental findings. The sequence $dI_5 \cdot dC_5$ shows only a small gel migration anomaly when repeated in phase with the turn of the DNA helix (k factor of 1.10 compared to 2.10 for $dA_5 \cdot dT_5$). This indicates that a spine of hydration might not be sufficient by itself to stabilize a B′ structure. However, the different hydrogen bonding pattern and charge distribution in $dI \cdot dC$ compared to $dA \cdot dT$ base pairs might also influence the spine of hydration in the minor groove. Thus, from these experiments, the role of a spine of hydration in DNA curvature cannot be ruled out.

(4) The remaining inference from the experiments described is that stacking interactions are the main driving force for B′-structure formation as suggested experimentally[47] and theoretically.[56] Optimal stacking in $dA_5 \cdot dT_5$ tracts is assumed to be dependent on a large propeller twist. To adopt the B′ structure, the 2-position of the purine has to be free.[47] The exocyclic groups at the 6-position of purine in the major groove might interact with one another to determine the extent of DNA curvature formed. According to this argument, the NH_2 group at the 6-position of dA would produce a large curvature compared to a smaller curvature for oxygen at the 6-position of dI. If this exocyclic group is reduced to H, the curvature is even smaller.[47] On the other hand, if the NH_2 group is enlarged by methylation, the curvature increases considerably.[68] This model provides an explanation for DNA curvature as an alternative to water spine formation in the minor groove. Of course, both models do not exclude one another, since both structural influences may operate synergistically.

The hypothesis can be made that the ability of a dA_n tract to adopt a B′ structure with large propeller twist, and hence to confer intrinsic curvature, is primarily determined by the stability of the stacking interactions between the adjacent base pairs.

[68] S. Diekmann, *EMBO J.* **6**, 4213 (1987).

[3] Global Features of DNA Structure by Comparative Gel Electrophoresis

By Donald M. Crothers and Jacqueline Drak

Introduction

The polyacrylamide gel electrophoretic mobility of DNA fragments is affected by systematic deviations from linear molecular shape. As a general consequence, comparative electrophoresis measurements can be used

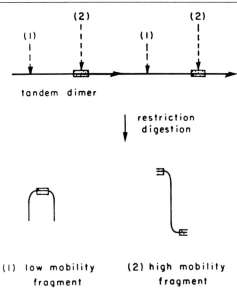

FIG. 1. Construction of circularly permuted variants of a DNA sequence differing in conformation because of the location of a bending locus, shown by the box.[1] The tandem dimer is cleaved with two different enzymes (1) and (2), yielding fragment (1), which is strongly bent and therefore of low gel mobility, and fragment (2), which is nearly linear because the bend is near the end, and is therefore of high mobility. (Reprinted with permission from *Nature*.)

to characterize bends in the DNA double helix, including their position, relative direction, and relative magnitude. Because the overall shape of a molecule containing two or more bends depends on the helical phasing between them, it is also possible to determine accurately the helical repeat of the sequence between them.

The first experiment which systematically exploited this idea was that designed to identify the locus responsible for sequence-directed bending of DNA molecules.[1] The underlying concept is diagrammed in Fig. 1. DNA fragments of equal length are excised from tandem repeats of a single sequence by cleavage with different restriction enzymes which cut only once in the sequence of interest. The result is a set of molecules with circularly permuted versions of the same sequence. If the molecule contains a structural feature which causes the helix axis to bend at a specific locus, then the members of the set will differ in shape: fragments in which the bend is near the center of the molecule will be most curved in overall appearance, whereas those in which the bend is at the end will be nearly linear.

[1] H.-M. Wu and D. M. Crothers, *Nature* (*London*) **308**, 509 (1984).

The electrophoretic consequence of increased overall curvature is a reduction in mobility μ on polyacrylamide gels. This observation, supported by experiments on a wide variety of systems, can be qualitatively justified by the theory of Lumpkin and Zimm.[2] Let Q be the total effective charge on the molecule; ζ is the frictional coefficient for translational movement of the DNA molecule along a "tube" in the gel; h_x is the projection on the field direction of the end-to-end vector of the DNA molecule contained in the tube; and L is the DNA contour length. The simplifying assumptions of Lumpkin and Zimm[2] lead to the prediction that the mobility (the velocity along the field direction divided by the magnitude of the electric field) is

$$\mu = \frac{Q}{\zeta} \langle h_x^2/L^2 \rangle \tag{1}$$

where the angled brackets represent the average over an ensemble of conformations. With the additional assumption that the shape of the molecule in the gel reflects its shape in solution, the reduced mean square projection $\langle h_x^2/L^2 \rangle$ is proportional to the average mean square end-to-end distance $\langle h_r^2 \rangle = \langle h_x^2 + h_y^2 + h_z^2 \rangle$. Hence one expects the mobility of curved molecules to be less than that of straight ones because their end-to-end distance is smaller. We leave for later the problem posed by the relationship between DNA shape in the gel and in solution.

Structural Features Elucidated by Comparative Electrophoresis

An effective simple rule for comparative electrophoresis is that it is sensitive to parameters which affect the mean square end-to-end distance for molecules of nearly invariant contour length L. Figure 1 shows that the end-to-end distance is sensitive to bend position. Figure 2 illustrates the dependence on helical phasing and relative bend direction when two bends are combined in the same molecule.[3,4] Variation of the end-to-end distance as linker length and hence helical phasing are changed results in a periodic dependence of the electrophoretic mobility on phasing, with a maximum at the trans isomer and a minimum for the cis isomer. The number of base pairs in the linker required to yield the cis isomer depends on both the relative direction of the two bends and the helical repeat of DNA in the linker. One can also design constructs which test the ability of calibrating bends of known magnitude to compensate for a bend of unknown size. Hence the primary structural features which can be studied

[2] O. J. Lumpkin and B. H. Zimm, *Biopolymers* **21**, 2315 (1982).
[3] S. S. Zinkel and D. M. Crothers, *Nature (London)* **328**, 178 (1987).
[4] J. Drak and D. M. Crothers, *Proc. Natl. Acad. Sci. U.S.A.* **88**, 3074 (1991).

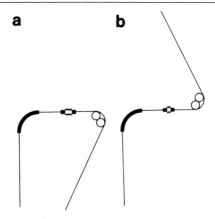

FIG. 2. Diagram showing the cis (a) and trans (b) isomers of molecules containing an A tract bend (heavy line) and a protein-induced bend. The dimeric CAP protein is indicated by two circles.[3] The isomers differ by half a helical turn in the length of the linker region between the two bends. (Reprinted with permission from *Nature*.)

by this method include bend position, relative bend direction, relative bend magnitude, and DNA helical repeat. Changes in DNA flexibility could also in principle affect comparative electrophoresis,[5] but that possibility has not yet been systematically explored.

Experimental Methods

Comparative electrophoresis measurements are performed using standard nondenaturing electrophoretic techniques, making them readily accessible to all molecular biological laboratories. Molecules are generally in the size range of 100 to 1000 base pairs (bp); the sensitivity to curvature is usually reduced outside of this range. DNA mobilities in polyacrylamide gels are more sensitive to DNA curvature than is found for agarose gels,[6] a fact which has been exploited by Anderson[5] to develop a two-dimensional gel method for detecting curved molecules in a large background of fragments of normal shape. Sensitivity of the mobility to curvature generally increases with increasing gel percentage.[6] However, as discussed in detail below, the use of a higher gel percentage (above about 6%) to maximize detection sensitivity may introduce new complications because mobility can depend on other features of the three-dimensional shape in molecules

[5] J. N. Anderson, *Nucleic Acids Res.* **14**, 8513 (1986).
[6] J. C. Marini, S. D. Levene, D. M. Crothers, and P. T. Englund, *Proc. Natl. Acad. Sci. U.S.A.* **79**, 7664 (1982).

containing more than one bend.[4] Decreased temperature can also lead to enhancement of the mobility anomaly in intrinsically curved DNA molecules.[7-10] However, this effect is probably due to an increase in the magnitude of curvature at lower temperature,[9] rather than to an effect intrinsic to the gel. For accurate measurements it is advisable to thermostat the gel, using, for example, the constant temperature gel apparatus manufactured by Hoefer Scientific (San Francisco, CA). For experiments on curved DNA molecules lacking a bound protein we generally use an acrylamide–bisacrylamide ratio of 29 : 1 to 39 : 1 (w/w). In cases where dissociation of a bound ligand is not a potential problem, full-strength TBE buffer [89 mM Tris–borate, 2 mM ethylenediaminetetraacetic acid (EDTA), pH 8.2] is a suitable choice for running the gel. Dissociation of electrostatically bound ligands such as proteins can often be reduced by diluting the gel buffer to $\frac{1}{2}$ or $\frac{1}{4}$. Our experiments follow standard DNA electrophoresis protocols in prerunning the gels: before loading samples, gels are run at constant voltage until the current is constant, within a few milliamperes, to ensure reproducibility of the electrophoresis conditions. For most acrylamide percentages this should take from 30 to 90 min.

Comparative electrophoresis measurements are simple in execution; the primary challenge is found in design and construction of an appropriate set of molecules whose comparison provides answers to the questions posed. In the following sections we consider examples which illustrate the general principles.

Bend Position: Locus of A Tract Bends

Figure 3 shows the results reported by Wu and Crothers[1] which identified periodically repeated A tracts as the source of DNA curvature in kinetoplast DNA; subsequent work has shown these tracts to be the dominant general origin of sequence-directed DNA bending. A 241-bp DNA fragment which was decidedly anomalous in electrophoretic mobility was cloned in tandem as in Fig. 1, then digested with a series of enzymes. Figure 3 shows the electrophoretic mobility as a function of the position of the ends of the molecule. Extrapolation of the data to the maximum gel mobility (the minimum position in the inverted distance of migration curve shown in Fig. 3), corresponding to a cut near base pair 140, identified the bending locus with its periodically repeated tracts of homopolymeric

[7] J. C. Marini, P. N. Effron, T. C. Goodman, C. K. Singleton, R. D. Wells, R. M. Wartell, and P. T. Englund, J. Biol. Chem. **259**, 8974 (1984).
[8] S. Diekmann and J. C. Wang, J. Mol. Biol. **186**, 1 (1985).
[9] S. D. Levene, H.-M. Wu, and D. M. Crothers, Biochemistry **25**, 3988 (1986).
[10] H.-S. Koo, H.-M. Wu, and D. M. Crothers, Nature (London) **320**, 501 (1986).

a

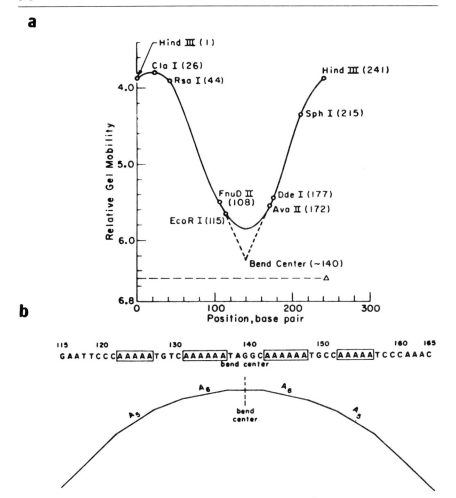

b

FIG. 3. Identification of the bending locus in a curved DNA.[1] (a) Experimental variation of the mobility of circularly permuted DNA sequences carrying a bending locus. The extrapolated minimum in the curve, corresponding to a maximum in the mobility, reflects positioning of the ends of the molecule at base pair 140. Hence, this site should correspond to the center of the bend. (b) DNA sequence at the bend locus, revealing A_{5-6} tracts repeated at 10-bp intervals. The curve represents the DNA helix axis, which is assumed to be deflected at each end of an A tract. (Reprinted with permission from *Nature*.)

dA · dT segments, each about half a helical turn long. Note that the minimum gel mobility is found for the fragment cut near base pair 20, which places the bend locus centered at 140 in the middle of the 241 bp fragment.

Given the symmetric disposition of four A tracts around position 140, the center of the bend must be located within a few base pairs of that locus, as is consistent with the extrapolation in Fig. 3. However, it should be recognized that the point of extrapolated maximum mobility is also affected by curvature elsewhere in the molecule, leading to possible errors of several base pairs in assigning the geometric center of the bend. For example, an experiment analogous to that in Fig. 3 for a larger fragment (423 bp), which contained a secondary bending locus 3' to the major site, identified position 148 rather than 140 as the bend center.[1] The theory of gel electrophoresis is not yet sufficiently well developed to provide a quantitative interpretation of data of this kind. However, a simple qualitative view identifies the extrapolated position of maximum mobility in Fig. 3 (corresponding to the minimum in the curve) with the point on the DNA helix axis at which a vector in the plane normal to the helix axis bisects the overall curvature angle of the fragment. Alternatively, one can simply associate the *minimum* mobility species with the circularly permuted sequence which minimizes the end-to-end distance of the DNA molecule.

Sequence Dependence of DNA Bending: Ligation Ladder Experiment

The bending locus experiment (Fig. 3) identified A tracts repeated in phase with the DNA helical screw as the source of DNA bending; overall curvature due to bends phased with the helical repeat had earlier been predicted by Trifonov and Sussman.[11] The requirement for phasing was confirmed by Hagerman[12] and Koo *et al.*,[10] who ligated synthetic oligonucleotides to form polymeric sequences with repeated motifs. In this approach, variation of sequence content and length allows independent variation of the motif and its phasing. The result of treatment of oligomers with DNA ligase is a ladder of products on an electrophoresis gel. Appropriate choice of ligase concentration and incubation time allows products from the monomer up to 20 or more repeats to be visualized on the gel. Hence, one can deduce the chain length for any band, and compare its mobility to that of "normal" DNA molecules which are ligated together to give a standard ladder. Blunt-ended 10-bp *Bam*HI linkers (New England Biolabs, Beverly, MA, sequence CGGGATCCCG) have been used for that pur-

---

[11] E. N. Trifonov and J. L. Sussman, *Proc. Natl. Acad. Sci. U.S.A.* **77,** 3816 (1980).
[12] P. J. Hagerman, *Biochemistry* **24,** 7033 (1985).

pose,[10] but other sequences can be substituted as long as it is verified by electrophoresis against restriction fragments or other standards that their electrophoretic mobility is indeed "normal."

The strategy used by Koo *et al.*[10] utilized double helical oligomers which have non-self-complementary ends, leading to ligation with a unique polarity. For example, the sequence

$$5'\text{-G G C A A A A A C G}$$
$$\phantom{5'\text{-}}\text{G T T T T T G C C C}$$

ligates only head-to-tail, yielding A_5 tracts phased every 10 bp. This approach is more general, although more costly in oligomer synthesis, than the strategy adopted by Hagerman,[12] in which self-complementary duplex oligomers having dyad symmetry were employed.

Comparison of the mobility of curved DNAs with the internal standards is usually expressed as the retardation coefficient R_L, defined as the ratio of the apparent chain length (judged by its electrophoretic mobility in comparison to DNA standards) to the actual chain length of the molecule. Values of R_L greater than 1 are generally associated with DNA curvature. However, we recommend caution in the use of this assumption, especially when R_L is less than about 1.2, because other factors such as helix rise, diameter, and flexibility can clearly affect gel mobility.

That A tracts have a preferred direction of bending was shown decisively by the experiment of Koo *et al.*[10] in which repeated A tracts were placed at a spacing of 1.5 helical turns. Such DNA molecules were found to be entirely normal in electrophoretic mobility, ruling out the possibility that the A tracts act as flexible hinges able to bend in a plane but without a preferred direction within the plane. Repetition of bends having a preferred direction at half-integral multiples of the helical repeat places alternate bends on opposite sides of the helix, yielding a zigzag structure which is straight at the level detected by electrophoresis. However, when the repetition occurs at integral multiples of the helical repeat, the overall molecular shape is curved, and the mobility is anomalously slow. Figure 4 illustrates plots of R_L against length for molecules in which the A tract bends are in and out of phase.

It is now clear that the dominant overall source of sequence-directed DNA bending is provided by A tract sequences, although their effects can be modulated by other dinucleotide sequences.[13] For convenience and simplicity in presenting underlying principles we shall neglect these second order contributions unless otherwise stated.

[13] A. Bolshoy, P. McNamara, R. E. Harrington, and E. N. Trifonov, *Proc. Natl. Acad. Sci. U.S.A.* **88**, 2312 (1991).

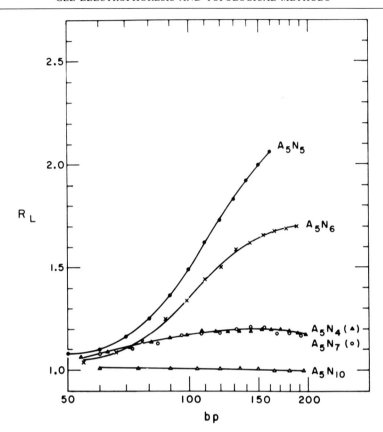

FIG. 4. Dependence of the ratio R_L of apparent to real DNA chain length on DNA size for different phasings of A tract bending sequences.[10] The anomaly in electrophoretic mobility is measured by deviations of R_L from 1; note that these are maximal for molecules containing the repeated sequence A_5N_5, in which the A tracts are repeated at 10-bp intervals. The anomaly is smaller but still pronounced for repeats of A_5N_6, for which the 11-bp repeat is also close to the DNA helical repeat. However, A_5N_4 and A_5N_7 are nearly normal in mobility. A_5N_{10} multimers are normal in electrophoretic mobility, indicating that A tracts separated by 1.5 helical turns do not cause DNA to bend. (Reprinted with permission from *Nature*.

Bend Magnitude: Functional Dependence of Electrophoretic Anomaly on Curvature

Most DNA molecules migrate in polyacrylamide gels according to their size, with little or no correction for curvature, even though it cannot be expected that they are rigorously straight.[13] The underlying reason for this is the quadratic rather than linear dependence of the electrophoretic anomaly $R_L - 1$ on curvature, which yields little variation in mobility with curvature as long as the curvature is small. A direct experimental demonstration of this dependence was provided by Koo and Crothers[14]; in view of the concerns offered by Hagerman[15] we provide here some additional clarifying comments.

There are two parts to the experiment: (1) establishing the calibration between curvature and mobility anomaly and (2) using the calibration and experimental data on the mobility of specific sequences to test quantitative models for A tract bending. As is customary for uniformly curved molecules of sequence repeat approximately equal to the helical repeat, it is assumed throughout that mobility anomalies reflect primarily the average curvature in solution, which in turn was assumed by Koo and Crothers[14] to be due solely to the presence of A tracts.

Sequences studied in an effort to accomplish objective (1) differed in the A tract density, varying from one per helical turn to one every four helical turns. As shown in Fig. 5, the deviations from normal electrophoretic mobility are small unless there is at least one A tract per two helical turns. When R_L is plotted against the square of the curvature, a linear dependence results; this is a purely empirical relationship, which does not rely on assumptions about molecular rigidity. The result of this experiment was consolidated in an equation which expressed the electrophoretic anomaly in terms of the calibration equation

$$R_L - 1 = [(9.6 \times 10^{-5})L^2 - 0.47](\text{relative curvature})^2 \qquad (2)$$

where R_L is measured in an 8% polyacrylamide gel, 29:1 acrylamide–bisacrylamide and L is DNA length in base pairs; Eq. (2) is valid for molecules in the size range 120–170 bp. The relative curvature is 1 for reference molecules of repeated sequence $(\text{GGGCA}_6\text{CGGCA}_6\text{C})_n$. We recommend that ligated oligomers of this sequence be included as internal standards when Eq. (2) is used to estimate curvature.

Absolute curvatures of A tract-containing molecules can be estimated from Eq. (2) by calculating first the relative curvature, which is obtained by solving Eq. (2) after inserting the experimental value of R_L. The absolute curvature is then the product of the relative curvature and the absolute curvature of the reference molecules under the appropriate ionic conditions. The most accurate value presently known for the absolute curvature is $18° \pm 3°/\text{A}_6$ tract (the corresponding sequence consists of ligated repeats of the 21-bp oligomer of relative curvature 1 above[16]); this value is, however, significantly modulated (up to 25%) by changing the sequence be-

[14] H.-S. Koo and D. M. Crothers, *Proc. Natl. Acad. Sci. U.S.A.* **85**, 1763 (1988).

[15] P. H. Hagerman, *Annu. Rev. Biochem.*

[16] H.-S. Koo, J. Drak, J. A. Rice, and D. M. Crothers, *Biochemistry* **29**, 4227 (1990).

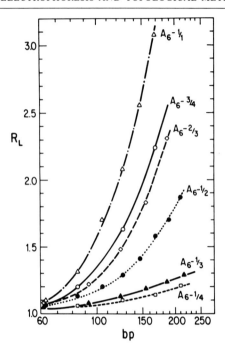

FIG. 5. Dependence of electrophoretic ratio R_L on length for multimers of differing A tract density.[10] The fractional numbers indicate the number of A tracts per helical turn in the multimers.

tween the A tracts (D. M. Crothers and T. Haran, unpublished results). Changing the ionic composition of the electrophoresis buffer leads to detectable changes in R_L values,[17] but because of Eq. (2) one infers that only modest changes in A tract curvature result, when compared to the experimental error in the absolute curvature. This conclusion is also supported by the rotational relaxation experiments of Levene et al.[9] Elevated temperature can, however, greatly reduce curvature.

Testing models for the parameters governing DNA curvature, objective (2), requires comparison of the predicted average curvature for a set of molecules with their observed mobility anomaly, taking account of the calibration equation [Eq. (2)]. As in all comparable calculations, average helix deflection angles from the model are used to calculate the average helix curvature. It is not assumed that the molecules are rigid in the electrophoretic comparison of their properties; thermal bending fluctuations occur in these as in all DNA molecules. In this comparison, it is vital

[17] S. Diekmann, Nucleic Acids Res. 15, 247 (1987).

that a uniform treatment be given to sequence repeat and the helical twist variable, which together control the phase match and clearly affect electrophoretic mobility.[4,10,12] Koo and Crothers[14] did this by comparing only molecules with sequence repeat of 10 bp, normalizing all calculated curvature and experimental electrophoretic properties to a control sequence of repeat A_6N_4. Improving on this approximation requires a large set of oligomers of differing sequence repeat (see below). The relative mobilities of all A tract-containing sequences could be predicted with a small parameter set which expresses the deflection of the DNA helix axis at each end of the A tract. The success of this simple model supports the validity of the underlying assumption, namely, that electrophoretic anomalies reflect the average systematic curvature of DNA at A tracts, and it further encourages the view that alterations of DNA flexibility need not be invoked as modulators of electrophoretic mobility in A tract-containing molecules.

An important factor in experimental design which derives from these considerations is that sensitivity of the electrophoretic mobility to sequence elements which produce curvature is greatest if they are introduced into molecules which are already curved. This principle, a simple consequence of the dependence of mobility anomaly on the *square* of overall curvature, should be taken into consideration when designing comparative electrophoresis experiments.

Cruciforms and Holliday Junctions

The widespread applicability of comparative electrophoresis methods can be illustrated by the studies of Lilley and colleagues on the four-helix junction, called the Holliday[18] junction, which results from cruciform formation in double helical DNA. Gough and Lilley[19] demonstrated that DNA is bent at a cruciform by using heteroduplex hybridization to construct molecules containing stable cruciforms. Molecules 400 bp in size showed extremely anomalous electrophoretic mobility, with R_L values above 4 in 12% polyacrylamide gels. The extent of the electrophoretic anomaly was found to vary with position of the junction, reaching a maximum when the junction is in the middle of the molecule.

Duckett et al.[20] extended this concept in a series of experiments which provided strong support for an X-shaped structure for the junction in Mg^{2+},

[18] D. M. J. Lilley,

[19] G. W. Gough and D. M. J. Lilley, *Nature (London)* **313**, 154 (1985).

[20] D. R. Duckett, A. I. H. Murchie, S. Diekmann, E. von Kitzing, B. Kemper, and D. M. J. Lilley, *Cell (Cambridge, Mass.)* **55**, 79 (1988).

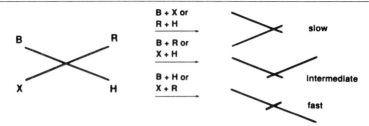

FIG. 6. Consequences of cleavage of two arms of an X-shaped Holliday junction.[20] In the model the helical arms are stacked to form colinear helices 80 bp in length, which are disposed in a manner which is neither parallel nor orthogonal. The six possible pairwise restriction digests thus may give three types of product, which will have different gel mobilities, the more acute angles giving more retarded migration. In this way the arms of the X were assigned as shown. (Reprinted with permission from *Cell.*)

with preferential association between specific arms. Figure 6 shows the constructs, in which the arms, B, R, X, and H have distinctive sequences. Cleavage of any single arm from this structure by restriction enzyme treatment yields four structures which migrate more rapidly than the intact junction but which are indistinguishable electrophoretically from each other. However, pairwise cleavage of the arms in all six combinations yields three different electrophoretic species. As shown in Fig. 6, these are thought to reflect pairs of arms separated by an acute angle (the slowest moving species with the largest bend angle), separated by an obtuse angle (the intermediate mobility species), or linearly related (the fastest moving molecules). Thus, although the junction is potentially 4-fold symmetric, there is clearly a preferential association between the linearly related pairs of arms. Further experiments showed that the preference is related to sequence at the junction, possibly reflecting preferential stacking of helical segments.[20]

Bend Direction: Geometry of A Tract Bends

A tracts repeated in phase with the DNA helical screw provide a convenient method for introducing curvature into any DNA molecule, thereby providing the ability to enhance or cancel the overall curvature of fragments containing a bend from another source, such as a helix defect or a bound protein. Whether an A tract bend will add constructively with a protein-induced bend depends on the relative direction in space of the two bends, which in turn depends on their individual directions relative to a local coordinate frame and on the helical twist between the corresponding coordinate systems. For example, a DNA bend toward the minor groove at the center of an A_6 tract will add constructively with a protein-induced bend which is toward the minor groove in a coordinate frame at its center whenever an integral number of helical turns separates the center of the A tract from the central axis of the protein binding site. Correspondingly, the two bends will tend to cancel when a half-integral number of helical turns separates them.

Some authors, for example, Hagerman,[15] have confused the distinction between the locus defining the bend *center*, as determined by the circular permutation experiment, and the *origin* (or translational location along the DNA molecule) of the reference coordinate frame which is used to describe the *direction* of the bend. For example, suppose that we know that a DNA sequence bends toward a protein, in an overall direction that it is equivalent to compressing the minor groove at the axis of approximate dyad symmetry which runs through the center of the DNA binding site. The intersection of the dyad axis with the DNA helix axis defines the origin of the reference coordinate frame. If the DNA curvature is greater on one side of the (imperfect) dyad axis than the other, then the position of the bend center will not coincide with the location of the origin of the reference coordinate frame. It follows logically that uncertainties in the *position* of the bend center do not necessarily confer uncertainty on the bend *direction,* and vice versa.

To draw conclusions about the direction of protein-induced bends, it was necessary first to establish the curvature direction of A tracts, which have subsequently become the primary standard for comparative measurements. Note that the direction of curvature can be defined relative to any arbitrary local coordinate frame in the helix. For example, a bend toward the minor groove in a coordinate frame located in the center of an A tract is equivalent in global geometry to a bend toward the A-containing strand about 2.5 bases away in the 3' direction. This is a consequence of the helical rotation of the reference coordinate frame as it is translated along the helix. It is important to recognize that comparative electrophoresis measurements can be used to establish *global* features of the bend, such as its approximate center and its direction relative to an arbitrarily defined coordinate frame, but this method is not generally able to specify precisely in which coordinate frame the bend is actually located or what its underlying structural basis is.

Symmetry Considerations

The plane of A tract bending can be defined relative to a coordinate frame at an arbitrary position in the tract. For simple symmetry arguments it is convenient to place its origin at the center of an A_{5-6} tract. Now consider two possible limiting cases: The bend direction is either toward one of the grooves (the positive or negative roll directions) or toward one of the backbone strands (the positive or negative tilt directions) at that coordinate position. Now imagine that the A and T strands are interchanged in the tract. This is equivalent to rotation of the segment about an axis in the reference coordinate frame which runs between the major and minor grooves, namely, the approximate dyad axis of the DNA double helix. If the A tract bend is toward one of the grooves at its center, its direction will be unchanged by this operation, but if it is toward one of the backbone strands, rotation will shift the direction by 180°.

Koo *et al.*[10] exploited this logic by constructing molecules having alternating A tracts and T tracts in phase with the helical repeat. The electrophoretic mobility of these molecules was nearly the same as those in which the A tracts were all on one strand. Because coherent addition of the bends is retained after the strand interchange operation, one can conclude that the bend is primarily toward one of the grooves in the coordinate frame at the center of the A tract. Had the bend been toward one of the strands, 180° rotation of every other A tract would have created a zigzag structure lacking overall curvature.

The experiments of Koo *et al.*[10] contained some indications that the A tract structure is not perfectly dyad symmetric about its central axis, but the sequences studied by Hagerman[21] provided decisive evidence to that effect. Comparison of the sequences A_4T_4 and T_4A_4 repeated in phase with the helical screw showed that the former are highly curved, whereas the latter are nearly normal in mobility, facts which are not compatible with full dyad symmetry of A tract structure. However, the observations can be accommodated by a slight change in the direction of the bend.[14,22] One way to visualize this is to retain the view that bending is toward one of the grooves, but to shift (and rotate) the position of the reference coordinate frame away from the center of the A tract by a little less than one-half base pair toward its 3' end. The consequence is to move the bends of the two A tracts closer together in A_4T_4 and further apart in T_4A_4:

$$N\ A\ A\ A\ A\ T\ T\ T\ T\ N\quad\quad N\ T\ T\ T\ T\ A\ A\ A\ A\ N$$
$$\uparrow\quad\quad\uparrow\quad\quad\quad\quad\uparrow\quad\quad\quad\quad\uparrow$$

In the illustration shown the two bends in A_4T_4 are about 120° apart in relative helical phasing, so they have a substantial resultant and hence add together effectively. However, they are approximately 180° apart in T_4A_4, resulting in their mutual cancellation.

[21] P. J. Hagerman, *Nature (London)* **321,** 449 (1986).
[22] L. E. Ulanovsky and E. N. Trifonov, *Nature (London)* **326,** 720 (1987).

Because of full dyad symmetry of the sequence, the resultant bend in A_4T_4 must be directed toward one of the grooves in a coordinate frame located on the dyad axis between the A and T tracts. This fact can be used to test further the conclusion that the A tract bend is toward one of the grooves in a reference frame close to the center of the tract. The sequences studied by Koo and Crothers[14] serve this purpose, and also illustrate the principle of *interdigitation* of bends of different kinds:

IHA:

G G G C A A A A A C G C C A A A A T T T T G C C G C G G G C C
         ↑ ←——— 10 bp ———→ ↑

OHA:

G G G C A A A A A C G G G C G G C C A A A A T T T T G C C G C
         ↑ ←——————— 15 bp ———————→ ↑

Alternate bends directed toward the same groove in the sequence IHA should be in phase but out of phase in OHA because of the approximate 1.5 helical turn spacing. This is the result found experimentally: ligation of the IHA sequence yields molecules of highly anomalous mobility, whereas the OHA sequence behaves normally.[14]

Absolute Bend Direction: Coupling to Known Standard

As we have seen, application of simple symmetry and geometric principles to the sequences studied by Koo et al.[10] and Hagerman[21] established that A tract bends behave as though they are directed toward either the major or the minor groove, at a coordinate locus slightly displaced, by about $\frac{1}{2}$ bp, to the 3′ side of the center of the A tract. Zinkel and Crothers[3] resolved the residual ambiguity in absolute direction by placing an A tract in the same DNA molecule as another bend whose direction could reasonably be assumed to be known. Chosen for this purpose was the catabolite activator protein (CAP)–DNA complex, for which it was assumed that DNA curves around the protein in a direction toward the minor groove in a reference coordinate frame located at the position of the approximate dyad axis of the DNA binding site; subsequent crystallographic work has validated this assumption.[23]

Because the experiment must distinguish only between the major and minor groove direction, the two bends studied can be some distance apart. To understand this, it is useful to consider the control experiment reported by Zinkel and Crothers,[3] in which two CAP bends were combined. When two identical bends are repeated in tandem, an integral number of helical

[23] S. C. Schultz, G. C. Schields, and T. A. Steitz, *Science* **253**, 1001 (1991).

turns must separate their centers in the cis isomer (Fig. 2). The spacing found was 84–85 bp, which could correspond to 9 turns at a helical repeat of 9.4, 8 turns at a helical repeat of 10.6, or 7 turns at a helical repeat of 12.1 bp/turn. Because CAP does not significantly unwind DNA,[24] the only acceptable value in this set is 10.6. As the bends are moved farther apart, the choice among the possible values of the helical repeat becomes less obvious because the two outlying values move closer to the consensus value of 10.5–10.7.

For the combination of a CAP bend with A tracts, Zinkel and Crothers[3] found that the G at the central axis of the kinetoplast DNA bend sequence[1]

A A A A A T G T C A A A A A A A T A G G C A A A A A A A T G C C A A A A A
                                    ↑

is located about 101–102 bp from the CAP site dyad axis in the cis isomer. This site is displaced by half-integral numbers of helical turns from the centers of the adjacent A tracts. Therefore, if we select the option in which A tract bends are directed toward the minor groove at their centers, then the overall bend must be directed toward the major groove at the central G. Because the CAP bend is toward the minor groove at the dyad axis, this would correspond to a half-integral number of helical turns, or 9.5 turns at a helical repeat of 10.6 to 10.7. Conversely, if the A tract bend is toward the major groove, the overall bend must be toward the minor groove in the coordinate frame at the central G, corresponding to 10 turns with a helical repeat of 10.1 to 10.2. Only the 10.6 to 10.7 values fall within the error margin of the result found for the very similar linker DNA sequence used for the two-CAP experiment. Hence the A tract bend direction must be toward the minor rather than the major groove in a coordinate frame near the center of an A_5 or A_6 tract.

Experiments by Zinkel[25] and Drak and Crothers[4] have revealed an important potential artifact in electrophoretic characterization of molecules containing two bends. At higher gel percentages (>6%) the maximum and minimum gel mobilities do not correspond to the trans and cis isomers, respectively, because electrophoretic mobility becomes dependent on the handedness of the superhelical shape possessed by the intermediate isomers[4] (see below). It is advisable to perform comparative electrophoresis experiments as a function of gel percentage for constructs of this kind. Experience so far[4,25] indicates that the correct cis and trans isomers are identified at low (3.5–6%) gel percentages.

The accuracy of the relative bending experiment can be improved by

[24] A. Kolb and H. Buc, *Nucleic Acids Res.* **10,** 473 (1982).
[25] S. S. Zinkel, Ph.D. Thesis, Yale University, New Haven, Connecticut (1989).

moving the unknown and reference bends closer together, which was done for the CAP–A tract combination by Gartenberg and Crothers.[26] They found that an A_6 bend adds optimally with the CAP bend when it is placed so that its center is 22 nucleotides to the 5' side of the CAP dyad axis. This is the position predicted for an A tract which bends toward its minor groove in a coordinate frame shifted by about 1 bp to the 3' side of its center. The difference between this result and the 0.5-bp 3' shift deduced from the symmetry properties of the DNA sequences discussed above corresponds to the likely error margin of the experiment, especially considering the possibility of slight unwinding or curvature contributions from non-A-tract sequence elements.[13]

Comparison with Crystallographic Structures

The direction of the A tract bend deduced from comparative electrophoresis measurements is not in agreement with crystallographic studies of A tract-containing oligonucleotides.[27–29] These show a variable direction of bending, but no structure has a bend directed primarily toward the minor groove at its center. On the other hand, we have recounted here a number of ways in which the A tract bend direction has been checked against other molecules. Furthermore, NMR measurements indicate that some critical distances across the minor groove are shorter in solution than in the crystal.[30,31] The effects of adding other ions such as Mg^{2+} to the electrophoresis buffer do not seem to change the direction substantially.[17] In this vein, a question might be raised about the influence of crystal packing forces[29] and organic solvents in the crystallization mixture in altering or destabilizing the solution structure of A tracts; clarifying this controversial problem will require further work.

Helical Repeat: A Tracts and Mixed Sequences

We illustrate here two methods for determining the DNA helical repeat by comparative electrophoresis. The basic idea is to find the DNA sequence repeat which matches the helical repeat, so that bends are phased at integral multiples of the helical repeat. Such constructs can be recog-

[26] M. R. Gartenberg and D. M. Crothers, *Nature (London)* **33,** 824 (1988).

[27] H. C. M. Nelson, J. T. Finsh, B. F. Luisi, and A. Klug, *Nature (London)* **330,** 221 (1987).

[28] M. Coll, C. A. Frederick, A. H.-J. Wang, and A. Rich, *Proc. Natl. Acad. Sci. U.S.A.* **84,** 8385 (1987).

[29] A. D. DiGabriele, M. R. Sanderson, and T. A. Steitz, *Proc. Natl. Acad. Sci. U.S.A.* **86,** 1816 (1989).

[30] J. G. Nadeau and D. M. Crothers, *Proc. Natl. Acad. Sci. U.S.A.* **86,** 2622 (1989).

[31] D. M. Crothers, T. E. Haran, and J. G. Nadeau, *J. Biol. Chem.* **265,** 7093 (1990).

GCAAAAAACGG	(11.00)
GCAAAAAACGGGCAAAAAACGGGCAAAAAACG	(10.67)
GGGCAAAAAACGGCAAAAAAC	(10.50)
GCAAAAAACGGCAAAAAACGGGCAAAAAACG	(10.33)
GCAAAAAACGGCAAAAAACGGCAAAAAACG	(10.00)
CGGGATCCGTCGACCATCTGT	(marker)

Fig. 7. Sequences of the oligonucleotides used in the ligation ladder experiment.[4] The numbers in parentheses indicate the corresponding sequence repeat. Top strands only are shown (written in the $5' \rightarrow 3'$ direction); bottom strands have the same length as their complementary top strand, but their ends are shifted by 3 bases, so that the duplexes have a 3-base $5'$ end overhang for the ligation reaction.

nized by comparative electrophoresis because the end-to-end distance is minimized when bends are in phase. A disadvantage of the method is that it is restricted to temperature and ionic conditions which are suitable for electrophoresis.

Consider the series of oligonucleotides shown in Fig. 7, which, when hybridized with their complementary sequences to yield a three-base overhang, can be ligated into curved multimers. All have A_6 tracts with closely related sequences separating them, but they differ in the period of overall sequence repeat, which varies from 10.0 to 11.0. Exact match between sequence repeat and helical repeat yields a planar curve without writhe, but any slight mismatch will produce right- or left-handed superhelical shapes. Because the end-to-end distance is minimized for the planar curve, one expects a maximum in the electrophoretic anomaly (minimum in the mobility) when the repeats match.

Figure 8 verifies this behavior. For different lengths and gel percentages one finds a maximum in the electrophoretic anomaly R_L at an interpolated value of the helical repeat of 10.34 ± 0.04. This value refers to the overall helical repeat of A_6 tracts interspersed with 4–5 bp of "normal" B-DNA sequence in TBE buffer at 22°. Unlike molecules with two separate bends (see below), these roughly circular molecules show no systematic dependence of mobility on the handedness of superhelical writhe even at very high gel percentages.[4]

Having an accurate value of the helical repeat of A tracts makes it possible to determine the helical repeat of arbitrary DNA sequences which separate two such bends. For instance, in an experiment designed to determine the helical repeat of B-DNA in solution[4] we constructed a set of isomeric molecules, each containing two separate bends produced by runs of adenines repeated every 10.5 bp (Fig. 9). The phasing between the bends was varied over one helical turn in steps of 2 bp by insertion of sequences of length n, $n + 2$, ..., $n + 10$ bp between them. After

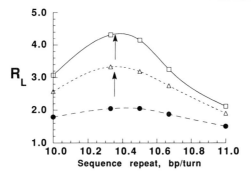

FIG. 8. Illustrative results of the ligation ladder experiments for multimers of the sequences in Fig. 7 which contain 15 A tracts.[4] The different polyacrylamide gel percentages used are ●, 5%; △, 8%; and □, 12% (w/v). Note that the magnitude of the electrophoretic anomaly, measured by $R_L - 1$, increases with gel percentage, but the interpolated position of the optimal sequence repeat, indicated by the vertical arrow, does not change significantly with gel percentage. From this experiment it is concluded that the average DNA helical repeat in the sequences in Fig. 7, taken equal to the sequence repeat which maximizes R_L, is 10.34 ± 0.04 bp/turn.

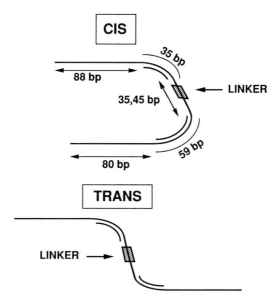

FIG. 9. Diagram of the cis and trans isomers of the set of DNA molecules containing two A tract bends (curved double lines) separated by a DNA segment whose variable length (35–45 bp) is modulated by the linker DNA segment of length 10–20 bp.[4]

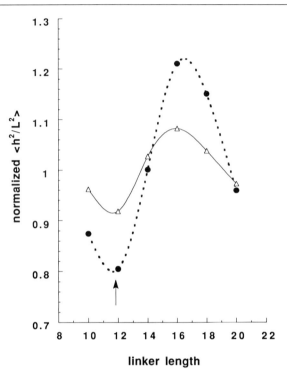

FIG. 10. Comparison[4] between the calculated normalized mean square end-to-end distance (●) and the normalized experimental mobility variations (△) of the molecules in Fig. 9. The calculations assumed 10.35 bp per turn in the A-tract bend regions and 10.50 bp per turn in the DNA segment that separates them. Experimental mobilities were measured on a 5% polyacrylamide gel. The arrow corresponds to the linker length of the cis isomer. From this and similar experiments it was concluded that the helical repeat h_B of the DNA segment between the bends is 10.49 ± 0.05 bp/turn.

identifying the cis and trans isomers experimentally on a 5% polyacrylamide gel, we used Monte Carlo simulations[16] to calculate the mean square end-to-end distances in the isomers. In these simulations the only variable was h_B, which was adjusted to bring about a match between the linker lengths corresponding to maxima and minima in the calculated mean square end-to-end lengths $\langle h_r^2 \rangle$, on one hand, and the experimentally observed linker lengths corresponding to trans and cis isomers, on the other. We varied h_B in steps of 0.02 bp/turn, calculated the normalized mean square end-to-end distance for all the isomers in the set, and then compared the variation of $\langle h_r^2 \rangle$ with the periodic variation in normalized experimental mobilities on a 5% gel. The results of such a comparison are illustrated in Fig. 10. A narrow range of h_B values (10.49 ± 0.05 bp/turn) was found to

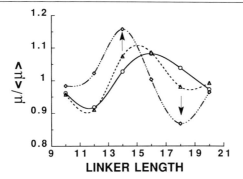

FIG. 11. Anomalous dependences of R_L on linker length which appear at higher percentages of polyacrylamide.[4] The gel percentages are ○, 5%, △, 8%; and ◇, 16%. The arrows indicate the isomers flanking the cis isomer whose mobilities are, respectively, dramatically increased and decreased in high percentage gels.

give good agreement between the phases of the experimental and theoretical curves; we discuss below the difference in their amplitudes.

In the lower gel percentage experiments, the linker length-dependent variation of the observed mobilities was in good qualitative agreement with the calculated values of $\langle h_r^2 \rangle$, as predicted by Eq. (1), with a helical repeat h_B in the expected range. However, anomalies became evident at increased gel percentage, as demonstrated in Fig. 11. Our experiments revealed that the determination of the *apparent* isomeric form depends on the gel percentage, an effect that is more pronounced for molecules with larger bends. We traced this phenomenon to the ability of high percentage gels ($\geq 6\%$) to discriminate between DNA molecules having right- or left-handed intrinsically superhelical shapes.

The sensitivity of our assay could be improved by cloning a larger isomeric set, in which the linker length is varied in steps of 1 bp instead of 2 bp. In addition, in the case of short sequences, incorporation of multiple repeats should amplify the effect of helical twist variations. This approach can also be used to study the change in helical twist produced on protein binding to the appropriate sequence, assuming that the protein does not strongly bend DNA and that the protein–DNA complex is stable during electrophoresis.

Bend Direction and Magnitude: Platinum-Induced Kink

The use of A tracts to characterize the direction and magnitude of other DNA bends can be illustrated with the experiment reported by Rice *et al.*[32] on the kink induced in DNA by the intrastrand cross-link produced

---

[32] J. A. Rice, D. M. Crothers, A. L. Pinto, and S. J. Lippard, *Proc. Natl. Acad. Sci. U.S.A.* **85**, 4158 (1988).

		Pt–Pt DISTANCE (base pairs)	Pt–A$_5$ DISTANCE (base pairs)
TCTCCTTCTTGG̑TTCTCTTCTC + GGAAGAACCAAGAGA	AA*AACGCCG AGAGTTTTTGCGGCAGA	32	13.5
	CAA*AACGCC AGAGGTTTTTGCGGAGA	32	14.5
	CCAA*AACGC AGAGGGTTTTTGCGAGA	32	15.5
	GCCAA*AACG AGAGCGGTTTTTGCAGA	32	16.5
	CGCCAA*AAC AGAGGCGGTTTTTGAGA	32	17.5
	GGCCG AGACCGGCGAGA	27	—

FIG. 12. Sequence of oligomers coupled to form DNA fragments used for the phase-sensitive determination of the relative bend direction at A tracts and the Pt-induced DNA kink.[32] The upper strand in each duplex is written in the 5' → 3' direction. The asterisk indicates the center of the A tract. The bottom fragment containing no A tract was coligated with the platinated fragment to yield control multimers with 27 bp (~2.5 helical turns) between the sites of platination, leading to out-of-phase bends and nearly normal electrophoretic mobility.

by the reaction of adjacent guanosine N-7 atoms with the antitumor agent cis-diamminedichloroplatinum(II). Because of the square planar geometry and cis coordination of the two guanosines at a single Pt, one expects a kink toward the major groove. The platinated oligonucleotide shown in Fig. 12 was coligated with A tract-containing molecules, in combinations which altered the phasing between Pt and the center of the A tract. Optimal addition between A tract- and Pt-induced bends is found when they are separated by 15.5 bp, the result expected for a Pt bend toward the major groove and an A tract bend toward the minor groove near its center. It was also observed that one Pt adduct in two helical turns produces about the same curvature as one A tract in one helical turn, indicating that the Pt-induced kink is about twice the size of the A tract bend.

Bend Direction, Relative Magnitude, and Change in Helical Repeat:
    Combined Parameters for Bulge Defect

The geometry of the DNA bend induced by a single bulge defect was studied by Rice and Crothers[33] through synthesis of an extensive set of comparison molecules, which we consider here for illustration of the

[33] J. A. Rice and D. M. Crothers, Biochemistry 28, 4512 (1989).

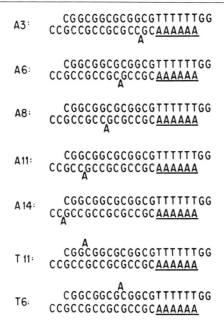

FIG. 13. Sequences used to measure the direction of bending of a bulged A relative to an A$_6$ tract. The spacing between the A tract and the bulge is varied over one helical turn. The sequences are named according to the number of base pairs separating the 5' end of the A tract from the bulge. The top strand in each duplex is written in the 5' → 3' direction.

underlying design principles. Figures 13 and 14 show the sequences. The molecules in Fig. 13 were designed in order to investigate the direction of the bulge bend relative to the A tract bend. Ligation and electrophoresis of the products showed that the anomaly is largest for isomer A6, in which a ¾ helical turn separates the center of the A$_6$ tract from the bulge. This is the result expected for an A tract bend toward the minor groove at its center and a bulge bend away from the strand containing the extra base. Bending toward one of the strands by the bulge bend was independently confirmed by the observation that when the bulge bend was placed on the other strand, it must be shifted in phase by ½ helical turn (T11) in order to add optimally with the A tract bend. (Note that this experiment provides yet another independent verification of the conclusion that the A tract bend is toward a groove at its center.) The magnitude of the bulge bend angle (approximately equal to the bend of one A tract) was estimated by adjusting it to produce calculated molecular curvatures which, in conjunction with Eq. (2), yielded the experimentally observed mobility anomalies.

The series of molecules in Fig. 14, with and without the bulge, provide a basis for estimating the change in helical repeat caused by the bulge.

MONOMER LENGTH	SEQUENCE
19	CCGCCGCGGCGTTTTTTGG CCGGCGGCGCCGCAAAAAA A
20	CGCGGCGCGGCGTTTTTTGG CCGCGCCGCGCCGCAAAAAA A
21	CGGCGGCGCGGCGTTTTTTGG CCGCCGCCGCGCCGCAAAAAA A
22	CGGCGGCGCGGCCGTTTTTTGG CCGCCGCCGCGCCGGCAAAAAA A

FIG. 14. Sequences used to study the helical periodicity of molecules containing an A tract bend and a bulge defect. The top strand in each duplex is written in the 5′ → 3′ direction.

This is in essence another version of the experiment shown in Fig. 8, but with poorer resolution because of the smaller number of sequences compared. The optimal electrophoretic anomaly was estimated at a helical repeat of 10.4 for the molecules lacking the bulge and 10.75 nucleotides (counting the bulge) for those which contain it. As the only difference between these two sets is one bulge per two helical turns, we ascribe an unwinding of $(10.75 - 10.4) \times 2$ or 0.7 ± 0.15 bp to the effect of the bulge. In other words, the absence of a T paired to the bulged A produces unwinding relative to the perfect helix by about 25°. Greater accuracy is possible in these experiments at the cost of synthesis and ligation of longer oligomers to provide more sequence lengths which are near integral multiples of the helical repeat in the range 10.4–10.8.

Some Reflections on Theory

Comparative electrophoresis has been developed as a largely empirical subject, with only general guidelines from theory. The fortuitous discovery of bent DNA molecules[1,6] provided a logical starting point for bootstrapping the method, with gradual extension to other systems. Because charged polymers moving through a gel under the influence of an electric field present an extraordinarily complex system, there is no generally accepted full theoretical treatment of the problem. Although there are dissenters from our point of view,[34] we prefer theories for polyacrylamide gel electrophoresis of molecules in the relevant size range (100–1000 bp) which begin from the concept of reptation, or motion of DNA molecules

[34] C. R. Calladine, H. R. Drew, and M. J. McCall, *J. Mol. Biol.* **201**, 127 (1988).

in a tortuous, tubelike path defined by the fibers of the gel.[2,35] However, recent experimental and theoretical developments have made it clear that one should not consider that the gel forms a rigid matrix, with a "tube" path determined solely by migration of the leading segment in the chain.[4,36] If that were so, molecules with a bend in the middle would follow the same tube, and have the same end-to-end distance as molecules which have the bend closer to the end, contrary to what has been observed. Furthermore, without "memory" of solution conformation in the gel, right- and left-handed superhelical shapes could not be distinguished by the gel.

We believe that a plausible model for electrophoresis of DNA is one in which the leading segment of the DNA chain determines the path among the gel fibers, but the gel is sufficiently elastic[36] to allow distortion of the shape of the tube in a way that restores some of the preferred solution shape. On the basis of this model one would expect that molecules which are curved would be less so in the gel than in solution, because DNA is also elastic. This is in agreement with the observations in Fig. 10, which show that varied phasing between two bends modulates the mobility by an amount that is less than would be expected from the solution A tract bend angle and application of Eq. (1). It will obviously be possible to adjust the bend angle to reduce the amplitude of the calculated mobility variation to that seen experimentally; such calculations of the apparent A tract bend angle inside the gel matrix are planned.

[35] L. Lerman and H. L. Frisch, *Biopolymers* **21,** 995 (1982).
[36] S. D. Levene and B. H. Zimm, *Science* **245,** 396 (1989).

[4] Use of Denaturing Gradient Gel Electrophoresis to Study Conformational Transitions in Nucleic Acids

By Ezra S. Abrams and Vincent P. Stanton, Jr.

Introduction

We use the term melting to denote, for double-stranded DNA (dsDNA), the temperature- or solvent-induced transition from a double-stranded, helical conformation to a single-stranded, random coil conformation. Melting is influenced by (among other things) solvent, DNA concentration, base composition, and sequence. Denaturing gradient gel electrophoresis

(DGGE)[1] is an experimental method in which solvent is varied to exploit the sensitivity of melting to changes in sequence. DNA fragments which differ by a single base (e.g., an dA to dT transversion), or even by a single methyl group, can be separated. In this chapter we briefly review relevant principles of DNA melting, illustrate the use of computer programs which greatly aid in the design and interpretation of experiments, provide protocols for procedures, and describe the versatility of DGGE as a tool for detecting various covalent modifications to double-stranded nucleic acids. Although we emphasize work on dsDNA, the principles apply to dsRNA or RNA–DNA hybrids.

Denaturing gradient gel electrophoresis is technically simple: DNA fragments are electrophoresed through a polyacrylamide gel which contains a concentration gradient of chemical denaturant. To obtain conditions where dsDNA is close to melting, the gel is immersed in a heated bath of electrophoresis buffer. Denaturant concentration and bath temperature are selected so that DNA fragments will partially melt during electrophresis. Sequence-dependent conformational changes are detected as variations in migration distance arising from the effect of partial melting on electrophoretic mobility: partially melted molecules are severely retarded in the gel.

Aside from technical ease, DGGE has other attributes which can be exploited. (1) Gel composition, gradient range, temperature, and voltage can easily be varied to resolve molecules with different properties. (2) There is an empirically tested theory and a set of computational tools which predict melting properties for DNA of known sequence, allowing prospective design of experimental conditions and facilitating the interpretation of data. (3) With the aid of existing computer programs experiments can be planned so that all—or very nearly all—possible single base changes in a short [<300 base pairs (bp)] sequenced segment of DNA can be detected. (4) Multiple samples (>100) can be analyzed under nearly identical conditions, and the relative amount of DNA in different bands can be accurately quantitated. (5) Ethidium bromide staining permits detection of DNA at the nanogram per band level; much lower amounts of DNA or RNA can be detected using radiolabeled molecules. (6) Individual species in complex mixtures, including genomic DNA digests, can be examined by several methods, including electrotransfer to nylon membranes with subsequent probe hybridization. (7) The method can be combined with conventional length-dependent separation to yield two-dimen-

---

[1] DGGE, Denaturing gradient gel electrophoresis; dsDNA, double-stranded DNA; T_m, midpoint of the melting transition; T_e, effective temperature in the gel; T_b, temperature of a DGGE tank; π, probability that a base pair is in a helical state; PCR, polymerase chain reaction; K_d, equilibrium constant for strand dissociation; ΔT, displacement in a DGG; MSR, most stable region.

sional gels where hundreds of distinct fragments are resolved. acrylamide matrix reduces diffusion, so at the end of the exp molecules remain separated and available for recovery and furthei   ,.
(9) Because DGGE does not depend on spectroscopy, optically dense, UV-absorbing or light-sensitive solutes can be introduced into the gradient, often at very high concentration, and their effects on DNA melting assessed. (10) DGGE separates molecules on the basis of melting properties rather than by length or mass. In favorable cases a particular base change can be detected in DNA fragments of widely different lengths.

How Denaturing Gradient Gel Electrophoresis Works

Our view of the physical basis of DGGE separations rests on a well-established theory of DNA melting and a tentative model for how melting affects the mobility of DNA molecules migrating through the pores of a fixed, immobile matrix of cross-linked acrylamide. As the temperature or concentration of denaturant in the solvent is raised, the melting of dsDNA is not continuous but stepwise: discrete segments of DNA, termed melting domains, undergo the helix-to-random coil transition in a predictable order (see Computational Tools, below). As DNA molecules migrate through a denaturing gradient gel (DGG) they are exposed to increasing concentrations of denaturant, and domains of progressively greater stability melt. In molecules with more than two domains (most molecules longer than 200 bp) partially melted intermediates are formed prior to strand dissociation. These intermediates may have three arms (when melting occurs at one end) or four arms (when melting occurs at both ends); the computation (see below) distinguishes only between linear and nonlinear molecules. Figure 1A shows the progression of partially melted intermediates for a hypothetical DNA fragment with three melting domains. Molecules formed when melting first occurs at an interior domain, to produce an internal single-stranded bubble bounded by helix (Fig. 1A, structure **II**), may be pictured as having four arms.

The mobility of a partially melted molecule is less than that of its fully helical counterpart. For a fully duplex DNA molecule to transit a gel pore, both ends of the molecule must pass through the same pore. Likewise, for a partially melted, three- or four-armed molecule to progress through the gel, each of the arms must eventually move through the same pore. We imagine that the spatial distribution of arms in partially melted molecules hinders this process.[2]

---

[2] An analogy is a person trying to navigate through the interpersonal pores in a densely crowded room. Horizontal extension of the arms, a small redistribution of mass, has a large effect on mobility.

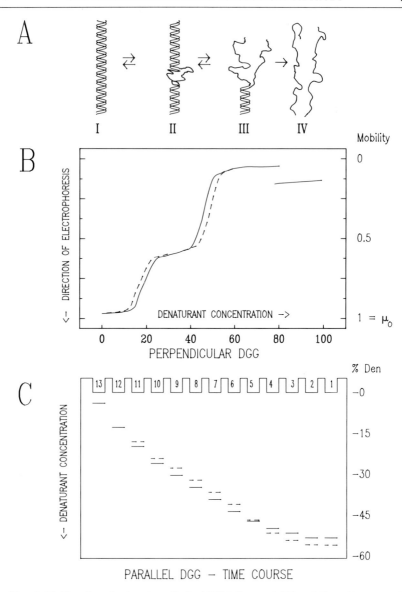

Fɪɢ. 1. Melting domains in a hypothetical DNA fragment (A) and the pattern expected when the fragment is run on either a perpendicular (B) or parallel (C) DGG. (A) and (B) are drawn so that the conformations shown in (A) correspond to the horizontal regions of the mobility curve in (B) directly below. (A) The temperature-dependent conformation of a DNA fragment with three melting domains is shown. (**I**) Fully helical duplex, low temperature form; (**II**) melting of a small, internal segment (domain one) produces a single-stranded bubble; (**III**) melting of the second domain results in a three-armed molecule; (**IV**) at high

Parallel and Perpendicular Gradients

Although it is possible to form thermal gradients in polyacrylamide gels,[3,4] we use a system in which the denaturing environment is determined by the concentration of urea and formamide in a gel maintained at elevated temperature. Heat and chemical denaturants have similar effects, including dependence on base composition, with respect to DNA melting.[5,6] The combined denaturing effect can be described by a single term, T_e, the effective temperature[7]:

$$T_e = T_b + x(\% \text{ denaturant}) \tag{1}$$

where T_b is the bath temperature, and x is a constant ($\sim$0.3°/1% denaturant) relating the effects of urea and formamide to temperature.[5,8,9]

There are two configurations for DGGE: the gradient of denaturant can be parallel or perpendicular to the electric field. We illustrate our

[3] D. R. Thatcher and B. Hodson, *Biochem. J.* **197**, 105 (1981).
[4] R. M. Wartell, S. H. Hosseini, and C. P. Moran, Jr., *Nucleic Acids Res.* **18**, 2699 (1990).
[5] K. Nishigaki, Y. Husimi, M. Masuda, K. Kaneko, and T. Tanaka, *J. Biochem. (Tokyo)* **95**, 627 (1984).
[6] H. Klump and W. Burkart, *Biochim. Biophys. Acta* **475**, 601 (1977).
[7] L. S. Lerman, S. G. Fischer, I. Hurley, K. Silverstein, and N. Lumelsky, *Annu. Rev. Biophys. Bioeng.* **13**, 399 (1984).
[8] L. S. Lerman, S. G. Fischer, D. B. Bregman, and K. J. Silverstein, *Proc. 2nd SUNYA Conv. Discipline Biomol. Stereodyn.* **1**, 459 (1989).
[9] S. Kogan and J. Gitschier, *Proc. Natl. Acad. Sci. U.S.A.* **87**, 2092 (1990).

temperature single strands are formed. In contrast to melting of domains one and two, strand dissociation is in strong disequilibrium, as indicated by the arrows (see text). (B) Pattern expected when the fragment shown in (A) is electrophoresed in a perpendicular DGG. The sample is applied across the top of the gel and migrates vertically downward. Melting of domains one and two results in the mobility transitions at 15–20% and 45–55% denaturant, respectively. Strand dissociation is at 80% denaturant; there is a discontinuity where single strands are formed. The dashed lines indicate the shift in the mobility transition resulting from a mutation in domain one or two. The mobility of the unmelted helix is defined as μ_0 equals 1. (C) Pattern expected when the fragment shown in (A), and a fragment with sequence changes in both domains one and two, is loaded at regular intervals, with the earliest loading at lane 1, on a parallel DGG. The solid line indicates the wild-type fragment, and the dashed line the variant. At the gel position reached by fragments run for a short period of time (lane 13), $T_e < T_m$ of domain one. Fragments run for a longer period of time (lane 11) have reached a concentration of denaturant sufficient to melt domain one; the variant fragment with a *less stable* domain one is retarded relative to the wild type. As the fragments migrate further into the gel (lanes 5–4) domain two melts; the variant fragment with the *more stable* domain two catches up with (lane 5) and passes (lane 4) the wild-type fragment. At slightly higher denaturant concentration the fragments have essentially zero velocity (lanes 2–1) and appear to stop moving.

description of the two configurations by contrasting the electrophoretic behavior of the hypothetical DNA fragment shown in Fig. 1A. In a perpendicular DGG, denaturant concentration increases across the gel from left to right, and the sample is applied across the top of the gel in a single well. As DNA moves downward, each molecule migrates vertically through a constant concentration of denaturant. Figure 1B shows the expected migration of the hypothetical fragment shown in Fig. 1A when electrophoresed on a perpendicular DGG. At the less denaturing side of the gel T_e is well below that required for even partial melting. Fully helical molecules (Fig. 1A, **I**) migrate in a length-dependent manner (Fig. 1B, left) as in conventional nondenaturing gel electrophoresis; we define the mobility of the helical form as μ_0 equals 1. Melting of the least stable internal segment (domain one) results in a mobility decrease owing to the presence of an internal, single-stranded region (Fig. 1A, **II**); mobility drops from 1.0 to 0.6 as denaturant increases from 18 to 25%. For any melting domain, there is a temperature interval over which both the melted and unmelted forms (forms **I** and **II** for domain one, Fig. 1A) are present in significant concentration. This interval corresponds to the steeply sloped regions of perpendicular gels, where mobility reflects the (rapidly shifting) average concentration of the two forms. At 25% denaturant the equilibrium between species **I** and **II** (Fig. 1A) is heavily weighted toward **II**.

Between 25 and 40% denaturant, where domain one has melted but before the second least stable segment (domain two) melts, the partially melted molecule again travels at a nearly constant velocity, affected slightly by an increase in viscosity owing to the increasing concentration of chemical denaturants. Melting of domain two produces a three-armed molecule (Fig. 1A, **III**), resulting in a second sharp decrease in velocity. Strand dissociation (Fig. 1A, **IV**) at 80% denaturant produces a discontinuity in the otherwise smooth melting curve.

The effect of a change in the stability of domain one or two, as might be caused by a base substitution, is shown by the dashed lines (Fig. 1B). A substitution which lowers stability in domain one reduces slightly the concentration of denaturant at which melting occurs, and hence shifts the mobility transition leftward, but it does not affect mobility at lower denaturant concentration, where domain one is completely helical, or at higher denaturant concentration, where domain one is completely melted. Similarly, a change which increases stability in domain two shifts the mobility transition rightward but has no effect on mobility at denaturant concentrations below 40% or above 55%, where domain two is either completely helical or completely melted.

Perpendicular DGGE is a useful first step in the analysis of a DNA fragment. The number of melting domains and their melting temperatures are revealed, providing a guide to the choice of optimal conditions for parallel DGGE.

In parallel DGGE the denaturant concentration increases from the top to the bottom of the gel. Typically wells are formed at the top of the gel, allowing the simultaneous analysis of similar samples. Figure 1C shows the pattern expected when the hypothetical fragment shown in Fig. 1A (wild type) and a variant with base substitutions in domains one and two is run on a parallel DGG for varying lengths of time, starting in lane 1 (longest running time). The initially helical molecules migrate at the same, virtually constant velocity until domain one begins to melt over a narrow range of denaturant (between lanes 12 and 11, Fig. 1C). Fragments differing by a single base in domain one start melting at a different concentration of denaturant, with consequent onset of retardation at different depths of migration. In Fig. 1C the wild-type fragment (illustrated by the solid line in Fig. 1B,C) begins to melt at approximately denaturant, whereas the less stable domain one variant (e.g., a dG · dC to dA · dT transition; dashed line in Fig. 1B and C) begins to melt at 12% denaturant.

After domain one has melted the resolved fragments migrate through the gel at reduced, roughly equal velocity (lanes 9–6). The fragment with the more stable domain one is at a higher concentration of denaturant. As domain two begins to melt (lanes 5 and 4; ~45% denaturant) mobility drops to virtually zero. The fragment with the more stable domain one undergoes this transition first because of its greater depth in the gradient. The variant fragment, with a less stable domain one but a more stable domain two, undergoes the mobility transition at a slightly higher concentration of denaturant; the variant fragment thus catches up with (lane 5, Fig. 1C) and passes (lane 4, Fig. 1C) the wild-type fragment because of the later onset of domain 2 melting. Both fragments are then virtually stopped, and further electrophoresis has little observable effect (lanes 2–1, Fig. 1C). In practice it can take over 100 hr for fragments to reach the denaturant concentration at which a second or third domain melts, because molecules with a melted region over 200 bp move very slowly (≤1 mm/ hr) in 8% acrylamide gels.

Stability of Partially Melted Molecules

The usefulness of DGGE for resolving partially melted molecules is limited by the eventual occurrence of strand dissociation at sufficiently denaturing conditions. Intramolecular melting, which is independent of DNA concentration (as in domains one and two, Fig. 1A) differs from bimolecular melting (strand dissociation), which does depend on concentration. It appears that intramolecular melting reactions are at equilibrium:

during electrophoresis the rates of melting and reannealing are equal and fast relative to movement through the gel, with some rare exceptions.[10,11] In contrast, complete melting of a DNA fragment to produce single strands is not reversible. The mobility of single strands is almost always greater than the mobility of the partially duplex species from which they were formed. Hence, the equilibrium between associated and dissociated strands that might prevail in a uniform bounded volume (a closed system) is thrown off balance by the removal and dilution of the separated strands. We call the region of a fragment which is in a duplex state just prior to strand dissociation the most stable region (MSR).

The differences between intramolecular and bimolecular melting have important practical consequences. First, DGGE does not generally resolve molecules that differ in the MSR. That is, DGGE resolves partially melted molecules; strand dissociation leads to a loss of resolution. GC clamping (see below) was devised to allow detection of mutations in the MSR. Second, in perpendicular gels the mobility curve is discontinuous after melting of the MSR, and the melting pattern at the strand dissociation temperature is markedly influenced by voltage.[12] Third, persistence in the gel of the partially melted molecule depends on the rate of strand dissociation at the T_e at which the fragment is arrested. Partially duplex molecules with a melted region over 200 bp are dramatically retarded in a DGG (velocity <1 mm/hr in 8% acrylamide gels). If the rate of strand dissociation is very low, the partially melted fragment will remain intact in the gel for a long time (>12 hr in an electric field of 8 V/cm for some fragments). On the other hand, if the rate of strand dissociation is significant (see below) migration of single strands away from the zone of the gel containing the arrested fragment will result in gradual loss of a detectable band; this loss may be affected by sequence changes in the MSR. The rate of strand dissociation will be affected by sequence changes in the MSR.[13]

Practical Concepts of Denaturing Gradient Gel Electrophoresis

Some of the major properties of DGGE are presented next. Mobility is determined by melting behavior, not fragment length. Two fragments of different lengths which share the same least stable domain(s) will

[10] S. A. Kozyavkin, S. M. Mirkin, and B. R. Amirikyan, *J. Biomol. Struct. Dyn.* **5,** 119 (1987).
[11] A. Suyama and A. Wada, *Biopolymers* **23,** 409 (1984).
[12] E. S. Abrams, S. E. MacMillan, N. Gabra, E. Schmitt, and L. S. Lerman, unpublished observations (1991).
[13] N. Cariello, Ph.D. Thesis, Massachusetts Institute of Technology, Cambridge, (1988).

retard at the same concentration of denaturant. Conversely, two fragments of the same length, but with least stable domains of different stability, will retard in the gradient at different denaturant concentrations.

Covalent modifications which alter the stability of a domain will be detected only if that domain melts. In almost all fragments larger than 1000 bp there are multiple domains; the range of domain stabilities is such that a partially melted species almost invariably occurs before strand dissociation. Sequence variants in high melting domains may not be detected because of the time required for the slowly moving partially melted species to migrate to the T_e at which these domains melt.

Sequence changes may increase or decrease domain stability, and multiple base changes may have additive or compensatory effects. The specificity of melting behavior arises from sequence-specific interactions: replacing a dG · dC base pair with a dC · dG base pair affects stacking interactions and consequently influences helix melting. Helix defects (non-Watson–Crick base pairs or internal loops of one or more unpaired bases) have large destabilizing effects. For each domain where melting affects fragment mobility, 50–90% of all possible single base changes will be detected,[14,15] and over 90% of helix defects will be detected.[16]

Because detection of helix defects is more efficient than detection of base substitutions, it is useful to form heteroduplexes (Fig. 7A). When two copies of a fragment which differ in sequence at one site are mixed, completely melted, and reannealed, four duplexes are created: the two original molecules (homoduplexes) and two molecules formed by complementary strands from each of the parental species (heteroduplexes). Each of the heteroduplexes will contain a unique helix defect.

There is an optimal set of electrophoresis conditions for detection of variants in each melting domain of a fragment. For domains of different stabilities optimal conditions may vary considerably, as illustrated in Fig. 1. DGGE is simplified for a given fragment when the region of interest lies in a single domain.

Finally, bands tend to sharpen as they retard because molecules at the trailing edge of a band are moving faster than molecules at the leading edge. As a result DGGE appears to be less sensitive to loading conditions (volume of sample loaded, ionic strength of sample) than other polyacrylamide gel electrophoresis procedures.

[14] M. Gray, A. Charpentier, K. Walsh, P. Wu, and W. Bender, *Genetics* **1271**, 1391 (1991).
[15] R. M. Myers, S. G. Fisher, L. S. Lerman, and T. Maniatis, *Nucleic Acids Res.* **13**, 3131 (1985).
[16] E. S. Abrams, S. E. Murdaugh, and L. S. Lerman, *Genomics* **7**, 463 (1990).

Computational Tools

Lerman and co-workers have prepared several computer programs that are useful in planning DGGE experiments for dsDNA fragments of known sequence.[7,17] The program MELT calculates the melting properties of a DNA fragment, whereas the effects of melting on fragment mobility in a gel can be estimated using the programs MUTRAV and SQHTX. MELT calculates the temperature (T) at which each base pair in a known sequence has a probability, π, of being in a helical (as opposed to a disordered) state, as well as the number of base pairs melted and the value of the dissociation constant (K_d) for strand dissociation at a series of temperature intervals selected by the operator. We refer to the temperature at a π value of 0.5 as the melting temperature, T_m, of a domain. For any DNA fragment a plot of T versus base pairs for a fixed value of π is termed a melt map.

Figure 2A shows melt maps (π 0.5) for a 350-bp *Hin*fI fragment from the human β-globin gene (solid line) and the same sequence with a 110-bp G,C-dense sequence (a GC clamp) attached to the promoter proximal end (dashed line). Figure 2B shows melt maps for part of the same *Hin*fI fragment (bp 50–350) with a 110-bp GC clamp attached to the promoter distal end. Melting contours for π values of 0.05, 0.5, and 0.95 are plotted. The sequences have been aligned on the x axis so that the promoter distal end of the β-globin sequence is bp 460 in each case. Melting domains appear as horizontal lines in the melt map; domains separated by no more than 5 or 10 bp can differ in stability by over 20° (compare the GC clamp to the adjacent low melting domain). The pattern of melting domains is a property of the entire fragment: adding a GC clamp to one end of a 350-bp fragment (Fig. 2A) alters melting in much of the fragment. A characteristic inverse relation between domain length and the temperature interval over which π varies from 5 to 95 helicity (melting bandwidth) is seen in Fig. 2B; only for long domains with narrow melting bandwidths is the 50% contour an accurate guide to melting.

Interpretation of the melt map provides valuable information about the physical structure of a molecule as it traverses a DGG. Consider the fragment in Fig. 2B: at 78° bp 190 to 435 are over 95% melted, bp 435 to 475 are 5–95% melted, bp 475–525 are much less than 5% melted, and bps 525–575 are between 5 and 50% melted. Except for the partially melted tail at bp 525–575 the average conformation of this molecule at 78° is similar to that of species **III** in Fig. 1A.

There is substantial evidence from electrophoretic experiments that

[17] L. S. Lerman and K. Silverstein, this series, Vol. 155, p. 482.

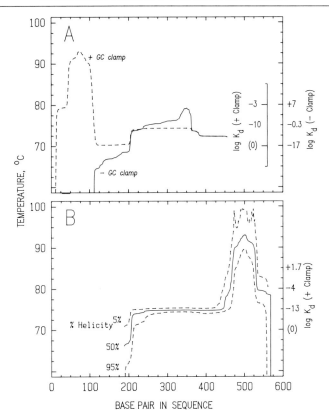

FIG. 2. Melt maps, showing the temperature (T) at which each base pair in a known sequence has a probability π of being in a helical state. (A) Melt maps for $T = T_m$ ($\pi = 0.5$) for a 350-bp *Hin*fI fragment from the human β-globin gene (solid line) and the same fragment with a 110-bp GC clamp at the promoter proximal end (dashed line). The sequences have been aligned on the x axis so that the globin fragment is bp 110–460. The value for log K_d at 70, 75, and 80° for both the clamped and unclamped fragments is shown. Where log K_d is below -20, a value of 0 is plotted. The melt maps for the clamped fragment are replotted in Fig. 3A. (B) Melt maps ($\pi = 0.05$, 0.5, and 0.95) for the β-globin *Hin*fI fragment (less the first 60 bp) with a GC clamp at the promoter distal end. The fragment has been aligned on the x axis so that globin DNA of the same sequence has the same coordinates as in (A). The steep domain boundary at bp 200 is present in the unclamped globin fragment in (A). Values for log K_d at 70, 75, and 80° are shown at right.

the melting behavior predicted by MELT is accurate.[15,18,19] Similar calculations made independently, using different computer programs (based on the same theory), are in quantitative agreement with results obtained by

[18] R. M. Myers, S. G. Fischer, T. Maniatis, and L. S. Lerman, *Nucleic Acids Res.* **13**, 3111 (1985).
[19] S. G. Fischer and L. S. Lerman, *Proc. Natl. Acad. Sci. U.S.A.* **80**, 1579 (1983).

optical spectroscopy,[20] although inverted repeats caused small, unpredicted effects.[21] Close correspondence of the sharp domain boundaries predicted by MELT with experimental results has been demonstrated.[19] For example, when 15 known single base mutations (transitions and transversions) in a 550-bp fragment of λ phage were studied by DGGE, the 9 mutations in the region predicted by MELT to be the least stable domain of the fragment were detected; 6 mutations in more stable domains were not detected.[19] The agreement between the calculated domain boundary and the domain boundary inferred from the experimental results was within 2 bp.

The MELT program can be used to predict accurately the incremental effect of sequence changes on melting behavior or, conversely, infer the type of base substitution from the measured gradient displacement.[22] The effects of helix defects on melting can also be calculated, though less precisely; accounting for non-Watson–Crick base pairs would require, at a minimum, calculating the destabilizing effect of all 104 possible triplets with a non-Watson–Crick base pair in the center, assuming that analysis of nearest neighbor interactions is sufficient for an accurate calculation. It is known that all non-Watson–Crick base pairs destabilize the helix, and that the magnitude of the destabilization is greater than that caused by a nucleotide substitution.[23-25]

The MELT program calculates stability assuming 20 mM cation, similar to the buffer used in DGGE. Whereas T_m varies sharply with Na$^+$ concentrations from 10 to 195 mM, the variation in domain structure is more modest.[26,27] Thus the MELT program can be used, cautiously, to predict domain pattern (but not T_m) for gels run with monovalent cation in the range 10 to 200 mM.

The MELT calculation is not useful at temperatures where strand dissociation occurs at a significant rate. A rough guide to the upper temperature at which the MELT program is dependable[17] (and the temperature at which strand dissociation occurs in a gel) is the temperature at which log K_d is about -5; at temperatures where log K_d is greater than -5, the melt map is valid only for solutions of very high DNA concentration, unobtainable with DGGE. We suggest use of an equilibrium constant

[20] R. D. Blake and S. G. Delcourt, *Biopolymers* **26**, 2009 (1987).
[21] C. R. McCampbell, R. M. Wartell, and R. R. Plaskon, *Biopolymers* **28**, 1745 (1989).
[22] S. G. Fischer and L. S. Lerman, *Proc. Natl. Acad. Sci. U.S.A.* **77**, 4420 (1980).
[23] S. Ikuta, K. Takagi, R. B. Wallace, and K. Itakura, *Nucleic Acids Res.* **15**, 797 (1987).
[24] H. Werntges, G. Steger, D. Riesner, and H. Fritz, *Nucleic Acids Res.* **14**, 3773 (1986).
[25] F. Aboul-ela, D. Koh, and I. Tinico, *Nucleic Acids Res.* **13**, 4811 (1985).
[26] O. Gotoh, A. Wada, and S. Yabuki, *Biopolymers* **18**, 805 (1979).
[27] R. D. Blake and P. V. Haydock, *Biopolymers* **18**, 3089 (1979).

(which varies quite sharply with temperature) in the absence of rates for strand dissociation.

For the unclamped β-globin HinfI fragment (Fig. 2A, solid line), log K_d is -5 at 72.5°; we infer that the large region of the fragment shown by the melt map to have a T_m above 72.5° could not be scrutinized by DGGE. In the clamped fragment (Fig. 2A, dashed line), log K_d is -5 at 78°; therefore, the entire β-globin sequence is accessible to inspection by DGGE.

It would be convenient in planning experiments to relate the melting of a fragment of known sequence to its mobility in a DGG, information not provided directly by the MELT calculation. The programs MUTRAV and SQHTX[17] make predictions about the relative mobility of molecules in a DGG. SQHTX calculates the effect on mobility of a change in stability at each base pair in a molecule for a series of gel running times. As mutations are frequently detected as helix defects in heteroduplex molecules, we normally use SQHTX to assess the effect of a non-Watson–Crick base pair on retardation.[16] The program uses, at each base pair, a uniform change in destabilization; we use a value of $-50°$ (actual values may be somewhat more or less). The SQHTX output complements the MELT calculation by including a quantitative estimate of the effect of partial melting on mobility in a DGG. The SQHTX algorithm assumes that a linear gradient of denaturant is equivalent to a linear gradient of temperature [Eq. (1)], and that the mobility of partially melted molecules is given by

$$\mu_t = \mu_0 \, e^{-(L/Z)} \tag{2}$$

where μ_t is the mobility of fragment at temperature t; L is the total number of base pairs melted at t; μ_0 is the mobility of the helical fragment ($L = 0$); and Z is a constant, taken as 75 bp.

Lerman et al.[7] proposed Eq. (2) based on arguments which may apply more to three-armed than four-armed molecules.[28] Equation (2) has not been rigorously tested but is consistent with the available data.[5] In principle, Eqs. (1) and (2) allow a determination of the mobility versus time profile for any fragment for which μ_0 is known. The value of μ_0 depends on fragment size, acrylamide concentration, and slightly on viscosity changes arising from denaturant gradient.

For a series of gel running times, SQHTX reports displacement: the difference between the temperature the reference molecule would reach and the temperature the variant would reach after the same time. Figure 3B is a plot of displacement (ΔT) versus base pairs at five different gel running times for the clamped β-globin molecule shown in Fig. 2A (dashed

[28] K. R. Shull, E. J. Kramer, and L. J. Fetters, Nature (London) **345,** 790 (1990).

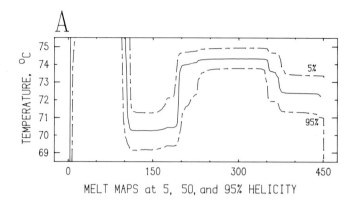

A

TEMPERATURE, °C

MELT MAPS at 5, 50, and 95% HELICITY

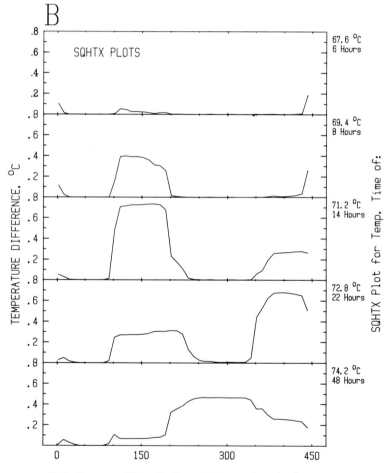

B

SQHTX PLOTS

TEMPERATURE DIFFERENCE, °C

SQHTX Plot for Temp, Time of:

67.6 °C
6 Hours

69.4 °C
8 Hours

71.2 °C
14 Hours

72.8 °C
22 Hours

74.2 °C
48 Hours

450 bp Fragment: 100 bp GC clamp + 350 bp β-Globin HinfI Fragment

line). For a variant molecule with a helix defect, ΔT will always be positive: the heteroduplex molecule will always occur at a lower effective temperature in the gradient than the wild type.

The melt map for the β-globin fragment (with expanded y axis) is shown in Fig. 3A, with curves for π values of 0.05, 0.5, and 0.95. Optimal resolution between wild-type and variant fragments (maximal ΔT) is a function of running time, but the relation varies depending on which melting domain the sequence change is located in. At short running times resolution is seen only at the extreme ends of the fragment. Separation of the wild-type fragment from variants with a mutation in the least stable domain (bp 100–140) requires shorter running times than are needed for resolution of fragments with a mutation in the domain between bp 175 and 325. The times listed to the right of each SQHTX plot in Fig. 3B are for a μ_0 of 1; for a fragment of this size we estimate[12] a μ_0 of 0.2. Detection of variants in the domain with a T_m of 74.5° (bp 200–350) requires running times longer than 22 hr, perhaps impractical using a parallel DGG with a starting T_e of 60°. There are three options: (1) use a parallel gel with a starting T_e just below that required to melt the domain (this requires careful empirical adjustment); (2) use a perpendicular gel (the high denaturant side is equivalent to a parallel gel at long running times, but unfortunately

---

FIG. 3. Comparison of melt maps and SQHTX plots. (A) The melt map for a 350-bp *Hin*fI fragment from the human β-globin locus with a GC clamp (the same sequence is shown by the dashed line in Fig. 2A) for π values of 0.05, 0.5, and 0.95. The x axis is expanded relative to Fig. 2. The melt map shows the pattern of melting domains; the mobility of the fragment in a DGG must be inferred. (B) SQHTX plots show the effect of a variant at each base pair in the fragment as a function of gel running time. The effect of a variant was simulated with a destabilization parameter of $-50°$, typical of the effect expected for a non-Watson–Crick base pair. The starting temperature for the calculation was 60°, and μ_0 (mobility of the unmelted fragment) was set at 1. The displacement [ΔT = (temperature of the wild-type fragment) − (temperature of the fragment with a mismatch at a base pair)] is plotted versus base pairs in the fragment. The temperature the wild-type fragment has reached at five different gel running times is shown at right. We assume a linear relation between temperature and position in a DGG; a displacement of 0.2° is equivalent to 2 mm in a typical DGG. A mutation in the GC clamp (bp 10–100) has no effect on mobility: the displacement is 0 at all running times. The largest separation in the gel between the wild-type fragment and a fragment with a mismatch in the low melting domain (bp 100–200) is predicted to occur when the wild-type molecule reaches a concentration of denaturant at which T_e equals 71.2°; at this temperature a mismatch in the domain at bp 200–350 has no effect, and a mismatch in the domain at bp 350–450 has a small effect. After 48 hr of electrophoresis, the wild-type molecule has reached a concentration of denaturant where T_e is equal to 74.2°. Molecules with a mismatch in the low melting domain (bp 100–200) are no longer resolved from wild type. Characteristically, a mismatch at the ends of the fragment is detectable at very early gel running times.

resolution in this region of a perpendicular gel is usually low); (3) the most general solution to the problem of fragments with multiple melting domains requiring different gel conditions and very long running times is to trim the fragment ends, retaining the region of interest. When the ends of the fragment can be chosen with some freedom [as when the polymerase chain reaction (PCR) is used] considerable benefit can often be gained by manipulation of melting domains.

Planning Experiments

How does one plan a DGGE experiment? The answer depends on (1) the availability of sequence information, (2) the completeness of sequence scrutiny necessary, and (3) the suitability of PCR products for the biological problem to be addressed.

If sequence information is not available, comprehensive scrutiny is unlikely. If the goal is a scan for sequence variants, the approach of Gray et al.[14] (see below) is useful when a large number of fragments are to be screened. Gray and co-workers[14] surveyed a large number of mutants at the *rosy* locus of *Drosophila melanogaster*. Restriction-digested genomic DNA was separated by DGGE and electroblotted to nylon filters, which were hybridized with labeled probes from the 7-kilobase pair (kbp) *rosy* locus. Mutations in 100 of 130 (77%) strains with rosy phenotypes were detected, with 36 of 43 sequenced mutations being single base changes. Lerman et al.[29] estimated that 60% of possible single base changes in the 35.4-kbp human β-globin cluster could be detected by analyzing DNA digested separately with two restriction enzymes. Thus, theoretical considerations suggest and experimental data prove that an impressive number of sequence variants can be detected in unclamped fragments without forming heteroduplexes.

Detection of variants is aided when the melting characteristics of the unsequenced fragment can be examined using perpendicular gels. The number and properties of domains can be gleaned before or after endonuclease restriction and used to select conditions for parallel DGG experiments. Polymorphic fragments will split into two bands as the variant domain melts (see Fig. 1B). This approach allows inspection of the full range of denaturing conditions (and therefore all assayable domains) in a comparatively short experiment. Because relatively nondenaturing condi-

---

[29] L. S. Lerman, K. Silverstein, and E. Grinfeld, *Cold Spring Harbor Symp. Quant. Biol.* **51**, 285 (1986).

tions prevail at one end of the gel one can identify polymorphic fragments on the basis of size. See Refs. 15, 30, and 31 for examples.

If sequence information is available there are two important considerations. (1) Is a comprehensive survey of the sequence, with near 100% likelihood of detection of any base change, necessary (as in screening an mRNA or cDNA for mutations), or is detection of some sequence variants sufficient, as in a search for polymorphisms? (2) Are the molecules to be studied amenable to analysis after PCR amplification? If so, fragment end points can be chosen with few constraints. Standard PCR is not suitable, however, for studies of DNA methylation or in experimental settings where the error rate of *Thermus aquaticus* (*Taq*) DNA polymerase is unacceptable.

If comprehensive sequence scrutiny is desired, careful manipulation of melting domains and use of GC clamps will be necessary. A MELT calculation is the usual first step. An optimal melt map is shown in Fig. 2B; the sequence of interest is confined to a single, flat melting domain bounded by a second domain with much higher T_m. In the clamped fragment, the T_m of the domain of interest should be lower (by at least 0.5–1°) than the temperature at which log K_d equals -5. Such a melt map can be obtained for almost all sequences with a low temperature melting domain of 150–400 bp. For extremely GC-rich sequences not amenable to clamping, two other solutions to the insensitivity of DGGE to variants in the MSR have been proposed. Smith *et al.*[32] used a solution melting step; Collins[33] used dITP-substituted DNA to lower the melting temperature of the GC-rich domain.

Several methods for attaching GC clamps to DNA fragments have been proposed[15,16]; the most convenient incorporates the clamp into the PCR primer.[31,34] If PCR-amplified DNA is not suitable, GC clamps can be added to genomic restriction fragments (or PCR products) by the heteroduplex extension procedure.[16] If a long sequence is to be scanned, a set of overlapping fragments must be designed to ensure that each nucleotide is placed in an assayable melting context. In favorable cases, up to five fragments which melt at different concentrations of denaturant can be analyzed simultaneously in one lane; variants in each fragment can be detected

[30] S. P. Cai and Y. W. Kan, *J. Clin Invest.* **85**, 550 (1990).

[31] V. C. Sheffield, D. R. Cox, and R. M. Myers, *in* "PCR Protocols: A Guide to Methods and Applications," (M. A. Innis, D. H. Gelfand, J. J. Sninsky, and T. J. White, eds.), p. 206. Academic Press, San Diego, 1990.

[32] F. I. Smith, T. E. Latham, J. A. Ferrier, and P. Palese, *Genomics* **3**, 217 (1988).

[33] M. Collins, personal communication (1991).

[34] V. C. Sheffield, D. R. Cox, L. S. Lerman, and R. M. Myers, *Proc. Natl. Acad. Sci. U.S.A.* **86**, 232 (1989).

unambiguously.[12,35] With restriction fragments, manipulation of melting domains is limited, and comprehensive scrutiny of a long sequence is not ensured.

The utility of optimizing fragment end points and adding GC clamps is illustrated by comparison of the melt maps shown in Fig. 2A (dashed line) and Fig. 2B (solid line). The fragment shown in Fig. 2B has a single low melting domain of over 200 bp; mutations throughout this region can be seen at a single gel running time. The inclusion of the low-melting tail of the *Hin*fI fragment (bp 110–210, solid line Fig. 2A) in the fragment shown in Fig. 2B is highly undesirable. If present, the low-melting tail would greatly increase (from 3 to >12 hr[12]) the time required in a parallel DGG before the fragment reaches a T_e of 74°.

About one-third of all the autoradiographic bands observed in the study by Gray *et al.*[14] were noticeably broad ("fuzzy"). In the sequence shown in Fig. 3A, the lowest melting domain is in the middle of the molecule, bounded on either side by more stable regions. Such "middle melter" molecules tend to form fuzzy bands, and the degree of broadening appears to be related to the difference between the melting temperature of the low melting domain relative to the boundary domains on either side. However, fuzzy melting behavior does not impede detection of single base changes.[12,36]

The DGGE–Southern blotting method is also useful for studying methylation in genomic DNA. Methylation of a single nucleotide can cause a mobility shift of the same magnitude as a base substitution.[37] Figure 4 shows part of a DGGE–Southern blot in which *Hae*III or *Hin*fI restricted human genomic DNA from five unrelated individuals was hybridized with a 4-kilobase (kb) single-copy probe. The apparent polymorphism(s) are due to differential methylation of the variant fragments, as shown by comparing genomic DNA with PCR-amplified DNA (Fig. 4).

Where many samples of the same sequence are to be analyzed and the determinant domain is not too long, the experimental procedure can be simplified somewhat by electrophoresis in a gel with a uniform concentration of denaturant.[38,39] The concentration of denaturant used depends on where the mobility difference between the two fragments is maximal as

[35] M. D. Traystman, M. Higuchi, C. K. Kasper, S. E. Antonarakis, and H. H. Kazazian, Jr., *Genomics* **6,** 293 (1990).

[36] M. Gray, personal communication (1991).

[37] M. Collins and R. M. Myers, *J. Mol. Biol.* **198,** 737 (1987).

[38] E. Hovig, B. Smith-Sorensen, A. Brogger, and A. L. Borresen, *Mutat. Res.* **262,** 63 (1991).

[39] M. Lichten, C. Goyon, N. P. Schults, D. Treco, J. W. Szostak, J. E. Haber, and A. Nicholas, *Proc. Natl. Acad. Sci. U.S.A.* **87,** 7653 (1990).

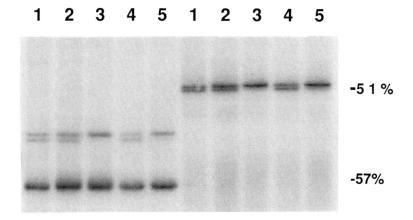

FIG. 4. Autoradiogram of a DGGE–Southern blot. Genomic DNA (10 μg) from Epstein-Barr virus (EBV) immortalized lymphocyte lines of five unrelated subjects was restricted with *Hae*III (left) or *Hin*fI (right). Digested DNAs were electrophoresed for 21 hr at 100 V on a 24 cm DGG (20–80% denaturant). The gel was denatured with alkali, neutralized, and electrotransferred to a Nytran membrane (Schleicher & Schuell, Keene, NH). The membrane was hybridized to a radiolabeled 4-kb single-copy probe from an intron of the human GNAI2 gene, washed [0.5% sodium dodecyl sulfate (SDS), 0.5× SSC at 65° for 1 hr], and autoradiographed for 4 days. The portion of the autoradiogram shown corresponds to 48–59% denaturant. The sharp, flat bands are characteristic of DGG. Note the smearing of the *Hae*III bands at 57% denaturant, a frequent band morphology. Two polymorphisms in apparent linkage disequilibrium are evident. Alternatively, the same polymorphic site may be present in both fragments. The polymorphic *Hin*fI fragment was cloned and sequenced, primers were made starting at the *Hin*fI termini and the fragment was amplified from genomic DNA of individuals 1–5. Extensive testing of the PCR products on perpendicular and parallel gels using gel conditions similar to those in the original experiment failed to reveal any polymorphism. We conclude that the variation in the autoradiogram is not due to base changes but almost certainly to methylation. The fragment contains multiple CpG dinucleotides, potential methylation sites.

estimated from a perpendicular DGG (e.g., 15% denaturant for domain one of the fragments shown in Fig. 1B).

Procedure

We describe and evaluate equipment and procedures that have been used successfully. It is our experience that the casting and running of a DGG require comparable effort but less care than a DNA sequencing gel. Many detailed procedures and comments on equipment are given by Myers *et al.*[40]

[40] R. M. Myers, T. Maniatis, and L. S. Lerman, this series, Vol. 155, p. 501.

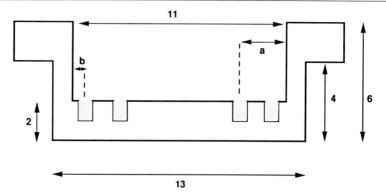

FIG. 5. Hangers to support the Hoefer upper buffer chamber in a tank are cut from ½ inch acrylic; an example is drawn to scale with critical dimensions in centimeters. The four cutouts (stippling) are 1 cm deep and 0.7 cm wide and accommodate support ribs in the buffer chamber. Spacing (a = 0.6 cm; b = 2.4 cm) is to the center of the cutouts.

Gel Equipment

Denaturing gradient gels are generally run as vertical gels, completely immersed in a tank of 60° buffer. Except for rare, very (dA · dT)-rich sequences which melt below 60° it is usually more convenient to maintain the bath at 60° and vary the range of chemical denaturant. A complete specialized apparatus for casting and running DGGs is available from CBS Scientific (Del Mar, CA). We use SE600 series cassettes from Hoefer Scientific (San Francisco, CA) which are also convenient for conventional polyacrylamide gel electrophoresis. This equipment is versatile, easy to set up, and allows the running of multiple gels easily. After pouring the gel(s) and loading the samples, the cassette is attached to the upper buffer chamber (SE6054), which is suspended in the tank with plexiglass hangers (Fig. 5). A wide variety of combs, plates, and spacers are available from Hoefer; we use 16 or 24 cm long spacers of 0.75 mm thickness for general work; thicker spacers (we have tried up to 3 mm) can be used for preparative gels. Each upper buffer chamber can hold two cassettes, and each cassette can hold two gels when an eared sandwich plate is placed between two standard plates.

Tanks and Thermoregulation

The requirement for a constant, homogeneous thermal environment is met by immersing the gel in a tank of electrolyte (the gel running buffer). The tank must be well stirred and the temperature accurately regulated for consistent results. Three different setups that work are the following:

(1) Immersion circulators, which combine heating, thermoregulation, and stirring in one device, are available from several manufacturers, for example, HETO Model 01 TT 620 (ATR, Inc., Laurel, MD) and Fisher (Pittsburgh, PA) Isotemp immersion circulator Model 730. An instrument that can be vertically adjusted to various depths (HETO) is best. These instruments are compact, easy to install and operate, and relatively inexpensive. The HETO unit can maintain the bath temperature within ±0.1° during an overnight run. (2) A mercury thermoregulator driving a standard controller (Cole-Parmer, Chicago, IL, Cat. Nos. N-02149-72 and N-02149-20) is accurate, inexpensive, and easily adjusted. Two quartz infrared immersion heaters (400 W each) are connected to the controller. Optimal thermoregulation occurs when the thermal sensor is close (2″) to the heat source. (3) A gel box and heat exchanger (Hoefer) are needed for use with the SE600 cassettes. For DGGE, pump hot water (1–2 gallons/min) through the heat exchanger.[35,39] A correction for the difference in the temperature of the circulating water and the temperature of the tank is needed.

We recommend glass tanks (CBS Scientific) for use with systems (1) and (2). Tanks of 14 × 38 × 38 (or 29) cm (width × length × height) are convenient for running 24 (or 16) cm long gels when using the HETO controller. If a mercury regulator and quartz immersion heaters are used, a smaller tank will suffice [14 × 30 × 38 (or 29)]. Plexiglass tanks are commercially available (Green Mountain Lab Supply), and aquarium tanks from pet stores will often work, although the dimensions of such tanks are generally inconvenient. The anode is usually a length of platinum wire which runs down the side of the tank.

Careful temperature monitoring is important with all three systems, particularly system (3) where, owing to the smaller buffer volume and limited heat exchange, heating from the electric field is significant. An ASTM 20C thermometer, with 0.1° gradations from 57° to 65°, is convenient.

Systems (1), (2), and (3) require a separate pump (e.g., a peristaltic pump with an output of 40–60 ml/min) to circulate buffer from the lower to the upper buffer chamber and thereby prevent the formation of pH differences. The Hoefer upper buffer chamber is provided with such a port so buffer can return to the tank. For systems (2) and (3), mixing can be accomplished either with a shaft (propeller blade) mixer or a heavy duty magnetic stirrer. Conventional magnetic stir bars will flip out of the rotating field, thereby ruining the experiment.

The high temperature and long running times typical of DGGE result in significant evaporation of running buffer. To reduce this, cover the buffer with $\frac{1}{2}$ inch diameter hollow polypropylene spheres or Styrofoam packing peanuts and then cover the tank with plastic wrap. We replace the buffer after 5 to 15 gels.

The essential components of a DGGE system are described above. In addition, tedious waiting for the tank buffer to warm from room temperature to 60° can be avoided by using an auxiliary heater [1000 W Vycor glass heater; Corning (Corning, NY), No. 16790]. For very long runs (e.g., 40-hr runs to resolve 50-kb molecules) it is necessary to replenish the buffer during the run. A liquid level sensor and relay (Cole-Parmer, Cat. Nos. L-07187-54 and L-07188-50) driving a peristaltic pump assures constant liquid level. Denaturing gradient gels are often run for 8 to 16 hr, so experiments may end during the night. It is convenient to have a timer (with at least a 10 A rating) to turn power supplies and heaters off. Bands do not diffuse appreciably over several hours if the gel is kept in the cassette.

Reagents. The majority of DGGE experiments reported in the literature use the buffer and gel solutions developed by Fisher and Lerman.[40,41]

$20 \times$ TAE running buffer: 0.8 M Tris base, 0.4 M sodium acetate, 20 mM Na_2EDTA, pH 7.4. For 4 liters, use 388 g Tris base, 218 g $CH_3CO_2Na \cdot 3H_2O$ (131.25 g anhydrous sodium acetate), and 30 g Na_2EDTA. Dissolve in 3.5 liters of distilled water, add glacial acetic acid ($\sim$120 ml) to pH 7.8, allow to cool to room temperature, adjust to pH 7.4, and bring to volume.

40% (w/v) Acrylamide–bisacrylamide (29 : 1)

Formamide, urea: commercially available molecular biology grade formamide and urea can be used without further purification

Denaturant solutions contain $1 \times$ TAE, acrylamide, and denaturant. Denaturant concentration is specified as percent of an arbitrary standard: 100% denaturant is defined as 40% (v/v) formamide plus 7.0 M urea. Prepare stocks of 0 and 100% denaturant with acrylamide and mix to make intermediate concentrations as needed. The acrylamide concentration can be adjusted to the size of the fragments being examined; 8% affords adequate resolution of 0.15- to 1-kb fragments. Denaturant–acrylamide stocks should be filtered through 0.45-μm nylon filters (formamide will dissolve many of the standard filter materials) and will last for at least 3 months when stored at 4°.

Ammonium persulfate: a 10% solution in water (APS) can be stored at 4° for 2 weeks. N,N,N',N'-Tetramethylethylenediamine (TEMED) is electrophoresis grade.

Loading buffer: sucrose- or glycerol-based load buffers can be used

[41] S. G. Fischer and L. S. Lerman, *Cell (Cambridge, Mass.)* **16,** 191 (1979).

Casting Gels

1. Assemble the Hoefer cassette on the casting stand as recommended by the supplier, applying a thin coat of petroleum jelly along the outer half of the spacers; failure to do so usually results in severe distortion of bands near the spacers.

2. Make up equal volumes of denaturant solutions corresponding to the two end points of the gradient by mixing 0 and 100% stocks; chill on ice. The total volume of denaturant solution should equal the volume between the plates. Alternatively, one can fill the plates to 90% of capacity with denaturing gradient gel and then top the gel with 0% denaturant. This allows clear demarcation of the beginning of the gradient and prevents urea and formamide from leaching into the sample wells. The Hoefer cassettes with 0.75 × 16 or 24 cm spacers require a total gradient volume of 16.4 or 25 ml to fill the gel plates within about 1.5 cm of the top.

3. Add catalysts [1/100th volume 10% (APS), 1/200th volume TEMED] to the denaturant solutions and mix by gently swirling. Pour the denser solution into the downstream (mixing) chamber of a two-chamber gradient maker (e.g., Hoefer SG30; similar devices are offered by other vendors) with a 10 × 3 mm stir bar in the mixing chamber. Slowly open the valve between the two chambers to fill the channel with denaturant solution; holding the gradient maker at an angle facilitates removal of air bubbles. Close the stopcock, and add the less dense solution to the other chamber of the gradient maker.

4. Place the gradient maker on a magnetic stirrer, connect to a peristaltic pump (0.8 mm ID tubing, with Luer fittings), open the internal and external channels, and turn on the peristaltic pump. A 20-gauge needle attached to the end of the 0.8 mm tubing will fit between plates with 0.75-mm spacers. (A peristaltic pump is not necessary: for gravity flow, use a larger diameter tubing and set the gradient maker just above the top of the gel.) Adjust the flow so that the gradient pours over a period of 5 (ESA) to 10 (VPS) min.

5. When the solution in the gradient maker has run out add 0% denaturant to the mixing chamber and fill the gel cassette. Insert the comb at an angle to avoid trapping air bubbles.

6. After the gel has polymerized remove the comb and immediately flush the wells with 1× TAE.

7. Load the gel and fasten the cassette to the upper buffer chamber as described by the supplier. Place the upper buffer chamber/gel assembly in the tank. The Hoefer upper buffer chamber has two ports for cassette attachment. When only one gel is run the unused port is closed with a plexiglass baffle. For a time course experiment it is necessary to load

wells while the gel remains in the tank; this is easily accomplished with microcapillary (sequencing) pipette tips or a Hamilton syringe.

The procedure for casting perpendicular gels, which have a single long well, is slightly different. Assemble the gel plates as above, then rotate the cassette 90°, so that the spacers are now horizontal. In the space where the comb would normally go, insert a spacer which has been trimmed to around 13.5 cm. This spacer is inserted to a depth of 1 cm, with one end (lightly coated with Celloseal) flush against the lower horizontal spacer. Use two or three small binder clamps to hold the spacer in place and prevent leaking. Insert the outlet of the gradient maker in the space between the 13.5 cm long spacer and the upper horizontal spacer. Pour the gradient as described above. After the gel has polymerized stand the cassette upright and fill any remaining space at the low denaturant end of the gel with 0% denaturant solution, so that when the 13.5 cm long trimmed spacer is removed the 0% denaturant forms one side of the well. Alternatively, if several samples are to be analyzed on the gel, make a small Teflon comb with several narrow (~2.5 mm) wells and insert it into the 0% gel. A small amount of each sample is loaded into one lane at the 0% end of the gel, and the remainder is mixed and loaded in the long well. This allows identification of the different species in a mixture on the basis of size.

Transfer of DNA from Gels to Membranes

Transfer of DNA fragments resolved by DGGE to nylon membranes allows repeated probing of the same DGG fractionation. The following procedure was devised by Gray et al.[14]

1. After electrophoresis wash the gel for 5 min in several changes of water to remove urea and formamide.
2. Stain the gel with ethidium bromide (1 μg/ml) for 10 min and examine under UV illumination for electrophoresis problems (e.g., buffer leak through the side spacer resulting in lane distortion). Photographic documentation is useful in interpreting subsequent autoradiograms.
3. The DNA can be denatured before or after electrotransfer. We find that duplex DNA is more uniformly and reproducibly transferred than denatured DNA. Nonetheless, satisfactory results have been obtained with either procedure. (a) To denature DNA before transfer, bathe the gel in 0.5 M NaOH for 5 min, neutralize in 0.5 M Tris (pH 8.0) for 5 min, and then soak in transfer buffer (20 mM Tris, 1 mM Na$_2$EDTA, pH 8.0) for 10 min. (b) If DNA is to be denatured on the nylon membrane after transfer, just soak the gel in transfer buffer for 10 min.
4. Assemble a cassette for electrophoretic transfer. Most commercial

electrotransfer apparatuses include a fenestrated plastic cassette to hold the gel and filter together. Submerge the cassette, a Scotchbrite pad, a piece of filter paper, and the labeled nylon filter in transfer buffer. Pick up the gel with a piece of adherent filter paper and lay it (gel side down) over the nylon filter. Remove any trapped air bubbles. Add a second Scotchbrite pad or more sheets of filter paper to make a tight sandwich when the other side of the plastic cassette is closed. Quickly lift the sandwich into the transfer tank with the nylon filter anodal to the gel. The brand of nylon filter used can dramatically affect the results. We have achieved most consistent success with Hybond N Plus (Amersham, Arlington Heights, IL).

5. Transfer by electrophoresis at 0.6–2 A for 2 hr. (The transfer buffer will warm by up to 15°.)

6. Rinse the filter in $2 \times$ SSC or $2 \times$ SSPE for 5 min and bake 2 hr.

7. Prehybridize and hybridize the membrane using standard Southern blot conditions.[42]

For transfer from 4% gels, which tend to be sticky, a negatively charged membrane such as Biodyne C (Pall) can be interposed between the gel and the membrane to which the DNA binds. During electroblotting the DNA appears to traverse the negatively charged membrane without hindrance.[12]

Using GC Clamps to Manipulate Melting Domains

As discussed above, synthetic or cloned GC-rich domains can be added to the ends of DNA fragments to create melting conditions favorable for DGG analysis. In general, 45-bp clamps seem to work consistently, whereas shorter clamps are less predictable. (A short clamp may produce an attractive melt map but it is important to consider the equilibrium constant of strand dissociation at the fragment arrest temperature as well.)

There are three main approaches to clamping. In the first approach, clone each fragment of interest into a plasmid with a GC clamp adjacent to a polylinker; a composite fragment (insert + GC clamp) is excised and run on a DGG. Several plasmids with GC clamps adjacent to polylinkers have been constructed.[16,40,43] Figure 6 shows the salient sequences of pCH1, pCH50, and pCH75, three GC clamp plasmids with different polylinkers.

In a second approach, Sheffield et al.[31,34] showed that addition of a 40-nucleotide GC-dense sequence to the 5' end of a PCR primer is a conve-

[42] G. M. Church and W. Gilbert, *Proc. Natl. Acad. Sci. U.S.A.* **81,** 1991 (1984).
[43] M. J. Tymms, B. McInnes, P. Alin, A. W. Linnane, and B. F. Cheetham, *Genet. Anal. Tech. Appl.* **7,** 53 (1990).

A

pCH1: 150 bp GC Clamp in the HincII site of pTZ19U

```
ACGACTCACT ATAGGGAAAG CTTGCATGCC TGCAGGTCCC GCGCCCCCCG    50

TGCCCCCGCC CCGCCCGCCG CGCCCCCCGT GCCCCCGCCC CGCCCGCCGC   100

GCCCCCCGTG CCCCCGCCCC GCCCGCCGCG CCCCCCGTGC CCCCGCCCCG   150

CCCGCCGCGC CCCCCGTGCC CCCGCCCCGC CCGCCGGACT CTAGAGGATC   200

CCCGGGTACC GAGCTCGAAT TCACTGGCCG TCGTTTTACA ACGTCGTGAC   250

TGGGAAAACCC   260
```

pCH50: 60 Bp GC Clamp in the HincII site of pTZ19U

```
ACGACTCACT ATAGGGAAAG cttgcgcgct agcgctgata tcaaaactgC    50

AGGTCCCGCG CCCCCCGTGC CCCCGCCCCG CCCGCCGCGC CCCCCGTGCC   100

CCCGCCCGCCC GCCGACTCT AGAGGATCCC CGGGTACCGA GCTCGAATTC   150

ACTGGCCGTC GTTTTACAAC GTCGTGACTG GAAAACCC   188
```

pCH75: 60 Bp GC Clamp in SphI site of pTZ18U

```
ACGACTCACT ATAGGGAAtt cgccatggcc ccagatctgg ggatcctctA    50

GAGTCGACCT GCAGGGTCGG CGGGCGGGCG GGGGCACGGG GGGCGCGGCG   100

GGCGGGGCGG GGGCACGGGG GGGGCGGGAC CCAAGCTTGG CACTGGCCGT   150
```

FIG. 6. Sequence (A) and restriction sites (B) of the polylinker and GC clamp of phagemids pCH1, pCH50, and pCH75. These phagemids are based on the pTZ vectors [D. A. Mead, *Protein Eng.* **1**, 67 (1986)]. In (A) the strand of DNA packaged in the pseudovirion is shown. The GC clamp is a head-to-tail array of a 29-bp monomer (indicated by the double underline in pCH1). Owing to the cloning strategy the ends of the clamp sequence vary. pCH1 is pTZ19U with a 150-bp GC clamp (bp 39–186) in the HincII site of the polylinker; pCH14 is a similar plasmid, but with a 60-bp GC clamp. pCH50 is pTZ19U with a 60-bp GC clamp (bp 56–114) in the HincII site and with the HindIII to PstI fragment replaced by a new polylinker (shown in lowercase) made in our laboratory. pCH75 is pTZ18U with a 60-bp GC clamp (bp 64–128) in the SphI site and with the EcoRI to XbaI fragment replaced by the EcoRI–XbaI fragment (shown in lowercase) from pGC1 (R. M. Myers, T. Maniatis, and L. S. Lerman, this series, Vol. 155, p. 501).

B

Enzyme	Sequence	Site in		
		pCH1	pCH50	pCH75
AccI	GTmkAC	—	—	53
BamHI	GGATCC	196	124	41
BglII	AGATCT	—	—	33
BssHII	GCGCGC	—	24	—
Eco47III	AGCGCT	—	31	—
EcoRI	GAATTC	217	145	16
HincII	GTyrAC	—	—	53
HindIII	AAGCTT	18	18	133
KpnI	GGTACC	205	133	—
NcoI	CCATGG	—	—	23
NheI	GCTAGC	—	28	—
PstI	CTGCAG	30	47	59
SalI	GTCGAC	—	—	53
SmaI	CCCGGG	201	129	—
SphI	GCATGC	24	—	—
SstI	GAGCTC	211	139	—
XbaI	TCTAGA	190	118	47

FIG. 6. (continued)

nient way to attach a clamp without an intervening cloning step (Fig. 7A). The GC-dense tail of the primer is not homologous to the template, and it does not interfere with the PCR. We do not know if multiplex PCR is possible with GC clamp primers. Several different sequences have been used as GC clamps in this procedure,[9] and the reaction conditions reported do not differ significantly from standard PCR conditions. Cleaner results are obtained from GC clamp oligonucleotides purified by gel electrophoresis or high-performance liquid chromatography (HPLC).

As a third approach, Abrams et al.[16] present a method for directly attaching a GC clamp to a genomic restriction fragment or an unclamped PCR product. The procedure, as applied to restriction enzyme-digested genomic DNA, is outlined in Fig. 7B (see Ref. 16 for details). Single-stranded probes of high specific activity are required for analysis of genomic DNA. The following procedure is recommended for PCR amplified DNA. (This procedure follows the scheme shown in Fig. 7B.)

1. Clone the PCR-amplified sequence of interest into a GC clamp vector.

2. Excise a restriction fragment (GC clamp + insert) from the plasmid and label the end near the GC clamp with $[\gamma^{-32}P]$ATP and T4 polynucleotide kinase. For example, after cloning a PCR product into the BglII site of pCH75 (Fig. 6), digest the plasmid with EcoRI and treat with T4 kinase and $[\gamma^{-32}P]$ATP. Heat-inactivate the kinase and digest the plasmid with HindIII. Purify the labeled restriction fragment by gel electrophoresis in

A

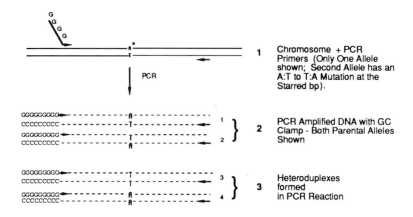

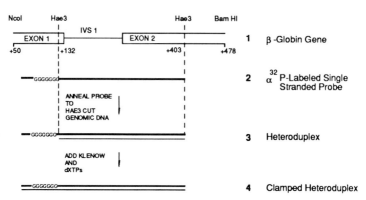

1   Chromosome + PCR
    Primers (Only One Allele
    shown; Second Allele has an
    A:T to T:A Mutation at the
    Starred bp).

2   PCR Amplified DNA with GC
    Clamp - Both Parental Alleles
    Shown

3   Heteroduplexes
    formed
    in PCR Reaction

B

1   β -Globin Gene

2   α ^{32}P-Labeled Single
    Stranded Probe

3   Heteroduplex

4   Clamped Heteroduplex

FIG. 7. Attaching a GC clamp using (A) clamp–primer PCR or (B) heteroduplex extension. (A) Attaching a GC clamp using a GC-tailed PCR primer. (1) One of the two PCR primers (indicated by the arrows) has a 40-nucleotide GC-rich 5′ tail. The tail is not complementary to the target sequence, indicated by the thin lines. Only one chromosome is shown. (2) Amplification generates two PCR products, one for each chromosome. The two alleles differ at a single site. (3) During the PCR procedure heteroduplex products are formed by product strand annealing. The extent of product strand annealing varies markedly during the reaction. (B) Strategy for attaching a GC clamp to a genomic restriction fragment. (1) Part of the human β-globin gene spanning the first intron is shown with pertinent restriction sites. Numbering is relative to the cap site, which is set as the zeroth base pair. Digestion with

low-melt agarose using TAE running buffer. If 5–10 µg of plasmid DNA is labeled the probe fragment (typically 300–600 bp) can be seen after ethidium bromide staining. Excise the band, dilute with 800 µl TEN (10 mM Tris-HCl, pH 8.0, 1 mM EDTA, 0.1 M NaCl), and melt the agarose (70°, 15 min). Typically 1–4 × 10^6 disintegrations/min (dpm) is recovered. After dilution this solution can be used directly in the hybridization step below.

3. Hybridize the labeled probe with PCR product in the following mixture: 3 µl (1800 dpm) probe; 15 µl of a typical PCR reaction (10–100 ng DNA); 4.2 µl KH buffer (0.5 M NaCl, 0.1 M Tris-HCl, pH 7.4); 0.42 µl 1 M dithiothreitol (DTT); 600 ng M13 viral DNA (see below); and water to a volume of 42 µl. Heat to 100° for 4 min and cool slowly (>40 min) to 30°.

4. For the clamp fill-in reaction, add the following to each hybridization reaction: 0.5 µl of 1 M MgCl$_2$; 0.6 µl of a solution with each of the four dXTPs at 25 mM; 0.9 µl KH buffer; 5.5 µl water; and 1.5 µl of Klenow (5 units/µl, New England Biolabs, Beverly, MA). After 10 min at 22° heat to 65° for 10 min and then stop by adding 15 µl of loading buffer and 2 µl of 0.5 M EDTA. Load 15–20 µl on a DGG of appropriate composition.

Comments

A low salt buffer is used for hybridization to ensure that the GC clamp melts. Heating at 100° for 4 min should inactivate Taq polymerase. For the clamp fill-in reaction a DNA polymerase with a 3′ → 5′-exonuclease activity is used to remove the nontemplate specified base often present at

---

HaeIII yields a 272-bp fragment. A wild-type copy of this HaeIII fragment was cloned adjacent to a 60-bp GC clamp, using the clamp vector pCH14 (see Fig. 6A), to give pCH14G1. (2) ^{32}P-Labeled single-strand probe is prepared from pCH14G1 by primer extension. The primer extension protocol yields a single-stranded molecule with precisely defined 5′ and 3′ ends. The probe is 375 nucleotides long; the 3′ 272 nucleotides are homologous to the β-globin HaeIII fragment, and the 5′ 100 nucleotides are homologous to the GC clamp and vector sequences. (3) When denatured, HaeIII-cut genomic DNA is annealed with the probe, a partially duplex structure is formed (the thick line represents the probe and the thin line the genomic DNA). The 5′ part of the probe (with the clamp) is present as a single-stranded tail, while the globin homologous sequences are present in a duplex, with one strand from the probe and the second strand from the genomic restriction fragment. Because the probe contains the wild-type β-globin sequence, a helix defect will be present if the genomic fragment contains a sequence variant. (4) The unclamped, partial duplex structure shown in (3) is converted to a clamped, fully duplex structure by treatment with Klenow fragment of DNA polymerase and dXTPs. The 3′ end of the genomic strand is the primer for DNA synthesis.

the 3' termini of PCR-amplified DNA.[44] When such PCR products are used in the heteroduplex extension procedure, this nontemplated base is the primer for the clamp fill-in reaction (Figure 7B, step 3). If the base is not removed, a helix defect will be present after the clamp fill-in reaction.

Single-stranded M13 viral DNA acts as a capture for the unlabeled strand of the (originally duplex) probe, which would otherwise compete with PCR molecules for the labeled probe strand. A cloned mutant PCR product can be used as a positive control. Such fragments can be routinely obtained from cloned PCR-amplified DNA.

Heteroduplex extension is similar to other procedures for detecting mutations in restricted genomic DNA by solution hybridization.[40,45–47] The sample of interest is digested with a restriction enzyme, mixed with a high specific activity single-stranded probe (^{32}P-labeled RNA probes can be used[48,49]), denatured, and allowed to anneal. The probe–genomic hybrids are then analyzed on a DGG. In general, 1–10 μg of genomic DNA per lane is used with high specific activity probes, which are difficult to prepare and relatively unstable owing to ^{32}P decay-induced strand breakage.

Using Polymerase Chain Reaction Products for Denaturing Gradient Gel Electrophoresis

Although PCR products have greatly simplified most aspects of sample preparation for DGGE, there are drawbacks aside from the loss of information regarding methylation. One concern with PCR-based DGGE is the error rate of *Taq* DNA polymerase, usually given as roughly 10^{-4}/base. PCR-induced mutations in melted domains will usually alter fragment mobility, especially when the novel mutant strands form heteroduplexes with wild-type strands. If DGGE is being used to identify rare mutations the error rate of *Taq* polymerase may be unacceptable. Use of thermolabile polymerases with lower error rates has proven useful in such circumstances.[50,51]

Another problem arises when a region containing two or more polymorphisms is amplified by PCR. PCR amplification of a heterozygous locus

[44] M. A. Innis, D. H. Gelfand, J. J. Sninsky, and T. J. White, (eds.), *in* "PCR Protocols: A Guide to Methods and Applications." Academic Press, San Diego, 1990.

[45] M. Collins, S. F. Wolf, L. L. Haines, and L. Mitsock, *Electrophoresis* **10**, 390 (1989).

[46] W. W. Noll and M. Collins, *Proc. Natl. Acad. Sci. U.S.A.* **84**, 3339 (1987).

[47] G. E. Riedel, S. L. Swanberg, K. D. Kuranda, K. Marquette, P. LaPan, P. Bledsoe, A. Kennedy, and B.-Y. Lin, *Theor. Appl. Genet.* **80**, 1 (1990).

[48] N. Takahashi, K. Hiyama, M. Kodaira, and C. Satoh, *Mutat. Res.* **234**, 61 (1990).

[49] F. I. Smith, J. D. Parvin, and P. Palese, *Virology* **150**, 55 (1986).

[50] P. Keohavong and W. G. Thilly, *Proc. Natl. Acad. Sci. U.S.A.* **86**, 9253 (1989).

[51] N. F. Cariello, P. Keohavong, A. G. Kat, and W. G. Thilly, *Mutat. Res.* **231**, 165 (1990).

produces four dsDNAs: the two parental alleles and two heteroduplexes (Fig. 7A). If the two alleles of the heterozygous locus differ at more than one site a novel type of PCR error called "jumping PCR"[44] will occur. The error results from the apparent failure of *Taq* polymerase to complete replication of all template strands during each round of PCR. Ends of some of the incomplete strands will lie between two polymorphic sites. In subsequent rounds the incomplete strands will act as long primers, anneal to heterologous templates, and be extended. The result will be new recombinant molecules which generally include all possible combinations of the starting polymorphic sites. The quantity of the various species is influenced by the distance between the polymorphisms and by the PCR conditions.[52]

Applications

In this section we briefly review some of the areas in which DGGE has proven useful and discuss possible new applications of the technique. The major use of DGGE to date has been to scan genes or parts of genes to detect sequence variants. The most extensive efforts have been devoted to detection of variants in the human β-globin[30] and Factor VIII genes[9,35] (only recent papers are cited). Denaturing gradient gel electrophoresis has been used to look for variants in genomic DNA from maize[47] and drosophila,[14] and in dsRNA viruses.[49] Lichten *et al.*[39] used DGGE to detect non-Watson–Crick base pairs present in unresolved recombination intermediates in yeast. Both frameshifts and base substitutions are seen.[51] One study which compared the efficiency of different methods for variant detection concluded that DGGE (including the use of clamps) was preferable to cleavage of helix defects by RNase A or chemicals.[53]

Comparison of DGGE to the recently described single-stranded conformation polymorphism (SSCP) technique[54,55] has shown the SSCP method to be comparable at variant detection when natural (unclamped) DNA fragments, chosen without regard to their melting properties, are studied: 14 small DNA fragments (<300 bp), each containing a sequenced transition or transversion, were studied with both methods. Seven of the 14 pairs of fragments were separated by SSCP, but only two by DGGE.[56] The appar-

[52] V. P. Stanton, Jr., unpublished observation (1991).
[53] B. D. Theophilus, T. Latham, G. A. Grabowski, and F. I. Smith, *Nucleic Acids Res.* **17,** 7707 (1989).
[54] M. Orita, H. Iwahana, H. Kanazawa, K. Hayashi, and T. Sekiya, *Proc. Natl. Acad. Sci. U.S.A.* **86,** 2766 (1989).
[55] Y. Suzuki, M. Orita, M. Shiraishi, K. Hayashi, and T. Sekiya, *Oncogene* **5,** 1037 (1990).
[56] H. Aburatani and V. P. Stanton, Jr., unpublished observation (1991).

ent low efficiency of DGGE (in this particular set of fragments) is largely due to mutations not lying in appropriate melting contexts. In principle GC clamping should increase the efficiency of DGGE to nearly 100%. As sequence information is accumulated ever more rapidly from human and other genomes, the advantages of DGGE will become more apparent because of the sequence-dependent computational tools on which the technique depends for maximal effectiveness.

Fisher and Lerman developed a two-dimensional (2-D) DNA gel system in which fragments are first resolved by length using conventional agarose or acrylamide gel electrophoresis, and then by melting properties in the second dimension. Using 2-D gels, EcoRI-digested DNA from Escherichia coli or an E. coli λ lysogen was analyzed. Approximately 300 distinct spots were visible in ethidium bromide-stained gels, and at least four of the seven EcoRI fragments of λ could be seen in DNA from the lysogen strain.[41] Poddar and Maniloff have used ethidium bromide-stained 2-D gels to estimate genome size in mycoplasmas.[57]

Uitterlinden and co-workers electroblotted 2-D gels and hybridized the resulting filters with radiolabeled probes.[58,59] When HaeIII-digested human genomic DNA was separated on a 2-D gel and probed with an oligonucleotide homologous to a VNTR core sequence, approximately 625 spots could be resolved. Two unrelated individuals typed with the VNTR probe differed at 150 spots, demonstrating the analytical power of the method.

Denaturing gradient gel electrophoresis is a powerful method for isolating mutant molecules after in vitro mutagenesis.[43,60] Some information as to the nature of mutations may be gained by comparing the retardation level of the mutant and wild-type molecules. The DGGE method can also be used to detect and quantitate error rates of DNA polymerases. With DGGE, PCR products can be examined directly, without the need for cloning. Thus, there is no need to worry about the effect of in vivo repair processes or PCR-induced damage, which inhibits replication in vivo. Keohavong and Thilly used DGGE to quantitate the PCR error rate of different DNA polymerases. The error rate per nucleotide was roughly 10^{-6} for phage T4 polymerase, 3×10^{-5} for modified T7 polymerase (Sequenase) and 2×10^{-4} for Taq polymerase.

DNA melting is influenced by methylation. Collins and Myers[37] methylated DNA restriction fragments in vitro using site-specific methylases

[57] S. K. Poddar and J. Maniloff, Nucleic Acids Res. 17, 2889 (1989).
[58] A. G. Uitterlinden, P. E. Slagboom, D. L. Knook, and J. Vijg, Proc. Natl. Acad. Sci. U.S.A. 86, 2742 (1989).
[59] A. G. Uitterlinden and J. Vijg, Trends Biotechnol. 7, 336 (1989).
[60] R. M. Myers, L. S. Lerman, and T. Maniatis, Science 229, 242 (1985).

and showed that unmethylated, hemimethylated, and fully methylated fragments could be resolved by DGGE. The effect on helix stability of methylation at a single site (e.g., 5mC or 6mA) is similar to the effect due to a sequence change. The development of electrotransfer techniques has simplified the study of genomic digests, and as a result DGGE is a very powerful method for studying methylation at virtually any nucleotide.

Denaturing gradient gel electrophoresis can be used to compare the effect of non-Watson–Crick structures on DNA stability and to separate molecules on the basis of such structures. For instance, the 359 nucleotide long, single-strand circular RNA of PSTV forms a highly base-paired, rodlike structure. Melting of this structure to produce a single-stranded circle leads to a mobility transition in temperature gradient gel electrophoresis.[61] Random nicks (present in the purified viral RNA) reduce the T_m of the transition. A similar result was obtained by Fischer and Lerman[22] using randomly nicked λ DNA. Senior et al.[62] showed that unpaired, terminal bases (dangling ends) increase the T_m of a 6-bp duplex. Melting temperatures for $[d(GC)_3]_2$ and $[d(GC)_3TT]_2$ were found to be 49.2 and 54.2°, respectively. This increase in stability, attributed to stacking interactions, is similar to the increase in stability expected for an additional $dA \cdot dT$ base pair. DGGE could be used to examine the effect of structures such as nicks and dangling ends on long molecules; the melt map can be used to choose fragments where the terminal base pairs are part of a long domain (e.g., the T_m of the terminal base pairs is not markedly lower than the T_m of the adjacent melting domain). DGGE has been used to study melting in partially base-paired single-stranded DNA.[63] The method can be used to characterize changes in both domain structure and T_m resulting from the incorporation of non-Watson–Crick bases. We have used the PCR to synthesize dsDNA with inosine or 7-deazaguanosine.[12,33]

The ability of DGGE to detect changes in large molecules is evident in the work of Lyubchenko and Shlyakhtenko,[64] who used DGGs run at low temperature (35–55°) to study the effect of superhelical density on melting in the plasmids pAO3 and pUC19. For pAO3 toposiomers there is a linear relation between superhelical density and melting temperature for the lowest melting transition. In addition, fine structure in the mobility curves was attributed to the presence of cruciforms.[64]

In principle, most compounds which form covalent adducts with DNA or RNA should alter helix stability; an effect at least as large as that due

[61] D. Riesner, G. Steger, R. Zimmat, R. A. Owens, M. Wagenhofer, W. Hillen, S. Volbach, and K. Henco, *Electrophoresis* **10**, 377 (1989).
[62] M. Senior, R. A. Jones, and K. J. Breslauer, *Biochemistry* **27**, 3879 (1988).
[63] K. Nishigaki, Y. Husimi, and M. Tsubota, *J. Biochem.* (*Tokyo*) **99**, 663 (1986).
[64] Yu. L. Lyubchenko and L. S. Shlyakhtenko, *Nucleic Acids Res.* **16**, 3269 (1988).

to a methyl group would be expected. However, compared to enzymatic methylation, the sequence specificity of such reactions is usually low, so that treatment of any long fragment would generate a large number of different species, as does exposure of dsDNA to UV light, as assayed by DGGE.[65]

Adding a GC clamp by PCR[31,34] (p. 28, and Fig. 5A) has been used to detect mutations in the CFTR gene,[66,67] in the p53 gene,[68] and in a comprehensive study of the Factor VIII gene.[69,70]

Acknowledgments

We thank Dr. L. Lerman for encouraging us to undertake this review, for sage criticism on the manuscript, and for helpful advice. We thank William Fripp for help with computations, and S. MacMillan for help with experiments on middle melting. E. Schmitt and N. Gabra made helpful suggestions. We thank S. Stanton for help with Fig. 1, and M. Collins for communicating unpublished data. E.S.A. thanks Leonard Lerman for advice, support, and encouragement. V.P.S. thanks David Housman for support and encouragement, and for first interesting him in DGGE. Support from grants from the National Institutes of Health to Dr. L. Lerman (NIH 8R37HG00345) and Dr. D. Housman (NIH PO1HL41484 and NIH RO1HG00299) is acknowledged.

[65] N. F. Cariello, P. Keohavong, B. J. Sanderson, and W. G. Thilly, *Nucleic Acids Res.* **16,** 4157 (1988).

[66] M. Devoto, P. Ronchetto, P. Fanen, J. J. Orriols, G. Romeo, M. Goossens, M. Ferrari, C. Magnani, M. Seia, and L. Cremonesi, *Am. J. Hum. Genet.* **48,** 1127 (1991).

[67] M. Vidaud, P. Fanen, J. Martin, N. Ghanem, S. Nicolas, and M. Goossens, *Hum. Genet.* **85,** 446 (1990).

[68] A. L. Borresen, E. Hovig, B. Smith-Sorensen, D. Malkin, S. Lystad, T. I. Andersen, J. M. Nesland, K. J. Isselbacher, and S. H. Friend, *Proc. Natl. Acad. Sci. U.S.A.* **88,** 8405 (1991).

[69] M. Higuchi, S. E. Antonarakis, L. Kasch, J. Oldenburg, E. Economou-Petersen, K. Olek, M. Arai, H. Inaba, and H. H. Kazazian, Jr., *Proc. Natl. Acad. Sci. U.S.A.* **88,** 8307 (1991).

[70] M. Higuchi, H. H. Kazazian, Jr., L. Kasch, T. C. Warren, M. J. McGinniss, J. A.. Phillips, C. Kasper, R. Janco, and S. E. Antonarakis, *Proc. Natl. Acad. Sci. U.S.A.* **88,** 7405 (1991).

[5] Two-Dimensional Gel Electrophoresis of Circular DNA Topoisomers

By RICHARD BOWATER, FAREED ABOUL-ELA, and DAVID M. J. LILLEY

DNA Supercoiling

Circular double-stranded DNA may exist in a variety of isomeric states that differ in the number of times that one strand is linked with the other, measured by the linking number Lk. These species are isomers of the topology of the molecule and are generally called topoisomers. This is discussed in greater detail in [6] and [7] in this volume, and for a rigorous treatment of the topological properties of circular DNA the reader should consult Ref. 1.

If we perform an imaginary experiment in which a linear DNA molecule of N base pairs (bp) is ligated into a circle, then if this is accomplished without the application of torsion to the molecule, the resulting linking number will be

$$Lk = Lk° = N/h \qquad (1)$$

where h is the helical repeat under the conditions of the experiment. $Lk°$ defines the relaxed state of the circular molecule. Note that h can be varied by changing the environmental conditions such as ionic strength and temperature. It may also be altered by interaction with ligands, most notably the intercalators. Intercalating agents are planar aromatic compounds such as ethidium bromide or chloroquine that insert between base pairs, causing a local unwinding.

Returning to the imaginary cyclization experiment, if torsion is applied prior to ligation, helical turns may be added or removed from the DNA molecule, changing the linking number of the circular species. Thus, we can define a linking difference (ΔLk) as

$$\Delta Lk = Lk - Lk° = Lk - N/h \qquad (2)$$

The linking difference may be positive or negative; when negative the molecule is underwound and negatively supercoiled. When ΔLk is zero, the molecule is relaxed; on average the molecule is planar, and the helical period is given by N/h. However, if $\Delta Lk \neq 0$, the average geometry of the

---

[1] J. H. White, N. R. Cozzarelli, and W. R. Bauer, *Science* **241**, 323 (1988).

molecule must be distorted from the relaxed conformation. The linking difference is related to distortions of the structure of the molecule by[2]

$$\Delta Lk = \Delta Tw + Wr \tag{3}$$

Tw is the twist of the molecule, measuring the rotation of the strands about the duplex axis. In fact, this property can be further decomposed into rotations about the local axis (winding number) and the laboratory axis (surface twist),[1] but the simple description is adequate for a working understanding of two-dimensional gel electrophoresis. Wr measures the writhing of the duplex axis in three-dimensional space. Thus, the linking difference is freely partitioned between the torsional and flexural geometric distortions, constrained only by the algebraic sum which must remain constant, according to Eq. (3).

For small DNA circles, writhing is energetically very costly, and therefore such molecules remain largely planar. For larger molecules this is not the case, and as the molecule is supercoiled the helical axis becomes distorted from an average planar circle, and is distorted about some kind of superhelical axis. Two kinds of distortion are possible in theory: an interwound or plectonemic structure, where the superhelical axis is linear, and a toroidal structure, where the axis is a circle. Experimental data favor the plectonemic structure, where the duplex DNA is wound onto the surface of an imaginary capped cylinder.

DNA Topology and Electrophoretic Mobility in Gels

The geometric distortion of a supercoiled molecule means that the average shape of the molecule is changed as a result of its topology, and the result is altered frictional properties. This can be seen very clearly by agarose gel electrophoresis, illustrated in Fig. 1. This shows the result of electrophoresing a series of topoisomers of the 3658-bp plasmid pAT153 in 1% agarose. A ladder of species is observed, each member of which is a single topoisomer differing from its neighbors by a linking difference of ± 1. The differences in shape between the different topoisomers results in a progressive change in electrophoretic mobility; the more supercoiled the species, the faster the migration. The resolving power of agarose gels for topoisomers is impressive, but unfortunately the range is limited, so that the more highly supercoiled species comigrate as a broad band at the lower end of the gel. The resolution can be improved for smaller DNA molecules by the use of composite polyacrylamide–agarose gels.

[2] F. B. Fuller, *Proc. Natl. Acad. Sci. U.S.A.* **68**, 815 (1971).

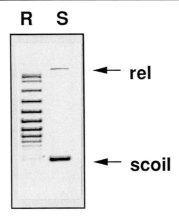

FIG. 1. Agarose gel separation of topoisomers of a 3.6-kilobase (kb) plasmid. A sample of pAT153 was incubated with topoisomerase I to induce partial relaxation of the DNA. This (track R), together with a sample of fully supercoiled DNA (track S), was electrophoresed on a 1% agarose gel. Note the ladder of topoisomeric species that is generated by topoisomerase treatment, extending in mobility from the fully supercoiled position (scoil) to the fully relaxed position (rel).

Supercoil-Dependent DNA Structure

As discussed in [8] in Volume 211 of this series, DNA supercoiling can stabilize an altered DNA structure that leads to a local change in twist. On formation of the new structure, the change in twist leads to a compensatory change in writhe according to Eq. (3), since the linking number of the molecule remains constant throughout. The change in writhe and consequent altered shape of the molecule leads to a change in the electrophoretic mobility; migration becomes slower due to the partial relaxation of writhing. Thus the molecule carrying the new structure will migrate more slowly than a molecule of the same linking difference that has not undergone the structural transition. By measuring the amount by which the migration is retarded, an estimate of the twist change for the transition may be made.

Two-Dimensional Gel Electrophoresis

The major problem with one-dimensional gel electrophoresis is the relatively limited range over which migration is a function of topology, as illustrated in Fig. 1. Two-dimensional gel electrophoresis[3] increases this range by a considerable degree, and it provides a simple way to reveal

[3] J. C. Wang, L. J. Peck, and K. Becherer, *Cold Spring Harbor Symp. Quant. Biol.* **47**, 85 (1983).

A. No structural transition:

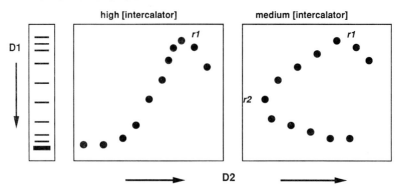

B. With structural transition:

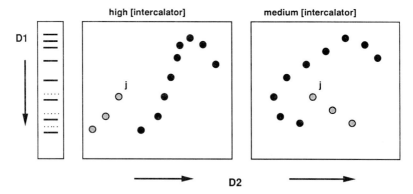

FIG. 2. Scheme to demonstrate the principle of two-dimensional gel electrophoresis. (A) Generation of two-dimensional gels from a circular DNA molecule that undergoes no structural transition. *Left,* One-dimensional gel electrophoresis (comparable to that in Fig. 1); *middle,* two-dimensional gel using a high chloroquine concentration in the second dimension; *right,* two-dimensional gel using a medium chloroquine concentration in the second dimension. (B) Equivalent to (A), except that we have now introduced a structural transition, leading to "jumps" of the topoisomer spots. See text for further details.

topology-dependent structural transitions and measure some parameters for the process.

The principle of two-dimensional gel electrophoresis is illustrated in Fig. 2. The basic idea of the gel may be grasped by considering first the result of studying the migration in the absence of any structural transition, shown in Fig. 2A. A complete set of topoisomers, from slightly positively supercoiled to highly negatively supercoiled, is electrophoresed from a single well of an agarose gel in the absence of intercalating agents (direction

D1). On the left is illustrated how this would appear on a conventional one-dimensional gel (comparable to Fig. 1). The gel is then soaked for several hours in an intercalating drug, usually chloroquine. The effect of the intercalation is a local unwinding, but because of the invariance of the linking number, the negative twist change must be compensated by a positive writhe [Eq. (3)], that is, the molecule becomes more positively supercoiled. The extent of positive supercoiling is directly proportional to the amount of chloroquine added. For the gel at center, sufficient chloroquine is added so that all the topoisomers, including those that were highly negatively supercoiled initially, become positively supercoiled. The gel is then electrophoresed for a second time, in a direction (D2) 90° to the first direction, with the same chloroquine concentration present in the running buffer.

The result is shown schematically in the middle of Fig. 2A. Those topoisomers with the lowest negative linking difference gain the greatest positive supercoiling in the second dimension; in other words, the topoisomers that were most negatively supercoiled (and therefore migrated fastest) in the first dimension become the least positively supercoiled (hence slowest) in the second dimension. Going from the right of the gel, the first few spots are topoisomers that were actually slightly positively supercoiled in the first dimension. We then pass through position r1, the position of relaxation (and therefore slowest migration in D1) in the first dimension. This is the position where $Lk = N/h$, where h is the helical periodicity in the absence of chloroquine. Moving further rightward the migration in D1 increases with $-\Delta Lk$, and this continues until a plateau is reached; this is the point at which the migration as a function of ΔLk has saturated in the first dimension.

The second dimension may be run with a lower concentration of chloroquine. This is often done to improve the resolution of topoisomers around the r1 position. The result is illustrated schematically by the right-hand gel of Fig. 2A. The migration of the topoisomer spots of smallest $-\Delta Lk$ is qualitatively similar to that of the previous gel, and of course position r1 is unchanged because this is determined by the first dimension. However, because the concentration of chloroquine is now lower, not all the topoisomers are positively supercoiled in the second dimension. The topoisomers of higher $-\Delta Lk$ remain negatively supercoiled throughout the experiment, although their negative supercoiling is proportionately reduced in the second dimension. At the point r2, $Lk = N/h'$ where h' is the helical periodicity in the presence of the prevailing chloroquine concentration, and hence the topoisomer at this position is relaxed in the second dimension. This topoisomer has the lowest mobility in the second dimension, and as we progress to higher values of $-\Delta Lk$ the mobility in the second dimension increases once more. For this reason the topoisomer spots appear to turn around at the r2 position.

Now we consider what happens when the molecule can undergo a transition to form an altered structure at some position, such as the formation of Z-DNA by a stretch of alternating cytosine–guanine residues. The local reversal of handedness of the purine–pyrimidine stretch gives rise to a local twist change, and hence a relaxation in the level of supercoiling. However, if the transition requires a threshold level of supercoiling to occur, the altered structure is absent in topoisomers of lower $-\Delta Lk$. The resulting two-dimensional gels are illustrated schematically in Fig. 2B. At low $-\Delta Lk$ the topoisomer spots migrate equivalently to those of Fig. 2A. However, above the threshold level of $-\Delta Lk$, the Z-DNA may form stably in the molecule, and hence there is a relaxation in the first dimension of the gel. Topoisomers carrying the Z-DNA migrate anomalously slowly in the first dimension (broken lines in the one-dimensional gel illustrated at left; Fig. 2B). However, the presence of the chloroquine in the second dimension effectively reduces the level of negative supercoiling; hence the Z-DNA is destabilized, and these topoisomers migrate with their normal mobility in D2. Thus, a structural transition is revealed by topoisomer spots having their normal second dimension mobility but a retarded mobility in the first dimension. This is seen as a discontinuous "jump," whereby the topoisomer spots describe a new line of slower mobility, beginning with topoisomer j. This is shown by the stippled spots of Fig. 2B.

From these gels three pieces of information may be obtained. First, by counting the topoisomer spots from position r1, the lowest value of $-\Delta Lk$ that supports the new structure (topoisomer j) can be measured. Second, by counting back from a given shifted topoisomer to the equivalent unshifted position with the same mobility in the first dimension, the twist change for the transition may be measured (an assumption is made that the presence of the new local DNA structure does not itself alter the overall gel mobility of the circular molecule—this may be more valid for some structures than others). With a reasonable separation of topoisomers, this can be measured with an accuracy of ± 0.2. Finally, by combining these two pieces of information, we may calculate the free energy of formation of the new structure, ΔG_f, according to

$$\Delta G_f = 1050 RT/N \times (L_j^2 - L_{j-t}^2) \tag{4}$$

where L_j is the linking difference of topoisomer j, and L_{j-t} is the linking difference of the unshifted topoisomer of equal mobility. The constant is

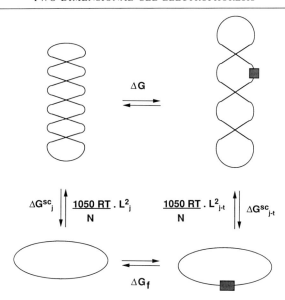

FIG. 3. Thermodynamic cycle for a local structural transition in a supercoiled DNA molecule. The reaction that is observed in the two-dimensional gel has a change in free energy of ΔG, and it comprises two components: formation of the local DNA conformation and relaxation of supercoiling owing to the change in twist. The formation of the new DNA structure has a free energy of ΔG_f. ΔG_j^{sc} is the free energy change for supercoiling topoisomer j (linking difference L_j), whereas ΔG_{j-t}^{sc} is the free energy of supercoiling topoisomer $j - t$ (linking difference L_{j-t}), where t is the local ΔTw for the structural change. The discontinuous jump in the gel occurs at topoisomer j, where ΔG is zero. Hence

$$\Delta G = 0 = \Delta G_{j-t}^{sc} - \Delta G_j^{sc} + \Delta G_f$$

and

$$\Delta G_f = \Delta G_j^{sc} - \Delta G_{j-t}^{sc}$$
$$\Delta G_f = (1050RT/N)(L_j^2 - L_{j-t}^2)$$

taken from Refs. 4–6. This results from the simple thermodynamic cycle shown in Fig. 3.

Other information may be forthcoming. From the sharpness of the spots, or any smearing between shifted and unshifted positions, some estimate of the rate of interconversion between the structural isomers may be obtained. If all the spots are sharp, the rate of interconversion must be

[4] R. E. Depew and J. C. Wang, *Proc. Natl. Acad. Sci. U.S.A.* **72,** 4275 (1975).
[5] D. E. Pulleyblank, M. Shure, D. Tang, J. Vinograd, and H.-P. Vosberg, *Proc. Natl. Acad. Sci. U.S.A.* **72,** 4280 (1975).
[6] D. S. Horowitz and J. C. Wang, *J. Mol. Biol.* **173,** 75 (1984).

faster than the electrophoretic equilibration in the pores of the gel. If the transition is a cooperative all-or-none process, then a single transition (like those illustrated schematically in Fig. 2) will be seen. However, it is possible that a transition can occur progressively with $-\Delta Lk$, as is seen with long Z-DNA tracts or in melting transitions, in which case a progressive retardation in the second dimension is seen.

Two-dimensional gels may also be used as a method for spreading out a wide distribution of topoisomers, for example, one obtained directly from cells grown under a particular set of conditions. This has been widely used in the study of the effects of transcription on the topology of cellular DNA[7,8] ([20] in this volume). The superhelix density of a narrower distribution of topoisomers may also be assigned by loading the distribution 1–2 hr ahead of the full distribution on a two-dimensional gel, using the same well; this is considerably easier than the older method that required several one-dimensional gels at different chloroquine concentrations and extensive densitometry.

Variables in Two-Dimensional Gel Electrophoresis

The conditions of gel electrophoresis can be varied to suit the structure it is intended to study. Because the perturbed structure is only present during the first dimension, it is not necessary to run the second dimension under any special conditions, beyond the choice of intercalator concentration. Potentially variable parameters include the following.

First Dimension

Different structures respond to environmental conditions in different ways. The following conditions are variable.

pH. Structures that require base protonation are more stable at lower pH. For example, the H-triplex requires cytosine protonation for the formation of C · G · CH$^+$ triple base pairs. The first dimension might therefore be electrophoresed in acetate buffer at pH 4.5.

Buffer Composition. Many perturbed structures require certain cations for their stability, such as the requirement for divalent metal ions for the folded conformation of cruciform structures.[9] Similarly, Z-DNA is strongly stabilized by hexamminecobalt(III), and this may be included

---

[7] L. F. Liu and J. C. Wang, *Proc. Natl. Acad. Sci. U.S.A.* **84,** 7024 (1987).

[8] H.-Y. Wu, S. Shyy, J. C. Wang, and L. F. Liu, *Cell* (*Cambridge, Mass.*) **53,** 433 (1988).

[9] D. R. Duckett, A. I. H. Murchie, S. Diekmann, E. von Kitzing, B. Kemper, and D. M. J. Lilley, *Cell* (*Cambridge, Mass.*) **55,** 79 (1988).

in the buffer for the first dimension. At the other extreme, some transitions, notably local helix melting of (A + T)-rich regions, are promoted by low ionic strength. To observe these transitions we find it sufficient to perform the first dimension in 45 mM Tris–borate, pH 8.3, although we have successfully carried out the electrophoresis at 10 mM Tris-HCl, pH 7.5, 0.5 mM ethylenediaminetetraacetic acid (EDTA), with a reduced voltage. In general with all such experiments we recirculate the electrophoresis buffer at rate exceeding 1 liter/hr. It is good practice to check the pH in the buffer compartments at the end of the electrophoresis, as nonuniform pH due to incomplete buffer recirculation will seriously affect the gels.

Temperature. Most transitions are temperature-dependent, some critically so. To study the thermal melting of (A + T)-rich sequences in supercoiled DNA we employ an apparatus for horizontal slab gel electrophoresis in which the gel is placed on top of a block through which water is circulated from a water bath. The temperature is measured by means of a platinum resistance element inserted into the gel itself.

Second Dimension

As explained above, the conditions of the second dimension do not affect the structural transition, but they can have a significant effect on the resolution of the topoisomers in different regions of the gel. The main variable in the second dimension is the concentration of chloroquine chosen. For optimal resolution of topoisomers around the r1 relaxed position, a relatively low concentration of chloroquine is used. For our standard buffer of 90 mM Tris–borate, pH 8.3, 1 mM EDTA (TBE) we use a concentration of 1–2 μg/ml chloroquine. These are good conditions for the study of transitions that occur at relatively low levels of supercoiling. For transitions that occur at higher values of $-\Delta Lk$, more chloroquine is required; we would use 5–20 μg/ml chloroquine in TBE. Under these conditions it becomes quite difficult to count topoisomers from the r1 position, and it is often necessary to perform a number of experiments using different sets of conditions.

A further technique is possible with two-dimensional gels when a perturbed structure is either sensitive or resistant to a particular probe. For example, a structure such as a cruciform may be especially sensitive to cleavage by a particular nuclease, whereas the unperturbed molecule is not. If the distribution is incubated with the enzyme prior to electrophoresis, only the topoisomers that have undergone the transition should be cleaved, and thus these topoisomers will be selectively absent from the resulting gel. This can be used to show a correspondence between topological and probing studies on a particular species.

Examples of Structural Transitions Studied by Two-Dimensional Gel Electrophoresis

We present three examples of structural transitions that may be studied using two-dimensional gel electrophoresis. This list is not exhaustive; for example, we have not included an example of the formation of the H-triplex structure by oligopurine · oligopyrimidine sequences[10]; the interested reader should refer to [9] in Volume 211 of this series.

Formation of Left-handed Z-DNA by (CG)₈ Sequence

Formation of Z-DNA[11] by alternating purine–pyrimidine sequences involves a change in the handedness of the helix. This leads to a local change in twist for a sequence of n bp of approximately

$$- \Delta Tw = n/h + n/12$$

where h the helical period of the alternating tract prior to Z-DNA formation. A two-dimensional gel showing Z-DNA formation in the sequence $(CG)_8$ is shown in Fig. 4. A sharp transition is seen at topoisomer $- \Delta Lk = 10$, with $- \Delta Tw = 3$. This corresponds to the cooperative formation of Z DNA by the complete $(CG)_8$ tract. The first dimension of this gel was electrophoresed at 4° in order to prevent any opening transitions in (A + T)-rich regions from occurring. In the second dimension, 1.8 μg/ml chloroquine was included in the electrophoresis buffer, and thus the topoisomers of greater $- \Delta Lk$ remain negatively supercoiled throughout.

Formation of Cruciform by Alternating Adenine–Thymine Sequences

Inverted repeats can form cruciform structures in supercoiled DNA,[12–14] with a twist change that is approximately equivalent to the loss of the normal right-handed helical period over the length of the sequence,

---

[10] V. Lyamichev, S. M. Mirkin, and M. D. Frank-Kamenetskii, *J. Biomol. Struct. Dyn.* **3**, 667 (1986).

[11] A. H.-J. Wang, G. J. Quigley, F. J. Kolpak, J. L. Crawford, J. H. van Boom, G. van der Marel, and A. Rich, *Nature (London)* **282**, 680 (1979).

[12] M. Gellert, K. Mizuuchi, M. H. O'Dea, H. Ohmori, and J. Tomizawa, *Cold Spring Harbor Symp. Quant. Biol.* **43**, 35 (1979).

[13] D. M. J. Lilley, *Proc. Natl. Acad. Sci. U.S.A.* **77**, 6468 (1980).

[14] N. Panayotatos and R. D. Wells, *Nature (London)* **289**, 466 (1981).

pCG8

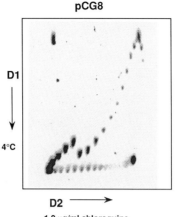

FIG. 4. Two-dimensional gel electrophoresis showing left-handed Z-DNA formation by a $(CG)_8$ sequence. Electrophoresis was carried out in $0.5 \times$ TBE buffer. The first dimension (D1) was performed at $4°$ in order to prevent helix opening in $(A + T)$-rich sequences in the plasmid. During the second dimension (D2), 1.8 μg/ml chloroquine was included in the electrophoresis buffer. Note the single, cooperative transition corresponding to complete formation of Z-DNA by the $(CG)_8$ sequence.

namely,

$$-\Delta Tw = n/h$$

where n bp is the length of the inverted repeat, and h the helical period prior to extrusion. pXG540 contains a sequence from an intron of the *Xenopus* globin $\alpha T1$ gene, which includes an $(AT)_{34}$ sequence. This is a 68-bp sequence of 2-fold symmetry that forms a cruciform structure with no kinetic barrier.[15] A two-dimensional gel is shown in Fig. 5. The second dimension of this gel was performed using 1.8 μg/ml chloroquine in TBE, and therefore half the distribution remains negatively supercoiled in the second dimension. The twist change is around -6, consistent with cruciform formation by the $(AT)_{34}$ sequence. The transition occurs discontinuously at a $-\Delta Lk$ of about 9, and this corresponds to a free energy of cruciform formation by the $(AT)_{34}$ sequence of 13.5 kcal/mol. This low ΔG is the reason why the cruciform is stable at a relatively low value of $-\Delta Lk$; indeed, the transition occurs on the positively supercoiled side of r2 in the two-dimensional gel.

The gel (Fig. 5) illustrates additional information that may be extracted from such data. The value of ΔTw measured for the cruciform transition

[15] D. R. Greaves, R. K. Patient, and D. M. J. Lilley, *J. Mol. Biol.* **185,** 461 (1985).

pXG540

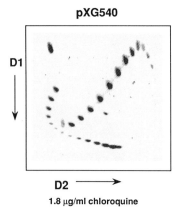

D1

D2 ⟶

1.8 µg/ml chloroquine

FIG. 5. Two-dimensional gel electrophoresis showing cruciform formation by an $(AT)_{34}$ sequence. Electrophoresis was carried out in TBE buffer, at room temperature. During the second dimension, 1.8 µg/ml chloroquine was included in the electrophoresis buffer. Note the cooperative transition at $-\Delta Lk = 9$ owing to cruciform formation by the $(AT)_{34}$ sequence.

was -5.8. This is rather low for a 68-bp sequence, for which we would expect

$$-\Delta Tw = 68/10.5 = 6.5$$

Indeed, by chance cruciform formation by an inverted repeat of exactly 68 bp of random sequence has been measured[16] to be just -6.5. As there can be no doubt that the $(AT)_{34}$ sequence does form a cruciform (e.g., it is cleaved by resolving enzymes as expected), we may ask what h would have to be to give a ΔTw of -5.8. The answer is 11.7 bp/helical turn; in other words, the $(AT)_{34}$ tract is considerably underwound prior to cruciform extrusion. This is consistent with an enhanced chemical reactivity of thymine bases in the sequence[17] and has led us to propose that the structure adopts a torsionally deformable alternating structure.[18]

Formation of Unpaired Regions by (A + T)-Rich Sequences at Elevated Temperature

(A + T)-rich sequences have a relatively low thermal stability and could potentially form stable opened regions or bubbles at low ionic strength and elevated temperature. The twist change would be very similar

[16] A. J. Courey and J. C. Wang, *Cell* (*Cambridge, Mass.*) **33**, 817 (1983).

[17] J. A. McClellan, E. Paleček, and D. M. J. Lilley, *Nucleic Acids Res.* **14**, 9291 (1986).

[18] A. Klug, A. Jack, M. A. Viswamitra, O. Kennard, Z. Shakked, and T. A. Steitz, *J. Mol. Biol.* **131**, 669 (1979).

pCollR315

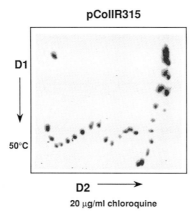

FIG. 6. Two-dimensional gel electrophoresis showing thermal melting of an (A + T)-rich sequence in pCollR315. Electrophoresis was carried out in 0.5× TBE buffer. The first dimension was performed at 50°. During the second dimension, 20 μg/ml chloroquine was included in the electrophoresis buffer, as a result of which all topoisomers were positively supercoiled in the second dimension. Note the multiple transitions corresponding to helix opening in the (A + T)-rich sequences.

to that for the formation of a cruciform by a sequence of the same length. Figure 6 shows a two-dimensional gel of a plasmid containing (A + T)-rich sequences that originated in ColE1, run in the first dimension in 45 mM Tris–borate, pH 8.3, at 50°. Clear transitions are evident that are consistent with bubble formation by opening of (A + T)-rich regions; their formation is markedly temperature-dependent and is suppressed by addition of salt to the first-dimension buffer. The transitions tend to be more continuous than those above, suggesting a progressive opening of a bubble as the molecule becomes more supercoiled. Three sequential transitions may be distinguished. The opening of the (A + T)-rich regions is consistent with their observed reactivity to single-strand-selective chemical agents, such as osmium tetroxide.[19] Prior modification of the topoisomer distribution with osmium tetroxide leads to a selective smearing of all the topoisomers beyond $-\Delta Lk = 11$, where the transitions begin, whereas those less supercoiled are completely unaltered. This shows that the two very different techniques of electrophoresis and chemical modification are actually studying the same events.

[19] J. C. Furlong, K. M. Sullivan, A. I. H. Murchie, G. W. Gough, and D. M. J. Lilley, *Biochemistry* **28**, 2009 (1989).

TABLE I
VOLUMES (μl) REQUIRED FOR PREPARATION OF TOPOISOMER DISTRIBUTIONS

	Final [EB] (μg/ml)									
	0	0.75	1.5	2.25	3.0	3.75	4.5	5.5	6.5	8.0
DNA[a]	9 μl	9	9	9	9	9	9	9	9	9
5 μg/ml EB	—	3	6	—	—	—	—	—	—	—
10 μg/mg EB	—	—	—	4.5	6	—	—	—	—	—
20 μg/ml EB	—	—	—	—	—	3.75	4.5	5.5	—	—
40 μg/ml EB	—	—	—	—	—	—	—	—	3.25	4
10× Buffer[b]	2	2	2	2	2	2	2	2	2	2
Water	8	5	2	3.5	2	4.25	3.5	2.5	4.75	4
Topoisomerase I	1	1	1	1	1	1	1	1	1	1

[a] A 300 μg/ml stock solution of supercoiled DNA is incubated with rat liver topoisomerase I, in the presence of a final concentration of ethidium bromide (EB) of between 0 and 8 μg/ml.

[b] The 10× buffer stock used for the incubation is 100 mM Tris, pH 8.0, 1 M NaCl, 200 mM EDTA. The concentrations of ethidium bromide may need to be changed if the incubation buffer is altered to suit a different topoisomerase, and some trial experiments may be required.

Preparation of Topoisomer Distributions

The first requirement for performing two-dimensional gel electrophoresis is a good distribution of topoisomers. This should cover the required range, that is, topoisomers of $\Delta Lk/Lk°$ from 0.01 to -0.08, and should have equal concentrations of all topoisomers. The distribution is generated by the action of a eukaryotic topoisomerase I on supercoiled DNA in the presence of a range of concentrations of ethidium bromide. We routinely employ a preparation of rat liver topoisomerase I prepared in our laboratory. It has a concentration of 10 mg/ml and gives a single band on a silver-stained polyacrylamide gel.

Ten 20-μl incubations with ethidium bromide concentrations between 0 and 8 μg/ml are set up according to Table I. After the addition of enzyme, the samples are incubated at room temperature for 60 min. The DNA is then precipitated with ethanol, sodium acetate at $-70°$, and the pellets redissolved in 20 μl of 10 mM Tris, pH 7.5, 0.1 mM EDTA. These solutions are stored at $-20°$.

Each incubation generates a Gaussian distribution of topoisomers about a mean, and the ethidium bromide concentration determines the position of the mean. We routinely examine a new set of distributions by electrophoresis in 1% agarose in TBE on a 10 × 10 cm gel, stained in 1 μg/ml ethidium bromide. From the intensity and positions of the distributions, it can be judged how to mix the separate distributions in order to produce one complete distribution of topoisomers covering the complete range of ΔLk. This is then ready for the two-dimensional gel electrophoresis.

Procedure for Two-Dimensional Gel Electrophoresis

A 1% agarose gel in the required running buffer is made by boiling the agarose in the appropriate buffer in a microwave oven until completely dissolved, followed by cooling to 60° before pouring. Our gels are 22 × 25 × 1 cm; the exact dimensions are not vital, but it should be approximately square, and the bigger the gel, the better will be the resolution of topoisomers. A single well is required at one corner (the top left as conventionally drawn), and once again the dimensions will affect the ultimate resolution. Although the well of a normal comb can be used, it is much better to generate a single circular well of about 2 mm diameter.

A sample of the complete topoisomer distribution containing 2.5 μg of DNA in a volume of 8 μl is required for the electrophoresis. The DNA concentration is quite critical as the window between overloading and underloading is narrow. To this is added 2 μl of a solution of 0.35 mg/ml Ficoll 70, containing 1 mg/ml of xylene cyanol. After vortex mixing the sample is applied to the single well, and electrophoresis carried out for 16–20 hr at 3.5 V/cm. The exact length of time will depend on the dimensions of the gel, the size of the plasmid, and the environmental conditions, and a preliminary experiment in one dimension may be necessary to judge this. The gel is disassembled and then soaked in TBE buffer containing the required concentration of chloroquine diphosphate for 6–10 hr in the dark. At the end of this period the gel is reassembled into a standard horizontal electrophoresis apparatus (temperature control is not important for the second dimension) such that the second dimension of electrophoresis is at 90° to the first. The running buffer for the second dimension is TBE containing the same chloroquine concentration that was used for the equilibration. Following electrophoresis in the second dimension the gel is stained in 1 μg/ml ethidium bromide in water for 1 hr, followed by extensive destaining in water at 4°. The extent of destaining required is proportional to the concentration of chloroquine that is used in the second dimension. The gel can then be photographed using a transilluminator providing 300 nm excitation of ethidium fluorescence. We normally photograph our gels with Kodak (Rochester, NY) Tri-X pan film, using red and green filters before the camera lens.

If the quantity of DNA that is available for two-dimensional gel electrophoresis is limited, these gels may be transferred to filters and probed with radioactively labeled DNA.[20]

[20] E. M. Southern, *J. Mol. Biol.* **98,** 503 (1975).

[6] Topological Structure of DNA Knots and Catenanes

By PETER DRÖGE and NICHOLAS R. COZZARELLI

Introduction

Knotted and catenated DNA molecules are frequent intermediate products of basic biological processes such as DNA replication, recombination, and topoisomerase action.[1,2] The determination of the complex structure of knots and catenanes provides important information about the processes that generate them and, in addition, allows conclusions about the structure of DNA.[2-6] Moreover, *in vitro* generated knotted or catenated DNA substrates of defined topology have been used successfully as tools to investigate the mechanisms by which, for example, enhancer elements function in the regulation of transcription and recombination.[7,8] The structural variety of knots and catenanes, however, is very rich.[9,10] Thus, in order to use a topological approach to the study of biological systems, it is crucial to determine unambiguously the topology of DNA molecules. In this chapter we describe briefly the basic concepts and methods by which the topological structure of DNA molecules is determined.

[1] A. Kornberg and T. A. Baker, "DNA Replication," 2nd Ed. Freeman, New York, 1992.
[2] S. A. Wasserman and N. R. Cozzarelli, *Science* **232,** 951 (1985).
[3] M. A. Krasnow, A. Stasiak, S. J. Spengler, F. Dean, T. Koller, and N. R. Cozzarelli, *Nature (London)* **304,** 559 (1983).
[4] S. A. Wasserman and N. R. Cozzarelli, *Proc. Natl. Acad. Sci. U.S.A.* **82,** 1079 (1985).
[5] S. A. Wasserman, J. D. Dungan, and N. R. Cozzarelli, *Science* **229,** 171 (1985).
[6] S. J. Spengler, A. Stasiak, and N. R. Cozzarelli, *Cell (Cambridge, Mass.)* **42,** 325 (1985).
[7] M. Dunaway and P. Dröge, *Nature (London)* **341,** 657 (1989).
[8] R. Kanaar, P. v. D. Putte, and N. R. Cozzarelli, *Cell (Cambridge, Mass.)* **58,** 147 (1989).
[9] D. Rolfsen, "Knots and Links." Publish or Perish Press, Berkeley, California, 1976.
[10] J. H. White, K. C. Millett, and N. R. Cozzarelli, *J. Mol. Biol.* **197,** 585 (1987).

METHODS IN ENZYMOLOGY, VOL. 212

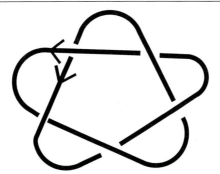

Fɪɢ. 1. Positive five-noded torus knot. The knot is portrayed as an embellished plane projection in which underpassing segments at crossings, or nodes, are omitted so that the three-dimensional structure of the knot can be reconstructed unambiguously. To determine the sign of nodes, a direction for traversing the molecule is chosen arbitrarily and directional arrows are fixed at each crossing. The arrows are shown here at only one of the nodes. If the overpassing arrow can be superimposed on the underpassing one by a clockwise rotation of less than 180°, then, by convention, the node is negative. If a counterclockwise rotation of less than 180° will align the arrows, then the node is positive. The knot depicted here is called a torus knot because it can be drawn without intersection on the surface of a torus. Because torus knots are completely regular, all the nodes are equivalent. The torus knot shown contains five (+) nodes. Its mirror image generated by interchanging underpassing and overpassing crossing segments contains five (−) nodes. Torus knots must have an odd number of nodes; the simplest member of the family contains three nodes and is called a trefoil.

Basic Definitions and Classifications

Knotting and catenation are true topological properties of DNA molecules. They can only be changed by breakage and reunion of the phosphodiester backbone of DNA, and not by any kind of structural deformation such as condensation or twisting. In this chapter, we will not directly consider another topological parameter, the widely known linking number (Lk), or its consequence, the supertwisting of the DNA double helix. In fact, in order to investigate the topological structure of knots and catenanes, the DNA backbone is usually nicked by DNase treatment to remove supercoils, because supertwisting interferes with the determination of knot and catenane structure. Nicking the DNA is superior to relaxation by topoisomerase treatment because the change from the conditions of relaxation to those in a gel, or during sample preparation for electron microscopy, can reintroduce supercoils. Nicks, gaps, or even the removal of one complete polynucleotide strand cannot change the topology of a knot or catenane.

It is convenient to portray DNA molecules and define their topological properties in plane projection as exemplified in Fig. 1. The DNA double helix in these drawings is represented by a single line, and when DNA is so flattened, segments cross to form what are termed nodes. The two types of nodes, positive (+) and negative (−), are defined by the convention described in the legend to Fig. 1. The underpassing segment at a node is identified in such drawings by removal of a short segment. To analyze an electron micrograph of a nicked knot, for example, the molecule is first redrawn with the minimum number of nodes. To do so, all accidental flopovers of segments occurring during sample preparation are removed. The path of the unfolded molecule is traced and, starting from any point of the molecule, an arrow is attached at each crossing to indicate the direction followed (Fig. 1). The sign of the nodes is then easily determined by reference to the convention.

It is often useful to determine the topological classification of a knot or catenane. This is most easily done by comparing the drawing of the molecule with an illustrated catalog. The most extensive one is given in a topology book by Rolfsen,[9] and a smaller but more accessible one can be found in the publication by White *et al.*[10] As an example, we illustrate the procedure with a knot that contains five nodes, such as the one shown in Fig. 1. By reference to the table, we find there are just four knots that contain five nodes; they are given the symbols 5_1, $\bar{5}_1$, 5_2, and $\bar{5}_2$. The subscript indicates the reference numbers in Rolfsen's table, and the bar means that it is the mirror image of the knot depicted there. It is easy to determine by inspection that the knot in Fig. 1 is $\bar{5}_1$. In more complex examples, classification is facilitated by the application of topological invariants and computer programs.[10,11]

Knots and catenanes can be classified into structurally distinct groups. The two that are most important biologically are called the torus and twist families. A member of the torus family can be drawn without intersection on the surface of a torus, a doughnut-shaped object, and an example is given in Fig. 1. It is the most regular knot or catenane structure possible. Thus, the knot depicted in Fig. 1 with the classification of $\bar{5}_1$ can also be called the (+) five-noded torus knot. The twist knot family has an interwound region, singly linked by two knot nodes. Thus, the knot 5_2 depicted in Fig. 2 is called a (−) five-noded twist knot. So many knots and catenanes produced in nature belong to these families because they can be derived either from the regular (+) winding of the two polynucleotide strands in the double helix or from the regular (−) winding of a DNA superhelix.

[11] J. H. Jenkins, M.A. Thesis, University of California, Berkeley (1990).

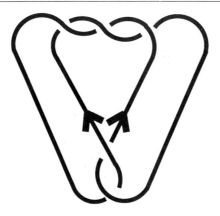

FIG. 2. Negative five-noded twist knot. The twist knot contains two types of nodes. These knots have an interwound, or plectonemic, region that contains a variable number of nodes—three in the example shown—and a single interlock containing two nodes that holds the plectonemic region in place. The example shown contains five ($-$) nodes, whereas its mirror image contains five ($+$) nodes. Its classification symbol is 5_2. The nodes are all of the same sign for twist knots with an odd number of nodes. For twist knots with an even number of nodes, the two linking nodes have one sign and the interwound nodes have the other sign.

Gel Electrophoresis

The two principal methods used for determining the topology of knots and catenanes are agarose gel electrophoresis and electron microscopy. Although only electron microscopy gives the complete stereostructure of individual molecules, electrophoresis provides a more rapid and quantitative overview of the population. It is best, therefore, to use both methods in combination.

Gel electrophoresis can resolve even complex mixtures of nicked knots or catenanes into discrete ladders. In the example shown in Fig. 3, right lane, the knots at each rung of the ladder have one more node than those on the rung immediately above. The electrophoretic mobility is therefore primarily determined by the number of irreducible nodes. Remarkably, the spacing between rungs decreases very little with increasing complexity, whereas for ladders of linking number topoisomers resolution rapidly reaches a limit at higher complexities. This extends the utility of gel electrophoresis for analyzing knots and catenanes. The smaller the size of a knotted or catenated DNA molecule, the higher the resolution, which, consequently, also shortens the time for electrophoresis. A lower limit, however, is set by the persistence length of DNA such that it is energeti-

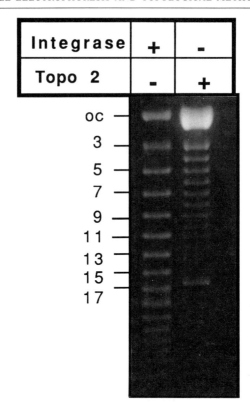

Integrase	+	-
Topo 2	-	+

FIG. 3. Separation of knots of different types and complexities by gel electrophoresis. Negatively supercoiled DNA of 7 kb was reacted with phage λ integrase (left lane) or T4 topoisomerase (right lane), nicked by DNase I treatment in the presence of ethidium bromide, purified, and separated on a high-resolution 0.8% agarose gel for about 24 hr. The DNA is visualized by ethidium bromide staining. Note that the complexity of the torus knots generated by integrase increases in steps of two nodes, whereas in the T4 twist knot ladder it increases in steps of one. Note also that twist and torus knots of the same complexity exhibit slightly different mobilities. Thus, the nine-noded twist knot, for example, migrates slightly ahead of the nine-noded torus knot. This difference in gel mobility becomes more apparent with an increasing number of knot nodes. The prominent band migrating between the 13- and 15-noded twist knot (right lane) represents residual supercoiled DNA. oc, Open circular unknotted DNA.

cally unfavorable for a molecule smaller than about 1 kilobase pairs (kb) to be knotted. Molecules of about 2 kb are the practical lower limit. Larger molecules can be resolved, but longer runs at lower voltage are needed. We have used 2-week runs for catenanes of total 24 kb in size.[12]

[12] H. W. Benjamin, M. M. Matzuk, M. A. Krasnow, and N. R. Cozzarelli, Cell (Cambridge, Mass.) 40, 147 (1985).

No band smearing was observed with the long runs. Buffer circulation, replenishment, and good temperature control are, of course, essential to obtain satisfactory results.

We cannot emphasize too strongly the need for preparing appropriate markers for electrophoresis. These will not only identify unambiguously the number of irreducible nodes in molecules, but can often further assist in determining aspects of their structure. For example, twist knots in Fig. 3 move slightly faster than torus knots with the same number of nodes.[13] Thus, by comparing the mobility of the DNA under investigation with that of known markers, one can obtain a very good first impression of the topological structure. Isolation of DNA from a preparative gel and analysis by electron microscopy are then necessary for confirmation.

Knot markers can most easily be prepared by topoisomerase treatment. Knotting of supercoiled DNA by T4 topoisomerase[14] yields a ladder of (−) twist knots spaced by one node (Fig. 3, right lane). Treatment of nicked DNA with a type 1 topoisomerase, such as *Escherichia coli* topoisomerase I, generates a ladder of random knots.[15] Intramolecular recombination performed by the phage λ integration system *in vitro* gives a ladder of (+) torus knots or (−) torus catenanes spaced by two nodes (Fig. 3, left lane).

The resolving power of gel electrophoresis is impressive. With small DNAs or long runs, stereoisomers with common node numbers can be separated. Thus torus, twist, and compound knots can be identified in the same sample.[13,16] Catenanes of four or more rings that differ only in linkage mode (i.e., straight chain or branched) can also be separated.[12] Two-dimensional gel electrophoresis in which the two dimensions are run under different conditions, or in which, between dimensions, the DNA is digested *in situ* with restriction enzymes, can resolve and identify very complex mixtures of linked forms.[16]

Electron Microscopy

Traditional techniques for electron microscopy of DNA can be useful for identifying molecules as being knotted or catenated, but they provide insufficient visual information to score reliably the sign of the nodes. The complete topological characterization of knots and catenanes required the introduction of the RecA protein coating method.[3,4] The cooperative

[13] P. Dröge and N. R. Cozzarelli, *Proc. Natl. Acad. Sci. U.S.A.* **86,** 6062 (1989).

[14] K. N. Kreuzer and C. V. Jongeneel, this series, Vol. 100, p. 144.

[15] F. B. Dean, A. Stasiak, Th. Koller, and N. R. Cozzarelli, *J. Biol. Chem.* **260,** 497 (1985).

[16] R. Kanaar, A. Klippel, E. Shekhtman, J. M. Dungan, R. Kahmann, and N. R. Cozzarelli, *Cell (Cambridge, Mass.)* **62,** 353 (1990).

binding of *E. coli* RecA protein to single- or double-stranded DNA yields a thick complex that, when stained and shadowed, allows a reliable identification of DNA overlay and underlay at each crossing point. Bacteriophage T4 encodes an analog of RecA, the UvsX protein, that also can be used for this purpose,[17] but RecA is easier to use. The technique can be difficult for beginners, and the following is a compendium of procedures that improve the chances for success.

First, it is worthwhile to train oneself in scoring molecules. An experienced observer can evaluate many more molecules than a novice. A simple way to gain this expertise is to practice with the simplest topological form of a knotted molecule, the three-noded trefoil. The signs of the three nodes of a trefoil must be the same, and thus it is easy to prepare a set of trefoils whose topological sign is unambiguous. A mask is then placed over each molecule by an independent observer so that only one node is exposed at a time. The nodes are scored and then the accuracy of the observer is evaluated. The training is continued until the reliability in node scoring is over 95%. If two independent and unbiased observers score molecules the same way, one can have great confidence in the result.

Second, we have found that there are a number of helpful aids in node scoring. (1) Because of the metal shadowing of the DNA, the overpassing segment is generally thicker than the underpassing one. (2) In following the stripes of the RecA helix or some other consistent feature of the molecule, the segment that shows continuity is on top. (3) If shadowing is predominantly unidirectional, the segment that disappears in the shadows is generally the underpassing one. (4) If a segment appears to jog at a node, it is the underpassing one.

Third, it is important to choose DNA of proper length. The smaller molecules that are the best for gel electrophoresis are not suitable for visualizing complex knots; in this case, one often observes just a clump. As a rough rule there should be at least 1 kb of DNA per node. The torus knots are so regular that the length requirement can be relaxed somewhat. Also, one can score more complex knots if the RecA coating of duplex DNA is done in the presence of the ATP analogue, ATPγS, because this procedure stretches out the DNA (see Fig. 4).

Fourth, there are a number of techniques for coating DNA with RecA. The easiest way, though, is to denature the DNA and coat a single-stranded knot or catenane.[4] Because RecA binds preferentially to single-stranded DNA, this method is much more likely to give complete coating (an example is shown in Fig. 4A). However, we have successfully coated duplex knots and catenanes. We prefer to bind RecA to duplex DNA in the presence of ATPγS (Fig. 4B).

[17] J. Griffith and T. Formosa, *J. Biol. Chem.* **260,** 4484 (1985).

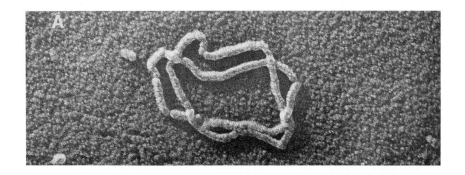

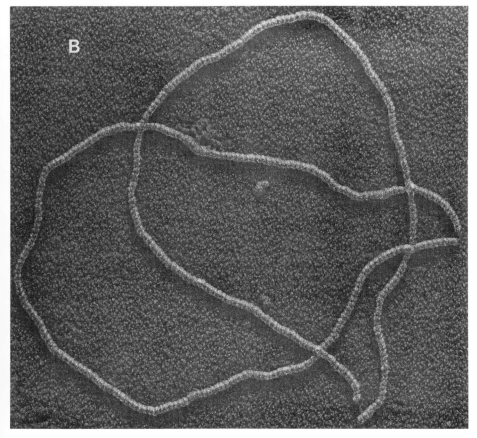

FIG. 4. Electron microscopy of RecA-coated four-noded catenanes. Shown are (+) right-handed four-noded catenanes coated with RecA by two different methods. In (A), the molecule was nicked, denatured by heat/glyoxal, and processed for RecA coating in the absence of ATP. In (B), double-stranded DNA was nicked and coated with RecA in the presence of ATPγS. Note that at identical magnifications, the molecule in B appears to be longer (about 150%) than that shown in (A), although the DNAs have the same size (4.3 kb). This is due to the unwinding and stretching out of the DNA helix by RecA.

Fifth, no matter what method is chosen, it is important to vary the concentration of RecA and of DNA in small increments and to be patient. The procedure sometimes succeeds or fails without obvious reasons. We have had the most difficulties if the DNA is impure or the RecA protein is degraded. Commercially available RecA protein works well. We also keep a single commercial preparation of ϕX174 single-stranded DNA for each newcomer to gain experience. The DNA is pure, already single stranded, and abundant, so that it is easy to get good results.

Sixth, the DNA that is spread for microscopy should be analyzed by gel electrophoresis. It is ideal to cut DNA bands from an ethidium bromide-stained gel. This not only enormously increases the fraction of desired DNA molecules but also allows a much higher frequency of molecules to be scored. Often, ambiguity in node scoring is removed when one knows that the molecule is a knot or catenane with a particular number of nodes. We have used a number of methods for isolating DNA from gels and found that using a commercial electroeluter is the most reliable.

The methods thus far commented on concern either knots or catenanes in which the absolute orientation of the DNA need not be established. However, for most catenanes the orientation of each ring must be determined for a complete topological description. We have used two methods for marking the orientation of RecA-coated catenanes. Orientation may be derived from an asymmetric partial DNA denaturation map. RecA protein is complexed to both duplex and denatured regions so that orientation and node overlay can be scored simultaneously.[4] Another technique involves the introduction of a site-specific nick into the DNA backbone.[4,18] This can be done by treatment with certain restriction enzymes in the presence of ethidium bromide or by treatment with enzymes that can nick DNA, such as the fd gene II product. The DNA is then partially denatured. The nick provides a preferential start site for denaturation, and an easily recognized long tail often results from a neighboring region high in A · T content. The determination of catenane orientation is far more difficult than just determining node sign, and it should be avoided whenever possible. Sometimes it is possible to design the experiment in such a way that, instead of a catenane, the diagnostic product is a knot or an intrinsically chiral catenane, such as the figure-eight catenane.

Other Methods for Determining Structure of Knots and Catenanes

If gel electrophoresis and electron microscopy of RecA-coated DNA are not suitable, other methods can sometimes be used to gain some topological information, however limited.

---

[18] E. Shekhtman, Ph.D. Thesis, University of California, Berkeley.

Induced Writhe of Catenanes

The catenation of two rings causes a mutual coiling of one DNA molecule about the other.[19] Thus, if the DNA of a catenane is nicked and religated, and one ring is cut away, then the remaining ring will have a linking number (Lk) other than that of relaxed DNA. This ΔLk will be positive if the winding of the catenane rings is right-handed and negative if it is left-handed. The magnitude of ΔLk is proportional to the number of windings of one ring around the other.[19] An advantage of this technique is that it can be applied to catenanes which are too highly intertwined to be analyzed by electron microscopy. One limitation is that it is most useful for regular catenanes, such as torus catenanes. However, determination of the handedness of the intertwining of catenanes can give a strong clue as to the processes that have generated them. For example, catenanes generated during the terminal stages of DNA replication are expected to be exclusively right-handed as a result of the unwinding of the parental DNA during replication fork movement.[20]

Products of Site-Specific Recombination

By far the most difficult aspect of DNA topology is the determination of the relative orientation of the two rings of a catenane. The relative orientation can be determined easily, however, if the DNA contains the sites for a site-specific recombination enzyme. For example, if the catenanes are right-handed parallel torus catenanes, they will be recombined by the Gin recombinase but not by the Tn 3 resolvase.[8,21] If the catenanes are right-handed and antiparallel, the specificity is reversed.

Stochastic Methods for Determining the Structure of a Regular Catenated Network

A large catenane with many rings such as that in kinetoplasts of parasites cannot be analyzed by traditional methods. The number of interlocks between catenanes can be determined, however, by analysis of the dimeric catenanes released from a partial restriction digest. In addition, the average number of rings linked to any ring can be determined from the relative amount of linear molecules, monomer circles, dimeric catenanes, trimeric catenanes, etc., released as a function of restriction enzyme digestion.[22]

[19] S. A. Wasserman, J. H. White, and N. R. Cozzarelli, *Nature (London)* **334**, 448 (1988).
[20] O. Sundin and A. Varshavsky, *Cell (Cambridge, Mass.)* **25**, 659 (1981).
[21] H. W. Benjamin and N. R. Cozzarelli, *J. Biol. Chem.* **265**, 6441 (1990).
[22] J. Chen and N. R. Cozzarelli, unpublished results (1991).

Conclusion

The application of a topological approach has proved to be a powerful tool in the study of biological systems. In this chapter we have briefly outlined the basic concepts and methods on which this approach is based. We have also pointed out some of the technical limitations and difficulties that one who chooses such an approach should be aware of. However, a clear advantage of this method is its broad area of application. This allowed us, for example, to study aspects of basic biological processes such as DNA replication, transcription, and DNA recombination at the molecular level. With the recent increasing success in the development of well-characterized *in vitro* systems, we are confident that a topological approach applied to these systems will further contribute to our knowledge about biological processes.

Section II

Probing DNA Structure *in Vitro*

[7] Probes of DNA Structure

By DAVID M. J. LILLEY

Introduction to Probing

Probing with enzymes and chemicals has become an important method for the study of DNA structure in the last decade. Often the first indication of local structural perturbation in DNA has come from enhanced local reactivity to either enzyme cleavage or chemical modification.[1-10] Probing methods can then be employed to characterize further the nature of the perturbed structure. Enzyme and chemical probing experiments have generated important insight into a number of altered DNA structures, including cruciform structures,[4,11-14] left-handed Z-DNA,[15-21] A tract-containing curved DNA,[22,23] H-triplexes,[24-28] G-tetrad structures,[29-31] and parallel-

[1] D. M. J. Lilley, *Proc. Natl. Acad. Sci. U.S.A.* **77**, 6468 (1980).
[2] N. Panayotatos and R. D. Wells, *Nature (London)* **289**, 466 (1981).
[3] C. C. Hentschel, *Nature (London)* **295**, 714 (1982).
[4] D. M. J. Lilley, *Nucleic Acids Res.* **11**, 3097 (1983).
[5] T. Kohwi-Shigematsu, R. Gelinas, and H. Weintraub, *Proc. Natl. Acad. Sci. U.S.A.* **80**, 4389 (1983).
[6] H. A. F. Mace, H. R. B. Pelham, and A. A. Travers, *Nature (London)* **304**, 555 (1983).
[7] J. M. Nickol and G. Felsenfeld, *Cell (Cambridge, Mass.)* **35**, 467 (1983).
[8] C. R. Cantor and A. Efstradiatis, *Nucleic Acids Res.* **12**, 8059 (1984).
[9] H. Htun, E. Lund, and J. E. Dahlberg, *Proc. Natl. Acad. Sci. U.S.A.* **81**, 7288 (1984).
[10] D. E. Pulleyblank, D. B. Haniford, and A. R. Morgan, *Cell (Cambridge, Mass.)* **42**, 271 (1985).
[11] D. M. J. Lilley and E. Paleček, *EMBO J.* **3**, 1187 (1984).
[12] G. W. Gough, K. M. Sullivan, and D. M. J. Lilley, *EMBO J.* **5**, 191 (1986).
[13] J. C. Furlong and D. M. J. Lilley, *Nucleic Acids Res.* **14**, 3995 (1986).
[14] P. M. Scholten and A. Nordheim, *Nucleic Acids Res.* **14**, 3981 (1986).
[15] C. K. Singleton, M. W. Kilpatrick, and R. D. Wells, *J. Biol. Chem.* **259**, 1963 (1984).
[16] W. Herr, *Proc. Natl. Acad. Sci. U.S.A.* **82**, 8009 (1985).
[17] B. H. Johnston and A. Rich, *Cell (Cambridge, Mass.)* **42**, 713 (1985).
[18] K. Nejedly, M. Kwinkowski, G. Galazka, J. Klysik, and E. Paleček, *J. Biomol. Struct. Dyn.* **3**, 467 (1985).
[19] J. K. Barton and A. L. Raphael, *Proc. Natl. Acad. Sci. U.S.A.* **82**, 6460 (1985).
[20] G. Galazka, E. Paleček, R. D. Wells, and J. Klysik, *J. Biol. Chem.* **261**, 7093 (1986).
[21] M. J. McLean, J. W. Lee, and R. D. Wells, *J. Biol. Chem.* **263**, 7378 (1988).
[22] A. Millgram-Burkoff and T. D. Tullius, *Cell (Cambridge, Mass.)* **48**, 935 (1987).
[23] J. G. McCarthy, L. D. Williams, and A. Rich, *Biochemistry* **29**, 6071 (1990).
[24] T. Kohwi-Shigematsu and Y. Kohwi, *Cell (Cambridge, Mass.)* **43**, 199 (1985).
[25] J. C. Hanvey, M. Shimizu, and R. D. Wells, *Proc. Natl. Acad. Sci. U.S.A.* **85**, 6296 (1988).
[26] H. Htun and J. E. Dahlberg, *Science* **241**, 1791 (1988).

stranded DNA,[32] as well as exploring more subtle variations occurring in B-DNA.[33-38] Chemical probing is particularly effective for the study of structural perturbation in supercoiled DNA, where nuclease probes cause problems because of the release of topological constraint with the first cleavage. Osmium tetroxide has recently been employed to reveal altered DNA structure in an alternating adenine–thymine tract in positively supercoiled DNA.[39] Chemical probes have also found application in the study of DNA structure in the vicinity of bound drug molecules[40-42] or proteins[37,43-45] and at base mismatches.[46,47] The latter is becoming important in the detection of single base mutations by means of heteroduplexed molecules. A considerable range of chemical probes is now available for the study of DNA structure; many of these agents are considered individually in the following chapters. The purpose of this chapter is to draw together some general features of the probes and their application.

Range of Probing Agents

Most of the enzymes employed in the study of DNA structure are nucleases, which probe the accessibility to the enzyme at each phosphodiester bond. Some, such as DNase I and S1 nuclease may be relatively

[27] B. H. Johnston, *Science* **241**, 1800 (1988).
[28] O. N. Voloshin, S. M. Mirkin, V. I. Lyamichev, B. P. Belotserkovskii, and M. D. Frank-Kamenetskii, *Nature (London)* **333**, 475 (1988).
[29] D. Sen and W. Gilbert, *Nature (London)* **334**, 364 (1988).
[30] W. I. Sundquist and A. Klug, *Nature (London)* **342**, 825 (1989).
[31] J. R. Williamson, M. K. Raghuraman, and T. R. Cech, *Cell (Cambridge, Mass.)* **59**, 871 (1989).
[32] J. Klysik, K. Rippe, and T. M. Jovin, *Biochemistry* **29**, 9831 (1990).
[33] C. Dingwall, G. P. Lomonossoff, and R. A. Laskey, *Nucleic Acids Res.* **9**, 2659 (1981).
[34] H. R. Drew, *J. Mol. Biol.* **176**, 535 (1984).
[35] H. R. Drew and A. A. Travers, *Cell (Cambridge, Mass.)* **37**, 491 (1984).
[36] H. R. Drew and A. A. Travers, *J. Mol. Biol.* **186**, 773 (1985).
[37] A. Spassky and D. S. Sigman, *Biochemistry* **24**, 8050 (1985).
[38] T. D. Tullius and B. A. Dombroski, *Science* **230**, 679 (1985).
[39] J. A. McClellan and D. M. J. Lilley, *J. Mol. Biol.* **219**, 145 (1991).
[40] K. R. Fox and G. W. Grigg, *Nucleic Acids Res.* **16**, 2063 (1988).
[41] M. J. Mclean, F. Seela, and M. J. Waring, *Proc. Natl. Acad. Sci. U.S.A.* **86**, 9687 (1989).
[42] B. M. G. Cons and K. R. Fox, *Nucleic Acids Res.* **17**, 5447 (1989).
[43] U. Siebenlist, *Nature (London)* **279**, 651 (1979).
[44] M. Buckle and H. Buc, *Biochemistry* **28**, 4388 (1989).
[45] S. Sasse-Dwight and J. D. Gralla, *J. Biol. Chem.* **264**, 8074 (1989).
[46] R. G. H. Cotton, N. R. Rodrigues, and R. D. Campbell, *Proc. Natl. Acad. Sci. U.S.A.* **85**, 4397 (1988).
[47] A. Bhattacharya and D. M. J. Lilley, *J. Mol. Biol.* **209**, 583 (1989).

sequence-neutral, whereas others, notably the restriction enzymes, may be highly sequence-specific. The restriction methylases may also be used in order to probe accessibility at particular locations. Enzyme probes have differing requirements for reaction conditions, but in general a given enzyme may have fairly strict requirements of buffer composition and conditions. Most enzymes will not tolerate extremes of ionic strength, pH, or temperature. Chemical probes, on the other hand, will normally withstand a wider range of conditions, and they can thus be more useful for the study of structural transitions in response to changing environmental conditions. For example, osmium tetroxide may be used over a wide range of conditions of cation chemistry and concentration, pH, temperature, and buffer composition. Table I summarizes the commonly used chemical probes of DNA structure. The list is not exhaustive, and other chapters in this volume should be consulted for further details. Moreover, Table I does not include many chemical agents, including such alkylating agents as the mustards[48] and [N,N-bis(2-chloroethyl)-N-nitrosourea],[49] that have not yet been employed to probe DNA structure. Chemical probes can approch perfect sequence neutrality, such as hydroxyl radicals that attack the deoxyribose groups of the backbone,[38] or can exhibit sequence selectivity. In some cases chemical cleavage agents have been given considerable sequence specificity by virtue of attached drugs or oligonucleotides.[50–52]

Benefits and Problems with Probing Experiments

The intrinsic problem with probing experiments is that the probing agent must react with the DNA structure, and may therefore perturb the object of study. At worst, the interaction might lead to the formation of a structure that did not exit in the absence of the probe. This is a particular concern with enzymes, where specificity arises because of protein–DNA interaction. For example, a single-strand-selective nuclease must selectively bind single-stranded DNA, and it could in principle stabilize structures that contain single-stranded elements. Such a concern might also extend to chemical probes that require intimate interaction with their targets, intercalation, for example. Such interactions could have the effect of catalyzing the formation of structures that are otherwise kinetically

[48] P. Brookes and P. D. Lawley, *Biochem. J.* **80,** 496 (1961).
[49] D. B. Ludlum, B. S. Krame, J. Wang, and C. Fenselau, *Biochemistry* **14,** 5480 (1975).
[50] J. H. Griffith and P. B. Dervan, *J. Am. Chem. Soc.* **108,** 5008 (1986).
[51] T. LeDoan, L. Perroualt, C. Hélene, M. Chassignol, and N. T. Thuong, *Biochemistry* **25,** 6736 (1986).
[52] H. Moser and P. D. Dervan, *Science* **238,** 645 (1987).

TABLE I
CHEMICAL PROBES USED TO STUDY DNA STRUCTURE[a]

Probe	Comments	Refs.
Haloacetaldehydes	A > C base, etheno adduct	b
Osmium tetroxide	T base, 5,6-diester adduct; cleaved with piperidine	c
Permanganate	T base, diol product; cleaved with piperidine	d
Diethyl pyrocarbonate (DEP)	A > G base; carbethoxylation; cleaved with piperidine	e
Formaldehyde	Cross-linking	f
Glyoxal	G base, etheno adduct	g
Glycidaldehyde	G base	h
Bisulfite	C base; deamination to dU	i
Hydroxylamine	C base	j
Dimethyl sulfate (DMS)	G base; N7 methylation; cleaved with piperidine	k
Methylene blue	Photooxidation of G base	l
Singlet oxygen	Cleavage reported at DNA kinks	m
Ethylnitrosourea	Phosphate ethylation	n
Carbodiimide	Single-strand-specific T and G base modification	o
Ozone	Strand scission at (A + T)-rich sequences	p
Psoralens	Supercoiling-dependent cross-linking	q
EDTA-Fe	Deoxyribose cleavage by HO· generated uniformly	r
MPE-Fe	Binding of complex to DNA; strand scission	s
Cu(o-phen)	Attack of deoxyribose leads to backbone scission	t
Transition ions	Binding or cleavage as function of ion stereochemistry	u
Uranyl ion	Photooxidation of deoxyribose	v

[a] This list and the references given are not intended to be comprehensive but are a guide to the probes available. For further information the reader should consult other chapters in the volume.

[b] D. M. J. Lilley, *Nucleic Acids Res.* **11,** 3097 (1983); T. Kohwi-Shigematsu, R. Gelinas, and H. Weintraub, *Proc. Natl. Acad. Sci. U.S.A.* **80,** 4389 (1983); K. Kayasuga-Mikado, T. Hashimoto, T. Negishi, K. Negishi, and H. Hayatsu, *Chem. Pharm. Bull.* **28,** 923 (1980); J. A. Sechrist III, J. R. Barrio, N. J. Leonard, C. Villar-Palasi, and A. G. Gilman, *Science* **177,** 279 (1972).

[c] D. M. J. Lilley and E. Paleček, *EMBO J.* **3,** 1187 (1984); G. C. Glikin, M. Vojtiskova, L. Rena-Descalzi, and E. Paleček, *Nucleic Acids Res.* **12,** 1725 (1984).

[d] S. Sasse-Dwight and J. D. Gralla, *J. Biol. Chem.* **264,** 8074 (1989).

[e] N. J. Leonard, J. J. McDonald, R. E. L. Henderson, and M. E. Reichmann, *Biochemistry* **10,** 3335 (1971).

[f] J. D. McGhee and P. H. von Hippel, *Biochemistry* **16,** 3279 (1977).

[g] K. Nakaya, O. Takenaka, H. Horinishi, and K. Shibata, *Biochim. Biophys. Acta* **161,** 23 (1968); D. M. J. Lilley, in "The Role of Cyclic Nucleic Acid Adducts in Carcinogenesis and Mutagenesis" (B. Singer and H. Bartsch, eds.), IARC publication 70, p. 83. IARC, Lyon, France, 1986.

[h] B. L. van Duren and G. Loewengart, *J. Biol. Chem.* **252,** 5370 (1977); Y. Kohwi, *Carcinogenesis* **10,** 2035 (1989).

[i] G. W. Gough, K. M. Sullivan, and D. M. J. Lilley, *EMBO J.* **5,** 191 (1986); R. Shapiro, B. Braverman, J. B. Louis, and R. E. Servis, *J. Biol. Chem.* **248,** 4060 (1973).

[j] B. H. Johnston and A. Rich, *Cell (Cambridge, Mass.)* **42,** 713 (1985); C. M. Rubin and C. W. Schmid, *Nucleic Acids Res.* **8,** 4613 (1980).

[k] A. M. Maxam and W. Gilbert, this series, Vol. 65, p. 499.

[l] T. Friedmann and D. M. Brown, *Nucleic Acids Res.* **5,** 615 (1978).

inaccessible. Even if this is not a problem, there is a more subtle concern, that of a driven equilibrium. If there is a dynamic equilibrium between two DNA structures, reaction of one form with a probe will drive the equilibrium into that conformation, and one might be led into overestimating the proportion of that structure, which in reality might be quite small.

For these reasons probing studies should be performed with care. The problems of structural perturbation can be largely overcome by working under "single-hit" conditions, using short times and low reagent concentrations. It is dangerous to rely on data from a single probe, and use of a variety of methods is most likely to avoid confusion. At the least, use of several different probes is advisable. The targets for these might be different; for example, if osmium tetroxide is used to probe thymine conformation, complementary information could be sought on adenine reactivity by use of bromoacetaldehyde or diethyl pyrocarbonate. Moreover, the positive result with some probes might be a failure to react (e.g., the inaccessibility of guanine N7 in a Hoogsteen pairing), and thus if the position of equilibrium is displaced, it will be driven in the opposite direction. In addition to the comparison of results using a number of different probes, it is advisable to obtain results from completely unrelated techniques, such as gel electrophoretic mobility. It is the agreement between a number of such different methods that is ultimately most convincing evidence for a particular structure.

Despite the potential problems with the use of probing methods, there are also substantial advantages. The most important is definition of the region of DNA that is structurally altered. Two-dimensional gel electro-

[m] M. E. Hogan, T. F. Rooney, and R. H. Austin, *Nature (London)* **328,** 554 (1987).

[n] U. Siebenlist and W. Gilbert, *Proc. Natl. Acad. Sci. U.S.A.* **77,** 122 (1980).

[o] P. Hale and J. Lebowitz, *J. Virol.* **25,** 305 (1978); P. Hale, R. S. Woodward, and J. Lebowitz, *Nature (London)* **284,** 640 (1980).

[p] K. Sawadaishi, K. Miura, E. Ohtsuka, T. Ueda, K. Ishizaki, and N. Shinriki, *Nucleic Acids Res.* **13,** 7093 (1985).

[q] R. R. Sinden, S. S. Broyles, and D. E. Pettijohn, *Proc. Natl. Acad. Sci. U.S.A.* **80,** 1797 (1983).

[r] T. D. Tullius and B. A. Dombroski, *Science* **230,** 679 (1985).

[s] P. B. Dervan, *Science* **232,** 464 (1986).

[t] D. S. Sigman, D. R. Graham, V. D'Aurora, and A. M. Stern, *J. Biol. Chem.* **254,** 12269 (1979); C.-H. B. Chen and D. S. Sigman, *Proc. Natl. Acad. Sci. U.S.A.* **237,** 1197 (1986).

[u] J. K. Barton, L. A. Basile, A. Danishefsky, and A. Alexandrescu, *Proc. Natl. Acad. Sci. U.S.A.* **81,** 1961 (1984).

[v] P. E. Nielsen, C. Jeppersen, and O. Buchardt, *FEBS Lett.* **235,** 122 (1988); C. Jeppersen and P. E. Nielsen, *Nucleic Acids Res.* **17,** 4947 (1989).

phoresis can reveal the existence of a topology-dependent structural transition in a circular DNA molecule (see [5] in this volume), but it gives no indication about which sequences are participating in the structural change. Probing methods can provide this information, and this is why the combination of these methods should be applied if possible.

Resolution of Different Probes

Probing can be carried out at the resolution of a single nucleotide in many cases. A number of chemical probes, such as osmium tetroxide and diethyl pyrocarbonate, generate base adducts that are labile to alkali, so that they may be cleaved using piperidine just as in the Maxam–Gilbert[53] reactions. Their sites of modification are readily revealed at single-nucleotide resolution by electrophoresis on sequencing gels. Other probes, such as the haloacetaldehydes, generate base adducts that are not themselves alkali-sensitive, but confer reactivity to a second modification reaction such that the secondary product may be cleaved with piperidine.[54] In earlier experiments, such adducts were frequently revealed by virtue of the sensitivity of the linearized DNA to cleavage by S1 nuclease.[5,11,55] The resolution of such an experiment is much lower, but there is one advantage; the double-stranded DNA may be electrophoresed in agarose, such that any modification within several kilobase pairs is revealed, and thus the sequence of almost an entire plasmid is probed in a single experiment.

Probing DNA Inside Cells

In recent years it has become possible to probe DNA structure within cells. This has two immediate advantages. First, we can ask questions about the structures adopted by DNA inside the cell. The great majority of studies of DNA structure are performed *in vitro,* and there is still a dearth of evidence for a biological role for many perturbed DNA structures. Second, one can use structural transitions in cellular DNA as a probe of cellular conditions, such as the level of DNA superhelical stress.

There are two ways in which this can be achieved. Enzymes can be employed to cleave or modify DNA inside the cell. Because proteins cannot be introduced into the cell, they must be synthesized there, using recombinant DNA methodology. In this way methylases[56] and resolving

---

[53] A. M. Maxam and W. Gilbert, this series, Vol. 65, p. 499.

[54] T. Kohwi-Shigematsu and Y. Kohwi, *Biochemistry* **29,** 9551 (1990).

[55] D. M. J. Lilley, *Nucleic Acids Res.* **11,** 3097 (1983).

[56] A. Jaworski, W.-T. Hsieh, J. A. Blaho, J. E. Larson, and R. D. Wells, *Science* **238,** 773 (1988).

enzymes[57] have been used to reveal the presence of Z-DNA and cruciform structures, respectively, in living cells. Of course, as these demonstrations have exclusively been confined to sequences that were introduced for the purpose, this still does not address the real biology, but that is another question. Some chemicals can be introduced into cells from the external medium in sufficient concentration to react with cellular DNA. Permanganate has been used to reveal open complex formation at promoters *in situ*,[45] whereas osmium tetroxide has demonstrated the existence of Z-DNA,[58,59] cruciforms,[60] and H-triplexes (E. Paleček, personal communication) inside bacterial cells.

[57] N. Panayotatos and A. Fontaine, *J. Biol. Chem.* **262**, 11364 (1987).
[58] P. Boublikova and E. Paleček, *Gen. Physiol. Biophys.* **8**, 475 (1989).
[59] A. R. Rahmouni and R. D. Wells, *Science* **246**, 358 (1989).
[60] J. A. McClellan, P. Boublikova, E. Paleček, and D. M. J. Lilley, *Proc. Natl. Acad. Sci. U.S.A.* **87**, 8373 (1990).

[8] Probing DNA Structure with Osmium Tetroxide Complexes *in Vitro*

By Emil Paleček

Introduction

Osmium(VIII) tetroxide is considered to be the most reliable reagent available for the cis-hydroxylation of alkenes.[1] Osmium tetroxide adds to the double bonds of the alkenes to produce osmate esters in which osmium is in a formal oxidation state of $+6$. These products can be hydrolyzed to the corresponding *cis*-diols. Formation of the osmate esters is accelerated in the presence of tertiary amines.[1,2] The reaction product formed in the absence of the ligands is a dimeric monoester (Fig. 1a) as shown by single-crystal X-ray analysis.[3] In the presence of a suitable ligand the reaction takes a different path leading to a hexacoordinate product (Fig. 1b). The hexacoordinate esters are more stable to both hydrolysis and exchange reactions than are the products formed in the absence of ligands (Fig. 1a). Single-crystal X-ray analysis of $(OsO_4)_2 \cdot$ hexamethylenetetraamine and $OsO_4 \cdot$ quinuclidine indicates a change in OsO_4 geometry from a tetrahedral to a trigonal bipyramidal one on coordination to a ligand nitrogen.[4] The

[1] M. Schröder, *Chem. Rev.* **80**, 187 (1980).
[2] S. F. Kobs and E. Behrman, *Inorg. Chim. Acta* **128**, 21 (1987).
[3] R. J. Collin, J. Jones, and W. P. Griffith, *J. Chem. Soc., Dalton Trans.*, 1094 (1974).
[4] W. P. Griffith, A. C. Skapski, K. A. Woode, and M. J. Wright, *Inorg. Chim. Acta* **31**, L413 (1978).

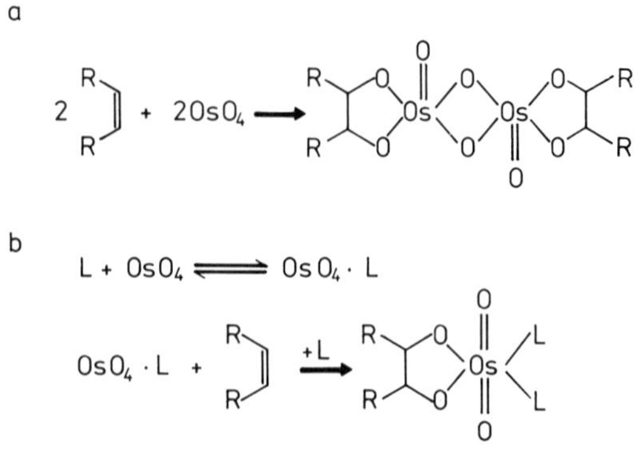

a

b

$$L + OsO_4 \rightleftharpoons OsO_4 \cdot L$$

FIG. 1. Reaction of osmium tetroxide with alkenes: (a) in the absence of a suitable ligand; (b) in the presence of a ligand. R, Alkyl side chain; L, tertiary amine ligand.

aromatic tertiary amines coordinate one OsO_4 if it is sterically allowed; the osmium–nitrogen bond is unusually long. Osmium tetroxide has been widely used in electron microscopy as a fixative and staining agent for biological tissues.

Osmium Tetroxide in Nucleic Acid Chemistry

In the absence of ligands osmium tetroxide reacts (similarly to $KMnO_4$) with thymine in single-stranded oligonucleotides and DNA to yield mainly *cis*-thymine glycol (5,6-dihydroxy-5,6-dihydrothymine) (Fig. 2a).[5–7] In addition to the thymine glycol, formation of cytosine glycol and 5,6-dihydroxycytosine in DNA was recently reported.[8] The *cis*-thymine glycols are produced as a result of γ-irradiation of DNA and also represent a minor product of UV-irradiation of DNA.[9,10] Endonucleases capable of recognizing and cleaving DNAs damaged by radiation and by osmium tetroxide treatment were found in prokaryotic and eukaryotic cells.[10–12] It was shown that 1.5 to 2.1 thymine glycols per M13 phage DNA corresponded to one lethal hit in a normal *Escherichia coli* cell.[13] Mutations

[5] K. Burton and W. T. Riley, *Biochem. J.* **98,** 70 (1966).

[6] M. Beer, S. Stern, D. Carmalt, and K. H. Mohlhenrich, *Biochemistry* **5,** 2283 (1966).

[7] K. Frenkel, M. S. Goldstein, N. Duker, and G. W. Teebor, *Biochemistry* **20,** 750 (1981).

[8] M. Dizdaroglu, E. Holwitt, M. P. Hagan, and W. F. Blakely, *Biochem. J.* **235,** 531 (1986).

[9] P. V. Harikaran and P. A. Cerutti, *Biochemistry* **16,** 2791 (1977).

[10] I. F. Ness and J. Nissen-Meyer, *Biochim. Biophys. Acta* **520,** 111 (1978).

[11] T. P. Brent, *Biochemistry* **22,** 4507 (1983).

[12] H. L. Katcher and S. S. Wallace, *Biochemistry* **22,** 4071 (1983).

[13] R. C. Hayes, L. A. Petrullo, H. Huang, S. S. Wallace, and J. E. LeClerc, *J. Mol. Biol.* **201,** 239 (1988).

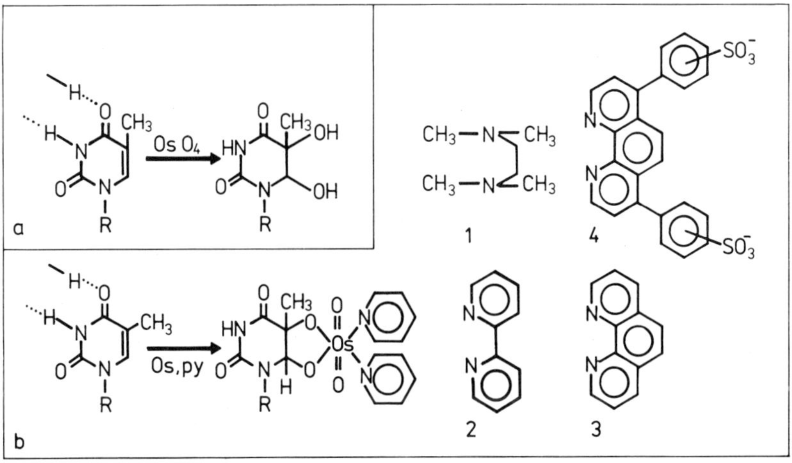

FIG. 2. (a) Reaction of osmium tetroxide (alone) with thymine. (b) Formation of the adduct between osmium tetroxide, pyridine, and thymine along with some tertiary amines which can replace pyridine in the osmium tetroxide complex; 1, tetramethylethylenediamine (TEMED); 2, 2,2'-bipyridine (bipy); 3, 1,10-phenanthroline (phe); 4, bathophenanthrolinedisulfonic acid (bpds); dotted bonds show hydrogen bonding in the Watson–Crick base pair. Only final reaction products are shown.

occurred preferentially at cytosine sites, and a common base change caused by osmium tetroxide treatment of DNA was the C → T transition.

In the presence of suitable tertiary amines [such as pyridine (py) and 2,2'-bipyridine (bipy)] osmium tetroxide forms stable complexes with pyrimidine bases (Figs. 2b and 3). The ligands may enormously increase the rate of formation of the esters. For example, the reaction of thymidine is increased 100-fold in the presence of pyridine and 10,000-fold in the presence of 2,2'-bipyridine (bipy).[14] At alkaline pH osmium(VI) ester complexes are hydrolyzed. In single-stranded nucleic acids OsO_4, pyridine (Os–py) and Os–bipy react with pyrimidine bases.[15] The reaction with thymine residues in poly(dT) was about 10-fold faster than that with cytosine in poly(dC). A slow modification of guanine residues was recently observed in poly(dG).[16,17] Owing to their good stability the hexacoordinate

[14] E. J. Behrman, *in* "Science of Biological Specimen Preparation" (J.-P. Revel, T. Barnard, and G. H. Haggis, eds.), p. 1. SEM Inc., Chicago, 1984.

[15] C.-H. Chang, M. Beer, and L. G. Marzilli, *Biochemistry* **16**, 33 (1977).

[16] E. Paleček, P. Boublikova, F. Jelen, A. Krejcova, E. Makaturova, K. Nejedly, P. Pecinka, and M. Vojtiskova, *in* "Structure and Methods, Volume 3: DNA and RNA" (R. H. Sarma and M. H. Sarma, eds.), p. 237. Adenine Press, Schenectady, New York, 1990.

[17] F. Jelen, P. Karlovsky, E. Makaturova, P. Pecinka, and E. Paleček, *Gen. Physiol. Biophys.* **10**, 461 (1991).

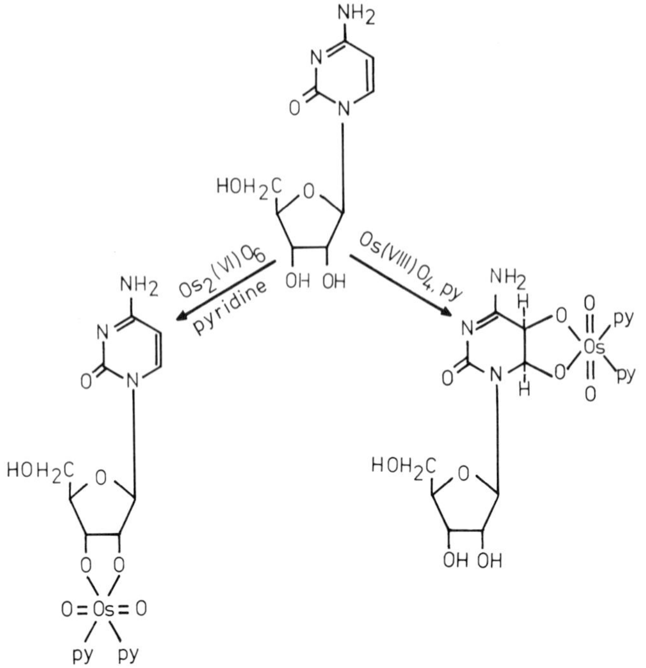

Fig. 3. Formation of oxoosmium(VI) nucleoside sugar esters [as a product of the reaction of cytidine monophosphate with potassium osmate(VI) and pyridine] and of heterocyclic osmate esters [resulting from the reaction of the same nucleotide with osmium tetroxide(VIII) and pyridine].

complexes were utilized in the 1970s for the introduction of a heavy metal stain with the intention of developing an electron microscopic method of DNA sequencing.[18]

$Os_2O_6py_4$ or potassium osmate, $K_2OsO_2(OH)_4$, in the presence of pyridine (or 2,2'-bipyridine), reacts with the sugar residue (Fig. 3) of any one of the usual ribonucleosides.[19,20] In the dinucleoside monophosphates ApU and UpA the osmate group is bonded only to the terminal ribose residue (Fig. 3).[20] Our preliminary results (E. Palecek, F. Jelen and T. Jovin, unpublished) suggest that this reaction can be applied to modification of RNA. Osmium (VIII) reagents react rapidly [in addition to the reaction with the 5,6-double bond of pyrimidine bases (Fig. 2)] with the isopentenyl-

[18] M. D. Cole, J. W. Wiggins, and M. Beer, *J. Mol. Biol.* **117**, 387 (1977).
[19] F. B. Daniel and E. J. Behrman, *J. Am. Chem. Soc.* **97**, 7352 (1975).
[20] F. B. Daniel and E. J. Behrman, *Biochemistry* **15**, 565 (1976).

adenine group[21] and oxidize thiobases such as 4-thiouridine.[5,22] Os–py and Os–bipy were used to prepare isomorphous heavy atom derivatives of tRNA for X-ray diffraction analysis.[23–26]

Osmium Tetroxide as Probe of DNA Structure

At the beginning of the 1980s we searched[27,28] for a chemical probe of the DNA structure that (1) would react specifically with single-stranded and distorted double-stranded regions in DNA under conditions close to physiological ones, (2) would form DNA adducts sufficiently stable even after removal of the unreacted agent, (3) would make it possible to detect a very small fraction of bases modified by the probe, and (4) would be potentially applicable for the DNA structure studies directly in cells and virus particles. We have found that the osmium tetroxide–pyridine reagent (Os–py) fulfills these requirements.[27–33]

Structural Specificity of the Osmium Probe

Single-crystal X-ray analysis of the osmium tetroxide bispyridine ester of thymine[34] and 1-methylthymine[35] suggested that the ester conformation is that of a half-chair with C-6 substantially out of plane and C-5 essentially in the plane. Formation of the *cis*-ester results in a substantial elongation of the C-5–C-6 and C-4–C-5 bonds. A deviation of the exocyclic substitu-

[21] J. A. Ragazzo and E. J. Behrman, *Bioinorg. Chem.* **5**, 343 (1976).

[22] K. Burton, *FEBS Lett.* **6**, 77 (1970).

[23] R. W. Schevitz, M. A. Navia, D. A. Bantz, G. Cornick, J. J. Rosa, M. D. H. Rosa, and P. B. Sigler, *Science* **177**, 429 (1972).

[24] F. J. Suddath, G. J. Quigley, A. McPherson, D. Sneden, J. J. Kim, and A. Rich, *Nature (London)* **248**, 20 (1974).

[25] J. D. Robertus, J. E. Ladner, J. T. Finch, D. Rhodes, R. S. Brown, B. F. C. Clark, and A. Klug, *Nature (London)* **250**, 546 (1974).

[26] J. J. Rosa and P. B. Sigler, *Biochemistry* **13**, 5102 (1974).

[27] E. Paleček, *Bioelectrochem. Bioenerg.* **8**, 469 (1981).

[28] E. Lukasova, F. Jelen, and E. Paleček, *Gen. Physiol. Biophys.* **1**, 53 (1982).

[29] E. Paleček and M. A. Hung, *Anal. Biochem.* **132**, 236 (1983).

[30] E. Lukasova, M. Vojtiskova, F. Jelen, T. Sticzay, and E. Paleček, *Gen. Physiol. Biophys.* **3**, 175 (1984).

[31] G. C. Glikin, M. Vojtiskova, L. Rena-Descalzi, and E. Paleček, *Nucleic Acids Res.* **12**, 1725 (1984).

[32] D. M. Lilley and E. Paleček, *EMBO J.* **3**, 1187 (1984).

[33] E. Paleček, *in* "Highlights of Modern Biochemistry" (A. Kotyk, J. Skoda, V. Paces, and V. Kostka, eds.), p. 53. VSP International Science Publishers, Zeist, The Netherlands, 1989.

[34] S. Neidle and D. I. Stuart, *Biochim. Biophys. Acta* **418**, 226 (1976).

[35] T. J. Kistenmacher, L. G. Marzilli, and M. Rossi, *Bioinorg. Chem.* **6**, 347 (1976).

ents from the plane of the 1-methylthymine residue was observed. The Os=O bonds are in a trans orientation with O=Os=O deviating from linearity (this deviation might be due to electronic effects involving the strong π bonding Os=O group and Os · O in the equatorial plane). It may be expected that the ability of the thymine osmate ester to form hydrogen bonds with adenine in DNA might be decreased, but no definite conclusion can be drawn from the results obtained.[34,35]

Os–py (and Os–bipy) react with single-stranded DNA[16,33] but in the B-DNA double helix the target C-5–C-6 double bond of thymine (Fig. 2) (located in the major groove) is not accessible to the bulky electrophilic osmium probe.[30] Local changes in the helix geometry may render this double bond accessible to the out-of-plane attack of the Os–py molecule on the base π orbitals. It has been shown that such changes may include single-base mismatches and bulges,[36–40] base unstacking in the four-way junctions at low ionic strengths,[41,42] distortions in the vicinity of single-strand interruptions,[28,30] local changes in twist in $(A–T)_n$ sequences,[43] and premelting of AT-rich sequences in supercoiled DNAs.[44]

Os–py has been applied to study local supercoil-stabilized DNA structures *in vitro* (see Refs. 16, 32, and references therein) as well as parallel-stranded DNA,[45,46] DNA–drug interactions,[47–51] etc. In local supercoil-stabilized DNA structures it reacts site specifically with bases in the formally single-stranded cruciform loop,[32,41,52,53] triplex loop,[53–57] displaced

[36] A. Bhattacharya and D. M. J. Lilley, *J. Mol. Biol.* **209,** 583 (1989).

[37] A. Bhattacharya and D. M. J. Lilley, *Nucleic Acids Res.* **17,** 6821 (1989).

[38] R. G. Cotton, N. R. Rodrigues, and R. D. Campbell, *Proc. Natl. Acad. Sci. U.S.A.* **85,** 4397 (1988).

[39] R. G. H. Cotton and R. D. Campbell, *Nucleic Acids Res.* **17,** 4223 (1989).

[40] H. Buc, D. M. J. Lilley, and M. Buckle, personal communication (1990).

[41] J. A. McClellan and D. M. J. Lilley, *J. Mol. Biol.* **197,** 707 (1987).

[42] D. R. Duckett, A. I. H. Murchie, S. Diekmann, E. von Kitzing, B. Kemper, and D. M. J. Lilley, *Cell (Cambridge, Mass.)* **55,** 79 (1988).

[43] J. A. McClellan, E. Paleček, and D. M. J. Lilley, *Nucleic Acids Res.* **14,** 9291 (1986).

[44] J. C. Furlong, K. M. Sullivan, A. I. H. Murchie, G. W. Gough, and D. M. J. Lilley, *Biochemistry* **28,** 2009 (1989).

[45] T. M. Jovin, K. Rippe, N. B. Ramsing, R. Klement, W. Elhorst, and M. Vojtiskova, *in* "Structure and Methods, Volume 3: DNA and RNA" (R. H. Sarma and M. H. Sarma, eds.), p. 155. Adenine Press, Schenectady, New York, 1990.

[46] J. Klysik, K. Rippe, and T. M. Jovin, *Biochemistry* **29,** 9831 (1990).

[47] M. J. McLean, A. Dar, and M. J. Waring, *J. Mol. Recognit.* **1,** 184 (1989).

[48] M. McLean, F. Seela, and M. J. Waring, *Proc. Natl. Acad. Sci. U.S.A.* **86,** 9687 (1989).

[49] A. Schwartz, L. Marrot, and M. Leng, *Biochemistry* **28,** 7975 (1989).

[50] M.-F. Anin and M. Leng, *Nucleic Acids Res.* **18,** 4395 (1990).

[51] M. Leng, *Biophys. Chem.* **35,** 155 (1990).

[52] D. M. J. Lilley, *Chem. Soc. Rev.* **18,** 53 (1989).

[53] E. Paleček, *Crit. Rev. Biochem. Mol. Biol.* **26,** 151 (1991).

half of the homopyrimidine strand in triplex DNA,[58,58a] and with bases at the B–Z junctions.[53,59–61]

Detection of Osmium Binding Sites

Reaction sites of the Os–py and Os–bipy probes in DNA chains can be detected in a number of ways (Fig. 4). Perhaps the most important is the analysis at single-nucleotide resolution, which is based either (a) on the labilization of the sugar–phosphate backbone at the site of the osmium adduct to cleavage with hot piperidine[59,62,63] as in the Maxam and Gilbert nucleotide sequencing technique,[64] (b) on the ability of the adduct to terminate the transcription,[56] or (c) on the ability of the adduct to stop DNA primer extension *in vitro*.[65] If a structural distortion sensitive to osmium modification is expected within a recognition site of a particular restrictase, the site-specific modification can be manifested by inhibition of the restriction cleavage.[60,66–69] Osmium binding sites can also be detected by nuclease S1 cleavage of the linearized DNA followed by electrophoresis in a nondenaturing gel.[31,32,55,60] Osmium modification of large DNA regions results in a partial or full relaxation of the supercoiled DNA molecule (accompanied by changes in the DNA electrophoretic mobility) owing to the formation of a denaturation bubble which can be visualized

[54] M. Vojtiskova and E. Paleček, *J. Biomol. Struct. Dyn.* **5**, 285 (1987).

[55] R. D. Wells, D. A. Collier, J. C. Hanvey, M. Shimizu, and F. Wohlrab, *FASEB J.* **2**, 2939 (1988).

[56] H. Htun and J. E. Dahlberg, *Science* **241**, 1791 (1988).

[57] M. Vojtiskova, S. Mirkin, V. Lyamichev, O. Voloshin, M. Frank-Kamenetskii, and E. Paleček, *FEBS Lett.* **234**, 295 (1988).

[58] J. Bernues, R. Beltran, J. M. Casasnovas, and F. Azorin, *EMBO J.* **8**, 2087 (1989).

[58a] J. Bernues, R. Beltran, J. M. Casasnovas, and F. Azorin, *Nucleic Acids Res.* **18**, 4067 (1990).

[59] B. H. Johnston and A. Rich, *Cell (Cambridge, Mass.)* **42**, 713 (1985).

[60] K. Nejedly, M. Kwinkowski, G. Galazka, J. Klysik, and E. Paleček, *J. Biomol. Struct. Dyn.* **3**, 467 (1985).

[61] T. M. Jovin, D. M. Soumpasis, and L. P. McIntosh, *Annu. Rev. Phys. Chem.* **38**, 521 (1987).

[62] T. Friedmann and D. M. Brown, *Nucleic Acids Res.* **5**, 615 (1978).

[63] G. Galazka, E. Paleček, R. D. Wells, and J. Klysik, *J. Biol. Chem.* **261**, 7093 (1986).

[64] A. M. Maxam and W. Gilbert, this series, Vol. 65, p. 499.

[65] A. R. Rahmouni and R. D. Wells, *Science* **246**, 358 (1989).

[66] E. Paleček, P. Boublikova, K. Nejedly, G. Galazka, and J. Klysik, *J. Biomol. Struct. Dyn.* **5**, 297 (1987).

[67] E. Paleček, P. Boublikova, G. Galazka, and J. Klysik, *Gen. Physiol. Biophys.* **6**, 327 (1987).

[68] E. Paleček, P. Boublikova, and K. Nejedly, *Biophys. Chem.* **34**, 63 (1989).

[69] K. Nejedly, R. Matyasek, and E. Paleček, *J. Biomol. Struct. Dyn.* **6**, 161 (1988).

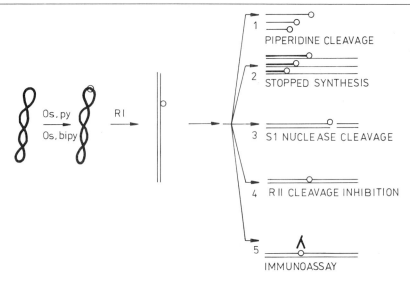

FIG. 4. Treatment of supercoiled DNA with osmium tetroxide–pyridine (Os–py) or with osmium tetroxide–2,2'-bipyridine (Os–bipy) and mapping of the reaction sites. To map the osmium binding sites DNA is cleaved by restrictase RI followed by: (1) cleavage with hot piperidine and sequencing using the principles of the Maxam and Gilbert technique; (2) *in vitro* transcription or DNA primer extension terminated at modified bases in the template followed by nucleotide sequencing; (3) digestion with nuclease S1 and electrophoresis on a nondenaturing gel; (4) digestion with another restrictase (RII), whose restriction site occurs at the place where a structural distortion is expected (e.g., at the B–Z junction as shown in Fig. 5); here, osmium modification of a base in the restriction site is manifested by inhibition of restriction cleavage; (5) immunoassay using polyclonal or monoclonal antibodies against the DNA–Os–py or DNA–Os–bipy adducts.

in the electron microscope without any special treatment (e.g., with glutaraldehyde) to fix the bubble.[31,70] Recently antibodies against DNA modified with Os–py or Os–bipy have been generated which can be applied for the detection of the adducts.[71] More details on the antibodies will be given elsewhere in this volume in the chapter on the use of osmium complexes to probe DNA structure *in situ*.[72]

Destabilization of DNA Double Helix by Osmium Tetroxide–Pyridine Reagent

In an Os–py reagent osmium tetroxide is usually applied at 1–2 mM concentrations, whereas the concentration of pyridine is higher by two orders of magnitude. At low ionic strengths, longer reaction times, and

[70] M. Vojtiskova, J. Stokrova, and E. Paleček, *J. Biomol. Struct. Dyn.* **2**, 1013 (1985).
[71] E. Paleček, A. Krejcova, M. Vojtiskova, V. Podgorodnichenko, T. Ilyina, and A. Poverennyi, *Gen. Physiol. Biophys.* **8**, 491 (1989).
[72] E. Paleček, this volume [17].

higher pyridine concentrations in the Os–py reagent, the initial attack of the probe on the DNA molecule may be followed by the formation and propagation of a denaturation bubble. Under properly chosen reaction conditions [e.g., 1 mM osmium tetroxide, 1–2% (v/v) pyridine, 15 min at 26° in 0.2 M NaCl, 25 mM Tris-HCl, 5 mM ethylenediaminetetraacetic acid (EDTA), pH 7.8] secondarily induced changes are negligible. Under such conditions (chosen in relation to the structure analyzed, i.e., pH 5–6 for a protonated DNA triplex), site-specific modification in the cruciform loop,[41,52] in the protonated triplex loop,[53,55–57] and at the B–Z junctions[60] (Fig. 5) were observed in supercoiled plasmids (at single nucleotide resolution) with no sign of changes induced by the probe. Nuclease S1 did not cleave at solitary Os–py-modified thymines[69] suggesting that the presence of a single adduct in DNA did not induce substantial changes in its backbone conformation (recognized by nuclease S1).

Circular dichroism (CD) measurements of calf thymus DNA treated with Os–py (2 mM osmium tetroxide, 1% (v/v) pyridine, DNA at a concentration of 300 μg/ml in 0.15 M NaCl, 0.015 M sodium citrate, pH 7.0) for several hours showed no changes in the CD bands.[30] Marked changes in both positive and negative CD bands and a steep increase in the amount of modified bases occurred under the given conditions only after more than 24 hr of reaction. In agreement with the CD results polarographic measurements showed only an increase in peak II (characteristic of double-stranded DNA) and no peak III (characteristic of single-stranded DNA) after 24 hr of treatment (Fig. 6b, curve 4). Peak III appeared, however, after 50 hr of Os–py modification (Fig. 6b, curve 5). The ability of pyridine to destabilize the DNA double helix was observed not only in the presence of osmium tetroxide[28,31,70] but also in its absence.[73,74] A large amount of data[16,33,52,54] (see Ref. 53 and references therein) suggests that Os–py is a suitable probe of the DNA structure, but the reaction conditions must be carefully chosen if reliable data are to be obtained. On the other hand, by using Os–bipy instead of the Os–py reagent (see below) the above-mentioned problems can be avoided.

Osmium Tetroxide Complexes

Substituting monodentate pyridine in Os–py reagent by bidentate 2,2′-bipyridine results in more stable DNA adducts and allows working with osmium tetroxide and the ligand at equimolar concentrations.[16,67,68] Different ligands have been tested for their ability to modify site specifically the cruciform loop and B–Z junctions in supercoiled plasmids (Table I). It has been shown the OsO$_4$ complexes with tetramethylethylenediamine (TEMED) or bathophenanthrolinedisulfonic acid (bpds) (Fig. 2) can site

[73] R. Bowater and D. M. J. Lilley, personal communication, 1991.
[74] E. Paleček, unpublished results.

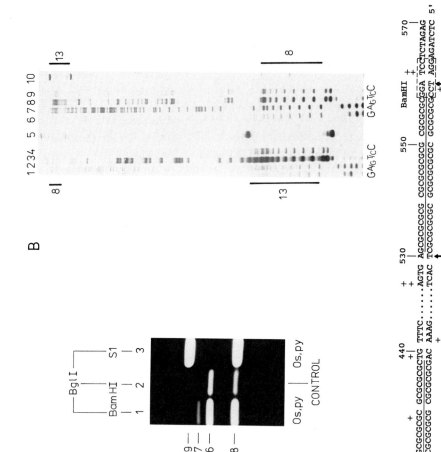

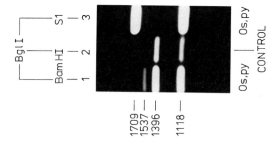

specifically modify the above-mentioned structures at millimolar and submillimolar concentrations. The adducts resulting from DNA treatment with Os–bipy, Os–bpds, or Os–TEMED labilize the sugar–phosphate backbone in a way similar to the Os–py adducts, and their reaction sites in the polynucleotide chain can be detected by hot piperidine cleavage[74] and sequencing.

A complex of osmium tetroxide with 1,10-phenanthroline (phe) showed a lower specificity toward the local DNA structures compared to Os–py, Os–bipy, Os–bpds, and Os–TEMED.[16,75,76] After its initial reaction with bases in the cruciform loop or B–Z junction it also reacted at other sites of the supercoiled and linear DNA molecules. Similar results were obtained from DNA modification with Os–bpds in the presence of chloroquine, suggesting that the Os–phe complex interacted with DNA by at least two modes, one of them probably being intercalation.[76] Os–phe may become useful in DNA footprinting.

Negative results with other compounds (Table I) do not mean that the given osmium tetroxide complex does not react at all with DNA; some of these compounds produced positive reaction when applied at higher concentrations. The results obtained with Os–py, Os–bipy, Os–bpds, and Os–TEMED suggest that these chemicals can be categorized as single-strand-selective probes of DNA structure[16,53] similar to diethyl pyrocarbo-

[75] P. Boublikova and E. Paleček, *Gen. Physiol. Biophys.* **6,** 475 (1989).
[76] P. Boublikova, M. Vojtiskova, J. Stokrova, and E. Paleček, unpublished results.

FIG. 5. Detection of structural distortions at the boundary between right-handed DNA and left-handed $(C–G)_n$ segments in a supercoiled plasmid. Four micrograms of supercoiled pPK2 DNA in 0.2 M NaCl, 25 mM Tris-HCl (pH 7.8) was treated with 1 mM osmium tetroxide, 2% pyridine for 15 min at 26° in a total volume of 100 μl. The reaction was terminated by ethanol precipitation. (A) The Os–py-treated DNA was digested with *Bgl*I followed by *Bam*HI (lane 1) or by nuclease S1 (lane 3) (no nuclease S1 cleavage at solitary Os–py-modified thymines was detected). Lane 2 contains fragments derived by cleavage of unmodified DNA with *Bgl*I plus *Bam*HI. (B) DNA was cleaved by *Hin*dIII and *Eco*RI. The *Eco*RI–*Hin*dIII fragment was separated by polyacrylamide gel electrophoresis and recovered from the gel. The 5′-ends were labeled with polynucleotide kinase and [^{32}P]dATP. Strands were separated, precipitated, dried, and redissolved in 100 μl of 1 M piperidine and incubated at 90° for 45 min. Piperidine was removed by lyophilization (three times repeated). The samples were dissolved in 20 μl of 80% formamide/Tris–borate, EDTA buffer and 3 μl aliquots heated at 90° for 2.5 min before loading on the 8% polyacrylamide sequencing gel next to Maxam and Gilbert sequencing reactions of the unmodified fragment. (C) Map of the Os–py-modified bases in the supercoiled pPK2 DNA. The lowest modification level shown here (+) amounts to 2–3% and the medium one (+ +) up to 5% of the maximum (↑). Alternating CG sequences are labeled with a solid line, and the *Bam*HI recognition sequences are boxed (dashed line). [Adapted from K. Nejedly, R. Matyasek, and E. Paleček, *J. Biomol. Struct. Dyn.* **6,** 261 (1988) with permission.]

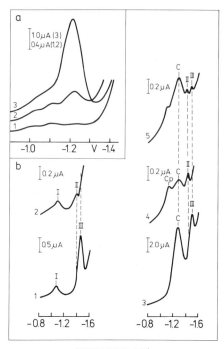

POTENTIAL (V)

FIG. 6. Voltammograms (polarograms) of Os–bipy- or Os–py-modified DNA samples. (a) Adsorptive stripping voltammetry of poly(dT) at a concentration of 0.6 μg/ml. Curve 1, Unmodified poly(dT); curve 2, poly(dT) containing 0.9% and, curve 3, 10% of bases modified with Os–bipy. One-tenth millimolar poly(dT) was incubated with 0.4 mM osmium tetroxide, 0.4 mM 2,2′-bipyridine (curve 3) or with 0.1 mM reagent (curve 2), for 20 min at 0°. Waiting time, 90 sec; curve 3 was measured at an instrument sensitivity 2.5-fold lower than that of curves 1 and 3. (F. Jelen, P. Pecinka, and E. Paleček, unpublished results, 1991.) (b) Differential pulse polarograms of Os–py-modified and unmodified calf thymus DNA. Curves 1 and 2, Unmodified DNAs: curve 1, thermally denatured DNA at a concentration of 31 μg/ml; curve 2, double-stranded (native) DNA at a concentration of 105 μg/ml. Curves 3–5, Os–py-modified DNA: curve 3, thermally denatured DNA, curves 4 and 5 double-stranded DNA; curves 3 and 4, 24 hr of modification; curve 5, 50 hr of modification; curves 1 and 3, DNA at a concentration of 31 μg/ml; curves 2 and 4, DNA at a concentration of 105 μg/ml; curve 5, DNA at a concentration of 207 μg/ml. For DNA modification, double-stranded DNA at a concentration of 300 μg/ml and denatured DNA at a concentration of 100 μg/ml were incubated with 2 mM osmium tetroxide, 1% pyridine at 26° for the times indicated above. Samples were dialyzed before the analysis. [Adapted from E. Lukasova, M. Vojtiskova, F. Jelen, T. Sticzay, and E. Paleček, *Gen. Physiol. Biophys.* **3,** 175 (1984) with permission.] Electrochemical measurements (both a and b) were performed with a Model PAR 174A Polarographic Analyser (EG & G Princeton Applied Res., Princeton, NJ). The three-electrode system included a Model 303 stationary mercury drop electrode as a dropping electrode (b) or as a hanging mercury drop electrode (a) plus a Pt wire as the auxillary electrode and a saturated calomel electrode K 77 as the reference electrode. The background electrolyte was 0.3 M ammonium formate with 50 mM sodium phosphate, pH 6.9.

TABLE I

ABILITY OF OSMIUM TETROXIDE–LIGAND COMPLEXES TO PRODUCE SITE-SPECIFIC MODIFICATION OF ColE1 CRUCIFORM IN SUPERCOILED PLASMID

Ligand with OsO$_4$	Cruciform modification[a]
Pyridine	+[b]
4,4'-Bipyridine	−
2,2'-Bipyridine	+
2,2'-Dithiobipyridine	−
8-Hydroxy-5-quinoline	−
8-Hydroxy-5-quinolinesulfonic acid	−
8-Hydroxy-7-iodo-5-quinolinesulfonic acid	−
1,10-Phenanthroline	+[c]
Bathophenanthrolinedisulfonic acid	+
2,9-Dimethyl-1,10-phenanthroline	−
5-Nitro-1,10-phenanthroline	−
Tetramethylethylenediamine	+

[a] +, Supercoiled pColIR215 DNA was treated with 2 mM osmium tetroxide, 2 mM 2,2'-bipyridine for 10 min at 37°. Site-specific modification was detected using nuclease S1 (after linearization with *Pst*I). −, No site-specific modification observed under the given conditions.

[b] The presence of at least 0.5% pyridine was necessary.

[c] In addition to cruciform modification, reactions at other sites were also detected.

nate, bromo- and chloroacetaldehyde, glyoxal, hydroxylamine, and other probes reacting preferentially with single-stranded DNA. The reactivity of Os–phe toward DNA is rather unusual and does not seem to be similar to any other probe of DNA structure so far available.

Probing of Local (Open) DNA Structures in Supercoiled Plasmid

Preparation of Osmium Tetroxide Solution

Osmium tetroxide is a toxic (oral LD$_{50}$ in rats, 14 mg/kg), volatile, pale yellow solid. It is soluble in water, benzene, carbon tetrachloride, chloroform, ethanol, and ether. It melts at 40° and has an appreciable vapor pressure at room temperature. The vapor affects the eyes and mucous membranes. Ample precaution to protect eyes, nose, and mouth is essential; contact with the skin (which is stained black) should also be avoided. To prepare an osmium tetroxide solution the ampoule of osmium tetroxide is usually carefully opened and dropped immediately into a flask with water. We found this procedure neither safe nor economical.

Materials

Ampoule of osmium tetroxide (0.25–1 g)
Liquid nitrogen
Ampoules

Procedure. An ampoule of osmium tetroxide is dipped into liquid nitrogen until the yellow crystals turn into powder (this usually takes a few minutes). Then the ampoule is opened and its contents are dissolved in water, or aliquots of about 10–50 mg are safely distributed into empty ampoules which are immediately closed and put again in liquid nitrogen prior to their sealing. If the entire amount of osmium tetroxide is dissolved in water, small volumes of the solution (2–10 ml) are distributed into vials with ground glass stoppers and kept at $-70°$ prior to their use (under these conditions solutions can be stored for months). If the solid substance is distributed into ampoules small volumes of fresh solution can be prepared when necessary. The solutions are calibrated using the following λ (ε) values[2]: 304 nm (1202), 297 nm (1442), 282 nm (1738), 275 nm (1733), 250 nm (3116). Usually 20–40 mM osmium tetroxide solutions are prepared. An osmium tetroxide solution stored at $4°$ should not be used after 2 weeks because of its instability. Working with small volumes of millimolar solutions of osmium tetroxide or its complexes is relatively safe; however, one must avoid contact with eyes and ingestion of the solution.

Modification of Supercoiled DNA

Local DNA structures such as cruciforms, left-handed DNA, and triplexes are stabilized by DNA supercoiling, and a certain threshold superhelical density is required for their extrusion. Thus, the superhelical density and the content of superhelical DNA in the analyzed sample should be controlled. Ionic strength, nature, and pH of the buffer should be chosen in agreement with the conditions under which the given local DNA structure is sufficiently stable. Modification of supercoiled DNA with Os–py or Os–bipy under the conditions given below does not introduce detectable amounts of strand breaks into the DNA molecule.

Reagents

10 μl of 5 mM osmium tetroxide
10 μl of 5 mM 2,2'-bipyridine (or 10% pyridine)
10 μl of 0.15 M sodium phosphate, pH 7.8

Procedure. Combine the above reagents in a 1.5-ml Eppendorf snap-cap tube, close the cap, and mix. Add 20 μl supercoiled DNA at a concentration of about 100 μg/ml (in 50 mM sodium phosphate, pH 7.8) and mix. Incubate at 26° for 20 min. Cool to 0° on ice and precipitate the DNA with ethanol as follows: Add 10 μl of 2.5 M ammonium acetate and 100 μl of 95% ethanol. Close the cap and mix. Chill at −70° for 8–10 min (dry ice–ethanol bath). Centrifuge at 12,000 g for 5 min (Eppendorf, 4°). Remove the supernatant with a Gilson pipette. If traces of the reaction mixture might interfere with the following steps (e.g., if inhibition of restriction cleavage is used to detect site-specific DNA modification, Figs. 4 and 5), extract the pellet with 70% ethanol and/or with diethyl ether: Add 100 μl of 70% (v/v) ethanol (ether) to the precipitate, close the cap, and mix for 1 min (Vortexer). Centrifuge at 12,000 g for 5 min. Remove the supernatant and dry under reduced pressure. Add 20 μl distilled water, close the cap, and redissolve the DNA (Vortexer).

In recent years analysis of the DNA Os–py (or Os–bipy) reaction sites at single-nucleotide resolution (Fig. 5B) has prevailed over other detection methods (Fig. 4). The former analysis is very useful as it can provide a detailed picture of the local DNA structure studied. On the other hand, it does not yield information about structural changes at different sites of the supercoiled DNA molecule. Such information can be obtained by using nuclease S1 cleavage combined with (native) agarose gel electrophoresis (Fig. 4). By this technique clusters of DNA–Os–py adducts can be reliably detected simultaneously at distant sites of the DNA molecule. On the other hand, solitary adducts in DNA strands may remain undetected.[66,67,69]

Details of the reaction conditions should be chosen depending on the detection method used for the DNA site-specific modification. For example, if osmium binding sites are to be detected at single-nucleotide resolution using the principles of Maxam and Gilbert sequencing (Fig. 4), a 5–15 min modification using the above conditions may be sufficient. If nuclease S1 cleavage followed by native gel electrophoresis with ethidium bromide staining is applied in further steps, a longer incubation with the reaction mixture (e.g., 30 min) may yield better results.

Determination of Amount of Osmium Adducts in DNA

Several methods for the determination of osmium in organic compounds are available.[6,15,77–83] They usually require relatively large amounts

[77] G. H. Ayres and W. N. Wells, *Anal. Chem.* **22,** 317 (1950).

[78] R. C. Mallett, S. J. Royal, and T. W. Steele, *Anal. Chem.* **51,** 1617 (1979).

[79] M. Balcerzak and Z. Marczenko, *Microchem. J.* **30,** 397 (1984).

[80] B. Keshavan and P. Nagaraja, *Microchem. J.* **31,** 124 (1985).

[81] A. T. Gowda, N. M. M. Gowda, and H. S. Gowda, *Microchem. J.* **31,** 378 (1985).

[82] C. F. Pereira and J. G. Gomez, *Analyst* **110,** 305 (1985).

[83] K. Smolander, M. K. Kauppinen, and R. Mauranen, *Analyst* **113,** 273 (1988).

of the substance to be analyzed and require decomposition of the substance prior to the (usually colorimetric) reaction. These techniques are either laborious or only approximate. The electronic absorption spectra of DNA and RNA adducts have been published.[15] DNA–Os–bipy adducts produce a band at about 310 nm which can be used to monitor the reaction of Os–bipy with DNA. The extinction coefficient is 6200 M^{-1} cm^{-1} for DNA (containing 50% pyrimidine bases) and 12,400 M^{-1} cm^{-1} per pyrimidine moiety. Adducts of DNA and RNA with Os–py and Os–bipy are reduced at the mercury electrode and produce catalytic currents at negative potentials.[21,24,29,84] By means of polarographic (voltammetric) methods (Fig. 6) DNA adducts can be determined at very low concentrations (by adsorptive stripping voltammetry starting from about 5 ng/ml of fully modified DNA).[29] The volume of the analyzed sample can be reduced to 5–20 μl.[85]

Osmium–Pyridine and Osmium–Bipyridine as Reagents for Thymine in Maxam and Gilbert Nucleotide Sequencing

Osmium tetroxide itself,[86] as well as Os–py,[62] were among the first chemical agents applied in nucleotide sequencing. Friedmann and Brown[62] showed that treatment of double-stranded DNA (at low ionic strength) with 3.1% osmium tetroxide, 6.3% (v/v) pyridine followed by treatment with hot piperidine resulted in chain cleavage at the modified thymidine residues. This procedure has found little application in nucleotide sequencing, most probably because of the influence of the DNA secondary structure (which should, under the given conditions, be disrupted to some extent; see section on destabilization of DNA double helix) on the accessibility of the T residues to the reagent. If, however, the DNA is gently denatured prior to treatment with reagents at substantially lower concentrations, very good results can be obtained. For instance, treatment of single-stranded DNA with 1 mM osmium tetroxide, 1% pyridine (or 1 mM bipy) for 4–16 min at 24° results in a specific modification of thymine residues[17] in Maxam and Gilbert[64] sequencing. At short reaction times Os–bipy gives better results.[17]

Concluding Remarks

In recent years chemical probes have become one of the most efficient tools for local supercoil-stabilized DNA structure research.[16,53] Among these probes osmium tetroxide represents a very important reagent mainly

[84] E. Paleček and F. Jelen, *in* "Charge and Field Effects in Biosystems" (M. J. Allen and P. N. R. Usherwood, eds.), p. 361. Abacus Press, Tonbridge, 1984.

[85] E. Paleček, *Anal. Biochem.* **170,** 421 (1988).

[86] E. D. Sverdlov, G. S. Monastyrskaya, and E. I. Budowskii, *Mol. Biol.* (*U.S.S.R.*) **11,** 116 (1977).

because of its ability to form various complexes which can be applied in DNA structure research *in vitro* and also in studies of DNA structure in cells. So far only commercially available ligands have been applied to produce osmium tetroxide complexes for probing DNA structure. It might be worthwhile to develop osmium tetroxide complexes tailored to specific local DNA structures and to determine the smallest structural changes in the DNA double helix sufficient for the reaction of the given complex to occur. The application of osmium tetroxide complexes to study DNA structure *in situ* will be discussed elsewhere in this volume.[72]

Acknowledgments

I wish to express my gratitude to Dr. F. Jelen for his help with preparation of the manuscript, and to Drs. K. Nejedly and M. Vojtiskova for critical reading.

[9] Detection of Non-B-DNA Structures at Specific Sites in Supercoiled Plasmid DNA and Chromatin with Haloacetaldehyde and Diethyl Pyrocarbonate

By T. Kohwi-Shigematsu and Y. Kohwi

Haloacetaldehyde

Reaction of Haloacetaldehyde with DNA

Both bromoacetaldehyde and chloroacetaldehyde react with the exocyclic amino group and the N-1 of adenine or N-3 of cytosine to generate $1,N^6$-ethenoadenine or $3,N^4$-ethenocytosine derivatives.[1-3] These reactions occur when adenine and cytosine bases are not hydrogen bonded,[4,5] and thus haloacetaldehyde is specific for unpaired bases (reviewed in Ref. 6). Although much less reactive than adenine and cytosine, guanine

[1] N. K. Kochetkov, V. N. Shibaev, and A. A. Kost, *Tetrahedron Lett.* **22**, 1993 (1971).

[2] J. R. Barrio, J. A. Secrist III, and N. J. Leonard, *Biochem. Biophys. Res. Commun.* **46**, 597 (1972).

[3] K. Kayasuga-Mikado, T. Hashimoto, T. Negishi, K. Negishi, and H. Hayatsu, *Chem. Pharm. Bull* **28**, 932 (1980).

[4] K. Kimura, M. Nakanishi, T. Yamamoto, and M. Tsuboi, *J. Biochem.* (*Tokyo*) **81**, 1699 (1977).

[5] T. Kohwi-Shigematsu, T. Enomoto, M. Yamada, M. Nakanishi, and M. Tsuboi, *Proc. Natl. Acad. Sci. U.S.A.* **75**, 4689 (1978).

[6] N. J. Leonard, *CRC Crit. Rev. Biochem.* **15**, 125 (1983–1984).

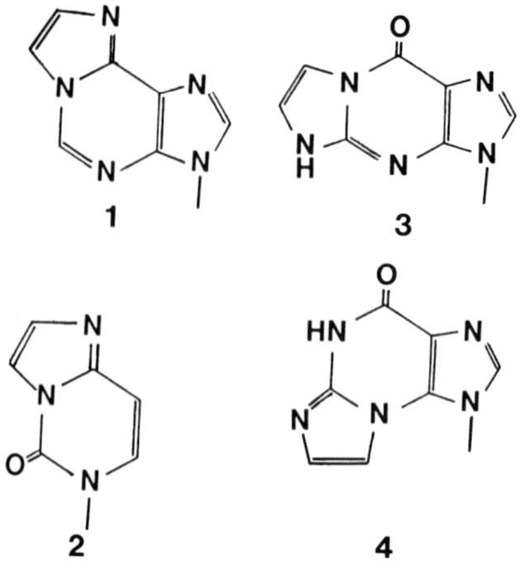

FIG. 1. Haloacetaldehyde reaction with DNA. Haloacetaldehyde reaction products of adenine (**1**), cytosine (**2**), and guanine (**3** and **4**) residues.

residues also react with haloacetaldehyde at the exocylic amino group and the N-1 or N-3 position.[7-10] The reaction products are shown in Fig. 1.

Advantage of Haloacetaldehyde as Probe to Detect Non-B-DNA Structures

The reaction specificities of haloacetaldehyde remain unaltered under a wide range of reaction conditions. Reactions at specific sites in supercoiled plasmid DNA can still be detected, for example, at pH 5–8, ionic concentrations as high as phosphate-buffered saline (PBS), and reaction temperatures up to 50°. This flexibility of choice in the reaction conditions allows detection of various non-B-DNA structures that form under specific conditions. Furthermore, various metal ions can be added to the reaction buffer to examine metal-induced DNA structures.[11,12] Because of very high reaction specificities to unpaired adenine and cytosine residues, the chemical reaction pattern at a single-nucleotide resolution remains the same for DNA modified with either 0.1 μl or 5 μl per 100 μl reaction volume.[11] Yet the reagent is so reactive that even a single unpaired base, over several kilobases of DNA, can be clearly detected. In addition to their use in *in*

[7] P. D. Sattsangi, N. J. Leonard, and C. R. Frihart, *J. Org. Chem.* **42**, 3292 (1977).
[8] F. Oesch and G. Doerjer, *Carcinogenesis* **3**, 663 (1982).
[9] B. Singer, S. J. Spengler, F. Chavez, and J. T. Kuśmierek, *Carcinogenesis* **8**, 745 (1987).
[10] J. T. Kuśmierek, W. Folkman, and B. Singer, *Chem. Res. Toxicol.* **2**, 230 (1989).
[11] Y. Kohwi and T. Kohwi-Shigematsu, *Proc. Natl. Acad. Sci. U.S.A.* **85**, 3781 (1988).
[12] Y. Kohwi, *Nucleic Acids Res.* **17**, 4493 (1989).

vitro studies using supercoiled plasmid DNAs, both bromoacetaldehyde and chloroacetaldehyde have been used successfully to detect reactable sites in chromatin when either whole cells or isolated nuclei are reacted with haloacetaldehydes[13-15] (see below).

Overview for Mapping of Haloacetaldehyde-Modified DNA Bases

Two methods to cleave DNA at the haloacetaldehyde-modified sites have been employed. One method employs nuclease S1, and the other employs chemicals used in Maxam–Gilbert sequencing reactions.

Nuclease S1 Method. The nuclease S1 method is used to locate halo-acetaldehyde-reactive sites over several kilobases of DNA (coarse mapping).[13] Once haloacetaldehyde reacts with unpaired DNA bases in supercoiled plasmid DNA or chromosomal DNA in either intact cells or nuclei, the modified residues can no longer base pair, even after the DNA is purified and digested with a restriction enzyme. For unmodified DNA, the DNA regions that were once unpaired in supercoiled plasmid DNA or in chromosomal DNA snaps back to the double-stranded structure after the DNA is purified and digested. Thus, only haloacetaldehyde-modified sites can be cleaved by subsequent treatment with the single-stranded-specific nuclease S1. The specific cleavage sites generated as a consequence of treatment with both haloacetaldehyde and nuclease S1 can then be mapped by cutting the DNA with an appropriate restriction enzyme, separating the fragments on an agarose gel, transferring the fragments to nitrocellulose filters, and probing with appropriately labeled DNA probes.

Alternatively, for supercoiled plasmid DNA, the DNA fragments with one end cut with both haloacetaldehyde and nuclease S1 and the other end cut with a restriction enzyme are separated and isolated from a native polyacrylamide gel and rerun on a polyacrylamide gel in 8.3 M urea, 100 mM Tris–borate, pH 8.3, and 2 mM ethylenediaminetetraacetic acid (EDTA) (urea-denaturing polyacrylamide gel). Multiple haloacetaldehyde-reactive sites are often detected within a given plasmid DNA population mixture and there might be more than one non-B-DNA structure formed in a given plasmid DNA population. This two-step process can help in identifying which haloacetaldehyde-reactive sites occur in the same plasmid DNA molecule and which do not. For this procedure, optimal nuclease S1 concentration and reaction time must be chosen such that it cleaves

---

[13] T. Kohwi-Shigematsu, R. Gelinas, and H. Weintraub, *Proc. Natl. Acad. Sci. U.S.A.* **80,** 4389 (1983).

[14] J. Bode, H.-J. Pucher, and K. Maass, *Eur. J. Biochem.* **158,** 393 (1986).

[15] T. Kohwi-Shigematsu and J. A. Nelson, *Mol. Carcinog.* **1,** 20 (1988).

DNA at one haloacetaldehyde-reactive site (site A) to generate a double-stranded break and also introduces nicks at other haloacetaldehyde-modified sites (sites B and C, for example). If haloacetaldehyde-reactive sites A, B, and C occur within the same plasmid DNA molecule, the purified, double-stranded DNA fragment [generated by restriction digestion at one end and nuclease S1 cleavage at the other end (at site A) containing sites B and C], should have nicks introduced by nuclease S1 at sites B and C. This can be revealed and mapped by the rerunning the isolated DNA fragment on subsequent denaturing acrylamide gel (see Refs. 16 and 17 for application of this approach).

Chemical Cleavage Method. The advantage of the chemical cleavage method over the nuclease S1 method is that it detects a single base modified with haloacetaldehyde. Furthermore, some haloacetaldehyde-modified regions do not assume the conformation recognized by nuclease S1.[11] Overall, the chemical cleavage method allows fine mapping of haloacetaldehyde-modified bases and aids in elucidating the exact structure adopted by a sequence of interest.

After isolating a one-end radiolabeled DNA fragment containing haloacetaldehyde-reactive sites from a native acrylamide gel, the DNA is subjected to Maxam–Gilbert chemical reactions.[18] After hydrazine/piperidine treatment of DNA, in addition to the expected cytosine ladder, new cleavages at haloacetaldehyde-modified adenine and guanine residues can be detected. After formic acid/piperidine treatment, new cleavages are detected for haloacetaldehyde-modified cytosines within the purine ladder. Similarly, dimethyl sulfate/piperidine treatment gives rise to new cleavages at haloacetaldehyde-modified adenine and cytosine. For most of the reaction the piperidine itself cleaves the DNA at haloacetaldehyde-modified sites, and it is easy to identify DNA bases that strongly react with haloacetaldehyde. However, for those DNA bases that have relatively weak reactivity with haloacetaldehyde, it is often difficult to distinguish the background cleavages caused by the piperidine reaction from cleavages at the haloacetaldehyde-modified sites. In many cases the intensity of bands increases when combined with hydrazine, formic acid, or dimethyl sulfate, and it is much easier to assign bands as new cleavages due to haloacetaldehyde modification when examined within given sequencing ladders. The reaction scheme is shown below.

[16] T. Kohwi-Shigematsu and Y. Kohwi, *Cell* (*Cambridge, Mass.*) **43**, 199 (1985).
[17] T. Kohwi-Shigematsu, T. Manes, and Y. Kohwi, *Proc. Natl. Acad. Sci. U.S.A.* **84**, 2223 (1987).
[18] A. M. Maxam and W. Gilbert, this series, Vol. 65, p. 499.

Reagent	Control DNA (expected cleavage)	Haloacetaldehyde-reacted DNA (new and enhanced* cleavage)
Hydrazine/piperidine	C	C*, A, G
Dimethyl sulfate/piperidine	G	G*, A, C
Formic acid/piperidine	G, A	G*, A*, C

Protocol 1: Preparation of Haloacetaldehydes

Precautions. Great care should be taken in handling haloacetaldehydes because they are mutagenic and probably carcinogenic.[19,20] All reactions with haloacetaldehyde should be performed wearing gloves and under a chemical hood. Haloacetaldehyde is detoxified in Clorox bleach.

Bromoacetaldehyde. Twenty-five grams of bromoacetaldehyde diethylacetal (Eastman Organic Chemical, Rochester, NY) is refluxed with 125 ml of 50% sulfuric acid (w/v) for 45 min in a round-bottomed flask under a chemical hood, and the solution is distilled. The boiling point of bromoacetaldehyde is 78°. The distillate is placed in a 50-ml round-bottomed flask and redistilled. The second distillate is fractionated into Eppendorf tubes and stored at −20° until use. For an alternative method, see Ref. 21.

Chloroacetaldehyde. Chloroacetaldehyde as a 50% (v/v) solution in water is commercially available from Fluka Chemical Corp. (Ronkonkoma, NY). This solution is doubly distilled and stored as described above. The boiling point of chloroacetaldehyde is 85–86°.

Both anhydrous chloroacetaldehyde and bromoacetaldehyde polymerize on standing, but the polymers revert to monomers on heating.

Protocol 2: Quick Assay to Confirm Haloacetaldehyde Reactivity

Supercoiled pBR322 plasmid DNA (10 μg) is either reacted or not reacted (control) with 2 μl of haloacetaldehyde in 100 μl buffer containing 50 mM sodium acetate at pH 5 for 30 min at 37°. DNA is precipitated with ethanol by adjusting the sodium acetate concentration to 0.3 M and adding 3 volumes of ethanol. Two micrograms of DNA is restricted with *Hind*III in a 30 μl medium salt restriction buffer for 30 min, the buffer is adjusted

[19] J. McCann, V. Simmon, D. Streitwieser, and B. N. Ames, *Proc. Natl. Acad. Sci. U.S.A.* **72**, 3190 (1975).

[20] B. L. van Duren, B. M. Goldschmidt, G. Loewengart, A. C. Smith, S. Melchionne, I. Seidman, and D. Roth, *J. Natl. Cancer Inst.* **63**, 1433 (1979).

[21] M. J. McLean, J. E. Larson, F. Wohlrab, and R. D. Wells, *Nucleic Acids Res.* **15**, 6917 (1987).

to S1 buffer (see Protocol 9), and 1 unit of nuclease S1 is added. The DNA sample is incubated at 37° for 10 min and precipitated with ethanol. DNA is resuspended in 10 μl of TE (10 mM Tris-HCl, pH 7.5, 1 mM EDTA). Five hundred nanograms of both haloacetaldehyde-modified and unmodified DNA treated with nuclease S1 are loaded on an 1% agarose gel next to *Hin*dIII-digested λ phage DNA size marker. The gel is stained with 1 μg/ml ethidium bromide. If haloacetaldehyde has reacted with pBR322 DNA, one can detect two sets of doublets, 3.0 and 3.2 kilobases (kb) and 1.2 and 1.4 kb, in addition to the parent DNA of 4.4 kb. These bands are derived from haloacetaldehyde-reactive sites at 3.0 and 3.2 kb clockwise from the *Hin*dIII site.

Protocol 3: Preparation of Supercoiled Plasmid DNA for Non-B-DNA Structural Analysis

To prepare supercoiled plasmid DNA suitable for chemical analysis, one needs DNA preparations with minimum nicks. DNA preparations contaminated with linear DNA cannot be used. A single bacterial colony of *Escherichia coli* DH5α is inoculated in 3 ml of LB medium (10 g Bacto-tryptone, 5 g of Bacto-yeast extract, 10 g of NaCl in 1 liter of water) containing the appropriate antibiotic and incubated at 37° for 7 hr with vigorous shaking. The culture is transfered to 500 ml of LB medium containing 10 mM Tris-HCl, pH 7.5, and antibiotic. This is incubated at 37° overnight with vigorous shaking. The following morning, cells are harvested by centrifugation at 4000 g for 10 min at 4° and lysed with the alkaline lysis procedure described in Ref. 22. Supercoiled plasmid DNA is purified by centrifugation to equilibrium in cesium chloride–ethidium bromide gradients twice. The DNA is purified from CsCl and ethidium bromide by successive extractions with CsCl-saturated 2-propanol and precipitated by addition of 3 volumes of 70% ice-cold ethanol. The DNA is immediately precipitated and resuspended in 400 μl of TE. Forty microliters of 3 M sodium acetate and 1 ml of absolute ethanol is added, and the DNA is reprecipitated. The pellet is washed with 70% ethanol by adding 1 ml of 70% ethanol, vortexing, and centrifuging. The pellet is then lyophilized, resuspended in autoclaved TE, and the concentration adjusted to 1 mg/ml. It is also fine to prepare plasmid DNA employing the chloramphenicol amplification method[22] as long as it is purified as described above. The chloramphenicol amplification method produces plasmid DNA with higher superhelical densities.

[22] T. Maniatis, E. F. Fritsch, and J. Sambrook, "Molecular Cloning: A Laboratory Manual." Cold Spring Harbor Laboratory, Cold Spring Harbor, New York, 1982.

*Protocol 4: Modification of Supercoiled Plasmid DNA
   with Haloacetaldehyde*

To examine the non-B-DNA structures adopted by the sequence of interest, a wide range of reaction conditions should be chosen, because the formation of non-B-DNA structures depends greatly on specific conditions. It is usually convenient to start with the following variations: (1) 50 mM Tris–maleate, pH 5, or sodium acetate at pH 5, (2) 50 mM Tris–maleate, pH 7, (3) 2 mM metal ions added to 50 mM Tris–maleate, pH 5, and (4) 2 mM metal ions added to 50 mM Tris–maleate, pH 7. Most likely one would detect haloacetaldehyde reactivity under one or more of these conditions. Because certain non-B-DNA structures are extremely sensitive to ionic strength, it is sometimes useful to employ 12.5 mM Tris–maleate together with varying Na^+ concentrations.[23,24]

Usually 20 μg of DNA is added to the reaction mixture described above to make a final volume of 100 μl. An aliquot of haloacetaldehyde is defrosted at 65° until it becomes nonviscous, and 2 μl is added to the reaction mixture. When the haloacetaldehyde is still viscous, it is polymerized and unreactive. The aliquot needs to be heated further until it is completely nonviscous, ensuring that the reagent is in the monomer form and reactive. The reaction is normally carried out for 1 hr at 37°. To stop the reaction, 7 μl of 4 M NaCl and 300 μl of absolute ethanol are added. The mixture is vortexed and centrifuged. The DNA pellet is resuspended in 200 μl of 0.3 M sodium acetate buffer at pH 5 to which 600 μl of absolute ethanol is added. The DNA is again pelleted and washed with 600 μl of 70% ethanol, vortexed, and recentrifuged. The haloacetaldehyde-modified DNA is lyophilized and resuspended in autoclaved TE to a concentration of 1 μg/μl and stored at 4°.

Protocol 5: Restriction Digestion and End-Labeling Methods

Four micrograms of haloacetaldehyde-modified DNA is restricted with an appropriate restriction enzyme in 20 μl of reaction buffer at 37° for 1 hr. When either a low- or medium-salt buffer is used for restriction, one can start the following labeling procedure immediately after restriction digestion. For a high-salt buffer, DNA must first be precipitated with ethanol and lyophilized before labeling.

3'-End-Labeling with Klenow Large Fragment. To the 20 μl solution described above, 1 μl of 10× medium-salt buffer, 4 μl of a 1 mM dNTP mixture (all deoxynucleotide triphosphates except for the radiolabeled

[23] T. Kohwi-Shigematsu and Y. Kohwi, *Biochemistry* **29**, 9551 (1990).
[24] J. Bode, Y. Kohwi, L. Dickinson, T. Joh, D. Klehr, C. Mielke, and T. Kohwi-Shigematsu, *Science* **255**, 195 (1992).

deoxynucleotide triphosphate), and 3 μl (30 μCi) of [α-^{32}P]dCTP or [α-^{32}P]dATP (3000 mCi/ml, either from ICN, Costa Mesa, CA, or NEN, Boston, MA), depending on the restriction site to be labeled, are added. One microliter of Klenow large fragment (1–2 units) (Boehringer-Mannheim, Indianapolis, IN, or BRL, Gaithersburg, MD) is added, and the reaction is allowed to proceed for 10 to 20 min at 37°.

5'-End-Labeling with Polynucleotide Kinase. To the 20 μl solution described above, 70 μl of TE, 10 μl of 10× calf intestinal phosphatase (CIP) buffer (0.5 M Tris-HCl, pH 9.0, 10 mM MgCl$_2$, 1 mM ZnCl$_2$, 10 mM spermidine), and 2 units of CIP are added. For DNA with 5' protruding ends, the reaction is incubated at 37° for 30 min, and an additional 2 units of CIP is added and incubated for 30 min further at the same temperature. For DNA with blunt ends or 3' protruding ends, the first incubation is at 37° for 15 min and then at 56° for additional 15 min. An additional 2 units of CIP is added, and the same incubation cycle is repeated. To the 100 μl reaction mixture, 10 μl of 10× STE (100 mM Tris-HCl, pH 8.0, 1 M NaCl, 10 mM EDTA) and 5 μl of 10% SDS (sodium dodecyl sulfate) are added and incubated at 68° for 15 min; the mixture is then extracted once with phenol–chloroform, once with chloroform, precipitated with ethanol, washed once with 70% ethanol, and lyophilized. The DNA is resuspended in 35 μl of TE, 5 μl of 100 mM dithiothreitol, 5 μl of either 10× kinase buffer for 5' protruding ends (0.5 M Tris-HCl, pH 7.6, 0.1 M MgCl$_2$, 50 mM dithiothreitol, 1 mM spermidine, 1 mM EDTA) or 10× kinase buffer for 3' protruding and blunt ends (0.5 M Tris-HCl, pH 9.5, 0.1 M MgCl$_2$, 50 mM dithiothreitol, 50% glycerol), and 5 μl of [γ-^{32}P]ATP (3000 mCi/ml, ICN). Two to five units of kinase (Pharmacia, Piscataway, NJ) is added, and incubation is continued at 37° for 30 min.

3'-End-Labeling with T4 Polymerase. The T4 polymerase method is convenient for labeling DNA at blunt ends and 3' protruding ends. First, a 1-min reaction at 37° with T4 polymerase is allowed in 20 μl of 33 mM Tris–acetate, pH 7.9, 66 mM potassium acetate, 10 mM magnesium acetate, 0.5 mM dithithreitol, 100 μg/ml of bovine serum albumin (BSA) (Sigma, St. Louis, MO, Fraction V). Second, the fill-in reaction is carried out for 15 min, at 37°, in the presence of 50 μCi of [α-^{32}P]dCTP and 100 μM each of cold dATP, dTTP, and dGTP. This reaction is chased with 100 μM cold dCTP for another 15 min.

The end-labeled DNA prepared as described above is adjusted to 100 μl with TE and separated from free nucleotides using spun-column chromatography with Sephadex G-50 fine resin.[22] The end-labeled DNA, collected in decapped Eppendorf tubes, are precipitated with ethanol, washed with 70% ethanol, and lyophilized. The DNA is further restricted with a restriction enzyme distal from the first restriction site to generate an end-labeled DNA fragment that contains sequences of interest. At this step, it is highly recommended to include spermidine at 1 mg/ml to assure complete digestion; without spermidine, the digestion is often incomplete.

Protocol 6: Purification of End-Labeled DNA Fragments

Native polyacrylamide gels, usually 6% or 8% acrylamide, 0.12% N,N'-methylene bisacrylamide, in 1 × TBE, are prepared using 30% acrylamide solution (National Diagnostics, Mannville, NJ), 2% N,N'-methylene bisacrylamide solution (National Diagnostics), and 5 × TBE stock (1 × TBE is 0.089 M Tris–borate, 0.089 M boric acid, 2 mM EDTA). We usually prepare gels of 75 ml volume. To the above solution are added 75 μl of TEMED (N,N,N',N'-tetramethylethylenediamine) and 750 μl of 10% ammonium persulfate before casting a gel.

The second restriction digestion is performed in a reaction volume of 20 μl. If a medium-salt or low-salt restriction enzyme buffer is used, 5 μl of loading buffer (50% glycerol in 1 × TBE with 0.25% each of xylene cyanol and bromphenol blue) is added, and the total of 25 μl is loaded directly onto the gel. In the case of high-salt buffer, the DNA needs to be precipitated with ethanol once and resuspended in 10 μl TE and 10 μl loading buffer. The gel is wrapped in plastic wrap and exposed to XAR film for 5–10 min to determine the location of the DNA fragment.

The acrylamide gel slice containing the desired DNA fragment is cut out and placed in an Eppendorf tube containing 800 μl of elution buffer (0.5 M ammonium acetate, 1 mM EDTA, and 1% SDS). DNA fragments that contain haloacetaldehyde-modified regions often exhibit a slower gel mobility than unmodified DNA fragments. It is important to excise acrylamide segments and not to leave out any portion of DNA fragments with different mobilities. The gel slice need not be crushed for DNA fragments smaller than 200 base pairs (bp). For DNA fragments of larger size, one can obtain better yields by crushing the gel slice. In this case, the crushed gel slices are transferred to end-sealed blue tips filled with a small amount of aquarium filter floss at the tips. The tops of the blue tips containing the crushed gel slice and 800 μl of elution buffer are sealed with Parafilm. The DNA samples in blue tips and Eppendorf tubes are then incubated at 37° overnight. The samples in the Eppendorf tubes are centrifuged at 10,000 g at 4° for 10 min, and the supernatant is transferred to new tubes. For elution of DNA samples from blue tips, the tips are first cut to allow the solution to pass through, and then the DNA solution is collected in Eppendorf tubes and centrifuged at 10,000 g at 4° for 10 min. The supernatant is recentrifuged, transferred equally into two tubes, and precipitated with 3 volumes of ethanol. The DNA is washed once with 70% ethanol and lyophilized.

As an alternative method to eluting DNA using the elution buffer, one

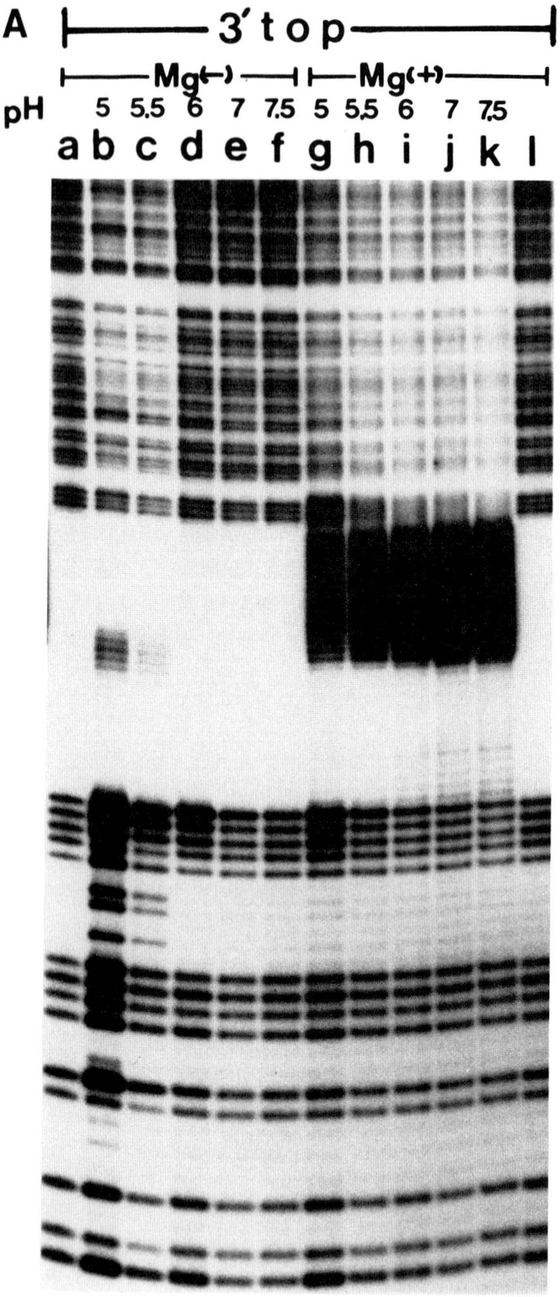

A
3′ top
Mg(−)
Mg(+)
pH
5 5.5 6 7 7.5 5 5.5 6 7 7.5
a b c d e f g h i j k l

B

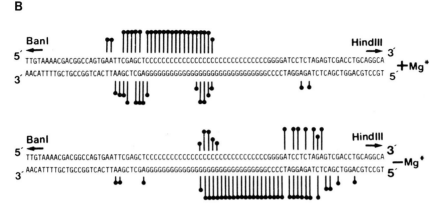

FIG. 2. (A) pH and Mg^{2+} dependency on triple-helical structure. Supercoiled plasmid pCATCG30 was treated with 2 μl of chloroacetaldehyde in 50 mM Tris–maleate buffer at pH values of 5, 5.5, 6, 7, and 7.5 in the presence or absence of 2 mM $MgCl_2$. The BanI–HindIII DNA fragment labeled at the 3' end of the HindIII site (top strand) was then isolated and subjected to Maxam–Gilbert chemical reaction with formic acid. Lanes a and 1 contain chloroacetaldehyde-untreated control DNA. pH had no effect on control DNA. (B) Chloroacetaldehyde-modified sites (top) in the presence of $MgCl_2$ and (bottom) in the absence of $MgCl_2$. (C) Probable structure adopted by the poly(dG)–poly(dC) sequence in supercoiled plasmid DNA and the corresponding hydrogen-bonding scheme for the base triplet in the presence (top) or absence (bottom) of 2 mM Mg^{2+}. ●, Hydrogen bonds in the non-B-DNA structure; ---, hydrogen bonds in the triplets. (From Kohwi and Kohwi-Shigematsu.[11])

can electroelute the DNA into a dialyzing tube. The DNA is extracted once with phenol and precipitated with ethanol. The DNA is then purified with an Elutip-dColumn (Schleicher & Schuell, Keene, NH). To isolate DNA from a large segment of an acrylamide gel, it is better to use this method then the elution buffer method. This is because contaminants from the acrylamide gel interfere with DNA mobility in the sequencing gels, and these can be removed by the electroelution/Elutip purification method. The extraction of DNA from a large acrylamide section with the elution buffer followed by purification with GENEclean II (BIO-101, La Jolla, CA) also works well. The purity of DNA is very important in obtaining a clean sequencing gel pattern.

Protocol 7: Chemical Cleavage of Haloacetaldehyde-Modified DNA

Lyophilized DNA samples divided equally in two separate tubes are labeled HZ for hydrazine reaction and the other FA for formic acid reaction. The step-by-step procedures for hydrazine and formic acid reactions are described in Ref. 22. There are some deviations. The temperature employed for chemical reactions is 15° instead of 20°. At each step of

FIG. 2. (*continued*)

precipitation of DNA with ethanol, there is no need to chill at $-70°$ before centrifugation. DNA is centrifuged at 10,000 g for 10 min at 4°. After the piperidine reaction, instead of lyophilization the sample is transferred to a new tube that contains 100 μl of 0.6 M sodium acetate at pH 5 in TE. The DNA is then precipitated with 3 volumes of ethanol. The DNA pellet is redissolved in 200 μl of 0.3 M sodium acetate and reprecipitated with ethanol, washed once with 70% ethanol, and lyophilized. To study the first 15–20 bp from the end-labeled site, lyophilization is recommended instead of successive ethanol precipitations. The DNA samples are resuspended in 90% formamide in $1 \times$ TBE loading buffer, heated at 95° for 5 min followed by quick chilling in ice, and loaded onto a urea-containing denaturing polyacrylamide gel.

Figure 2A shows a typical pattern after overnight exposure of the film at $-20°$. A poly(dG)–poly(dC) sequence of 30 bp in supercoiled plasmid DNA was reacted with chloroacetaldehyde. The data represent the chloroacetaldehyde reactivity of one strand containing the poly(dC) tract when specific non-B-DNA structures are formed depending on the presence or absence of Mg^{2+} at various pH values. The modification sites of both strands [for the chloroacetaldehyde reaction pattern for the other strand containing the poly(dG) tract, see Ref. 11] are summarized in Fig. 2B, and the structures proposed are shown in Fig. 2C.

More recently, we examined longer tracts of poly(dG)–poly(dC) sequences in supercoiled plasmid DNA and detected chloracetaldehyde reaction patterns different from that of either a dG · dG · dC or dC$^+$ · dG · dC triplex.[25] In the case of a 45 bp-poly(dG)–poly(dC) sequence at pH 5 in the presence of Mg^{2+}, chloroacetaldehyde reactivity in the poly(dC) tract was restricted to two major regions, which occur in intervals of 15 bp: the first region centers at 15 bp, the second region at 30 bp from the 3' end of the poly(dC) tract. For the poly(dG) tract, chloroacetaldehyde-reactive guanine residues were mapped to the guanines complementary to the chloroacetaldehyde-modified cytosines. A structure consistent with this chemical reaction pattern was proposed and shown in Fig. 3. It is of interest to note that similarly long poly(dG)–poly(dC) sequences stably form a typical dG · dG · dC triple structure in PBS (137 mM NaCl, 2.68 mM KCl, 8 mM Na_2HPO_4, 1.47 mM KH_2PO_4, pH 7.4) in the presence of Mg^{2+}.[26] Therefore, it is important to test many different ionic environments to study non-B DNA structure formation for a given sequence.

Haloacetaldehyde reactivity with guanine residues is detected only rarely. When a reaction buffer at pH 5 or higher is employed and if modified

[25] T. Kohwi-Shigematsu and Y. Kohwi, *Nucleic Acids Res.* **19**, 4267 (1991).
[26] Y. Kohwi and T. Kohwi-Shigematsu, *Genes & Dev.* **5**, 2547 (1991).

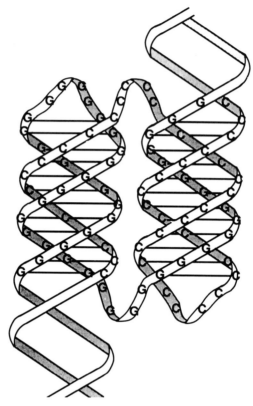

FIG. 3. Model proposed for an altered triple-helix structure: a fused dG · dG · dC and dC⁺ · dG · dC triplex structure. Hydrogen bonds within triple-helices are shown by horizontal lines connecting the three bases. (From Kohwi-Shigematsu and Kohwi.[25])

DNA is eluted from a native acrylamide gel at 37° overnight, hyperreactive guanine residues can be identified and mapped based on the appearance of new bands within a cytosine ladder (G-new bands). Typical example is the chloroacetaldehyde hyperreactivity at the center guanine residues within the poly(dG) tract when the poly(dG)–poly(dC) tract forms an intramolecular G · G · C triplex structure in supercoiled plasmid DNA. The guanine residues over the 5′ half of the dG tract are expected to be unpaired in the absence of Mg^{2+} under acidic pH. However, no G-new bands were detected. Instead guanine reactivity with chloroacetaldehyde was suggested by enhancement in cleavages at these guanines after formic acid/piperidine reaction.[11] These guanine residues are found to be very reactive with another chemical probe, diethyl pyrocarbonate, as shown in Figs. 6 and 7.

Important Points to Note

Haloacetaldehyde reactivities with adenines and cytosines are specific for these nucleotides when they are involved in non-B-DNA structures. However, for the analysis of the guanine residues involved in non-B-DNA structures, one should be careful in choosing the reaction conditions. The haloacetaldehyde reaction with DNA at pH 4.5 or below induces DNA cleavages at every guanine residue, which can be detected either with or without piperidine reaction (the lower the pH, the more cleavages are induced without piperidine reaction). These DNA cleavages occur when chemically modified DNA is subjected to incubation at 60–68° overnight at the stage of elution from acrylamide. When elution is done at 37°, very few, if any, cleavages are detected. This is presumably due to depurination induced at modified guanine residues at high temperature. To avoid background due to nonspecific guanine cleavages, a reaction buffer below pH 5 should not be used, or, if such pH conditions must be used, the DNA should not be heated to 50° or higher prior to the piperidine reaction. It is best to elute the DNA at 37–42° overnight.

Protocol 8: The Nuclease S1 Cleavage Method

The haloacetaldehyde-modified and unmodified supercoiled plasmid DNA are digested by an appropriate restriction enzyme, end-labeled at the restriction site by either Klenow large fragments or T4 polymerase (not by kinase), and suspended in 30 μl of nuclease S1 buffer (30 mM NaOAc, 100 mM NaCl, 2 mM ZnCl$_4$, pH 4.5). In the resulting linear DNA, only those DNA sequences that had been modified with haloacetaldehyde are digested by nuclease S1. It is recommended that the DNA is digested with varying nuclease S1 concentrations, for example, 0.05, 0.1, and 1 unit/μg of DNA, at 22° for 10 min. The DNA was precipitated with ethanol and immediately loaded onto a 6–8% native acrylamide gel next to a labeled DNA size marker. The appearance of fragments specifically from the nuclease S1/haloacetaldehyde treated DNA and not the control unmodified DNA reflect haloacetaldehyde-reactive sites in the original supercoiled DNA. To locate the reactive sites, it is necessary to map the haloacetaldehyde-reactive sites from several different restriction sites. Once this is done, specific DNA fragments are isolated from a native acrylamide gel, eluted as described in Protocol 6, and rerun on a sequencing gel. On a sequencing gel, it is important to run the DNA fragments next to the Maxam–Gilbert chemical sequencing ladder prepared by the control DNA fragment end-labeled at the same restriction site. By comparing the nuclease S1 cleavage patterns of haloacetaldehyde-modified DNA fragment with varying amounts of nuclease S1, one can tell which haloacetaldehyde reactive sites occur in the same DNA fragment.[16,17]

In Situ Probing of DNA Structure by Haloacetaldehyde

It is of great interest to be able to detect non-B-DNA structures within cells. Haloacetaldehyde reactions that detect non-B-DNA structures in supercoiled plasmid DNA can be used directly to probe DNA structure in chromatin in eukaryotic cells or plasmid DNA within bacterial cells.

Briefly, to study haloacetaldehyde-reactive sites in chromatin in eukaryotic cells, cells are directly related with haloacetaldehyde, followed by puriciation of genomic DNA. With nuclease S1, many kilobases of DNA are surveyed for haloacetaldehyde-reactive sites. In the case of plasmid DNA modified with haloacetaldehyde *in situ* in *E. coli* cells, the chemical cleavage method can be used directly on the purified plasmid DNA to examine whether a specific region of DNA forms a non-B-DNA structure.

Chromatin in Eukaryotic Cells

Protocol 9: In Situ Modification of Chromatin with Haloacetaldehyde

Bromoacetaldehyde is a better probe for *in vivo* detection of unusual DNA structures than chloroacetaldehyde owing to its stronger reactivity. Nevertheless, chloroacetaldehyde has also been successfully used in our laboratory (Ref. 15 and T. Kohwi-Shigematsu and Y. Kohwi, unpublished results). Cells from tissue culture plates are harvested and washed twice in standard saline citrate buffer (0.15 M NaCl, 15 mM sodium citrate, 10 mM Tris-HCl, pH 7.4) by centrifugation at 300 g for 5 min. The cells are then suspended in 150 mM NaCl, 1 mM EDTA, 30 mM Tris-HCl, 2 mM ZnCl$_2$, pH 6.8, at a DNA concentration of 1 mg/ml for reaction with haloacetaldehyde. Alternatively, nuclei are prepared by vortexing the cell suspension vigorously in reticulocyte standard (RS) buffer (10 mM Tris-HCl, 10 mM NaCl, 3 mM MgCl$_2$, pH 7.4)/0.5% Nonidet P-40. The nuclei are pelleted at 700 g for 10 min at 4° and suspended in 30 mM sodium acetate, 30 mM NaCl, 1 mM EDTA, 2 mM ZnCl$_2$, pH 5, at a DNA concentration of 1 mg/ml for subsequent reaction with haloacetaldehyde. To 1 ml of the cell or nuclear suspensions, 20–60 μl of haloacetaldehyde is added. Various haloacetaldehyde concentrations in this range should be tested for each cell type used. The mixture is swirled well and incubated at 37° for 15 min. Usually, 15 min of incubation is sufficient. However, we recommend including one more sample at a longer incubation time of 30 min. The reaction mixture is swirled occasionally during the incubation period.

To stop the reaction, the cell suspension is mixed with 4 ml of RS buffer Nonidet P-40, vortexed vigorously, and centrifuged at 700 g for 10 min at 4°. The supernatant is removed, and the nuclei pellet is vortexed and washed with the same buffer twice. For the nuclear suspension reacted with haloacetaldehyde, nuclei are pelleted at 700 g for 10 min, and the supernatant is removed and washed with RS buffer Nonidet P-40 twice. The genomic DNA is then purified by incubating the nuclei in 0.5% SDS, 10 mM Tris-HCl, pH 7.0, 10 mM EDTA containing 200 μg/ml of proteinase K for 6 hr at 37° and extracted once with phenol, once with a combination of phenol and chloroform (1 : 1), and once with only chloroform before the DNA is precipitated with ethanol. The DNA pellet is dissolved in 400 μl of TE, transferred to Eppendorf tubes, reprecipitated with ethanol, washed once with 70% ethanol, and lyophilized. The DNA is dissolved in TE to final concentration of 1 μg/μl.

Employing ZnCl$_2$ instead of MgCl$_2$ works well most of the time, producing genomic DNA with minimal nicks introduced by endonuclease activity within the cell. However, for cells that have high endonuclease activity, better results are sometimes obtained using 30 mM sodium acetate at pH 5 instead of 30 mM Tris-HCl at pH 6.8. It is important to note that, unlike the case of supercoiled plasmid DNA, the detection of haloacetaldehyde reaction sites in chromatin is not subject to change at varying pH in the range of pH 5–8 nor by the presence of metal ions, either MgCl$_2$ or ZnCl$_2$. Unpaired DNA bases in chromatin seem to be well fixed, possibly by specific protein–DNA interactions, whereas in plasmid DNA there is considerable flexibility as to which particular DNA sequences in the plasmid DNA react under the various chemical conditions used.

Protocol 10: Restriction Enzyme and Nuclease S1 Digestion of DNA

Ten micrograms of genomic DNA, either haloacetaldehyde-modified or unmodified control, in 60 μl of restriction buffer is digested with an appropriate restriction enzyme (at least 10 unit/μg of DNA) in the presence of spermidine for several hours to avoid partial digestion. The best choice of restriction site lies 2 kb from the region of interest. After restriction digestion, the linear DNA is subjected to nuclease S1 digestion. The restriction mixture is adjusted to 30 mM sodium acetate, 100 mM NaCl, 2 mM ZnCl$_2$, pH 4.5, with the use of 2 M sodium acetate, pH 4.5, 4 M NaCl, and 100 mM ZnCl$_2$. The sample is kept in ice until nuclease S1 is added. We suggest trying various nuclease S1 concentrations when genomic DNA is used. One unit (0.1 unit/μg of DNA), 5 units (0.5 unit/μg of DNA), or 10 units (1 unit/μg DNA) of nuclease S1 is added, and incubation proceeds at 37° for 10 min. The DNA is immediately precipitated with ethanol, washed once with 70% ethanol, briefly dried, and resuspended in loading buffer.

Protocol 11: Southern Transfer of DNA and Sequence Hybridization

Resulting DNA fragments are separated in an agarose gel with DNA size markers, namely, *Hin*dIII-digested λ phage DNA and *Hae*III-digested φX174 DNA. It is also convenient to prepare genomic DNA digested with several other restriction enzymes as additional size markers to facilitate more precise mapping. It is better to run a full-size agarose gel at least 20 cm in length to allow a good separation of DNA. The Southern transfer method is described in detail in Ref. 22. Instead of baking the nitrocellulose after transfer, we usually cross-link the DNA onto nitrocellulose using a UV cross-linker from Strategene (La Jolla, CA).

The nitrocellulose filter is then placed in a sealed plastic bag and soaked in 15 ml of hybridization buffer containing 7.4 ml of deionized formamide, 2.3 ml of 20× SSPE (7.2 M NaCl, 0.6 M sodium phosphate, and 20 mM NaEDTA, pH 7.0), 3.0 ml of 50% dextran sulfate, 1.5 ml of 10% SDS, 0.15 ml of 100× Denhardt's solution (1 g each of Ficoll, polyvinylpyrrolidone, and Sigma Fraction V BSA per 50 ml), and 0.75 ml of 10 mg/ml denatured salmon sperm DNA. The DNA probe is prepared by isolating a DNA fragment from an agarose gel: the gel is stained with ethidium bromide, and the band containing the DNA fragment of interest is cut and placed in a dialysis tube that contains a minimal amount of 1× TBE buffer. The DNA is then electrophoresed into a dialysis tube. The DNA is extracted once with phenol, once with chloroform, and precipitated with ethanol. To label the DNA the nick translation technique[22] is employed. After the reaction, the sample is adjusted to 100 µl with TE and purified using Sephadex G-50 spun-column chromatography as described above. It is essential to obtain a specific activity near 1 × 10⁸ counts/min (cpm)/µg DNA and use 1 × 10⁷ cpm for hybridization. Boiling the DNA probe at 100° should be avoided after purification via spun-column chromatography for it can cause background. Thus, the DNA probe is mixed in hybridization buffer preincubated at 68° for 30 min and further incubated at the same temperature for 30 min before hybridization. Hybridization is allowed to proceed overnight, and then the nitrocellulose is washed with 0.1% SDS, 0.1× SSC (1× SSC is 0.15 M NaCl, 15 mM sodium citrate, pH 7.4) at 65° for 1 hr. The filter is dried briefly, wrapped in plastic wrap, and exposed to XAR film at −80° for 1 to 2 days. Figure 4 shows typical results of such experiments.

The additional bands beside the expected parent band, indicating haloacetaldehyde reactivity at specific site, can be very faint. To enhance the intensity of the signal, an alkaline agarose gel[22] may be employed instead of a neutral gel, as nicks introduced by S1 nuclease at specific sites can be revealed by an alkaline gel and not with a neutral gel. Alkaline agarose gel electrophoresis works very well for supercoiled plasmid DNA. For genomic DNA, however, success depends on its quality; the DNA must be free of nicks introduced by endogenous endonucleases.

Plasmid DNA in *E. coli* Cells

Protocol 12: In Situ Modification of Plasmid DNA in E. coli Cells with Haloacetaldehyde

In *E. coli* cells, DNA is complexed with proteins. The effective, or unconstrained superhelical density inside cells can be estimated as approximately half that of the purified DNA.[27-35] To test non-B-DNA structure formation in *E. coli* cells, it is recommended to vary effective superhelical density inside cells. This can be done by adding chloramphenicol to an exponentially growing *E. coli* cells and examining haloacetaldehyde reactivity at various points of time after treatment with chloramphenicol. Here, we describe a procedure for detecting an intramolecular dG · dG · dC triplex in *E. coli* as an example.[36] Because this triplex structure is stabilized at neutral pH in the presence of Mg^{2+} (Ref. 11) and under physiological ionic strength (Ref. 26), it was expected that it could form easily inside cells in a relatively long dG tract, providing that sufficient unconstrained superhelical stress is available inside cells.

A one milliliter confluent culture of *E. coli* AB1157 cells (Ref. 37) transformed with plasmid DNAs containing varying lengths of poly(dG)–poly(dC) sequences was inoculated into the 500 ml M9 minimum

[27] R. R. Sinden, S. S. Broyles, and D. E. Pettijohn, *Proc. Natl. Acad. Sci. U.S.A.* **80,** 1797 (1983).

[28] D. R. Greaves, R. K. Patient, and D. M. J. Lilley, *J. Mol. Biol.* **185,** 461 (1985).

[29] L. J. Peck and J. C. Wang, *Cell (Cambridge, Mass.)* **40,** 129 (1985).

[30] D. M. J. Lilley, *Nature (London)* **320,** 14 (1986).

[31] J. B. Bliska and N. R. Cazzarelli, *J. Mol. Biol.* **194,** 205 (1987).

[32] J. A. Borowiec and J. D. Gralla, *J. Mol. Biol.* **195,** 89 (1987).

[33] R. R. Sinden and T. J. Kochel, *Biochemistry* **26,** 1343 (1987).

[34] G. Zheng, T. Kochel, R. W. Hoepfner, S. E. Timmons, and R. R. Sinden, *J. Mol. Biol.* **221,** 107 (1991).

[35] W. Zacharias, A. Jaworski, J. E. Larson, and R. D. Wells, *Proc. Natl. Acad. Sci. U.S.A.* **85,** 7069 (1988).

[36] Y. Kohwi, S. Malkhosyan, and T. Kohwi-Shigematsu, *J. Mol. Biol.* **223,** 817 (1992).

[37] P. Howard-Flanders, R. P. Boyce, and L. Theriot, *Genetics* **53,** 1119 (1966).

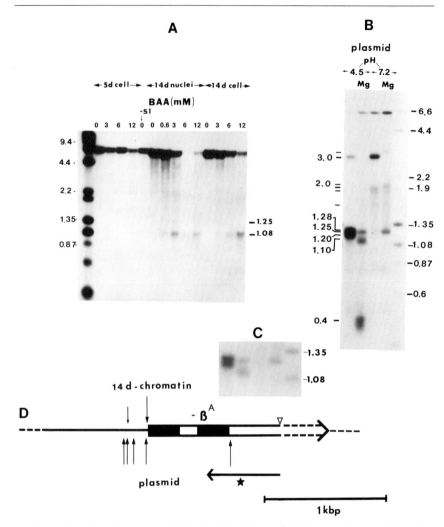

FIG. 4. Reaction of bromoacetaldehyde with chicken β^A-globin gene hypersensitive sites in chromatin and in recombinant supercoiled plasmid DNA. (A) Five-day (5d, the embryonic lineage) or 14-day (14d, the adult lineage) intact erthyrocytes (as well nuclei from 14-day erythrocytes) were isolated and allowed to react at 37° for 15 min with bromoacetaldehyde (BAA) at 0, 3, 6, and 12 μl per 100 μl of reaction volume. The DNA was purified, digested with *Hin*dIII, and treated with nuclease S1. The resulting DNA was separated on a 1.5% agarose gel and blot hybridized to the *Hin*dIII–*Bgl*II probe indicated by the arrow and star. For 5-day cells, the far right-hand lane represents a control (neither bromoacetaldehyde nor nuclease S1). No subband was detected for 5d cells. The major site for the reaction of bromoacetaldehyde in 14d nuclei or cells yields a subband of about 1.08 kb corresponding to a position about -10 bp from the mRNA cap site. A minor site that yields a subband of about 1.25 kb corresponds to a position at about -170 bp from the mRNA site. (B) Supercoiled plasmid HR16 DNA was incubated with 100 mM bromoacetaldehyde for 1 hr at

medium (6 g Na_2PO_4, 3 g KH_2PO_4, 0.5 g NaCl, 1 g NH_4Cl per liter) supplemented with 2 g/liter cassamino acids.[22] Cells were grown in the M9 minimum medium with shaking at 37° to a cell density corresponding to an OD_{550} of 0.4, and chloramphenicol was added to a final concentration of 150 $\mu g/ml$. The culture was further incubated at 37° for 2 hr. The bacteria were washed twice with PBS and suspended in 500 μl of PBS with 2 mM Mg^{2+}. A 40 μl sample of CAA in 500 μl of PBS was combined with 500 μl of bacterial suspension and the reaction mixture was incubated at 37° for 20 min. The bacteria were pelleted and washed twice with PBS. The plasmid DNA was isolated and purified as described in Protocol 3. The fine mapping for CAA-modified sites were done exactly as described for purified supercoiled plasmid DNA in Protocols 5–7. The CAA-modification patterns obtained from the direct probing of plasmid DNAs in *E. coli* cells are shown in Fig. 5 next to the corresponding *in vitro* modification patterns.

The linking numbers of extracted plasmid DNA at various points of time after chloramphenicol treatment were examined by separating the topoisomers in a 1% agarose gel in 1 × TBE buffer containing chloroquine (4 $\mu g/ml$). Optimal levels of chloroquine amounts and agarose concentration vary depending on the size of the plasmid DNA. We found that the DNA purified from *E. coli* cells treated with chloramphenicol for 1 to 2 hours had maximum superhelical density. During this period, extracted DNA was more supercoiled than that of untreated cells by approximately four turns. Only under this condition was an intramolecular dG · dG · dC triplex formation detected.[36]

Diethyl Pyrocarbonate

Reaction of Diethyl Pyrocarbonate with DNA

Diethyl pyrocarbonate reacts at the N-6 and N-7 positions of adenine and the N-7 position of guanine when these positions are especially accessi-

---

37° in the presence or absence of 2 mM $MgCl_2$ at pH 4.5 (50 mM sodium acetate) or pH 7.2 (50 mM Tris-HCl). After the reaction, the DNA was purified and digested first with *Hin*dIII and then with nuclease S1 at 0.5 unit/μg in the S1 buffer described in the text. The DNA was separated on a 1.2% agarose gel and blot hybridized to the *Hin*dIII–*Bgl*II probe. Markers in the extreme right-hand lane are *Hin*dIII-cleaved λ phage DNA and *Hae*III-cleaved φX174 fragments. The cluster of sites represented by subbands between 1.28 and 1.10 kb map to the 5′ side of the β^A-globin gene; the cluster of sites represented by the 3- and 2-kb fragments map in pBR322. The 0.4-kb subband is also seen by nuclease S1 in supercoiled plasmids and in chromatin. (C) A lighter exposure of the 1.10- to 1.35-kb region of (B). (D) Summary of sites of bromoacetaldehyde reaction for chromatin and plasmid DNA. ▽, *Hin*dIII cleavage site. (From Kohwi-Shigematsu *et al.*[13])

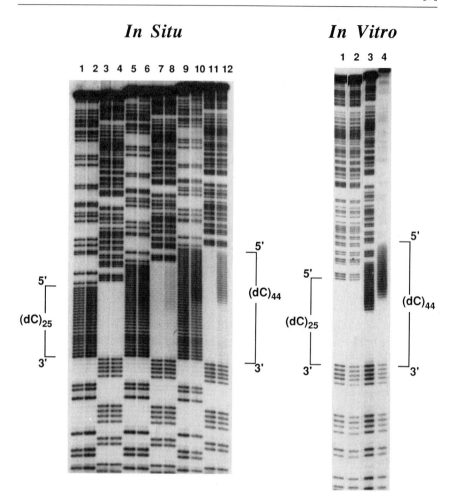

FIG. 5. Detection of a dG · dG · dC triplex structure in *E. coli* cells. Various lengths of poly(dG)–poly(dC) sequences were inserted between the *Sma*I and *Sac*I sites of the pUC13 plasmid by the G–C tailing method using terminal transferase.[25] (A) Plasmid DNA containing 25, 35, 44 bp of poly(dG)–poly(dC) sequences in *E. coli* cells were probed with chloroacetalde-hyde as described in the text. The *Hin*dIII–*Bgl*I DNA fragment with the 3' end radio-labeled at the *Hin*dIII restriction site was subjected to Maxam–Gilbert chemical reactions. The odd-numbered lanes are control DNAs, not treated with chloroacetaldehyde, and the even-numbered lanes are DNA treated with chloroacetaldehyde. Chloroacetaldehyde treated and untreated plasmid DNA containing a 25 bp (lanes 1–4), 35 bp (lanes 5–8), and 44 bp (lanes 9–12) of poly(dG)–poly(dC) sequences were subjected to either the hydrazine piperidine reactions (lanes 1, 2, 5, 6, 9, and 10) or to the formic acid/piperidine reaction (lanes 3, 4, 7, 8, 11, and 12). (B) Purified supercoiled plasmids containing various lengths of poly(dG)–

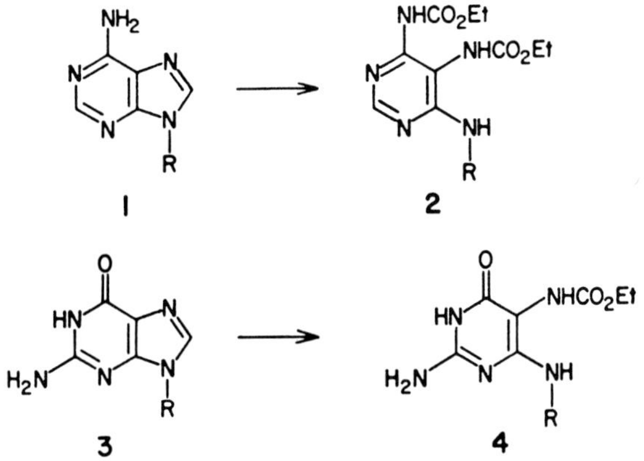

FIG. 6. Reaction of diethyl pyrocarbonate with adenosine and guanosine. Adenosine (1) reacts with diethyl pyrocarbonate to give the ring-opened dicarbethoxylated derivative (2) as the major product. Similarly, guanosine (3) reacts with diethyl pyrocarbonate to give the ring-opened product (4). (From Vincze et al.[41])

ble (Fig. 6),[38–41] and potential applications of diethyl pyrocarbonate in nucleic acid structure determination were suggested early on.[40] In RNA, purines are reactive with diethyl pyrocarbonate only if they assume a single-stranded but not a double-stranded conformation.[42] Thus, several laboratories examined whether it could be used as a probe to detect DNA conformational changes at the nucleotide level. Diethyl pyrocarbonate was found to react preferentially with purines in Z-DNA that are in the

[38] N. J. Leonard, J. J. McDonald, and M. E. Reichmann, Proc. Natl. Acad. Sci. U.S.A. 67, 93 (1970).
[39] N. J. Leonard, J. J. McDonald, R. E. L. Henderson, and M. E. Reichmann, Biochemistry 10, 3335 (1971).
[40] R. E. L. Henderson, L. H. Kirkegaard, and N. J. Leonard, Biochim. Biophys. Acta 294, 356 (1973).
[41] A. Vincze, R. E. L. Henderson, J. J. McDonald, and N. J. Leonard, J. Am. Chem. Soc. 95, 2677 (1973).
[42] D. A. Peattie and W. Gilbert, Proc. Natl. Acad. Sci. U.S.A. 77, 4679 (1980).

poly(dC) sequences were treated with 2 μl of chloroacetaldehyde in 100 μl of PBS containing 2 mM Mg^{2+} for 1 hr. The DNA was precipitated with ethanol twice and analyzed as described above. Control DNA containing a 25 bp of poly(dG)–poly(dC) sequence without the chloroacetaldehyde treatment is shown in lane 1. Chloroacetaldehyde-treated DNAs containing a 25 bp (lane 2), 35 bp (lane 3), and 44 bp (lane 4) of poly(dG)–poly(dC) sequences are shown. The DNAs were subjected to the formic acid/piperidines treatment. The regions for the dC tracts are indicated by brackets. (From Kohwi et al.[36])

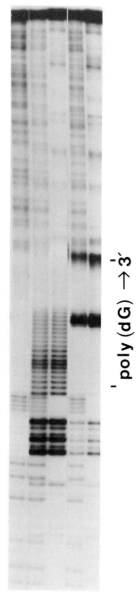

usual syn conformation[43,44] and single-stranded regions including cruciform loops[45,46] and B-DNA–Z-DNA junctions.[17] Diethyl pyrocarbonate also detects the structure (H-DNA) formed by homopurine–homopyrimidine sequences with mirror symmetry by reacting with the 5' halves of the purine strands that are not involved in the triple-helix formation.[47–50] The following is a step-by-step procedure for DNA structural analysis using diethyl pyrocarbonate as the probe.

Protocol 13: Modification of Supercoiled Plasmid DNA with Diethyl Pyrocarbonate

Four micrograms of supercoiled plasmid are dissolved in 200 μl of either 50 mM sodium acetate (pH 5) or 50 mM sodium cacodylate (pH 6 or 7, adjusted with KOH). Three microliters of diethyl pyrocarbonate (Sigma) is added, and the mixture is vortexed and incubated at 20° for 15 min. The sample is mixed once during incubation. Twenty microliters of 3 M sodium acetate and 600 μl of ethanol are added; the mixture is then vortexed and centrifuged at 10,000 g for 10 min. The DNA pellet is redissolved in 0.3 M sodium acetate in TE and reprecipitated with ethanol. The DNA is restricted with an appropriate restriction enzyme and labeled at either the 5' or 3' end as described in Protocol 5. The DNA is further digested with the second restriction enzyme, and the DNA fragment is

[43] W. Herr, *Proc. Natl. Acad. Sci. U.S.A.* **82**, 8009 (1985).
[44] B. H. Johnston and A. Rich, *Cell (Cambridge, Mass.)* **42**, 713 (1985).
[45] P. M. Scholten and A. Nordheim, *Nucleic Acids Res.* **14**, 3981 (1986).
[46] J. C. Furlong and D. M. J. Lilley, *Nucl. Acids Res.* **14**, 3995 (1986).
[47] H. Htun and J. E. Dahlberg, *Science* **243**, 1571 (1989).
[48] B. H. Johnston, *Science* **241**, 1800 (1988).
[49] O. N. Voloshin, S. M. Mirkin, V. I. Lyamichev, B. P. Belotserkovskii, and M. D. Frank-Kamenetskii, *Nature (London)* **333**, 475 (1988).
[50] J. C. Hanvey, J. Klysik, and R. D. Wells, *J. Biol. Chem.* **263**, 7386 (1988).

Fig. 7. Detection of triple-helical structures formed by a poly(dG)–poly(dC) sequence with diethyl pyrocarbonate. Supercoiled plasmid pCATCG30 was treated with 3 μl of diethyl pyrocarbonate in 200 μl of 50 mM sodium acetate at pH 5 in the presence (lanes 4, 5) or absence (lanes 2, 3) of MgCl$_2$. The samples were incubated at 20° for 15 min. The DNA was precipitated twice with ethanol. The *Ban*I–*Hind*III DNA fragment labeled at the 5' end of the *Hind*III site (bottom strand) was then isolated and subjected to either hydrazine/piperidine reaction (lanes 1, 2, 4) or piperidine reaction alone (lanes 3, 5). Lane 1 represents unmodified control DNA. Diethyl pyrocarbonate-reactive sites shown in lanes 2 and 3 are as follows:

3'-GGGGGGGGGGGGGGGGGGGGGGGGGGGGGGGGGGCCCCTAGGAGATCTCAGCTGGACGTCCGT-5'   (Mg^{2+}, pH 5)
          *   * * * * * *   *   *

where asterisks represent diethyl pyrocarbonate-reactive sites.

isolated from a native polyacrylamide gel. After eluting DNA from the gel (Protocol 6), it is resuspended in 100 μl of 1 M piperidine and heated at 95° for 5 min. The DNA is precipitated twice with ethanol, washed once with 70% ethanol, lyophilized, and resuspended in 90% formamide/1 × TBE loading buffer. The DNA sample is denatured at 90° for 5 min and electrophoresed through urea-containing denaturing polyacrylamide gels next to sequencing ladders generated by Maxam–Gilbert chemical reactions[18] as markers (Protocol 7).

Merit of Using both Haloacetaldehyde and Diethyl Pyrocarbonate

Haloacetaldehyde is specific for unpaired adenine and cytosine residues. However, its reaction with guanine residues is very slow at pH 5–8. On the other hand, diethyl pyrocarbonate is reactive with guanine and adenine residues that are involved in non-B-DNA structures. Therefore these two chemical probes complement each other when used in combination to elucidate the actual non-B-DNA structure. An example of such a case is demonstrated with the structure adopted by a poly(dG)–poly(dC) sequence in the presence or absence of Mg^{2+} as shown in Figs. 2 and 7.

Acknowledgments

This work was supported by grants from the National Institutes of Health (RO1-CA39681 to T.K.-S. and RO1-CA51377 to Y.K.) and by funds provided by the Cigarette and Tobacco Surtax Fund of the State of California through the Tobacco-Related Disease Research Program of the University of California (1KT98 to Y.K.), and an American Cancer Society Faculty Award to T.K.-S.

[10] Hydroxylamine and Methoxylamine as Probes of DNA Structure

By BRIAN H. JOHNSTON

Introduction

Of the chemical reagents that can be used for Maxam–Gilbert-style DNA sequencing, several are also useful as structural probes for DNA.[1-3] This chapter describes the use of two of these, hydroxylamine and its

---

[1] B. H. Johnston and A. Rich, *Cell* (*Cambridge, Mass.*) **42**, 713 (1985).
[2] W. Herr, *Proc. Natl. Acad. Sci. U.S.A.* **82**, 8009 (1985).
[3] E. Paleček, *EMBO J.* **3**, 1187 (1984).

derivative methoxylamine. They react with exposed cytosine residues in a manner that allows alkaline cleavage of the backbone. They are highly reactive with single-stranded (ss) DNA and practically unreactive with double-stranded (ds) DNA having a uniform structure, such as B- or Z-form DNA. However, they can react strongly at structural discontinuities in dsDNA, for example, junctions between B-DNA and Z-DNA[1] or between out-of-phase segments of Z-DNA.[4] These characteristics make hydroxylamine and methoxylamine highly sensitive detectors of non-B-DNA segments or single-stranded regions within predominantly B-form DNA.

Chemistry

Hydroxylamine is highly specific for cytosine residues at neutral and acid pH. The rate of reaction with free adenosine is only 1/200 the rate with cytidine, and with guanosine it is even less. Thymines remain unreactive at all pH values.[5] Taking advantage of this specificity, Rubin and Schmid[6] proposed its use as a substitute for hydrazine in the Maxam–Gilbert "C" reaction because it is less affected by air oxidation and less hazardous to handle. It is a strong nucleophile but a weak base (pK_a 6.5).[7] Consequently, at neutral pH, where the most biologically interesting DNA reactions take place, a considerable proportion of hydroxylamine exists as the free base, which is the reactive, nucleophilic form.

The reactions of hydroxylamine with cytosine residues are mainly responsible for its usefulness as a structural probe. However, under some conditions, hydroxylamine does react with bases other than cytosine. At pH 10, it reacts with uracil residues in RNA[8] or in DNA in which the cytosines have been deaminated (e.g., by treatment with nitrous acid), leading to release of the base and susceptibility of the backbone to scission by acid hydrolysis.[9]

In the presence of oxygen, treatment of DNA with low concentrations of hydroxylamine (0.01–0.1 M) results in reaction with all four bases and extensive cleavage of the polynucleotide chain.[5,10] These reactions are

[4] B. H. Johnston, G. J. Quigley, M. J. Ellison, and A. Rich, *Biochemistry* **30**, 5257 (1991).
[5] N. K. Kochetkov and E. I. Budovskii (eds), "Organic Chemistry of Nucleic Acids," Part B, p. 295 and 408. Plenum, New York, 1972.
[6] C. M. Rubin and C. W. Schmid, *Nucleic Acids Res.* **8**, 4613 (1980).
[7] D. M. Brown, *in* "Basic Principles in Nucleic Acid Chemistry" (P. O. P. Ts'o, ed.), p. 6. Academic Press, New York, 1974.
[8] D. Verwoerd, H. Kohlhage, and W. Zillig, *Nature (London)* **192**, 1038 (1961).
[9] H. S. Shapiro and E. Chargaff, *Biochemistry* **5**, 3012 (1966).
[10] E. Freese and E. B. Freese, *Biochemistry* **4**, 2419 (1965).

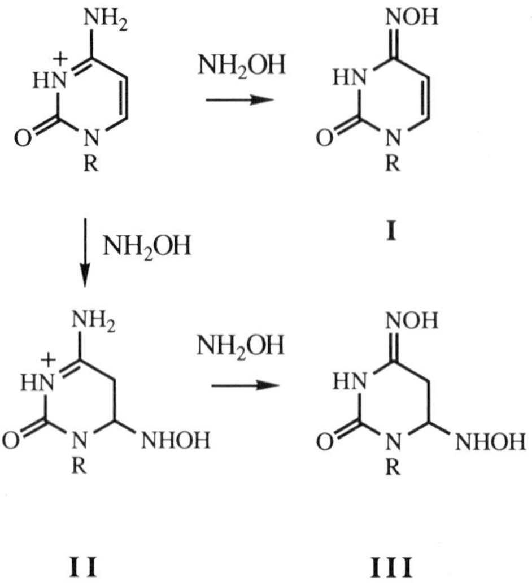

FIG. 1. Reactions of hydroxylamine with cytosine residues. See text for details.

inhibited at higher concentrations of hydroxylamine ($>0.1\ M$), by removal of oxygen, or by the addition of catalase or peroxidase, and they appear to be mediated by hydroxyl radicals produced through oxidation of hydroxylamine. The oxygen-dependent reactions are probably mostly responsible for the mutagenicity and other biological effects of hydroxylamine.[11,12a] (For earlier biological uses of hydroxylamine and methoxylamine, see Phillips and Brown[12b] and Kochetkov and Budovskii.[12c])

At the pH and hydroxylamine concentrations used for chemical sequencing or structural probing, these oxygen-dependent reactions are not significant. However, when hydroxylamine-reactive sites in DNA are visualized by backbone cleavage by piperidine, a faint guanine ladder is frequently seen. This ladder is probably mainly due to depurination during the reaction with hot piperidine, since it is seen to some extent when unmodified DNA is treated with piperidine.

Reaction with free cytosine in solution produces two major products: N^4-hydroxycytosine (Fig. 1, compound **I**) and N^4-hydroxy-6-hydroxylamino-5,6-dihydrocytosine (**III**) (see Kochetkov and Budovskii[5] and

[11] H.-J. Rhaese and E. Freese, *Biochim. Biophys. Acta* **155**, 476 (1968).
[12a] H.-J. Rhaese, E. Freese, and M. S. Melzer, *Biochim. Biophys. Acta* **155**, 491 (1968).
[12b] J. H. Phillips and D. M. Brown, *Prog. Nucleic Acid Res. Mol. Biol.* **7**, 349 (1967).
[12c] N. K. Kochetkov and E. I. Budovskii, *Prog. Nucleic Acid Res. Mol. Biol.* **9**, 403 (1969).

Brown[7] and references therein). The first product results from direct substitution of hydroxylamine for the exocylic amino group, and the second from addition across the 5,6-double bond (**II**), followed by substitution at the amino group by a second hydroxylamine molecule. The ratio of the concentrations of the two products [**III**]/[**I**] depends directly on pH, ionic strength, and hydroxylamine concentration and inversely on temperature.[13,14] Treatment of cytidine 5'-triphosphate with 7 M hydroxylamine at pH 6.5 and 30° yielded **III**, whereas treatment with 1 M hydroxylamine at pH 5 and 50° yielded **I** (see Kochetkov and Budovskii[5] and references therein). The saturation of the 5,6-double bond in **III** makes the base susceptible to ring opening and strand scission on treatment with hot piperidine; thus, relative production of the backbone-labilizing product **III** over product **I** is favored by high hydroxylamine concentration and high pH. The cytosine–hydroxylamine reaction is efficient in the pH range of 5 to 7, with a maximum at pH 6.0–6.2.[8] It falls off at low pH because the reactive free base form of hydroxylamine is converted to the protonated form; on the other hand, the decline in rate at high pH indicates that it is probably the protonated form of cytosine that reacts with hydroxylamine.[5,7] Protonation at the N-3 position is likely to increase the electrophilic character of the C-6 position.

5-Methyl- or 5-hydroxymethyl-substituted cytosines are much less reactive toward addition by hydroxylamine at the 5,6-double bond (**II**), and somewhat less reactive in the exchange reaction at the 4-position (see Brown[7] and references therein).

The foregoing general comments on the chemistry of hydroxylamine apply also to methoxylamine and other O-alkyl derivatives of hydroxylamine, except that the pH optimum for methoxylamine is 5.1 (versus 6.1 for hydroxylamine), and methoxylamine does not react with uracil even at alkaline pH.

Structural Specificity

B–Z Junctions

A striking demonstration of the sensitivity of hydroxylamine to helical structure is shown in Fig. 2,[15] which shows a sequencing gel of a sequence consisting of alternating $(CA)_{31}$ abutting a mixed sequence (plasmid

[13] N. K. Kochetkov, E. I. Budovskii, E. D. Sverdlov, V. N. Shibaev, R. P. Shibaeva, and G. S. Monastyrskaya, *Tetrahedron Lett.*, 3253 (1967).

[14] E. I. Budovskii, E. D. Sverdlov, R. P. Shibaeva, G. S. Monastyrskaya, and N. K. Kochetkov, *Mol. Biol. (U.S.S.R.)* **2**, 329 (1968).

[15] D. B. Haniford and D. E. Pulleyblank, *Nature (London)* **302**, 632 (1983).

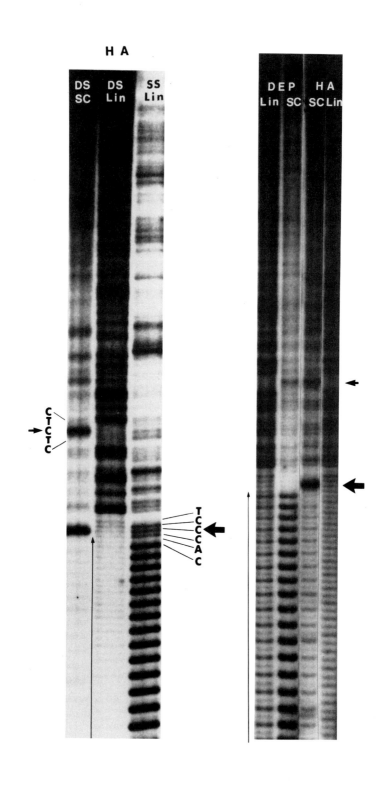

pDHf14[15]). The left-hand autoradiograph shows the sequence reacted with hydroxylamine in three conformations: the denatured form (right lane); the relaxed duplex, which is all B-DNA (middle lane); and the supercoiled form, in which the CA repeat is in the Z conformation and the adjacent sequence is B-DNA (left lane). The treatment of the denatured DNA was performed at reduced temperature and hydroxylamine concentration, yet all the cytosine residues are still highly reactive toward hydroxylamine (right lane). In B-DNA they are completely unreactive; the faint bands in the CA repeat in the middle lane are adenine residues. In the supercoiled (left) lane, the central cytosine of a string of three cytosines is highly reactive, defining the B–Z junction and demonstrating its high degree of localization by the fact that the cytosines on either side are almost unreactive. This sensitivity and precision of hydroxylamine as a locator of junctions between B-DNA and Z-DNA make it a valuable probe for mapping Z-DNA.

When hydroxylamine is used in combination with diethyl pyrocarbonate (DEP), which highlights the actual bases that have adopted the Z conformation[1,2] (see the chapters on Z-DNA[16] and DEP[17]), quite precise assignments of right- or left-handed conformations can usually be made. This is illustrated in the right-hand autoradiograph in Fig. 2, which shows reaction with DEP (left two lanes; note hyperreactivity of the CA repeat when supercoiled) alongside reaction with hydroxylamine (right two lanes). Another example is seen in Fig. 3, which is a sequencing gel of

[16] B. H. Johnston, this series, Vol. 211 [7].
[17] T. Kohwi-Shigematsu and Y. Kohwi, this volume [9].

FIG. 2. Autoradiographs of sequencing gels showing reaction of a plasmid containing $(CA)_{31}$ with hydroxylamine and diethyl pyrocarbonate. *Left:* An 8% polyacrylamide gel showing hydroxylamine reactivity in the region of the upper (5′) end of the CA repeat. When the reaction is performed on a denatured restriction fragment (SS Lin), the most reactive residues are cytosines within the repeat. When linearized intact plasmid is reacted (DS Lin), guanines are most reactive and the (faint) A and C bands of the repeat are of equal intensity. With supercoiled plasmid (DS SC) a single cytosine at the end of the repeat becomes hyperreactive, and the only bands seen within the repeat are faint adenines. *Right:* The same region, showing reactivity to both hydroxylamine and diethyl pyrocarbonate side by side. The plasmid is pDHf14,[1,15] a derivative of pBR322. SC, Highly supercoiled; Lin, linearized; DS, double-stranded; SS, single-stranded; DEP, diethyl pyrocarbonate; HA, hydroxylamine. The vertical lines span the CA repeat, large horizontal arrows mark the hydroxylamine-reactive cytosine at the B–Z junction on the 5′ side of the repeat, and small arrows indicate a secondary hydroxylamine-reactive base at the center of an alternating CT sequence. (Reproduced from Johnston and Rich[1] by permission.)

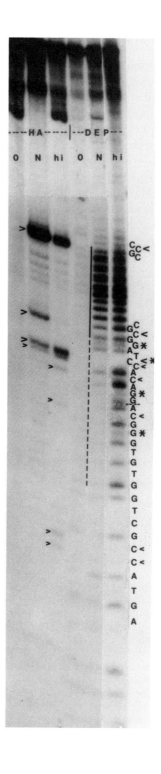

plasmid pLP32,[18] containing the sequence $(CG)_{16}$ inserted into the BamHI site of pBR322, in both linear and supercoiled forms. On moderate supercoiling ($\sigma \sim -0.055$), the CG insert is converted to Z-DNA as indicated by hyperreactivity to DEP (compare lane DEP-O, relaxed, with lane DEP-N, supercoiled). At the same time, single cytosines at either end of the CG insert become highly reactive to hydroxylamine (lane HA-N). The fact that cytosines adjacent to the reactive ones are virtually unreactive again demonstrates the ability of hydroxylamine to distinguish structural details at high resolution. When the superhelical density increases to approximately -0.1, the extent of Z-DNA expands to include sequences below the CG sequence. This extension is seen in the shift in position of one of the B–Z junction bands to cytosines some distance farther down in the gel (lane HA-hi, arrowheads), along with hyperreactivity to DEP extending into that region (lane DEP-hi). Similar shifts with supercoiling have been described for a natural sequence from pBR322.[19]

The appearance of DEP reactivity and hydroxylamine reactivity do not always go hand in hand. If a portion of a long repeated sequence capable of forming Z-DNA undergoes the transition to Z-DNA, the Z region will often be effectively delocalized over the repeated sequence. As a result, the DEP reactivity, which covers all the purines that are in the Z conformation, appears to extend over the entire sequence, but the two reactive cytosines at the B–Z junctions are "smeared out" to the point of being undetectable. Only when the region of Z-DNA extends over the entire repeated sequence do the B–Z junctions become locked at the ends of the sequence and appear as bands on a sequencing gel.[20a]

Another example of a structural distortion in the DNA helix that is sensitively detected by hydroxylamine is the junction between two seg-

[18] L. J. Peck, A. Nordheim, A. Rich, and J. C. Wang, Proc. Natl. Acad. Sci. U.S.A. 79, 4560 (1982).
[19] B. H. Johnston, J. Biomol. Struct. Dyn. 6, 153 (1988).
[20a] B. H. Johnston, W. Ohara, and A. Rich, J. Biol. Chem. 263, 4512 (1988).

FIG. 3. Reactivity of a plasmid containing $(C–G)_{16}$ (pLP32; Ref. 18) to hydroxylamine (HA) and diethyl pyrocarbonate (DEP). Hydroxylamine and DEP were used on intact plasmids in the relaxed state (O), at native (N), and at high (hi) superhelical densities. Cytosines strongly or slightly reactive to hydroxylamine when supercoiled are labeled with large and small arrowheads, respectively. The latter appear mostly at high levels of supercoiling. Bases in perfect purine–pyrimidine alternation (the CG insert) are within the solid vertical line, whereas those mainly alternating but with some out of alternation (asterisks) are along the dotted vertical line. The short horizontal dotted line in the sequence marks a potential Z–Z junction, where the phase of alternation changes. (Reproduced from Johnston and Rich[1] by permission.)

ments of Z-DNA which differ in the phase of their syn–anti alternation about the glycosidic bond; this has been called a Z–Z junction.[1] A recent study[4] of a model Z–Z junction showed that when a (C-C) · (G-G) sequence forms the Z–Z junction, the 5′-most cytosine is hyperreactive to hydroxylamine but the adjacent cytosine is not.

The large differential reactivity for single- versus double-stranded DNA makes hydroxylamine also a sensitive detector of H-DNA. This DNA conformation, formed by mirror-symmetric homopurine–homopyrimidine sequences (see elsewhere in this series[20b]), consists of a triple-stranded region, a single-stranded loop, and a longer apparently single-stranded region.[21] The cytosine residues on each of the single-stranded regions are hypersensitive to hydroxylamine, as are the junctions between the different parts of the overall structure.[22,23] In Fig. 4 is shown the reactivity of a $(C-T)_{18}$ sequence for hydroxylamine, methoxylamine, and also osmium tetroxide, which has a structural specificity similar to that of hydroxylamine except that it reacts with thymine residues.[1,3,24] Under conditions of pH 6 or below and negative supercoiling, cytosines in the middle and at one end become hyperreactive; they are, respectively, the loop and the junction between Hoogsteen and Watson–Crick base pairing at the end of the CT sequence. Thymines in between the cytosines are hyperreactive to OsO_4 under the same conditions.

Figure 4 also demonstrates the subtle differences that can sometimes be seen between hydroxylamine and methoxylamine. Lanes 7 and 8, reactivity with hydroxylamine at pH 6, show some reactivity in regions of the sequence that are expected to be involved in Hoogsteen pairing within the triplex, perhaps because a fraction of the molecules are in a dynamic equilibrium between the duplex and triplex conformations, which leaves them in a more reactive conformation during the interconversion. Methoxylamine shows no such reactivity under the same conditions (lanes 3 and 4), possibly reflecting either less sensitivity to subtle conformational changes (methoxylamine being a larger molecule) or perhaps a slightly different influence by the two probes on the position of equilibrium between the B and H conformations. The latter could be a consequence of the slightly different proportions of protonated and unprotonated species of the two probes at pH values near 6 (the pK_a of hydroxylamine is 6.5; that of methoxylamine is 5.8).[7]

Although the rate of the hydroxylamine reaction peaks at pH 6, it can

[20b] M. D. Frank-Kamenetskii, this series, Vol. 211 [9].

[21] V. I. Lyamichev, S. M. Mirkin, and M. D. Frank-Kamenetskii, *J. Biomol. Struct. Dyn.* **3,** 667 (1986).

[22] B. H. Johnston, *Science* **241,** 1800 (1988).

[23] H. Htun and J. E. Dahlberg, *Science* **241,** 1791 (1988).

[24] E. Paleček, this volume [8].

be used over a rather wide range of pH values with appropriate adjustments in the time of reaction. Figure 4b shows the effect of pH over the range of 4.5–8 on reactivity with H-DNA, which requires acid pH for stability. Despite the reduced efficiency, the reactivity characteristic of the single-stranded loop of H-DNA increases dramatically from pH 6.5 to 4.5 despite only a 2-fold adjustment in the incubation time.

Method

Preparation of Hydroxylamine Solution

The free amine is susceptible to oxidation and is prepared from the hydrochloride as needed. Dissolve 2.97 g of hydroxylamine hydrochloride (from Aldrich, Milwaukee, WI, or another supplier) in about 5 ml of deionized water in a 10-ml beaker. Titrate to the desired pH using diethylamine; for pH 6, about 3 ml is required. Bring the final volume to 10 ml with deionized water. The solution is usable for at least a week if stored at $-20°$.[6] Methoxylamine is prepared in essentially the same way; it is somewhat more stable than hydroxylamine.[7]

Modification Reaction

For a 100-μl reaction mixture, bring the desired amount of DNA to 25 μl in 10 mM Tris/1 mM EDTA or other compatible buffer. Add 75 μl of 4 M hydroxylamine and incubate until the desired level of modification is achieved. For visualization of about 250 nucleotides on a sequencing gel after backbone cleavage, an incubation of 15–20 min at 20–37° is typically appropriate. For low-resolution visualization of cleavages on an entire plasmid, for example, a lower level of modification will be required, so that the average number of cleavages is less than one per molecule. The reaction is terminated by adding 11 μl of 3 M sodium acetate plus 333 μl of ethanol, chilling 5–10 min, and centrifuging in a microcentrifuge for 5–15 min, depending on the DNA concentration. The supernatant is removed, 250 μl of 0.3 M sodium acetate is added, the pellet is redissolved by vortexing, and 750 μl of ethanol is added. The tube is chilled and centrifuged as before, the supernatant is removed, and the pellet is dried in a Speed-Vac concentrator (Savont Instruments, Hicksville, NY).

Cleavage of DNA Backbone and Gel Electrophoresis

To the modified, dried-down DNA add 90 μl of water (or more if an aliquot is to be saved for a no-piperidine control), mix to dissolve, and transfer to a screw-capped 1.5-ml microcentrifuge tube (Sarstedt,

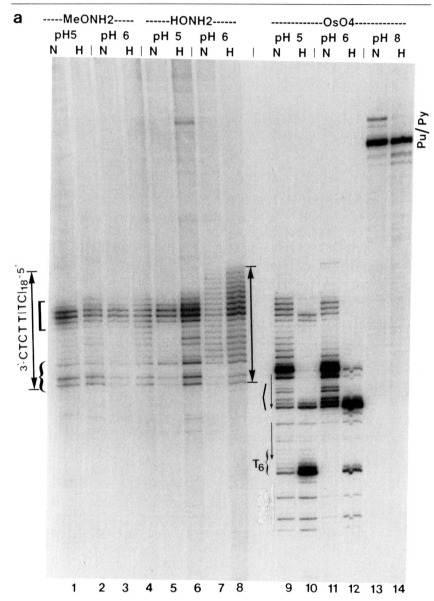

FIG. 4. Reactivity of H-DNA to methoxylamine (MeONH₂), hydroxylamine (HONH₂), and osmium tetroxide (OsO₄). Autoradiographs of sequencing gels are shown for a fragment of the human U1 gene containing the sequence $(C-T)_{18}$ (pGEM2-SB-OF[23]). (a) Reactivity of the H-DNA conformation under various conditions and its disappearance at pH 8. (b) The effect of pH on reactivity toward hydroxylamine, showing only the $(C-T)_{18}$ region. Superhelical density is indicated by N (native; $\sigma \sim -0.055$) and H (high; $\sigma \sim -0.12$). The

b   pH

4.5 5 5.5 6 6.5 7 8

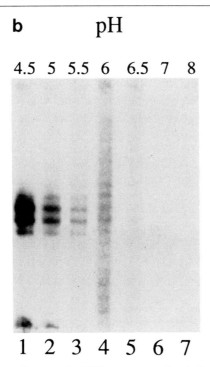

1 2 3 4 5 6 7

double-headed arrow exactly spans the $(C-T)_{18}$ sequence, the single-headed arrows indicate short AT-rich direct repeats, and Pu/Py indicates an alternating purine–pyrimidine sequence. Hyperreactive regions in the middle and at the lower (3′) end of the $(C-T)_{18}$ sequence are designated by heavy square brackets and heavy braces, respectively. An adjacent AT sequence hyperreactive to OsO_4 is indicated by an angled bracket. All reaction mixtures were preincubated in the appropriate buffer plus 1 mM EDTA for 15 min at 37° before the addition of the reagent to induce the H-DNA conformation. The reaction buffers, titrated to the indicated pH values were 50 mM sodium acetate for pH 4.5 to 5.5, 50 mM sodium cacodylate for pH 6 to 7, and 20 mM EDTA for pH 8. Osmium tetroxide reactions were initiated by adding pyridine (titrated to the appropriate pH with HCl) to 4% final concentration, followed by aqueous OsO_4 to 1%. OsO_4 reactions were terminated by three extractions with carbon tetrachloride at 0°, followed by ethanol precipitation. For the hydroxylamine pH series in (b), hydroxylamine was titrated to the various pHs with diethylamine. To compensate for the slower reaction rate at nonoptimal pH values, the time of incubation with hydroxylamine was increased for pH values below 6 as follows: pH 4.5, 35 min; pH 5.5, 20 min; pH 6 and 6.5, 15 min. (Reproduced from Johnston[22] by permission.)

Princeton, NJ). Using siliconized tubes will aid in recovering material after the cleavage step. Add 10 μl of piperidine (e.g., Fisher, Pittsburgh, PA), cap the tube tightly, vortex to mix, and place in a heating block or water bath at 90° for 15 min. Transfer the tubes to ice, remove the caps, and evaporate the piperidine solution in a Speed-Vac; this procedure takes a couple of hours with the heater on, or it can be left overnight with the heater off. Add 50–100 μl of distilled water, evaporate to dryness again, add another 50–100 μl of water, and once again evaporate to dryness. Add 5–20 μl of sequencing gel loading buffer (80% deionized formamide, 25 mM EDTA, 0.3% bromphenol blue, 0.3% xylene cyanol) and dissolve the DNA residue by pipetting up and down with a micropipettor. Check the recovery by drawing up the liquid into the pipette tip and holding it against a Geiger counter. Incubating the tube for some time at 37° or 55° may aid recovery. When the radioactivity is sufficiently in solution, spin out any suspended particulates and transfer the samples to new tubes. This procedure is helpful because particulates can bind significant proportions of the labeled DNA, leading to misleading estimates of the amount in solution. Load equal amounts of radioactivity for each sample on a sequencing gel of an appropriate acrylamide concentration, and electrophorese as usual. Transfer the gel to Whatman (Clifton, NJ) 3MM or similar paper, dry, and place on X-ray film, with an intensifying screen if desired.

Comments

In a structural probing experiment, the possibility of the probe altering the structure being examined must be taken into account. For hydroxylamine and methoxylamine, the principal cause for concern in this regard is the relatively high ionic strength of the reaction, a consequence of the relatively weak reactivity of the probes and the need to titrate them to achieve most pH conditions. One possible way (not yet validated) to reduce the ionic strength would be to start with the basic form of hydroxylamine, perhaps by passing the hydrochloride over an ion-exchange column or dissolving it in alkaline solution and extracting the free base, and then titrate with acid to achieve the desired pH. Another approach is simply to use a lower concentration of hydroxylamine and a longer reaction time.

Figure 5 shows the effects of concentration on the level of modification at pH 7 as detected by piperidine cleavability, using plasmid pLP324,[25] which contains the sequence $(C-G)_{12}$. Here, 2 M and even 1 M hydroxylamine allowed a clear visualization of the junction even using typical reaction conditions of 15 min at 22° (30° for 1 M hydroxylamine). At pH 6, where the intensity of the reaction is higher (compare lanes 6 and 10), reactivity at a concentration of 1 M is even clearer (not shown). In practice, most conformational transitions examined to date are much more heavily influenced by pH or supercoiling than by ionic strength. Nevertheless, it is advisable to include controls as to whether the ionic strength of

[25] L. J. Peck and J. C. Wang, *Cell (Cambridge, Mass.)* **40,** 129 (1985).

Reagent	DEP				hydroxylamine							
Concentration	sat.	sat.	sat.	sat.	3M	3M	2M	2M		1M	3M		3M
pH	7.0	7.0	7.0	7.0	7.0	7.0	7.0	7.0		7.0	6.0		6.0
pip inc'n , min	30	30	20	30	30	30	15	30		30	15		30
removal of modifying reagent	2X ppt	spin col.	chl+ col.	chl+ col.	2x ppt	spin col.	spin col.	spin col.		spin col.	spin col.		spin col.
	1	2	3	4	5	6	7	8		9	10		11

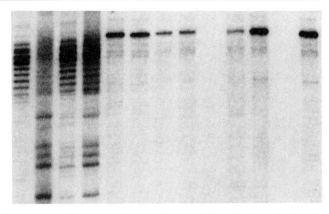

FIG. 5. Effects of various reaction parameters on the reactivity of hydroxylamine at a B–Z junction. The plasmid examined is pLP324,[25] which contains the sequence $(C-G)_{12}$, at a superhelicity of approximately -0.055. This sequence is in the Z conformation, as can be seen by the dark bands in lane 1 representing reaction of the guanine residues with diethyl pyrocarbonate (DEP). The single dark band in lanes 5–11 is due to hydroxylamine reacting with a cytosine at one of the B–Z junctions. The intensity of the reaction is highest for 3 M hydroxylamine at pH 6 (lane 10) and decreases with decreasing concentration (lanes 6–9) and with reaction at pH 7 (lanes 5–9). Equivalent results are seen whether the time of piperidine treatment ("pip inc'n") is 15 or 30 min and whether the reagent is removed by two ethanol precipitations ("2× ppt") or by passage through a P-10 (Bio-Rad, Richmond, CA) spin column ("spin col."). For completeness, the effects of cleavage time and removal procedure for the reaction with DEP are shown in lanes 1–4. For this reagent, precipitation (lane 1) is preferable to either a spin column (lane 2) or extraction with chloroform followed by a spin column ("chl + col.," lanes 3–4).

the hydroxylamine reaction appears to influence the conformational state of the DNA. This can be done by testing whether the reactivity pattern for another structure-sensitive probe, such as DEP or OsO_4, varies under high and low salt conditions, as shown in Fig. 3, lanes 4 and 5.[1]

An alternative method of removing the reagent is simply to pass the reaction mixture through a spin column.[26] The quality of the results is

[26] J. Sambrook, E. F. Fritsch, and T. Maniatis, "Molecular Cloning: A Laboratory Manual," 2nd Ed. Cold Spring Harbor Laboratory, Cold Spring Harbor, New York, 1989.

essentially equivalent, as can be seen by comparing lanes 5 and 6 of Fig. 5. For small numbers of samples, the spin column approach may be faster.

Figure 5 also demonstrates that cleavage of the backbone is complete after a 15-min treatment with 1 M piperidine at 90° (compare lanes 7 and 8). The standard 30-min cleavage used in the Maxam–Gilbert procedure is satisfactory but sometimes yields a slightly higher level of nonspecific cleavage.

A rapid alternative method for visualizing the sites of reaction with hydroxylamine, which eliminates the need for end-labeling the DNA, is the polymerase block approach. The unlabeled DNA is annealed to a labeled primer that is extended by a DNA polymerase such as the Klenow enzyme,[27] or else it is transcribed using a phage polymerase such as Sp6 or T7 in the presence of an α-labeled nucleoside triphosphate.[23] The products of the polymerase are electrophoresed on a sequencing gel in the presence of appropriate markers. This approach is described in [15] in this volume.[28]

Acknowledgments

This work was supported in part by National Institutes of Health Grant 1R29GM41423.

---

[27] S. Sasse-Dwight and J. D. Gralla, this series, Vol. 208, p. 146.
[28] H. Htun and B. H. Johnston, this volume [15].

[11] Using Hydroxyl Radical to Probe DNA Structure

By MARY ANN PRICE and THOMAS D. TULLIUS

Introduction

It has become increasingly clear that DNA is not monotonously B form in structure, merely acting as a repository of the sequence information for structurally more interesting RNA and protein molecules. We now know of DNA molecules with gross alterations from the B structure that are biologically important, such as the four-stranded Holliday intermediates in recombination and transposition and the G-quartets in telomeres. DNA also adopts conformations that deviate more subtly from the B form. An example is the conformation associated with intrinsically bent DNA. Small but potentially important sequence-dependent differences from the structural parameters of the ideal B form, expressed as variations in sugar pucker, helical twist, base pair roll, or minor groove width, are found in high-resolution structures of DNA determined by X-ray crystallography and NMR spectroscopy.

Structure determination at atomic resolution by crystallography or NMR requires large amounts of time and material, and to date it has been successful only for relatively small oligonucleotides. It would be a formidable, if not impossible, task to solve every DNA structure by these methods. An experimental approach involving DNA cleavage in solution represents a useful alternative to NMR and crystallography for the determination of DNA structure. DNA molecules of any length may be studied, there are few constraints on conditions, and experiments that give local structural information are quickly and easily performed.

Our laboratory has introduced the hydroxyl radical as a reagent for the study of the structure of DNA and DNA–protein complexes. The hydroxyl radical has several important features which distinguish it from other commonly used DNA cleavage agents. The hydroxyl radical is small compared to the enzyme deoxyribonuclease I, or even the chemical nucleases bis-1,10-phenanthrolinecopper(I) and methidiumpropylEDTA · Fe(II). The high reactivity of the hydroxyl radical leads to extremely low specificity in cleavage of DNA. The hydroxyl radical has no preference for cleavage of a particular base, nor does it exhibit sequence selectivity in cleavage. Because of these attributes, the hydroxyl radical has been used extensively as a reagent for high-resolution footprinting of protein–DNA complexes.

How, though, can a reagent that cuts DNA so nonspecifically be useful in the study of DNA structure? Most reagents that are used to determine aspects of DNA structure in solution work by reacting preferentially with DNA that is in an unusual conformation. Osmium tetroxide, for example, modifies unpaired (and perhaps unstacked) thymidines. Although the hydroxyl radical cleaves every nucleotide of almost any DNA molecule, we have found subtle, but reproducible, variations in the cleavage pattern that likely reflect structural details of the DNA. These variations in cleavage rate are small for mixed-sequence DNA in the B form but can be much larger for DNA of more unusual structure, such as bent DNA or DNA in a Holliday junction.

In this chapter we describe the application of the hydroxyl radical as a reagent for elucidating structural details of DNA. The use of the hydroxyl radical for footprinting complexes of protein with DNA has been the subject of two previous chapters in this series.[1,2] We have tried to make the present contribution complete in itself, but the reader may find these other chapters useful as well.

Chemistry of Reaction

We generate the hydroxyl radical by the Fenton reaction of hydrogen peroxide with the complex of iron(II) and EDTA:

$$[Fe(EDTA)]^{2-} + H_2O_2 \rightarrow [Fe(EDTA)]^- + OH^- + \cdot OH \tag{1}$$

An electron from Fe(II) splits hydrogen peroxide into the hydroxide ion and the neutral hydroxyl radical. Ascorbate is added to the reaction to reduce Fe(III) to Fe(II), which can then react with more peroxide. We use the EDTA complex of Fe(II) so that the hydroxyl radical is generated by a negatively charged metal complex that does not bind to DNA. The hydroxyl radical can also be generated by γ-radiolysis of water, yielding a DNA cleavage pattern identical to that produced by the Fenton reaction.[3]

The hydroxyl radical cleaves DNA by abstracting a hydrogen atom from a deoxyribose residue in the DNA backbone. Subsequently the sugar breaks down, leaving a single-nucleoside gap with predominantly 5'- and 3'-phosphate ends. A small amount of 3'-phosphoglycolate is formed as well. This product migrates faster than the corresponding 3'-phosphate and can be seen when high percentage polyacrylamide gels are used to separate the products of hydroxyl radical cleavage of 5'-radiolabeled DNA (see, e.g., Fig. 7).

Single-stranded gaps caused by hydroxyl radical treatment are detected by running the reaction products on a denaturing polyacrylamide gel. The products of the Maxam–Gilbert G or G + A sequencing reaction,[4] which also have phosphate ends, are run in parallel as size markers for sequence assignment. The concentrations of the cutting reagents are adjusted so that very few DNA molecules are gapped more than once. The hydroxyl radical cleaves DNA much less specifically than most DNA cleavage reagents, so the result of an experiment is a ladder of bands on a sequencing gel that corresponds to fairly even cutting at every position in the backbone.

[1] T. D. Tullius, B. A. Dombroski, M. E. A. Churchill, and L. Kam, this series, Vol. 155, p. 537.

[2] W. J. Dixon, J. J. Hayes, J. R. Levin, M. F. Weidner, B. A. Dombroski, and T. D. Tullius, this series, Vol. 208, p. 380.

[3] J. Hayes, L. Kam, and T. D. Tullius, this series, Vol. 186, p. 545.

[4] A. M. Maxam and W. Gilbert, this series, Vol. 65, p. 499.

DNA Substrate

Several sources of DNA are suitable for the experiment. Naturally occurring DNA sequences are often most easily studied when cloned into plasmids, since plasmid DNA is easy to prepare and to purify in large amounts and is convenient to manipulate. Hydroxyl radical cleavage experiments on the bent kinetoplast DNA of a trypanosome parasite were done on a restriction fragment of a plasmid in which was cloned kinetoplast minicircle DNA.[5] By cloning a DNA molecule into the polylinker of a vector, the sequence of interest can be excised by enzymes that singly cleave the parent molecule, and the resulting fragment can be uniquely end-labeled on either strand at either end by standard procedures.

If the goal of an experiment is to understand a particular structural motif, synthetic DNA molecules are useful for systematically testing the properties of many sequences. For example, a large number of sequences have been surveyed for their bending properties. To study these molecules by hydroxyl radical cleavage, we have found it useful to clone synthetic oligonucleotides.[6,7] Several tandem repeats of a given sequence are introduced into a polylinker site in a plasmid. Having more than one example of the sequence in a single restriction fragment gives a clear indication of the reproducibility of interesting features in the cleavage pattern. A further advantage is that by introducing several different test oligonucleotides into the same sequence context, the amount of cleavage of flanking DNA can be used as an internal control for the level of cutting, so that cleavage patterns of the different sequences can be compared quantitatively. Another consideration in hydroxyl radical cleavage experiments is the distance of the sequence of interest from the radioactive end-label. The ideal distance is 25–50 base pairs (bp). In such a situation DNA fragments of interest will not dissociate from the duplex during ethanol precipitation, and good resolution is achieved on denaturing gels. A polylinker offers flexibility in the choice of restriction site so that a fragment of the desired size can be obtained.

Certain unusual DNA structures cannot be cloned, for example, an immobile four-way junction. Here it is necessary to work with synthetic oligonucleotides directly.[8–10] We have also used synthetic oligonucleotides

[5] A. M. Burkhoff and T. D. Tullius, *Cell (Cambridge, Mass.)* **48,** 935 (1987).

[6] A. M. Burkhoff and T. D. Tullius, *Nature (London)* **331,** 455 (1988).

[7] M. A. Price and T. D. Tullius, manuscript in preparation (1992).

[8] M. E. A. Churchill, T. D. Tullius, N. R. Kallenbach, and N. C. Seeman, *Proc. Natl. Acad. Sci. U.S.A.* **85,** 4653 (1988).

[9] J.-H. Chen, M. E. A. Churchill, T. D. Tullius, N. R. Kallenbach, and N. C. Seeman, *Biochemistry* **27,** 6032 (1988).

[10] A. Kimball, Q. Guo, M. Lu, R. P. Cunningham, N. R. Kallenbach, N. C. Seeman, and T. D. Tullius, *J. Biol. Chem.* **265,** 6544 (1990).

in high-resolution studies of hydroxyl radical cleavage of duplexes that have been characterized by X-ray crystallography.

We next describe procedures common to the study of both restriction fragments and synthetic oligonucleotides. After, we provide experimental protocols specific to each type of substrate, with illustrative examples for each.

General Experimental Procedures

The DNA sample must be free of background nicking in order to analyze the results of a hydroxyl radical cleavage experiment. Care must be taken at all steps in the procedure to ensure that DNA is not exposed to metal ions, nucleases, or anything else that might nick the DNA. We use doubly deionized water in the preparation of all solutions, and also sterilize solutions by autoclaving or passage through 0.22-μm filters.

Preparation of Cutting Reagents

An aqueous solution of iron(II) is made by dissolution of ferrous ammonium sulfate (Aldrich, Milwaukee, WI) to give a concentration of 2–40 mM. Aliquots of the solution can be stored in a $-20°$ freezer, but some oxidation occurs. To prevent contamination of other experiments by iron, glassware that has been used to make iron solutions is used solely for that purpose. After washing, this glassware is rinsed with dilute HCl to remove residual metal. Dilutions of aqueous EDTA (Aldrich, gold label) are made from a stock solution of 0.2–0.5 M (pH 8.0). Iron(II) and EDTA solutions are mixed at a molar ratio of 1:2. Use of excess EDTA ensures that all the iron is chelated, since free ferrous ion could bind to DNA and cause undesired oxidative damage. The iron(II) and EDTA solutions can be premixed and frozen in aliquots, or mixed just before doing the experiment. A solution of sodium ascorbate (Sigma, St. Louis, MO) is made up at a concentration of 10–100 mM. Aliquots of this solution can be frozen. Hydrogen peroxide is diluted to the desired concentration from the 30% (w/v) solution (Aldrich) that is purchased from the supplier.

To achieve the proper amount of cleavage, namely, no more than one cut per DNA molecule, appropriate concentrations of reagents must be determined for each set of conditions and each length of DNA used. In 10 mM Tris-HCl, without any other hydroxyl radical scavengers present, final concentrations of 50 μM Fe(II) : 100 μM EDTA, 1 mM sodium ascorbate, and 0.03% hydrogen peroxide are good starting points. Because ascorbate is consumed in the reaction, if the initial concentration of

$[Fe(EDTA)]^{2-}$ is increased, the ascorbate concentration must also be raised to obtain more production of the hydroxyl radical. We use at least a 10-fold molar excess of ascorbate relative to $[Fe(EDTA)]^{2-}$.

The reaction can be performed in the presence of almost any buffer or salt. Organic buffers and solvents which contain carbon–hydrogen bonds are efficient scavengers of the hydroxyl radical. This point should be kept in mind when choosing reagent concentrations. Inorganic salts, however, do not scavenge the hydroxyl radical, and therefore any concentration can be used without affecting the amount of cutting.[1] The concentrations of salts in our experiments are usually 5–50 mM. The ability to vary salt concentration without affecting the cutting reaction is advantageous when studying DNA structure, since conformation often depends on the salt concentration.[11,12]

To test for the level of cutting under the desired experimental conditions, an untreated sample containing the same amount of radioactivity as a cleaved sample should be run on the sequencing gel as a control. A comparison of the intensities of the full-length bands after a short exposure to X-ray film is used to determine the fraction of full-length molecules remaining in a treated sample. Using the Poisson distribution[13] it can be shown that, if more than 70% of the molecules are uncut, then fewer than 5% contain more than one cut.

Cleavage Experiments Using DNA Restriction Fragments

Preparation and Manipulation of Plasmids

Plasmids are amplified and isolated by standard methods, usually the cleared-lysate method.[14] Plasmids are further purified by Sephacryl S-500 (gel filtration) and RPC-5 (ion-exchange) chromatography on a fast protein liquid chromatography (FPLC) system (Pharmacia, Piscataway, NJ). We have also purified supercoiled plasmid DNA using cesium chloride gradients, taking great care not to expose the DNA sample to light while it still contains ethidium bromide so as to avoid nicking. Small-scale preparations of plasmid DNA[14] have also been used, but efficient 5′-radiolabeling of such samples is difficult owing to RNA contamination.

DNA restriction fragments (made from 2–10 μg of plasmid DNA) are

---

[11] S. Diekmann, *Nucleic Acids Res.* **15,** 247 (1987).
[12] D. R. Duckett, A. I. H. Murchie, and D. M. J. Lilley, *EMBO J.* **9,** 583 (1990).
[13] M. Brenowitz, D. F. Senear, M. A. Shea, and G. K. Ackers, this series, Vol. 130, p. 132.
[14] T. Maniatis, E. F. Fritsch, and J. Sambrook, "Molecular Cloning: A Laboratory Manual." Cold Spring Harbor Laboratory, Cold Spring Harbor, New York, 1982.

labeled at the 3' or 5' end with ^{32}P by standard protocols, purified on a native polyacrylamide gel, and isolated from the gel slice by the crush and soak procedure.[14] Radiolabeled DNA samples are stored either in TE buffer (10 mM Tris-HCl, 1 mM EDTA, pH 8.0) or in 10 mM Tris-HCl (pH 8.0), at 1000–2000 counts/min (cpm) per microliter (as measured by a Geiger counter held next to the pipette tip containing 1 μl of the DNA solution). Storage of higher concentrations of radioactive DNA in buffer, or storage in water, leads to autodegradation of the DNA, giving a background that interferes with interpretation of the hydroxyl radical cleavage pattern.

Cleavage of DNA Restriction Fragments with Hydroxyl Radical

Cleavage experiments are usually performed on 10,000–50,000 cpm of a radiolabeled restriction fragment. The volume of the solution before addition of the cutting reagents is usually 70 μl, although any volume may be used. Cutting is initiated by adding 10 μl of each of the three reagents [iron(II) EDTA, hydrogen peroxide, and sodium ascorbate], each at 10× concentration, to the 70-μl DNA sample, as separate drops placed on the side of the reaction tube. The three drops are simultaneously added to the DNA solution, and the tube is tapped to mix the reagents with the DNA. The reaction is allowed to proceed for 1–2 min, after which 100 μl of a stop solution (135 mM thiourea, 135 mM EDTA, and 0.6 M sodium acetate) is added. The DNA is precipitated by adding 0.5 ml ethanol. The DNA is immediately pelleted by centrifugation, resuspended in 0.2 ml of 0.3 M sodium acetate, precipitated with ethanol, rinsed with 70% (v/v) ethanol, and lyophilized.

Gel Electrophoresis

Dried DNA samples are resuspended in 2–4 μl of formamide loading buffer, denatured by heating at 90° for 3–5 min, put on ice, and loaded on a denaturing polyacrylamide sequencing gel with dimensions of 0.35 mm × 30 cm × 40 cm. All the denaturing gels for which we show autoradiographs in this chapter had a ratio of acrylamide to bisacrylamide of 19 : 1 and contained 8.3 M urea and 1× TBE (100 mM Tris, 100 mM borate, and 2.8 mM EDTA). For the best resolution denaturing gels should be made from fresh reagents. Gels are electrophoresed in 1 × TBE running buffer. Gels must be prerun and run at a fairly high temperature (45–50°) to keep the DNA denatured.

When electrophoresis is complete, the gel is dried onto Whatman (Clifton, NJ) 3MM filter paper. Gels that are lower than 10% polyacrylamide can be transferred to filter paper, covered with Saran wrap plastic wrap, and dried. For gels that are greater than 10% polyacrylamide, it is easiest to transfer the gel to Saran wrap, place it on filter paper, and then dry the gel.[15] Dried gels are exposed to X-ray film (Kodak X-Omat, Rochester, NY) at room temperature for the best resolution, or at $-70°$ with an intensifying screen for fast results. A good exposure of a gel that contains 10,000 cpm per lane can be obtained in about 5 days at room temperature. Autoradiographs are analyzed by densitometry. We use a Joyce-Loebl Chromoscan 3 densitometer or a Molecular Dynamics (Sunnyvale, CA) Model 300E densitometer.

Results: Bent DNA

Our work on the structural basis of DNA bending offers a good example of the use of hydroxyl radical cleavage to analyze the structure of a DNA sequence cloned in a plasmid. Early evidence that DNA could be stably curved came from the observation of anomalously slow gel mobility of a restriction fragment of trypanosome kinetoplast minicircle DNA (kDNA).[16–18] Bent kinetoplast DNA has subsequently been studied by a variety of methods that are sensitive to the global curvature of the DNA, including electric birefringence[19] and electron microscopy.[20] Sequence determination showed that bent kDNA contained short runs of adenines phased at approximately 10-bp intervals.[16,18]

A restriction fragment containing bent kDNA was subjected to hydroxyl radical cleavage with the aim of revealing details of the structure of the adenine tracts at the nucleotide level.[5] We observed a remarkably sinusoidal cleavage pattern that was phased precisely with the adenine tracts. Each minimum in the pattern occurred at the 3' adenine in an adenine tract, whereas the maxima were at or near the 5' adenines of the tracts. A monotonic decrease in cleavage rate from a 5' to a 3' adenine was seen in each adenine tract.

The study of synthetic DNA oligonucleotides designed to test models for DNA bending has contributed greatly to our understanding of DNA curvature. The hydroxyl radical cleavage pattern of a cloned synthetic

[15] G. E. Shafer, M. A. Price, and T. D. Tullius, *Electrophoresis* **10**, 397 (1989).

[16] J. C. Marini, S. D. Levene, D. M. Crothers, and P. T. Englund, *Proc. Natl. Acad. Sci. U.S.A.* **79**, 7664 (1982).

[17] H.-M. Wu and D. M. Crothers, *Nature (London)* **308**, 509 (1984).

[18] S. Diekmann and J. C. Wang, *J. Mol. Biol.* **186**, 1 (1985).

[19] P. J. Hagerman, *Proc. Natl. Acad. Sci. U.S.A.* **81**, 4632 (1984).

[20] J. Griffith, M. Bleyman, C. A. Rauch, P. A. Kitchin, and P. T. Englund, *Cell (Cambridge, Mass.)* **46**, 717 (1986).

A

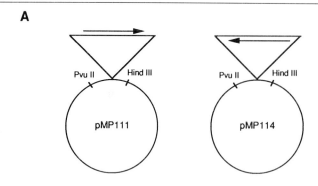

5′-AATTGGAGGGCAAAAATGGGCAAAAATGGGCAAAAATGGGCAAAAATGGTCC
     CCTCCCGTTTTTACCCGTTTTTACCCGTTTTTACCCGTTTTTACCAGGAATT-5′

FIG. 1. Hydroxyl radical cleavage pattern of $(A_5N_5)_4$. (A) Plasmid constructs. The DNA duplex shown is cloned into the *Eco*RI site of plasmid pUC18. Both orientations of the insert were selected. 3′-Labeling of the *Hind*III site of pMP111 radiolabels the adenine-rich strand; of pMP114, the thymine-rich strand. (B) Autoradiograph of hydroxyl radical cleavage products of 3′-labeled *Hind*III–*Pvu*II restriction fragments of pMP111 (lanes 1, 3, and 5) and pMP114 (lanes 2, 4, and 6). In lanes 1 and 2 are run untreated controls; in lanes 3 and 4, the products of Maxam and Gilbert G reactions; and in lanes 5 and 6, hydroxyl radical cleavage products. (C) Densitometric traces of lane 5 (upper scan) and lane 6 (lower scan). The film was scanned using a Molecular Dynamics Model 300E densitometer. Resulting line graphs were plotted using Microsoft Excel. Both scans are shown from 5′ to 3′ left to right and are marked with the sequence of the DNA. (D) 3′-Offset between minima in the cleavage pattern. The scan of the adenine tract is shown 5′ to 3′ left to right, whereas the thymine tract is shown 3′ to 5′, so that each base lines up vertically with its complement. Dots above the peaks indicate minima in the two cutting patterns.

DNA molecule containing phased adenine tracts is shown in Fig. 1. This fragment contains four repeats of d(GGGCA$_5$T) cloned into the *Eco*RI site of pUC18 (Fig. 1A).[7] After digestion of the plasmid with *Hind*III and *Pvu*II and radiolabeling at the 3′ end with ^{32}P, 30,000 cpm of the fragment [in 10 mM Tris-HCl (pH 8.0), 10 mM NaCl, and 5 mM magnesium acetate] was treated with 50 μM Fe(II) : 100 μM EDTA, 0.03% H_2O_2, and 1 mM sodium ascorbate (final concentrations). The reaction products were electrophoresed on an 8% denaturing polyacrylamide gel. From the autoradiograph it is obvious that the cutting pattern of a short adenine tract (Fig. 1B, lane 5), or of its complementary thymine tract (Fig. 1B, lane 6), differs from that of mixed sequence DNA. The densitometric scans show an almost sinusoidal cleavage pattern for the repeating runs of adenines (Fig. 1C,

B <u>1 2</u> <u>3 4</u> <u>5 6</u>

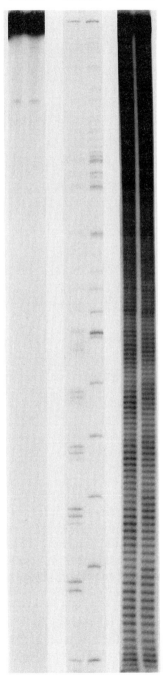

C

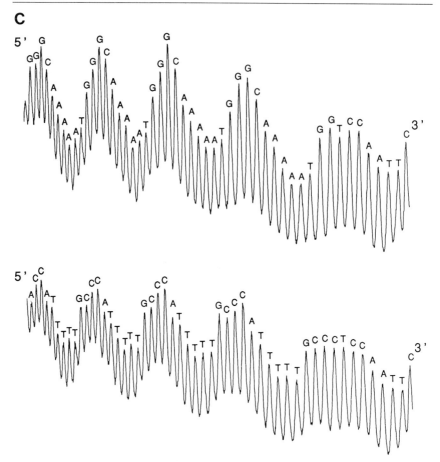

FIG. 1. (*continued*)

top scan), with the cutting decreasing from the 5' end to the 3' end of the adenine tract, very similar to the pattern seen with bent kinetoplast DNA.[5] The complementary strand also exhibits a modulated cleavage pattern (Fig. 1C, bottom scan), but the minima in the cutting patterns on the two strands (marked with dots) are offset in the 3' direction by a few nucleotides (Fig. 1D).[5] This offset suggests that the cleavage pattern reflects a property of the minor groove of adenine tract DNA. The deoxyriboses on opposite strands that are nearest each other across the minor groove are not members of a base pair, but instead are shifted relative to each other two or three nucleotides in the 3' direction. The decrease in cleavage by the hydroxyl radical most probably reflects steric occlusion arising from

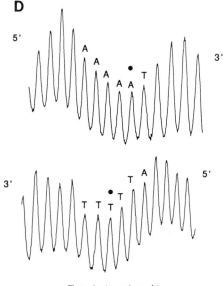

FIG. 1. (*continued*))

narrowing of the minor groove. Indeed, NMR[21] and X-ray fiber diffraction[22] experiments on poly(dA) · poly(dT), and several crystal structures of duplexes containing adenine tracts,[23–25] have revealed that adenine tracts have notably narrow minor grooves.

Subsequent cleavage experiments on adenine tract sequences that are not bent have shown that the hydroxyl radical cleavage pattern can distinguish bent from unbent DNA. Adenine tracts in certain nonbent molecules are cleaved by the hydroxyl radical at levels comparable to flanking, mixed sequence DNA. Thus, the hydroxyl radical senses structure rather than base composition.[6,7]

Cleavage Experiments Using Oligonucleotides

Preparation, Purification and Radioactive Labeling of Oligonucleotides

Oligonucleotides are synthesized on an automated DNA synthesizer (Biosearch 8600, Milliger/Biosearch, Burlington, MA) using phosphor-

[21] R. W. Behling and D. R. Kearns, *Biochemistry* **25**, 3335 (1986).
[22] D. G. Alexeev, A. A. Lipanov, and I. Y. Skuratovskii, *Nature (London)* **325**, 821 (1987).
[23] H. C. M. Nelson, J. T. Finch, B. F. Luisi, and A. Klug, *Nature (London)* **330**, 221 (1987).
[24] M. Coll, C. A. Frederick, A. H.-J. Wang, and A. Rich, *Proc. Natl. Acad. Sci. U.S.A.* **84**, 8385 (1987).

amidite chemistry. Purification is achieved either by chromatography on a Mono Q column using an FPLC system (Pharmacia), by use of an oligonucleotide purification cartridge (Applied Biosystems, Foster City, CA), or by electrophoresis on a preparative denaturing polyacrylamide gel (3 mm × 16 cm × 19.5 cm) (10–20% acrylamide, depending on the length of the fragment). The band containing the oligonucleotide is detected by brief UV shadowing of the gel and then excised. Precipitation of the purified DNA is carried out by addition of 5 volumes of ethanol to a solution of the DNA either in 0.3 M sodium acetate and 10 mM magnesium acetate or in 0.5 M sodium acetate. The DNA pellet is rinsed with 85% ethanol. An ethanol slurry of an oligonucleotide is usually stored at −70° for at least 1 hr before centrifugation to ensure efficient recovery of the DNA.

The oligonucleotide (300–500 pmol) is 3′- or 5′-radiolabeled by standard methods, albeit in a small volume (10 μl total). For 3′-radiolabeling using the Klenow fragment of DNA polymerase I, the DNA must be double-stranded and must contain a 5′ overhang, so the oligonucleotide must be annealed to an appropriate complementary strand before labeling. The labeling reaction is stopped by addition of 10 μl of 10 M urea, and the DNA is loaded directly onto a denaturing polyacrylamide gel (0.4 mm × 30 cm × 36 cm) for separation from unincorporated label and smaller oligonucleotides. To have a single population of labeled DNA, it is important to achieve good separation of the labeled full-length oligonucleotide, of length n, from any contaminating ($n - 1$)-mer. After elution from the gel slice, labeled DNA is stored either as an ethanol slurry (0.5 M sodium acetate, 5 volumes of ethanol) or at a concentration of 1 μM in TE buffer.

Before cleaving an oligonucleotide with the hydroxyl radical, it must be annealed to its complementary strand(s) to form the molecule of interest. Annealing is carried out in the buffers and salts to be used in the cleavage reaction. Because high salt concentration favors duplex over single strands, salt concentration is an important variable in achieving complete annealing. We use at least 30 mM NaCl or 2 mM magnesium acetate. The labeled strand is mixed with cold complementary strand(s), heated for 2–10 min at 70–90°, and then cooled slowly. We often try several ratios of cold to hot strand(s), and we always assay for duplex or junction formation by electrophoresis on a native polyacrylamide gel. Figures 2 and 6 show the results of typical annealing assays. We have found that a

---

[25] A. D. DiGabriele, M. R. Sanderson, and T. A. Steitz, *Proc. Natl. Acad. Sci. U.S.A.* **86,** 1816 (1989).

50-fold excess of cold complementary strand over labeled strand reliably leads to complete duplex or junction formation.

Cleavage of Oligonucleotides with Hydroxyl Radical

To 7 μl of annealed radiolabeled DNA, 1 μl of each cutting reagent is added. Because of the small volumes of the reagents, the droplets must be pushed into the DNA solution with the pipette tip after the last reagent is added. Reactions are allowed to proceed for 1–2 min and are stopped by addition of 1 μl of 0.05–0.5 M thiourea. It is important to establish that the reaction is completely stopped, so several thiourea concentrations should be tested. The oligonucleotide is dried by evaporation under reduced pressure (using a Speed-Vac concentrator) directly after the cutting reaction, because short DNA fragments are incompletely precipitated by ethanol.

As with a restriction fragment, the dried oligonucleotide is resuspended in 2–4 μl of formamide loading buffer, heated at 90° for 3–5 min, put on ice, and loaded on a denaturing polyacrylamide gel. Because the experiment requires the resolution of short DNA fragments on the gel, the products of hydroxyl radical cleavage of an oligonucleotide are run on a high percentage polyacrylamide gel (up to 25%).

Because elevated salt concentration favors annealing, a sample of an oligonucleotide often is high in salt. When such a sample is run on a denaturing polyacrylamide gel, the salt front can distort the shapes of bands containing DNA smaller than a 4-mer. One way to avoid this problem is to anneal the strands in as little salt as possible while still ensuring complete duplex or junction formation. Alternatively, if only a fraction of a sample is run in a lane, the concentration of salt is diluted and the salt front runs with shorter DNA fragments. The mobility of the salt front relative to DNA increases with the percentage of polyacrylamide in the gel. Gels up to 25% polyacrylamide can be used to make the salt front run with smaller DNA fragments.

When electrophoresis is complete, the gel is transferred from the glass plate to Saran wrap, placed on filter paper, and dried. Autoradiographs can be analyzed by either one- or two-dimensional densitometry.

Results

Adenine Tract DNA. The structures of several DNA duplexes have been determined by single crystal X-ray diffraction.[23-26] We wished to compare the hydroxyl radical cleavage pattern of one such duplex to

[26] K. Yanagi, G. G. Prive, and R. E. Dickerson, *J. Mol. Biol.* **217**, 201 (1991).

structural parameters obtained from the crystal structure to gain insight into the source of variations in the cleavage frequency of the hydroxyl radical. Our goal was to achieve high-resolution hydroxyl radical cleavage patterns so that the integrated intensity of each band could be determined accurately for comparison to the structural parameters. We performed a cleavage experiment on a DNA duplex containing an $A_6 \cdot T_6$ sequence that had been characterized crystallographically.[23] We synthesized two strands, each of which contained the twelve bases in the crystal (shown in boldface below) and four additional bases at the 5' end of the strand so that we could uniquely 3'-radiolabel either strand:

5'-GATC**CGCAAAAAAGCG**

GCGTTTTTTCGCTCGA-5'

Each strand is radiolabeled in turn at the 3' end (the A-rich strand with [α-^{32}P]dATP, and the T-rich strand with [α-^{32}P]dGTP), annealed to its unlabeled complement, and cleaved with the hydroxyl radical. Labeled DNA (70 pmol) is mixed with 3.5 nmol of unlabeled complementary strand in 30 mM NaCl, 10 mM Tris-HCl (pH 8.0) in a volume of 35 μl. The DNA is annealed by heating at 70° for 10 min and then cooling slowly by allowing the heating block to return to room temperature.

To test for duplex formation, control samples are run on a native gel. Samples are prepared by adding 3 μl of 6× native dye[14] to either 7 μl of annealed DNA or 7 μl (10 pmol) of labeled DNA with no added unlabeled strand. The native gel is 20% acrylamide (19:1, acrylamide–bisacrylamide) and 1× TBE. When a 50-fold excess of unlabeled strand is used in the annealing reaction, all of the labeled strand is contained in a single slowly migrating band on a native gel (Fig. 2, lanes marked DS). The 3'-labeled sample, to which no additional unlabeled strand was added, unexpectedly comigrates mostly with the duplex (Fig. 2, lanes marked SS). We suspect that this is due to the presence of the unlabeled strand from the radiolabeling reaction that was not separated from the labeled strand (which is the same length). In contrast, 5'-labeled DNA, which is never annealed to the complementary strand, runs almost entirely as a single rapidly migrating band (data not shown).

Hydroxyl radical cleavage reactions are carried out on samples containing 7 μl of annealed radiolabeled DNA, with the following final concentrations of cutting reagents: 50 μM Fe(II):100 μM EDTA, 1 mM ascorbate, and 0.03% H_2O_2. The reaction is stopped by addition of 1 μl of 50 mM thiourea. Reactions are carried out at both 0° and room temperature. Reactions at 0° are allowed to proceed for 2 min before stopping, and room temperature reactions for 1 min. The products of hydroxyl radical cleavage

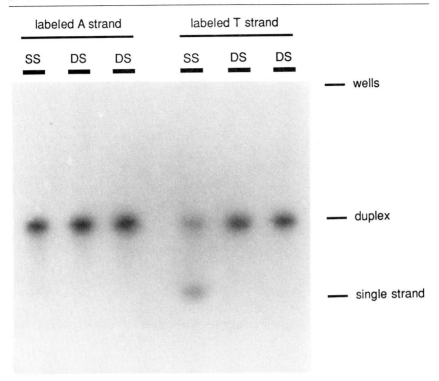

FIG. 2. Assay for duplex formation for an oligonucleotide containing the sequence d(CGCAAAAAAGCG). Lanes marked DS contain annealed DNA samples to which have been added a 50-fold excess of cold complementary strand to radiolabeled strand; lanes marked SS contain radiolabeled DNA with no added cold complementary strand. The two DS samples for each radiolabeled fragment were from different annealings that were run on separate sequencing gels.

are run on a 25% denaturing polyacrylamide gel. A typical autoradiograph is shown in Fig. 3.

Autoradiographs are analyzed by two-dimensional densitometry. The results we show here were obtained by scanning the autoradiograph with an Eikonix Model 1412 camera interfaced to a Hewlett-Packard Vectra computer. This gel scanning system was developed in the laboratory of Professor Gary Ackers (Dept. of Biochemistry, Washington University School of Medicine, St. Louis, MO). The resulting data file contains an array of optical density values for each pixel of the scan. We analyze this file using an image analysis program, IMAGE, written by Dr. Wayne Rasband (NIH), running on a Macintosh II computer. (We now scan and analyze autoradiographs with a Molecular Dynamics Model 300E

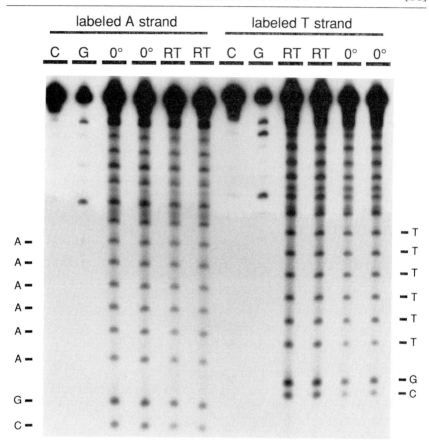

FIG. 3. Hydroxyl radical cleavage patterns of an oligonucleotide containing the sequence d(CGCAAAAAAGCG). Lanes marked C contain untreated controls; lanes marked G contain the products of a Maxam and Gilbert G reaction; lanes labeled 0° contain hydroxyl radical reaction products obtained at 0°; lanes marked RT contain the products of reactions carried out at room temperature. The sequence of the DNA in each set of lanes is indicated to the sides of the autoradiograph.

densitometer and its associated software). Using the program IMAGE, the optical density of a band is measured by drawing a rectangle around the band. In choosing where to draw a rectangle, it is easiest to differentiate between the film background and the various levels of intensity in the band when the image is displayed in pseudocolor. The program subtracts the background by picking the optical density value in the rectangle that occurs most often (the mode), and subtracting this value from the optical density value of each pixel in the rectangle. This makes it necessary to extend the rectangle into the area between lanes, so that sufficient background is included; therefore, only every other lane on the gel can be loaded with sample. The intensity of a band that is reported by the program is thus the sum of all optical density values in the rectangle after subtracting the background.

We normalize the intensity for each band in each lane by calculating the ratio of the intensity of a band to the total intensity of the lane. Data from 14 lanes of each strand at each temperature were averaged. Plots of the normalized cleavage intensities for both strands, obtained at room temperature, are shown in Fig. 4. As with the one-dimensional scan of the restriction fragment containing repeating adenine tracts (Fig. 1C), two-dimensional analysis of the cleavage pattern of the oligonucleotide shows a decrease in the hydroxyl radical cleavage frequency from 5′ to 3′ along the adenine tract. The minimum in the cleavage pattern of the thymine-rich strand is located more to the 3′ direction in the oligonucleotide compared to the restriction fragment (compare the plot in Fig. 4 with the bottom scan in Fig. 1C). This discrepancy may be due to differences in the sequences flanking the adenine tracts, or to the qualitative nature of the analysis of the cleavage pattern of the restriction fragment.

DNA Junctions. A more unusual DNA structure is the four-stranded Holliday junction, an intermediate in recombination reactions.[27] When a Holliday junction contains homologous DNA, branch migration is possible. This is a necessary process in the cell, but it makes the four-stranded DNA molecule too unstable to permit structural study by standard methods. This problem was overcome by the construction of immobile oligonucleotide models for Holliday junctions, which have fixed branch points.[28,29] The DNA junction J1, one of the first four-way junctions prepared and the most extensively studied, consists of four 16-mer oligonucleotides that anneal to form a branched DNA structure with four 8-bp-long duplex arms. This molecule has been characterized by physical methods including calorimetry, circular dichroism spectroscopy, and ¹H NMR spectroscopy.[30–33] These studies showed that the four strands of J1 anneal into a fully base-paired structure. However, an important question remained, that of the disposition of the four arms relative to one another in solution.

[27] R. Holliday, *Genet. Res.* **5**, 282 (1964).
[28] N. C. Seeman, *J. Theor. Biol.* **99**, 237 (1982).
[29] N. C. Seeman and N. R. Kallenbach, *Biophys. J.* **44**, 201 (1983).
[30] N. R. Kallenbach, R.-I. Ma, and N. C. Seeman, *Nature (London)* **305**, 829 (1983).
[31] N. C. Seeman, M. E. Maestre, R.-I. Ma, and N. R. Kallenbach, *Prog. Clin. Biol. Res.* **172A**, 99 (1985).
[32] D. E. Wemmer, A. J. Wand, N. C. Seeman, and N. R. Kallenbach, *Biochemistry* **24**, 5745 (1985).
[33] S.-M. Chen, F. Heffron, W. Leupin, and W. J. Chazin, *Biochemistry* **30**, 766 (1991).

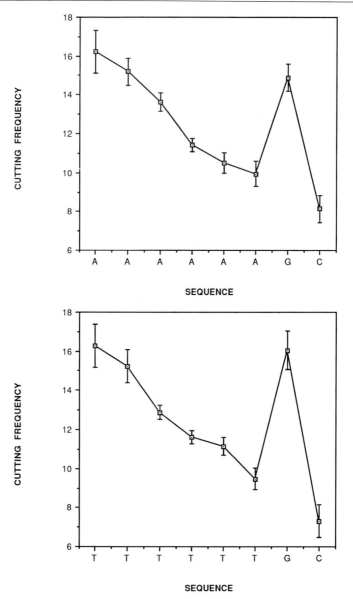

FIG. 4. Quantitative analysis of hydroxyl radical cleavage of the sequence d(CGCAAAAAAGCG) at room temperature. The cutting frequency, represented as the percentage of the intensity of a particular band relative to the sum of the intensities of the eight bands measured, is plotted versus the nucleotide sequence of each strand. Each plot is shown 5′ to 3′ left to right.

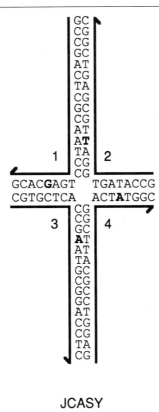

JCASY

FIG. 5. Sequence of the four-way DNA junction JCASY. The strands are numbered 1–4; the arrowhead on each strand indicates the 3′ end of the strand. For convenience in comparing results with previous work, the nucleotides on each strand are numbered so that the junction is flanked by positions 8 and 9 on each strand. Numbering begins at the 5′ end of each strand; the nucleotide at position 12 on each strand is shown in boldface.

Here we describe hydroxyl radical cleavage experiments on the molecule JCASY, which is identical to J1 except that two opposite arms have been extended by 7 bp (Fig. 5). We determined cleavage patterns for all four strands of JCASY. To prepare the four-way junction, one 5′-radiolabeled strand is combined with 50-fold molar excesses of the remaining three strands [2 pmol of labeled strand and 100 pmol of unlabeled strands in 50 mM Tris-HCl (pH 8.0), 10 mM MgCl$_2$] in a final volume of 7 μl. The sample is annealed by heating to 90° for 2 min and then cooling to room temperature. In a separate experiment each strand was annealed to its complement under identical conditions, to form the duplex control.

The annealed DNA is cleaved by the hydroxyl radical using a cutting reagent of the following composition: $100 \mu M$ Fe^{2+} : $200 \mu M$ EDTA, 0.15% H_2O_2, 1 mM ascorbate. After 2 min the reaction is stopped by addition of 1 μl of 0.5 M thiourea. To assay for junction formation, 7 μl of the reaction mixture is run on a 20% native polyacrylamide gel [19 : 1, acrylamide–bis-acrylamide; 1 × TAE-Mg (40 mM Tris, 200 mM acetic acid, 2 mM EDTA, 10 mM magnesium acetate, pH 8.0)]. Samples were prepared for native gel electrophoresis by addition of 4 μl of loading buffer (1 × TAE-Mg, 75% glycerol, 0.03% xylene cyanol, and 0.03% bromphenol blue). Again, a 50-fold excess of cold strand(s) leads to complete duplex or junction formation (Fig. 6). The remaining 4 μl of the cleavage reaction is dried on a Speed-Vac concentrator and electrophoresed on a 20% denaturing polyacrylamide gel (Fig. 7).

Figure 8 shows densitometric scans of the hydroxyl radical cleavage patterns of duplex controls and junction molecules for all four strands of JCASY. The nucleotide positions along each strand are numbered from the 5' end. Positions 8 and 9 of each strand surround the branch point. In JCASY, the greatest degree of protection from hydroxyl radical attack is seen at positions 8 and 9 in strands 2 and 4 (Fig. 8). In contrast, strands 1 and 3 do not exhibit protections at positions 8 and 9, but are weakly protected at positions 12 and 13. This protection pattern reveals that JCASY has 2-fold structural symmetry, not 4-fold symmetry as one would predict for a tetrahedral or square planar structure. The protection pattern is consistent with the Sigal–Alberts model for the Holliday junction,[34] in which the crossover positions of two of the strands are in the center of the molecule, and thus would be expected to be protected from attack. On the other two strands these positions lie on the outside, and should be exposed to attack (Fig. 9).[8] A Sigal–Alberts Holliday structure consists of two helical domains that are coaxial. The helices may be envisioned to run either parallel or antiparallel to one another. Both versions of the model predict that positions on the helical strands half a helical turn away from the branch point will fall in the interior of the molecule (Fig. 9). The observed protections at position 12 and 13 on the helical strands are thus consistent with the model.

Hydroxyl radical cleavage of a four-way junction thus shows that the structure is two-fold symmetric, planar or close to planar, with preferred crossover strands. It has subsequently been shown by experiments in other laboratories that the antiparallel Sigal–Alberts conformation is the preferred conformation.[12,35]

The strong protections at the branch point on strands 2 and 4, and the

[34] N. Sigal and B. Alberts, *J. Mol. Biol.* **71,** 789 (1972).
[35] J. P. Cooper and P. J. Hagerman, *J. Mol. Biol.* **198,** 711 (1987).

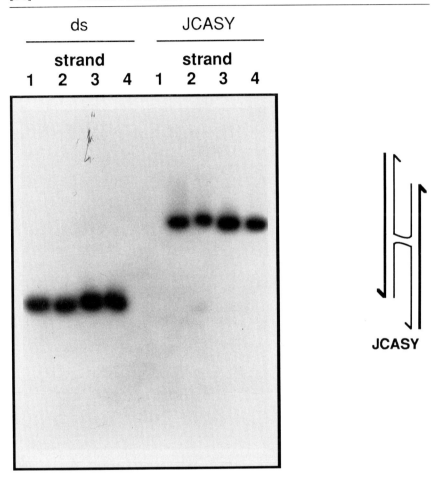

FIG. 6. Assay for formation of duplex and junction. Shown is the autoradiograph of a native polyacrylamide gel on which was electrophoresed products of annealing of the strands of JCASY. Lanes 1–4 show each of the four strands of JCASY annealed to its complement, to form a normal duplex. Lanes 5–8 show each of the four strands annealed to the other three strands (see Fig. 5) to form the junction.

absence of branch point protections on strands 1 and 3, demonstrates that strands 2 and 4 are the crossover strands (Fig. 8). The ability of the hydroxyl radical to detect crossover and noncrossover strands in a Holliday junction has been the most striking success of our application of the technique to Holliday junctions. Subsequent work employing a battalion of DNA cleavage reagents has not uncovered any that detect crossover structure with certainty. The hydroxyl radical protection patterns of

strand 4

control   ss   ds   JCASY

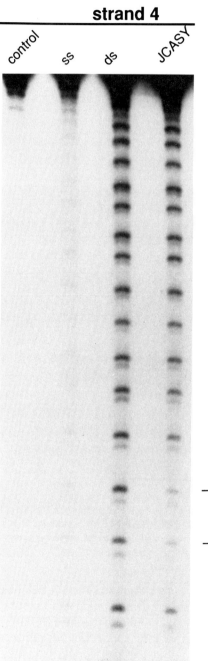

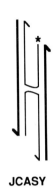

— 9

— 8

JCASY

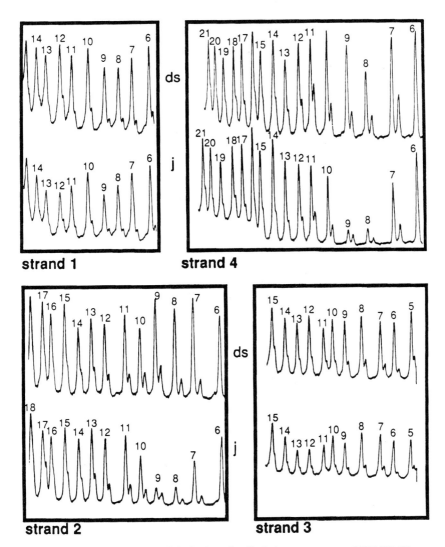

FIG. 8. Densitometric scans of the hydroxyl radical cleavage pattern of JCASY. The top scans in each box (marked ds) are control cleavage patterns of radiolabeled strands annealed to form duplex and treated with hydroxyl radical. Bottom scans (marked j) are of strands annealed to form the junction JCASY and treated with hydroxyl radical.

FIG. 7. Hydroxyl radical cleavage pattern of strand 4 of JCASY. The lane marked control contains an untreated sample; ds contains duplex treated with hydroxyl radical; JCASY contains the junction treated with hydroxyl radical. Positions 8 and 9 are marked to the side of the autoradiograph.

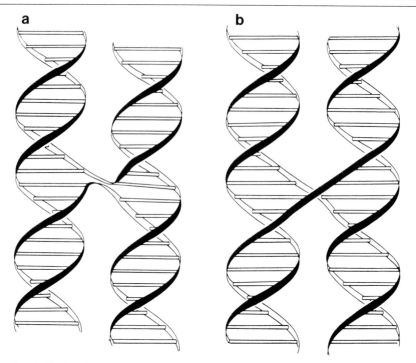

FIG. 9. Sigal–Alberts structures. (a) Antiparallel, or open, Sigal–Alberts structure. Here, the helical (noncrossover) strands run in opposite directions, that is, the 5′ end of one helical strand is at the top of the figure whereas the 5′ end of the other helical strand is at the bottom. (b) Parallel, or closed, Sigal–Alberts structure. In this structure the helical strands run in the same direction.

JCASY described here are virtually identical to those that we have published for the related molecule J1[8] and to those that we have obtained for a third related junction that has adjacent, rather than opposite, arms extended.[10]

Summary and Perspectives

The examples presented in this chapter show that the hydroxyl radical cleavage experiment yields high resolution, reproducible structural data for DNA in solution. Adenine tracts in both a bent restriction fragment and a short oligonucleotide duplex have a characteristic pattern of hydroxyl radical cleavage that is related to the unusual conformation adopted by these tracts. The four strands of an immobile Holliday junction show distinctive hydroxyl radical protections that report on the crossover configuration of the molecule. Not only does this technique provide structural data at the nucleotide level, it can easily be performed by any laboratory that is experienced in the methods of molecular biology. We hope that this description will encourage researchers to apply this approach to other DNA molecules.

Acknowledgments

We thank Amy Kimball for allowing us to use data on junction DNA and for assistance with the preparation of the manuscript, Ned Seeman and Neville Kallenbach for providing junction DNA molecules, and Gary Ackers for making available to us the two-dimensional gel scanning system. We are grateful to Sarah Morse for synthesis of oligonucleotides. We are indebted to Margaret Weidner, who did much of the original work in our laboratory on high-resolution cleavage analysis of DNA oligonucleotides. This work was supported by U.S. Public Health Service Grant GM 40894. T.D.T. is a fellow of the Alfred P. Sloan Foundation and the recipient of a Camille and Henry Dreyfus Teacher–Scholar award and a Research Career Development Award (CA 01208) from the U.S. PHS. M.A.P. acknowledges the support of a graduate fellowship from the National Science Foundation.

[12] Transition Metal Complexes as Probes of Nucleic Acids

By Christine S. Chow and Jacqueline K. Barton

Introduction

It has become increasingly clear that the one-dimensional base pair sequence of double-stranded DNA contains an abundance of three-dimensional structural polymorphism. Do proteins take advantage of this conformational variability in recognizing specific binding sites on the double helix? DNA-binding proteins may distinguish their targets not only through specific hydrogen bonding interactions with the base pairs but also through a series of specific electrostatic and van der Waals interactions with the sugar–phosphate backbone and base pair stack. Furthermore noncontacting base pairs, through bending, looping, and other mechanisms, can serve to orient helical nucleic acid regions with disparate elements within one protein or between proteins. In other words, the shape of a nucleic acid site is likely to be an important factor governing its recognition.

It therefore becomes important to unravel the relationship between DNA sequence and its three-dimensional structure. X-Ray crystallography has so far provided the highest resolution views of nucleic acid structures.[1]

[1] O. Kennard and W. N. Hunter, *Q. Rev. Biophys.* **22**, 327 (1989).

But the number of such structures with different sequences is sufficiently small that, from these structures alone, sequence-dependent variations in local conformation cannot be determined. Additionally, owing to considerations of crystal packing and the short lengths of the oligonucleotides crystallized, other methods are needed to complement the crystallographic studies in which a given sequence in the context of the polymer is examined. Enzymatic assays and, importantly, also chemical probes of the accessibility of sugars or base residues to different reactions have provided valuable tools to characterize these sequence-dependent variations in solution.

Determining the three-dimensional structure of an RNA molecule is still more complex. Here the majority of the polymer need not be helical in structure. Tertiary folding, single-stranded looped regions, as well as novel elements of hydrogen bonding and stacking all contribute in defining the structure of the RNA molecule. Crystallography[2] and NMR methods[3] have provided only a few clues as to the structures that an RNA polymer likely adopts. Moreover, with RNAs, fewer enzymatic and chemical probes are currently available.

In our laboratory we have designed a series of transition metal complexes which recognize their nucleic acid binding sites based on shape selection.[4] By matching the shapes and symmetries of the metal complexes to particular variations in local nucleic acid conformation, a family of molecules which target different DNA sites have been developed. The recognition of a site depends on the local conformation, or shape, rather than on the sequence directly. Indeed, based purely on such considerations of shape and symmetry, a high level of specificity may be achieved. The molecules prepared serve as a novel series of conformation-selective probes, and these may be utilized to map the topological variations in structure along the nucleic acid polymer.

The probes, shown in Fig. 1, are all derived from the parent tris(phenanthroline)metal complex. In this family we may vary the metal center so as to change the spectroscopic or reactive characteristics of the complex, and we may vary the bidentate ligands so as to alter the recognition characteristics of the complex. The complexes share several features which are essential to their application as shape-selective probes. First,

[2] S. H. Kim, J. L. Sussman, F. L. Suddath, G. J. Quigley, A. McPherson, A. H. J. Wang, N. C. Seeman, and A. Rich, *Proc. Natl. Acad. Sci. U.S.A.* **71**, 4970 (1974); E. Westhof, P. Dumas, and D. Moras, *J. Mol. Biol.* **184**, 119 (1985).

[3] D. J. Patel, L. Shapiro, and D. Hare, *Q. Rev. Biophys.* **20**, 78 (1987); C. Cheong, G. Varani, and I. Tinoco, Jr., *Nature (London)* **346**, 680 (1990).

[4] A. M. Pyle and J. K. Barton, *in* "Progress in Inorganic Chemistry: Bioinorganic Chemistry" (S. J. Lippard, ed.), Vol. 38, p. 413. Wiley, New York, 1990.

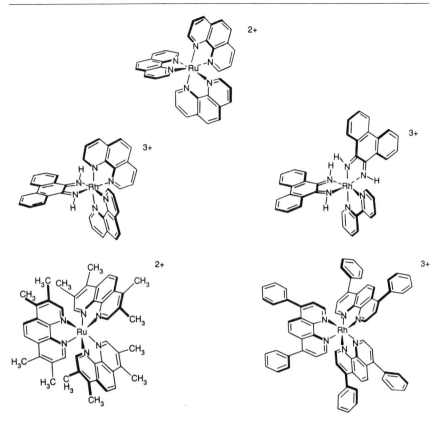

FIG. 1. Transition metal complexes used to probe nucleic acid structure (clockwise from the top): Δ-Ru(phen)$_3^{2+}$, Δ-Rh(phi)$_2$(bpy)$^{3+}$, Λ-Rh(DIP)$_3^{3+}$, Λ-Ru(TMP)$_3^{2+}$, Δ-Rh(phen)$_2$(phi)$^{3+}$.

the complexes are all coordinatively saturated and inert. There is no direct coordination of the metal center to the nucleic acid. Instead the recognition is based on a series of noncovalent interactions between the coordination complex and the polynucleotide. Second, if conclusions are to be drawn about the structure of the nucleic acid site based on a shape complementarity to the metal complex, the metal complex must be rigid in structure. The three-dimensional structures of the probes with octahedral coordination about the metal center are indeed extremely rigid and well defined. Third, we take advantage of the chirality of the tris-chelated octahedral complexes in specifying the symmetries of particular sites on the polymer.[5]

[5] J. K. Barton, *Science* **233**, 727 (1986).

Last, the probes described in this chapter all promote DNA strand cleavage at their binding sites upon photoactivation.[6] By coupling photoreactivity to the shape and symmetry constraints of the metal complex, a family of conformation-specific, and therefore site-specific, DNA-cleaving molecules are obtained. Photoactivated cleavage chemistry may also become useful in probing DNA structure and recognition inside cells.[7]

Extensive spectroscopic and biochemical studies[4,8–11] have shown that these metal complexes bind noncovalently to nucleic acids through two basic modes: interacalation and hydrophobic groove binding. The groove-bound mode appears to favor the minor groove of the helix, whereas intercalation appears for these complexes to originate from the major groove of the helix. For intercalation into a right-handed helix, it is the Δ isomer that is favored. In contrast, for the groove-bound mode, it is the complementary Λ symmetry that is favored for binding against the surface of the right-handed helix.

Nucleic acid cleavage chemistry can be extremely powerful in marking binding sites on the polynucleotide and generally in probing sensitively conformational variations that occur in solution under physiological conditions. To maximize the information that can be gleaned from these studies, information regarding the mechanism of strand cleavage is essential. The cleavage chemistry of the phenanthrenequinonediimine (phi) complexes of rhodium have been characterized in some detail.[12] Mechanistic studies indicate that cleavage results from the direct abstraction of the C-3′ H atom from the sugar by a delocalized excited state radical on the phi. In contrast to cleavage[13] by FeEDTA, no diffusible radical mediates the reaction. Hence site specificity in cleavage is observed. Also, in contrast to other cleaving reagents,[4] these intercalators probe directly the major groove of the helix; governed by the topology of the intercalative binding, a single-base 5′-asymmetry in

[6] M. B. Fleisher, K. C. Waterman, N. J. Turro, and J. K. Barton, *Inorg. Chem.* **25,** 3549 (1986).

[7] L. B. Chapnick, L. A. Chasin, A. L. Raphael, and J. K. Barton, *Mutat. Res.* **201,** 17 (1988).

[8] C. V. Kumar, J. K. Barton, and N. J. Turro, *J. Am. Chem. Soc.* **107,** 5518 (1985).

[9] J. K. Barton, J. M. Goldberg, C. V. Kumar, and N. J. Turro, *J. Am. Chem. Soc.* **108,** 2081 (1986).

[10] A. M. Pyle, J. P. Rehmann, R. Meshoyrer, C. V. Kumar, N. J. Turro, and J. K. Barton, *J. Am. Chem. Soc.* **111,** 3051 (1989).

[11] J. P. Rehmann and J. K. Barton, *Biochemistry* **29,** 1701 (1990); J. P. Rehmann and J. K. Barton, *Biochemistry* **29,** 1710 (1990).

[12] A. Sitlani, E. C. Long, A. M. Pyle, and J. K. Barton, *J. Am. Chem. Soc.* **114,** in press (1992).

[13] R. P. Hertzberg and P. B. Dervan, *Biochemistry* **23,** 3934 (1984).

cleavage is frequently observed. In addition, no secondary reactants such as dithiothreitol (DTT) or hydrogen peroxide are required; only ultraviolet light is required. The reaction does not even require oxygen. Finally, because the reaction originates on the sugar moiety, no base preference is inherent in the cleavage chemistry.

To a first approximation, all sugars react with equal efficiency, and therefore sites of cleavage can be considered to reflect directly sites of binding. The cleavage characteristics of the ruthenium-based photochemistry differ substantially from those of rhodium complexes. For the groove-bound probes, the site-specific hydrogen atom abstraction chemistry of the rhodium series is not productive.[14] Ruthenium(II) polypyridyls, however, have been well characterized[15] as efficient sensitizers of singlet oxygen in their excited state. Photolysis with visible light of ruthenium complexes bound to nucleic acids promotes strand cleavage in a reaction mediated by singlet oxygen.[16] Here clearly a diffusible species, $^1O_2$, mediates the reaction; rather than obtaining single sites of cleavage, a region (several bases) along the helix is reacted. The target of singlet oxygen chemistry moreover is the nucleic acid base. Piperidine treatment is therefore required to convert the lesion to a strand break. Because the chemistry targets the different bases with different efficiencies, guanine being most reactive, cleavage reactions must also be normalized against those of the parent $Ru(phen)_3^{2+}$ (where phen is phenanthroline), which binds the helix with little specificity.

In this chapter we discuss the utility and application of these different transition metal complexes in probing DNA and RNA structure in solution. We include a discussion of the sequence-neutral cleavage complex, $Rh(phi)_2bpy^{3+}$ (where bpy is bipyridine), a useful reagent for high-resolution photofootprinting,[17] as well as several conformation-specific tools to examine nucleic acid structure. These various ruthenium and rhodium complexes may be applied to detect subtle variations in B-DNA conformations[18,19] or to investigate global secondary structures of a polynucleotide such as DNA cruciforms,[20] left-handed Z-DNA,[21]

[14] Unpublished results in our laboratory.
[15] A. Juris, V. Balzani, F. Barigelletti, S. Campagna, P. Belser, and A. von Zelewsky, *Coord. Chem. Rev.* **84**, 85 (1988).
[16] H.-Y. Mei and J. K. Barton, *Proc. Natl. Acad. Sci. U.S.A.* **85**, 1339 (1988).
[17] K. Uchida, A. M. Pyle, T. Morii, and J. K. Barton, *Nucleic Acids Res.* **17**, 10259 (1989).
[18] A. M. Pyle, E. C. Long, and J. K. Barton, *J. Am. Chem. Soc.* **111**, 4520 (1989).
[19] A. M. Pyle, T. Morii, and J. K. Barton, *J. Am. Chem. Soc.* **112**, 9432 (1990).
[20] M. R. Kirshenbaum, R. Tribolet, and J. K. Barton, *Nucleic Acids Res.* **16**, 7943 (1988).
[21] J. K. Barton and A. L. Raphael, *Proc. Natl. Acad. Sci. U.S.A.* **82**, 6460 (1985).

and A-form DNA.[16,22] In addition, we describe how the secondary and tertiary structure of RNA may be examined using this methodology.[23] We discuss the different information that may be gained from these methods and as well point out some of the necessary precautions in their application. In general, studies with these complexes provide a unique and sensitive handle to probe elements of nucleic acid polymorphism in solution.

DNA and RNA Preparation, Purification, and Labeling

The plasmid DNA of interest is purified and digested with a restriction enzyme to linearize and give the desired protruding 5' terminus for dephosphorylation and end-labeling with T4 polynucleotide kinase and [γ-^{32}P]ATP[24] or a 3' terminus for labeling with [α-^{32}P]dATP and terminal deoxynucleotidyltransferase.[25] This labeling is followed by a second enzyme digest, yielding the desired end-labeled DNA fragment which is then purified on a nondenaturing polyacrylamide gel and isolated by electrophoretic elution using the Schleicher and Schuell (Keene, NH) Elutrap system. The linearized, labeled restriction fragments are then used directly for photocleavage experiments.

DNA oligonucleotides have been synthesized on a Pharmacia (Piscataway, NJ) Gene Assembler using the phosphoramidite method[26] and purified by reversed-phase high-performance liquid chromatography (HPLC). The oligonucleotides are 5' or 3' end labeled by the same procedures as with the restriction fragments. The labeled oligonucleotides are gel purified on denaturing polyacrylamide gels and electroeluted or purified by the Nensorb (Du Pont, Boston, MA) purification method.

Purified RNA samples are 3'-end-labeled with T4 RNA ligase and [5'-^{32}P]pCp[27] or dephosphorylated and 5'-end-labeled with T4 polynucleotide kinase and [γ-^{32}P]ATP. All the necessary precautions are taken in order to prevent contamination by RNases. Following the end-labeling procedure, the RNA samples are precipitated, resuspended and denatured in loading dye, and gel purified on a 40 cm long and 0.8 mm thick denaturing polyacrylamide gel (10–20%) for 8 to 12 hr at 600 V, visualized, and electroeluted. The labeled RNA fragments are stored at $-20°$ in 10 mM Tris-HCl (pH 7.5) and are renatured (heat in storage buffer to 70° for

[22] H.-Y. Mei and J. K. Barton, *J. Am. Chem. Soc.* **108**, 7414 (1986).
[23] C. S. Chow and J. K. Barton, *J. Am. Chem. Soc.* **112**, 2839 (1990).
[24] G. Chaconas and J. H. van de Sande, this series, Vol. 65, p. 75.
[25] R. Roychoudhury and R. Wu, this series, Vol. 65, p. 43.
[26] M. J. Gait, "Oligonucleotide Synthesis, a Practical Approach." IRL Press, Oxford, 1984.
[27] T. E. England, A. G. Bruce, and O. C. Uhlenbeck, this series, Vol. 65, p. 65.

10 min and slow cool to room temperature) prior to the photocleavage experiments.

The syntheses of ruthenium[16,17] complexes and of rhodium complexes containing phen[20] and phi[28] have been described. [Ru(TMP)$_3$]Cl$_2$ and [Rh(DIP)$_3$]Cl$_3$ (where TMP is tetramethylphenanthroline and DIP is diphenylphenanthroline) may be obtained from Bethesda Research Laboratories (Gaithersburg, MD). Other complexes may also be obtained in small quantities from the authors.

Photocleavage by Rhodium and Ruthenium Complexes

Metal Stock Solutions

Metal stock solutions are made to approximately 1 mM and can be stored frozen for several weeks. The concentrations are measured according to the given extinction coefficients. The stock solutions are diluted immediately prior to addition to the reaction mixtures.

Photolysis

Lamp Irradiations. Irradiations of the DNA/RNA reaction mixtures (typical volume is 20–50 μl) have been performed in either 0.6- or 1.7-ml siliconized polypropylene tubes. The open reaction tubes are fixed such that the reaction mixture is directly in the focal point of a 1000 W Hg/Xe lamp beam focused and filtered with a monochromater (Oriel, Stratford, CT, Model 77250) and a glass filter to eliminate light below 305 nm. The wavelengths typically used for cleavage experiments are 310 to 365 nm for the rhodium complexes and 442 nm for the ruthenium complexes ($\pm$6 nm).

Laser Irradiations. Irradiations of the DNA/RNA reaction mixtures can be performed using a He/Cd laser (Liconix, Sunnyvale, CA, Model 4200 NB, 442 nm, 22 mW). Open reaction tubes (0.6- or 1.7-ml siliconized tubes) are placed in a holder such that the reaction mixture is directly in the line of the laser light, which is filtered with a glass filter to eliminate the light below 400 nm.

Transilluminator. Footprinting experiments have been conducted using a transilluminating light box, which is commonly found in laboratories to examine ethidium-stained gels. Prepared reaction mixtures of 50-μl volumes are placed in shortened polypropylene tubes which have been cut down to approximately 7 mm high and 100 μl in volume to minimize the distance between the reaction mixtures and the UV light source. The samples are then placed in a pipette tip rack and irradiated by inverting

[28] A. M. Pyle, M. Chiang, and J. K. Barton, *Inorg. Chem.* **29**, 4898 (1990).

a UV transilluminator (Spectroline Model TR302 with a broad band of irradiation centered at 302 nm) on top of them. The UV filter of the transilluminator may be removed to increase light intensity. Samples should be less than 1 cm from the light source and irradiated for 20 min at room temperature.

Other Light Sources. Before using a new light source, one may test its photocleavage efficiencies with the metal complexes by measuring irradiation times needed for 50% conversion of form I to form II DNA on agarose gels. For high-resolution cleavage studies on polyacrylamide gels, irradiation times should be increased by 5–10 times for best results. Controls for photodamage (samples irradiated in the absence of metal) should always be included.

High-Resolution Mapping of Cleavage Sites on DNA and RNA Fragments

Gel Electrophoresis Techniques

DNA or RNA pellets from the cleavage reactions are dissolved in 1 to 4 μl of loading buffer and electrophoresed in TBE buffer [90 mM Tris, 90 mM boric acid, 2.5 mM ethylenediaminetetraacetic acid (EDTA), pH 8.3] on denaturing polyacrylamide gels (19:1 acrylamide–N,N'-methylenebisacrylamide, 8 M urea, TBE buffer, 0.08% ammonium persulfate). DNA loading buffer consists of 80% (v/v) deionized formamide, 50 mM Tris–borate (pH 8.3), 1 mM EDTA, and 0.025% (w/v) each xylene cyanol and bromphenol blue. RNA loading buffer consists of TBE buffer, 7 M urea, and 0.025% (w/v) each xylene cyanol and bromphenol blue.

Chemical Sequencing Methods

DNA reaction products are coelectrophoresed with Maxam–Gilbert sequencing reactions.[29] RNA reaction products are coelectrophoresed with Peattie–Gilbert sequencing reactions.[30] We find it desirable to employ chemical rather than enzymatic sequencing. Sequencing products comigrate with the cleavage reactions since both contain phosphate termini. Furthermore the sequencing reactions give an additional check on the labeled fragment as a substrate for cleavage.

[29] A. M. Maxam and W. Gilbert, this series, Vol. 65, p. 499.
[30] D. Peattie, *Proc. Natl. Acad. Sci. U.S.A.* **76,** 1760 (1979).

Autoradiography, Densitometry, and Data Processing

After removal from the glass plates, the gels are dried and subjected to autoradiography (Kodak, Rochester, NY, X-Omat AR) at $-60°$ with an intensifying screen or at ambient temperature without an intensifying screen. Autoradiographs are scanned on a laser densitometer (LKB, Uppsala, Sweden, Model 2222-020 Ultrascan XL and GelScan XL software).

Sequence-Neutral DNA Cleavage and Footprinting by $Rh(phi)_2(bpy)^{3+}$

The complex bis(phenanthrenequinonediimine)(bipyridyl)rhodium(III) [$Rh(phi)_2(bpy)^{3+}$] binds DNA avidly (log $K \geq 7$) and with photoactivation cleaves DNA efficiently in a sequence-neutral fashion.[18] As a result this metallointercalator can serve as a sensitive, high-resolution photofootprinting reagent.[17] Cleavage with this complex allows one to map with precision the binding sites of small molecules or large proteins on DNA. Because the complex intercalates into the helix, sensing both grooves, it can be used to footprint molecules bound in either groove.

Reaction Conditions

A typical reaction mixture for cleavage of DNA by $Rh(phi)_2(bpy)^{3+}$ in the absence of a DNA-binding molecule is as follows: 50 μl containing ^{32}P-end-labeled DNA fragment [$\sim$20,000–30,000 counts/min (cpm)], 5 μM base-paired DNA (concentration adjusted with sonicated calf thymus carrier DNA), and 5 μM $Rh(phi)_2(bpy)^{3+}$ (synthesized as described previously[28]; ε_{350}, 23,600 M^{-1} cm^{-1}) in Tris–acetate buffer (50 mM Tris, 20 mM sodium acetate, 18 mM NaCl, pH 7.0). Cleavage may also be performed in other buffers, which may be necessary to ensure proper binding of the agent to be footprinted. Samples are vortexed and centrifuged briefly, allowed to equilibrate for 15 min at ambient temperature, and irradiated at 310 nm on the lamp for approximately 5 to 10 min or with the transilluminator for 20 min.

Following irradiation, 1 μl of 9 mM calf thymus DNA and 25 μl of 7.5 M ammonium acetate are added to the reaction mixture, which is then heated to 90° for 2 min and precipitated by addition of 150 μl ethanol. This melting procedure is necessary to dissociate the metal complex fully. The DNA pellet is then rinsed twice with cold ethanol, dried, and gel electrophoresed.

Cleavage of DNA in the presence of a DNA-binding molecule is as follows: the reagent to be footprinted is bound to the DNA under the appropriate conditions followed by the addition of the rhodium complex and incubation at room temperature for 15 min. The final reaction mixture

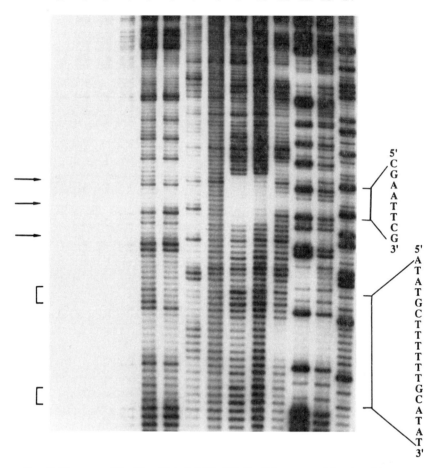

FIG. 2. Cleavage of the 3′-end-labeled *Hin*dIII–*Pvu*II fragment of the plasmid pJT18-T6 by the transition metal probes. Autoradiogram of the 8 *M* urea denaturing 8% polyacrylamide gel of the 245-bp fragment. All conditions for these reactions are as described in the text. Lane 1: Dark control, labeled fragment in the absence of metal complex or light. Lanes 2–4: Light controls, DNA fragment irradiated at 310 nm for 10 min, 313 nm for 1 min, and 442 nm for 25 min, respectively, without metal complex. Lane 4 has also been treated with piperidine. Lanes 5 and 6: Fragments after photolysis in the presence of Ru(TMP)$_3^{2+}$ and Ru(phen)$_3^{2+}$, respectively, and treatment with piperidine. The bracketed regions to the left of the gel indicate sites of preferential cleavage by Ru(TMP)$_3^{2+}$. Lane 7: Fragment after photolysis in the presence of Rh(phen)$_2$(phi)$^{3+}$. Irradiation was performed at 313 nm for 40 sec. Arrows to the left of the gel indicate specific, sharp sites of cleavage. Lane 8: Fragment after photolysis with Rh(phi)$_2$(bpy)$^{3+}$ indicating sequence-neutral cleavage by the reagent. Lanes 9 and 10: *Eco*RI footprinting with Rh(phi)$_2$(bpy)$^{3+}$. The following conditions were used: 1 m*M* CaCl$_2$, 50 m*M* Tris, 20 m*M* sodium acetate, 18 m*M* NaCl, pH 7.0, 5 μ*M* DNA nucleo-

contains 5 μM DNA and 5 μM Rh(phi)$_2$(bpy)$^{3+}$ (higher or lower concentrations can be employed but a ratio of 1 : 1 for the DNA–metal complex is optimal). Samples are then irradiated and ethanol precipitated, as above, followed by phenol–chloroform extraction and another ethanol precipitation and wash.

Discussion

An example of footprinting with Rh(phi)$_2$(bpy)$^{3+}$ is shown in Fig. 2. These DNA cleavage experiments were conducted on a 3'-end-labeled *Hin*dIII–*Pvu*II fragment of the plasmid pJT18-T6. The 245-base pair (bp) fragment[17] contains an *Eco*RI endonuclease binding site as well as a binding site for distamycin. Therefore the footprinting of both a protein bound primarily in the major groove and a small molecule bound in the minor groove may be illustrated. As seen in Fig. 2, sequence-neutral cleavage of the DNA is apparent in the presence of metal complex without other site-specifically bound molecules (lane 8). In the presence of 240 or 120 units of *Eco*RI and 1 mM CaCl$_2$ (lanes 9 and 10, respectively), however, a clear footprint is evident which covers base pairs G64 to T73. In the presence of higher protein concentrations one observes other regions of protection along the DNA fragment, reflecting the lower affinity binding sites for the protein. As shown in lane 11 of Fig. 2, with the minor groove-bound distamycin, a sharp footprint is obtained at the T$_6$ binding site for distamycin (base pairs 40–45) and also at weaker distamycin binding sites TTAA (66–69) and ATTA (72–75).

A rhodium–base pair ratio of 1 : 1 is important in order to maintain sequence neutrality in cleavage. At low rhodium–DNA ratios, the complex shows some sequence selectivity. Also, at high salt concentrations (greater than 100 mM NaCl, 10 mM MgCl$_2$, 10 mM CaCl$_2$, or 10 to 100 mM EDTA) some loss in sequence neutrality is apparent. However, the reaction with

---

tides, 5 μM metal, and 240 or 120 units *Eco*RI, respectively. The DNA fragment was incubated for 30 min at ambient temperature with *Eco*RI. The DNA carrier and metal complex were then added and the mixture incubated for another 15 min, followed by irradiation at 310 nm for 10 min. The samples were extracted with phenol–chloroform and precipitated as described in the text. The *Eco*RI footprint is shown in the upper bracketed region to the right of the gel. Lane 11: Distamycin footprinting with Rh(phi)$_2$(bpy)$^{3+}$. The following conditions were used: 50 mM Tris, 20 mM sodium acetate, 18 mM NaCl, pH 7.0, 17 μM distamycin, 10 μM DNA nucleotides, 5 μM metal. The DNA fragment was incubated for 10 min at ambient temperature with distamycin. The DNA carrier and metal complex were then added and the mixture incubated for another 5 min, followed by irradiation at 310 nm for 10 min. The distamycin footprint is shown in the lower bracketed region to the right of the gel. Lanes 12–14: Maxam–Gilbert sequencing reactions, G, G + A, and T + C, respectively.

Rh(phi)$_2$(bpy)$^{3+}$ is not inhibited by moderate concentrations of salts, EDTA, glycerol, or reducing agents which may be necessary to obtain a native interaction of the DNA with protein. The facts that no chemical activators are required for the cleavage and further that the rhodium complex and the photocleavage reaction are not perturbed by the presence of a wide range of reagents are clear advantages in applying this footprinting technique to a host of experiments, including those where metalloproteins may be involved.

Another advantage in using Rh(phi)$_2$(bpy)$^{3+}$ as a footprinting reagent lies in the high resolution of its cleavage patterns. The sharp pattern of cleavage is obtained because the complex is itself rigid and reacts with the polymer through direct strand scission at the binding site rather than through a reaction mediated by a diffusing species. The single-nucleotide resolution is consistent with the crystallographic results found for both distamycin[31] and *Eco*RI[32] and shows the applicability of this reagent for mapping molecules, big or small, bound in either groove of the helix. There are two caveats in the application of this complex for photofootprinting, however. Rh(phi)$_2$bpy^{3+} binds DNA with high affinity ($K \geq 10^7 \ M^{-1}$). Hence one must consider the competition of the rhodium complex for a given binding site with a weakly binding protein or small molecule. Second, the photoactivation, while advantageous from the standpoint of needing no chemical activators, requires some thought and consideration be given to light sources. Relatively unsophisticated light sources can be utilized; however, such sources must be placed close to the samples (to maximize the photon flux hitting the solution), and some ingenuity must be applied to work out the logistics for irradiations.

Mapping Major and Minor Groove Topologies of DNA with Rh(phen)$_2$(phi)$^{3+}$ and Ru(TMP)$_3$$^{2+}$

Used in concert, cleavage experiments with bis(phenanthroline)(phenanthrenequinonediimine)rhodium(III), [Rh(phen)$_2$(phi)$^{3+}$], and tris(tetramethylphenanthroline)ruthenium(II), [Ru(TMP)$_3$$^{2+}$], provide a strategy to

[31] M. L. Kopka, C. Yoon, D. Goodsell, P. Pjura, and R. E. Dickerson, *J. Mol. Biol.* **183**, 553 (1985); M. Coll, C. A. Frederick, A. H.-J. Wang, and A. Rich, *Proc. Natl. Acad. Sci. U.S.A.* **84**, 8385 (1987).

[32] C. A. Frederick, J. Grable, M. Melia, C. Samudzi, L. Jen-Jacobson, B.-C. Wang, P. Greene, H. W. Boyer, and J. M. Rosenberg, *Nature (London)* **309**, 327 (1984); J. A. McClarin, C. A. Frederick, B.-C. Wang, P. Greene, H. W. Boyer, J. Grable, and J. M. Rosenberg, *Science* **234**, 1526 (1986).

[33] One example of such an application of these probes is given in P. W. Huber, T. Morii, H.-Y. Mei, and J. K. Barton, *Proc. Natl. Acad. Sci. U.S.A.* **88**, 10801 (1991).

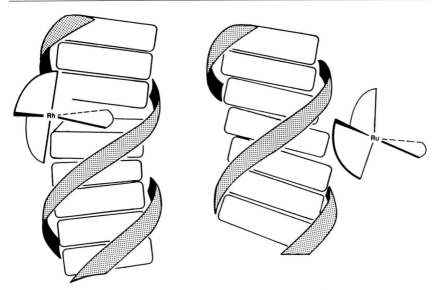

FIG. 3. Schematic illustration of intercalation by $Rh(phen)_2(phi)^{3+}$ (left) in open major grooves of DNA and of surface binding by $Ru(TMP)_3^{2+}$ (right) against wide and shallow minor grooves.

examine local variations in conformation within ostensibly B-DNA. With these reagents one can map aspects of the topologies of both major and minor grooves of the double helix such as whether the groove is widened or narrowed, whether the groove is opened in a smooth fashion or more sharply, and whether there are changes in the propeller twisting of base pairs or the tilting of bases.

These complexes bind to DNA through different modes and mark their binding sites using different cleavage chemistry. Figure 3 illustrates schematically the binding of these complexes to a double helix. Like $Rh(phi)_2(bpy)^{3+}$, $Rh(phen)_2(phi)^{3+}$ binds avidly to double helical DNA by intercalation,[12,18] and the intercalation appears to be in the major groove of the helix. Unlike $Rh(phi)_2(bpy)^{3+}$, however, $Rh(phen)_2(phi)^{3+}$ binds to DNA with sequence selectivity, recognizing preferentially two families of sites, 5'-pyrimidine–pyrimidine–purine–purine-3' steps, such as 5'-CCAG-3', and homopyrimidine–homopurine tracts. Enantioselectivity favoring the Δ isomer is observed at the 5'-pyrimidine–pyrimidine–purine–purine-3' steps but not along the homopyridine tracts. The recognition of these sites by the rhodium complex is based on shape selection.[18] Comparisons with crystallographic results on sequences targeted by $Rh(phen)_2(phi)^{3+}$ indicate that the sites preferentially bound are those which are more open in the major groove. It appears that intimate stacking of the phi ligand between the DNA base pairs necessitates sites in which the major groove is open so as to accommodate the bulk of the ancillary, overhanging phenanthroline ligands. Moreover the enantioselectivity in cleavage correlates well with the opening of sites owing to changes in propeller twisting.[19] Alternatively, at sites cleaved with no enantioselectivity, the openings may arise primarily through a negative inclination and/ or displacement of the DNA bases.

$Ru(TMP)_3^{2+}$ binds to the surface of the minor groove of DNA.[16,22] The complex has been shown to bind cooperatively to A-form helices of various base sequences yet exhibits little binding to B- or Z-form synthetic polymers. Some chiral discrimination in its binding is evident, favoring the Λ isomer. Spectroscopic results indicate no detectable intercalation by the complex, in contrast to the parent $Ru(phen)_3^{2+}$ which binds DNA through both the intercalative and groove-bound modes. The basis for the conformational selectivity of $Ru(TMP)_3^{2+}$ rests in its shape. The steric bulk of the methyl groups of $Ru(TMP)_3^{2+}$ not only precludes intercalation but also interferes with minor groove binding by the complex unless the minor groove is more open or shallow to accommodate the bulky methylated complex. The complex binds preferentially to the A form, since the A conformation contains a shallow and wide minor groove.

DNA Cleavage by $Rh(phen)_2(phi)^{3+}$

Reaction Conditions. A typical reaction mixture for cleavage of DNA by $Rh(phen)_2(phi)^{3+}$ is as follows: 50 μl containing ^{32}P-end-labeled DNA fragment (~20,000–30,000 cpm), 50 μM DNA base pairs (concentration adjusted with calf thymus carrier DNA), and 5 μM $Rh(phen)_2(phi)^{3+}$ (synthesized as described previously; ε_{362}, 19,400 M^{-1} cm^{-1})[28] in Tris–acetate buffer (50 mM Tris, 20 mM sodium acetate, 18 mM NaCl, pH 7.0) or 50 mM sodium cacodylate buffer (pH 7.0). The samples are vortexed and centrifuged briefly, allowed to equilibrate for 5 min at ambient temperature, and irradiated on the lamp at 313 nm for 30 sec to 3 min or at 365 nm for 5 to 10 min. The fragments are precipitated with 25 μl of 5 M ammonium acetate and 150 μl ethanol, and the resulting pellet is washed twice with 80 μl cold ethanol.

Cleavage conditions with oligonucleotides are as follows: 50 μl containing 500 μM nucleotide (200,000 cpm), 50 μM $Rh(phen)_2(phi)^{3+}$, and 25 mM sodium cacodylate (pH 7.0). The samples are irradiated for 7.5 min at 313 nm on the lamp. Following irradiation, an aliquot (20,000–30,000 cpm) is taken and lyophilized to dryness followed by resuspension in formamide loading dye.

DNA Cleavage by Ru(TMP)$_3$$^{2+}$

Reaction Conditions. A typical reaction mixture for cleavage of DNA with Λ-Ru(TMP)$_3$$^{2+}$ is as follows: 20 μl containing ^{32}P-end-labeled DNA fragment ($\sim$20,000–30,000 cpm), 100 μM DNA base pairs (concentration adjusted with sonicated calf thymus carrier DNA), 30 μM Λ-Ru(TMP)$_3$$^{2+}$ (synthesized and enantiomers separated as described previously; ε_{438}, 24,500 M^{-1} cm^{-1})[16,17] or Ru(phen)$_3$$^{2+}$ (synthesized as described; ε_{442}, 19,000 M^{-1} cm^{-1})[17] in buffer (5 mM Tris-HCl, 50 mM NaCl, pH 7.0), and 1.2 mM imidazole. The samples are vortexed and centrifuged briefly, allowed to equilibrate at room temperature for 15 min, and irradiated on the lamp or the laser at 442 nm for 25 min. Following irradiation, the fragments are precipitated with 4 μl of 5 M ammonium acetate and 120 μl ethanol. After washing twice with 80 μl cold ethanol and drying, the fragments are treated with 50 μl piperidine [10% (v/v) distilled piperidine in deionized water] at 90° for 30 min. The samples are immediately frozen following piperidine treatment and lyophilized for at least 1 hr. The samples are washed twice with 40-μl aliquots of water and lyophilized to dryness after each addition.

Discussion

Figure 2 illustrates cleavage by Rh(phen)$_2$(phi)$^{3+}$, Ru(phen)$_3$$^{2+}$, and Ru(TMP)$_3$$^{2+}$ on the pJT18-T6 restriction fragment. Differences both in the sites of strong cleavage and in the characteristic patterns of cleavage are evident with these complexes. Taken together the patterns of cleavage can be used to map aspects of the structure at these sites.

Reaction with racemic Rh(phen)$_2$(phi)$^{3+}$, shown in lane 7 of Fig. 2, leads to both single sharp sites of cleavage and a region where several bands of lesser intensity are evident. Sharp cleavage, for example, is seen at the *Eco*RI binding site, 5′-CGAATTCG-3′. This sharp cleavage therefore marks a site which opens abruptly in the major groove. Furthermore we find enantiomeric discrimination in cleavage at this site (data not shown), from which we can suggest that the site becomes open owing to a change in propeller twisting. The crystal structure of the dodecamer 5′-CGCGAATTCGCG-3′, the same internal sequence contained in this fragment, shows the greatest change in propeller twist at this same T*CGC* step; as a result in the crystal this site is indeed the most open in the major groove.[34] A different pattern of cleavage is seen across the T$_6$ tract. Here cleavage is evident at all sites except the last T to the 5′ side. The pattern

---

[34] R. Dickerson and H. R. Drew, *J. Mol. Biol.* **149,** 761 (1981).

of cleavage varies smoothly across the tract, and no enantiomeric discrimination is apparent (data not shown). Here, then, the pattern of cleavage marks an open region in the major groove of DNA and one which likely is not determined by propeller twist changes but instead by a gradual variation in the inclination and displacement of bases toward the major groove. Consistent with results from other laboratories, then, from these data we can conclude that the DNA bends smoothly across this T_6 tract.[35]

Cleavage results with $Ru(TMP)_3^{2+}$, though quite different from those seen with the rhodium complex, are actually quite complementary. Here, to obtain specific regions of binding by the complex, patterns of cleavage by $Ru(TMP)_3^{2+}$ (lane 5, Fig. 2) must be compared to those by $Ru(phen)_3^{2+}$ (lane 6, Fig. 2). The ruthenium complexes sensitize the formation of singlet oxygen, a diffusible species which reacts with the DNA bases with preference for guanine residues.[16] Piperidine treatment is necessary to achieve complete strand scission. As can be seen in Fig. 2, then, reactions with $Ru(phen)_3^{2+}$ produce cleavage at all guanine residues owing to the fact that singlet oxygen is being generated at largely uniform concentrations near the helix by nonselectively bound complex or by complex free in solution. Reactions with Λ-$Ru(TMP)_3^{2+}$ in the presence of imidazole (which is believed to help solubilize the complex) also yield a guanine-specific cleavage, but additional sites where Λ-$Ru(TMP)_3^{2+}$ is bound preferentially, creating locally higher concentrations of singlet oxygen, are apparent. In Fig. 2 it can be seen that such sites flank the bent T_6 tract at the two 5'-ATAT-3' sites.

A drawback in the application of $Ru(TMP)_3^{2+}$ as a probe rests in this need for comparison with $Ru(phen)_3^{2+}$. Histograms for the cleavage experiments may be constructed where the intensity of each band on the autoradiograph is converted to a relative probability of cleavage at a site.[16] The probability of cleavage by $Ru(TMP)_3^{2+}$ at each site is then obtained by subtraction of the $Ru(phen)_3^{2+}$ cleavage probability to normalize for the differential base reactivity of singlet oxygen.

Whereas $Rh(phen)_2(phi)^{3+}$ probes openings in the major groove, $Ru(TMP)_3^{2+}$ probes the shallowness of the minor groove. $Ru(TMP)_3^{2+}$ preferentially binds A conformations. But the binding of the complex to a site does not establish that the site is in the A conformation, only that the minor groove is sufficiently open and shallow to permit association of the bulky $Ru(TMP)_3^{2+}$. The minor groove of a canonical B-form helix is too narrow for detectable binding. An extended region of cleavage by $Ru(TMP)_3^{2+}$ without cleavage by the rhodium complex would point to a

---

[35] T. Morii, D. Campisi, J. K. Barton, unpublished results.

predominantly A-like conformation. In Fig. 2 we see no preferential cleavage by $Ru(TMP)_3^{2+}$ about the B-like *Eco*RI site (although some cleavage is evident to the 5' side at the sequence 5'-TTTC-3'). We do, however, see cleavage by $Ru(TMP)_3^{2+}$ flanking the bent T_6 tract, with cleavage by $Rh(phen)_2(phi)^{3+}$ at the center. This pattern suggests an opening in the major groove (not due to propeller twisting) in the center of the T_6 tract and the necessary openings of the minor groove above and below. Cleavage experiments with $Ru(TMP)_3^{2+}$ and $Rh(phen)_2(phi)^{3+}$ provide unique and complementary structural information concerning the minor and major grooves of DNA helices.

Shape-Selective Cleavage of Unusual DNA Structures by $Rh(DIP)_3^{3+}$

Tris(diphenylphenanthroline)rhodium(III), $[Rh(DIP)_3^{2+}]$, specifically targets unusual non-B-DNA structures such as Z-DNA and cruciforms. Thus, site-specific cleavage by the complex can be used to probe such altered conformations. The cleavage is sufficiently specific that mapping experiments are first conducted at low resolution to determine sites of altered conformation along the genome. Thereafter cleavage experiments, focused on the region of interest, may be conducted with single-nucleotide resolution to characterize the sites structurally.

Low-Resolution Mapping

A typical reaction mixture for cleavage of plasmid DNA with $Rh(DIP)_3^{3+}$ is as follows: 20 μl containing 200 μM in nucleotides of supercoiled plasmid DNA and 10 μM $Rh(DIP)_3^{3+}$ (synthesized as described previously[20]; ε_{296}, 116,000 M^{-1} cm^{-1})[4] in Tris–acetate buffer (50 mM Tris, 20 mM sodium acetate, 18 mM NaCl, pH 7.0). The samples are mixed by gently tapping the side of the tube. Centrifugation should be avoided because it may lead to precipitation of the metal complex. The samples are incubated for 1 min at room temperature and irradiated on the lamp at 313 nm for 20 to 60 sec. Immediately following irradiation, the samples are precipitated by adding 4 μl of 5 M ammonium acetate and 120 μl ethanol. The DNA pellets are washed twice with 80 μl ethanol and dried. Under these conditions, photolysis should yield partial to complete conversion to form II and form III DNA.

After resuspension in the appropriate enzyme reaction buffer, restriction enzyme is added to give linearization (total volume 20 μl). The DNA–enzyme mixture is digested at 37° for 1.5 hr. For subsequent digestion with S1 nuclease, adjust the pH of the reaction mixture to 6.0 with 12 μl Tris–acetate buffer (50 mM Tris, 20 mM sodium acetate, 18 mM NaCl, pH 5.0), add 4 μl ZnSO$_4$ to give a final concentration of 1 mM, add 2 μl

S1 (10–20 units), and incubate the sample for 20 min at 37°. The reaction is quenched with the addition of bromphenol blue dye containing 50% (w/v) sucrose, 0.5% (w/v) bromphenol blue, and 10 mM EDTA. Electrophoresis on 1% agarose gels (50 mM Tris–acetate, 20 mM sodium acetate, 18 mM NaCl, pH 7.0) resolves the fragments produced.

High-Resolution Mapping

A typical reaction mixture for cleavage of plasmid DNA for high-resolution mapping with Rh(DIP)$_3^{3+}$ is as follows: 20 μl containing 200 μM in nucleotides of supercoiled plasmid DNA and 10 μM Rh(DIP)$_3^{3+}$ in buffer (5 mM Tris-HCl, 50 mM NaCl, pH 7.0). The samples are irradiated on the lamp at 332 nm for 3.5 min (conditions may vary in order to convert approximately 50% form I DNA to cleaved species). Immediately following irradiation, the samples are precipitated by adding 4 μl of 5 M ammonium acetate and 120 μl ethanol. The DNA pellets are washed twice with 80 μl ethanol and dried. Following irradiation, the DNA is cut with an appropriate restriction enzyme (e.g., 15 units BglI, 1.5 hr at 37°). The samples are phenol extracted and ethanol precipitated. The linearized DNA is 5'- or 3'-end-labeled according to the published procedures.[24,25] After a second enzyme digest, the desired DNA fragment is isolated on a nondenaturing polyacrylamide gel. The purified labeled fragments are denatured in loading buffer and resolved on a denaturing polyacrylamide gel. The untreated labeled fragment is sequenced by the method of Maxam and Gilbert and coelectrophoresed.

Discussion

On photoactivation, Rh(DIP)$_3^{3+}$ specifically cleaves unusual DNA conformations and thus can be used to detect such structures on long DNA substrates. Low-resolution experiments are first used to determine the location of the structure, and whether the cleavage is single-stranded, double-stranded, and dependent on DNA superhelicity. In low-resolution experiments,[20,21] if distinct bands become evident after restriction without the addition of S1 nuclease the cleavage is primarily double stranded; single-stranded cleavage events are revealed by subsequent treatment with S1 nuclease.

Cleavage of Z-DNA by the complex and its less efficient cobalt analogue have been characterized.[21,36] Cleavage is found at alternating pyrimidine–purine stretches under conditions where the Z form is adopted, with

[36] M. R. Kirshenbaum, Ph.D. Dissertation, Columbia University, New York (1989).

TABLE I

SITES OF DOUBLE-STRANDED CLEAVAGE BY Rh(DIP)$_3^{3+a}$

DNA	Location of Rh(DIP)$_3^{3+}$ cleavage sites[b] (kb)	Location(s) of palindromic repeat(s) (kb)	Size of stem (bp)	Size of loop (bp)
pBR322	3.25 ± 0.07	3.0650	11	3
		3.2205	10	6
		3.1230	9	5
pColE1	0.09 ± 0.20	0.1005	13	5
φX174	2.25 ± 0.11	2.3300	13[c]	2
(rfII)	4.14 ± 0.11	3.9650	9	3
	3.07 ± 0.10	3.0085	8	8

[a] Adapted from Ref. 20.

[b] Averages were calculated from at least six experiments with at least two restriction enzymes, where the error given is 1σ.

[c] The palindromic sequence contains a single internal incorrect purine–pyrimidine pairing.

greater cleavage at the border of the stretch and with one strand being preferred. Supercoiled substrates have been used in these experiments.

Perhaps most easy to detect are the double-stranded cleavage sites which appear to mark sites of cruciform formation. Table I shows a variety of different cruciforms targeted by Rh(DIP)$_3^{3+}$.[20] These cruciforms vary in terms of the length, sequence, and the sequence flanking. What is targeted by the rhodium complex is the cruciform structure. Low-resolution experiments show double-stranded cleavage in the vicinity of the palindomic sites and furthermore the necessity of supercoiling (native superhelical density) for specific targeting. In a cruciform, the extrusion of the inverted repeat sequence into two intrastrand hydrogen-bonded segments relieves torsional strain associated with supercoiling.[37] If the plasmid DNA is first linearized with a restriction enzyme before treatment with the metal complex, no cleavage is apparent. If the supercoil stress is relieved, the structure recognized by the metal complex is no longer present. Recognition is of the structure, not the sequence, and the complex does not promote extrusion. Cleavage also varies as a function of magnesium concentration and temperature as would be expected based on the requirements for cruciform extrusion.[37] Different cruciforms are targeted with different efficiencies, however. High-resolution experiments[20] show

[37] D. M. J. Lilley, K. M. Sullivan, and A. I. Murchie, *Prog. Nucleic Acid Res. Mol. Biol.* **1,** 126 (1987).

the cleavage by the rhodium complex to occur directly flanking the cruciform stem and preferentially at the flank which is higher in AT content.

Other unusual conformations are also targeted by the complex, at least some of which correspond to conformations which have not yet been well characterized.[38] The fact that the complex cleaves at a particular site does not itself establish that a sequence adopts a cruciform or Z-site. Specific cleavage does, however, indicate an unusual non-B conformation and one which is neither A-like nor single-stranded. One may further identify the conformation targeted by observing the specific pattern of cleavage, whether double- or single-stranded, at several positions or a single site, and through the sequence itself (e.g., if it is an alternating purine–pyrimidine sequence or flanking a palindromic repeat). A disadvantage in the application of the complex lies in the fact that if no specific conformation is detected, after extensive irradiation some secondary photoreaction at 5'-GG-3'and 5'-GA-3' sites become apparent as does some covalent binding of the rhodium complex to the DNA.[36,39] Low irradiation levels are required to detect unusual conformations, and the prime advantage in the application of this probe lies in its rapid and sensitive detection of cruciforms (in particular because of the double-stranded cleavage) and other unusual conformations and their ready localization to regions on long plasmids.

Shape-Selective Cleavage of RNA by Metal Complexes

This same family of transition metal complexes can also be applied to probe RNA structure. It has been shown that photolysis of the metal complexes in the presence of RNA promotes cleavage at diverse sites on the RNA.[23] Photocleavage by the different complexes appears to involve similar patterns of recognition as seen with DNA.

Reaction Conditions

Cleavage reactions are similar to those for DNA fragments. The buffer conditions are identical; however, reaction times may vary, and the cold carrier is tRNA (100 μM nucleotides). Cleavage conditions for the rhodium complexes are as follows: 10 μM Rh(phi)$_2$(bpy)$^{3+}$ at 310 nm for 2 to 8 min on the lamp; 10 μM Rh(phen)$_2$(phi)$^{3+}$ at 365 nm for 10 min on the lamp; 2.5 μM Rh(DIP)$_3^{3+}$ at 313 nm for 6 min on the lamp. Cleavage with Ru(TMP)$_3^{2+}$ or Ru(phen)$_3^{2+}$ (2.5 μM) is performed at 442 nm for 20 min

---

[38] I. Lee, unpublished results.
[39] M. B. Fleisher, H.-Y. Mei, and J. K. Barton, *Prog. Nucleic Acid Res. Mol. Biol.* **2**, 65 (1988).

on the laser followed by aniline treatment: resuspend the modified RNA in 20 μl of 1.0 M aniline buffered at pH 4.5 with acetic acid and incubate at 60° in the dark for 20 min, freeze the reaction mixture at −70° and lyophilize to dryness, and wash the samples twice with 40-μl aliquots of water.

Discussion

Much of our understanding of how these complexes interact with RNA and hence how they may be applied in probing RNA structure comes from studies[23] carried out on tRNA^{Phe}, one of the few crystallographically characterized[2] RNA polymers, as well as on mutant[40] tRNA substrates. The sites targeted are consistent with the parameters for recognition observed on DNA helices. In the case of Rh(DIP)$_3^{3+}$ and in particular with Rh(phen)$_2$(phi)$^{3+}$, the novel recognition leads to highly specific sites of cleavage. The novel and specific sites cleaved provide a scheme therefore for the application of these reagents in cleavage experiments which serve as fingerprints to monitor the changes in structure that may arise on mutation or in binding alternate molecules to the RNA. Cleavage of tRNA^{Phe} by the two rhodium complexes is shown in Fig. 4.

Rh(DIP)$_3^{3+}$, for example, recognizes and cleaves the folded loop regions of a tRNA molecule. Similar to recognition of DNA cruciforms, the complex appears to recognize hydrophobic regions of the molecule which are not purely double- or single-stranded. However, it is not yet known how the complex specifically interacts with this region of the molecule. Of more interest, specific cleavage is also consistently observed adjacent to GU mismatches (3′ side of U) within double helical regions of RNA (see C70 in Fig. 4).[41] These reactions are unique, highly specific, and can provide a structural marker.

Recognition of RNA structure by Rh(phen)$_2$(phi)$^{3+}$ is of particular interest in that the complex does not simply interact with double- or single-stranded regions of the molecule, but rather with regions that exhibit tertiary interactions.[40,42] These include on tRNA^{Phe} the D-T loop region, the triply bonded sites where three bases are hydrogen bonded one to another, and the helix–loop junction regions of tRNA in which the major groove is accessible to the metal complex. We can understand the basis for the targeting of these specific sites of tertiary folding by the complex

[40] C. S. Chow, L. S. Behlen, O. C. Uhlenbeck, and J. K. Barton, *Biochemistry* **31**, 972 (1992).
[41] C. S. Chow and J. K. Barton, submitted for publication; C. S. Chow, Ph.D. Dissertation, California Institute of Technology, Pasadena, CA, 1992.
[42] C. S. Chow, K. M. Hartmann, S. L. Rawlings, P. W. Huber, and J. K. Barton, *Biochemistry* **31**, 3534 (1992).

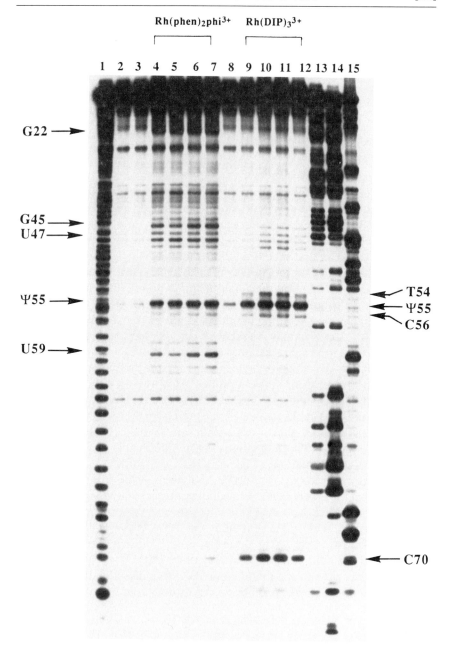

based on its recognition characteristics of DNA. One would not expect any interaction of $Rh(phen)_2(phi)^{3+}$ in the major groove of RNA because the helix is generally A form with a deep shallow major groove which is inaccessible to the metal complex. At the sites of triply bonded bases, the third base may provide an accessible surface for stacking in the helix from the major groove. $Rh(phen)_2(phi)^{3+}$ therefore provides an extremely useful reagent for targeting such sites of tertiary folding. Specific cleavage by the reagent alone does not establish such a tertiary interaction, however. Nonetheless it does allow for the characterization of the site as neither double helical nor in a single coiled form but instead folded in a manner which permits access for stacking into a helix from the major groove. The fact that cleavage is obtained with this reagent calls for additional experiments to conclude whether the site targeted participates in tertiary folding. Most importantly, this reagent can be very usefully applied in comparing native structures of RNA to mutant RNAs.[40] Because the complex has the ability to fingerprint specific (and generally few) regions of tertiary structure, cleavage patterns on a mutant RNA can serve to indicate structural changes resulting from the mutation.

Cleavage results with $Rh(phi)_2(bpy)^{3+}$ are generally similar to those obtained with $Rh(phen)_2(phi)^{3+}$, but less specific. The smaller $Rh(phi)_2(bpy)^{3+}$ recognizes additional sites such as helix–helix junctions of the type found in tRNA (acceptor stem and D-stem) or in 5 S rRNA (helices I and V).[42]

Selective cleavage by the ruthenium complexes on tRNA[23] is not extremely revealing. Although we have observed a preferential binding of the Λ isomer to RNA, the lack of selective cleavage by $Ru(TMP)_3^{2+}$ on tRNA suggests that perhaps the shortness of the helices in tRNA does not favor interaction with the metal complex. It is likely, however, that in regions of the RNA in which molecules can intercalate, one would see enhanced cleavage with $Ru(phen)_3^{2+}$ over $Ru(TMP)_3^{2+}$, which cannot in-

---

FIG. 4. Cleavage of tRNA^{Phe} by the rhodium probes. Autoradiogram of the 8 M urea denaturing 20% polyacrylamide sequencing gel of yeast tRNA^{Phe} ^{32}P-labeled at its 3' end. The concentrations of the metal complexes were 10 μM for $Rh(phen)_2(phi)^{3+}$ and 2.5 μM for $Rh(DIP)_3^{3+}$. Buffer conditions were 50 mM Tris, 20 mM sodium acetate, 18 mM NaCl, pH 7.0, for $Rh(phen)_2(phi)^{3+}$ and 5 mM Tris, 50 mM NaCl, pH 7.0, for $Rh(DIP)_3^{3+}$. Other buffer conditions may be used; with 10 mM Mg^{2+} and higher concentrations of metals, similar results are obtained. Lane 1: Alkaline hydrolysis. Lane 2: RNA control in the absence of metal complex or light. Lanes 3 and 8: Light controls; irradiations at 365 nm for 10 min and 313 nm for 8 min, respectively. Lanes 4–7: Specific cleavage by $Rh(phen)_2(phi)^{3+}$ at irradiation times of 4, 6, 8, and 10 min, at 365 nm. Lanes 9–12: Specific cleavage by $Rh(DIP)_3^{3+}$ at irradiation times of 2, 4, 6, and 8 min at 313 nm. Lanes 13–15: Sequencing reactions, A at 37°, A at 90°, and U, respectively. Arrows at left indicate $Rh(phen)_2(phi)^{3+}$ cleavage sites, and arrows at right show $Rh(DIP)_3^{3+}$ cleavage sites.

tercalate. This is seen in the case of tRNA in which $Ru(phen)_3^{2+}$ preferentially cleaves in the D-T loop region. Studies on larger RNA helices need to be conducted. The transition metal complexes certainly provide new, specific, and unique probes to examine RNA structures.

Acknowledgments

We are grateful to the National Institute of General Medical Science (GM33309 and GM08346) and to the Ralph M. Parsons Foundation for financial support of research.

[13] Probing DNA Structure with Psoralen *in Vitro*

By David W. Ussery, Robert W. Hoepfner, and Richard R. Sinden

Introduction: Psoralen and Its Photochemistry

Psoralens have been widely used as probes of DNA structure, RNA structure, and DNA–protein interaction. Bacterial and eukaryotic cells are permeable to psoralens which can photobind to DNA inside cells. Psoralen binds preferentially to DNA in cells. The binding of psoralens to DNA is sensitive to superhelical density, alternate DNA conformations, and protein association. Psoralens initially bind noncovalently to DNA by interacalation into the DNA double helix. The intercalative binding is sufficiently weak that relatively few psoralens will bind DNA, resulting in little perturbation of the topology and structure of DNA. On irradiation with near-UV light (320–400 nm), an intercalated psoralen can be photobound to an adjacent pyrimidine base, forming a monoadduct. When the adjacent base in the opposite strand is also a pyrimidine, interstrand cross-links can form in a second photochemical reaction (Fig. 1).[1] Cross-links covalently bind the two strands of the double helix together. The number of psoralens photobound to DNA can be carefully controlled by the incident exposure to UV light. Psoralens bind tightly to proteins and membranes but the binding is not dependent on irradiation with 360 nm light.[2,3] Psoralen monoadducts and cross-links can be easily assayed by a number of methods, which will be described or referenced in this chapter.

[1] R. S. Cole, *Biochim. Biophys. Acta* **217**, 30 (1970).
[2] S. Frederiksen and J. E. Hearst, *Biochim. Biophys. Acta* **563**, 343 (1979).
[3] J. D. Laskin, E. Lee, E. J. Yurkow, D. L. Laskin, and M. A. Gallo, *Proc. Natl. Acad. Sci. U.S.A.* **82**, 6158 (1985).

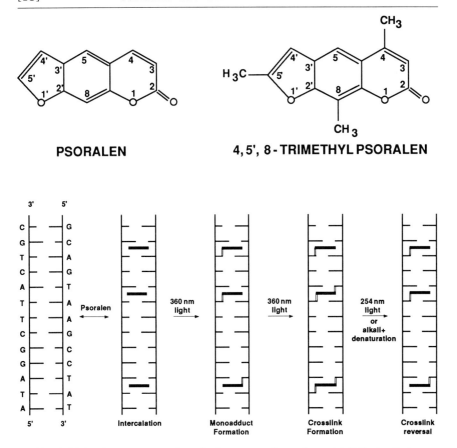

FIG. 1. Molecular structure and numbering system of psoralen and 4,5′,8-trimethylpsoralen. At bottom is shown the mechanism of reaction between psoralen and DNA. Psoralen intercalates between the stacked bases of the DNA double helix. This reaction is reversible and occurs in the absence of light. Although psoralen is shown intercalating preferentially between 5′-TA dinucleotides, psoralen intercalates relatively independently of base structure. On addition of 360 nm light, psoralen forms monoadducts to single strands. Following a second photochemical reaction, interstrand cross-links are formed. The preferred site for Me$_3$-psoralen cross-link formation is 5′-TA ≫ 5′-AT > 5′-GT [F. Esposito, R. G. Brankamp, and R. R. Sinden, *J. Biol. Chem.* **263,** 11466 (1988)]. If an adjacent pyrimidine is not present in the opposite strand a cross-link cannot form. Cross-links can be photoreversed by two different methods: photoreversal or treatment with alkali under denaturing conditions. Detailed conditions for the photobinding and cross-link reversal are given in the text.

Psoralen Derivatives

Several different derivatives of psoralen have been used to study DNA structure. Figure 1 shows the structure and numbering system of psoralen and 4,5′,8-trimethylpsoralen (Me_3-psoralen). Psoralen derivatives have been described that vary in solubility and reactivity. A number of different psoralens are commercially available including a large collection offered by HRI Associates, Inc. (2341 Stanwell Drive, Concord, CA 94520). Me_3-psoralen (which we routinely use in our laboratory) has a high quantum yield resulting in a rapid rate of photoaddition. However, it also photodestructs readily. The reactivity of Me_3-psoralen with supercoiled DNA has been well characterized.[4,5] One derivative, 8-methoxypsoralen (8-MOP), is long lived in solution, although less photoreactive than Me_3-psoralen. Psoralen derivatives have been described that are more soluble than psoralen or Me_3-psoralen.[6] The use of biotinylated psoralen derivatives for labeling nucleic acid hybridization probes has been reviewed.[7]

Psoralen Photobinding

Psoralens bind to DNA by interacalation between the stacked base pairs of DNA. The intercalative binding unwinds the DNA helix by about 28° per psoralen molecule.[8] On irradiation with 360 nm light, monoadducts are initially formed. Psoralens form cyclobutane bonds between the pyrone 3,4- or furan 4′,5′-double bonds of the psoralen and the 5,6-double bonds of pyrimidine bases. The photochemistry of psoralens has been reviewed.[6,9] There are specific psoralen derivatives (e.g., angelicin or isopsoralen) which will only form monoadducts. In addition, because the formation of monoadducts is quicker than the formation of a cross-link, a short laser pulse of light (15 nsec) will form only monoadducts.[10]

The formation of interstrand cross-links requires further irradiation with 360 nm light. According to Shi and Hearst, only furan-sided monoadducts can be driven to cross-links.[11] The formation of cross-links requires the presence of adjacent pyrimidines on opposite strands (see Fig. 1);

[4] J. E. Hyde and J. E. Hearst, *Biochemistry* **17**, 1251 (1978).
[5] R. R. Sinden, J. O. Carlson, and D. E. Pettijohn, *Cell (Cambridge, Mass.)* **21**, 773 (1980).
[6] J. E. Hearst, S. T. Isaacs, D. Kanne, H. Rapoport, and K. Straub, *Q. Rev. Biophys.* **17**, 1 (1984).
[7] C. Levenson, R. Watson, and E. L. Sheldon, this series, Vol. 184, p. 577.
[8] G. Wiesehahn and J. E. Hearst, *Proc. Natl. Acad. Sci. U.S.A.* **75**, 2703 (1978).
[9] G. D. Cimino, H. B. Gamper, S. T. Isaacs, and J. E. Hearst, *Annu. Rev. Biochem.* **54**, 1154 (1985).
[10] B. H. Johnston, M. A. Johnson, C. B. Moore, and J. E. Hearst, *Science* **197**, 906 (1977).
[11] Y. Shi and J. E. Hearst, *Biochemistry* **26**, 3792 (1987).

consequently, not all furan side monoadducts can form cross-links. Me$_3$-psoralen and 8-MOP have been shown to have a sequence specificity for 5'-TA $\gg$ 5'-AT $>$ 5'-GT.[12] The sequence specificity of several other psoralen derivatives has been studied.[13] In addition, although 5'-TA represents a preferential cross-link site, the rate of cross-linking can vary 3- to 4-fold depending on the composition of bases flanking the 5'-TA dinucleotide. Changes 3 base pairs (bp) away from the 5'-TA can result in significant changes in the reactivity of the 5'-TA.[12] Presumably, the reactivity of psoralen with DNA is sensitive to subtle changes in the twist, tilt, and roll of the bases as these parameters change with DNA sequence. Consequently, it may be difficult to predict theoretically the sequence specificity of the reactivity of psoralen with DNA.

Based on the crystal structure of a psoralen–thymine monoadduct, it has been suggested that a psoralen cross-link will introduce a stable 56° bend into DNA.[14] Two-dimensional NMR analysis of short cross-linked oligonucleotides (8 bp) has also suggested that cross-links bend DNA.[15] However, analyses of longer pieces of cross-linked DNA on polyacrylamide gels are not consistent with a stable bend introduced by the psoralen cross-link.[16,17]

Psoralen Photobinding Procedure

DNA is dissolved in TEN buffer [10 mM Tris, pH 7.6, 1 mM ethylenediaminetetraacetic acid (EDTA), 50 mM NaCl] at 50 μg/ml. Ten microliters of a saturated solution of Me$_3$-psoralen in ethanol is added per milliliter DNA solution [a 1% (v/v) solution]. This adds 6–7 μg Me$_3$-psoralen per milliliter. However, Me$_3$-psoralen is only soluble at about 0.6 μg/ml.[18] The use of a saturated solution minimizes variation in Me$_3$-psoralen concentration which might affect the rate and extent of photobinding. Before irradiation, DNA is typically equilibrated with Me$_3$-psoralen for 2 min to allow for intercalation. This length of time is most likely excessive since equilibration is probably quite rapid. DNA is placed in an open plastic petri dish (of varying size from 35 to 150 mm depending on the sample volume) to form a thin layer of solution less than 5 mm deep. Multiple small samples of DNA ($<$50–100 μl) are typically pipetted as drops onto a petri dish. If

[12] F. Esposito, R. G. Brankamp, and R. R. Sinden, *J. Biol. Chem.* **263**, 11466 (1988).

[13] V. Boyer, E. Moustacchi, and E. Sage, *Biochemistry* **27**, 3011 (1988).

[14] D. A. Pearlman, S. R. Holbrook, D. H. Pirkle, and S.-H. Kim, *Science* **227**, 1304 (1985).

[15] M. T. Tomic, D. E. Wemmer, and S.-H. Kim, *Science* **238**, 1722 (1987).

[16] R. R. Sinden and P. J. Hagerman, *Biochemistry* **23**, 6299 (1984).

[17] T. E. Haran and D. M. Crothers, *Biochemistry* **27**, 6967 (1988).

[18] S. T. Isaacs, C. J. Shen, J. E. Hearst, and H. Rapoport, *Biochemistry* **16**, 1058 (1977).

many samples need to be irradiated simultaneously they are placed in wells of a microtiter dish (24 or 96 well). The geometry of this configuration, however, is not identical to that for samples irradiated in an open petri dish, and this will affect the cross-link yield.

Samples in uncovered petri dishes are placed on ice in a 4° room about 4–5 cm below two General Electric BLB20 bulbs. A number of commercial long-wavelength lamps are available. However, a lamp can be easily made by obtaining standard "black light" bulbs from a good lighting store and placing them in a standard fluorescent light fixture. Two GE BLB20 bulbs at 4–5 cm produce an incident light intensity of approximately 1.2 kJ m^{-2} per minute as measured by a Blak Ray Long Wave UV Meter Model J-221 (Ultraviolet Products, Inc., San Gabriel, CA). The incident light intensity can be varied by changing the distance between the bulbs and the sample. Me$_3$-psoralen reacts and/or photodestructs within about 20 min of irradiation. Therefore, for long exposures readdition of Me$_3$-psoralen every 10–15 min is required to ensure a nearly linear rate of photobinding. The rate of reaction and photodestruction for 8-methoxypsoralen is much slower, and it needs to be replenished only every 60 min.

Reversal of Trimethylpsoralen Cross-links

Psoralen cross-links are reversible. There are two methods commonly used for cross-link reversal: photoreversal and alkali reversal. Photoreversal occurs between 240 and 310 nm and will initially leave monoadducts attached predominantly to the furan ring.[19,20] Alkali reversal leaves monoadducts attached through the pyrone ring.[21] This knowledge has been utilized to deliver a site-specific monoadduct, with the psoralen in a particular orientation (e.g., furan side or pyrone side photobound).[22]

Photoreversal. The following photoreversal procedure is routinely used in our laboratory. Samples (up to 20 μg DNA) are resuspended from an ethanol precipitate in 50 μl water. The samples are pipetted into caps of 1.5-ml polypropylene microcentrifuge tubes mounted (upside down) on a small piece of cardboard. Microcentrifuge tube caps allow greater recovery of small volumes compared to plastic petri dishes where significant liquid will adhere to the surface. The samples are placed 15 cm under a 254 nm ultraviolet (germicidal) UV lamp and irradiated for 10 min at room temperature with an incident light intensity of 0.54 kJ m^{-2} per

[19] G. D. Cimino, Y. Shi, and J. E. Hearst, *Biochemistry* **25**, 3013 (1986).
[20] Y. Shi and J. E. Hearst, *Biochemistry* **26**, 3786 (1987).
[21] Y. Shi, H. Spielmann, and J. E. Hearst, *Biochemistry* **27**, 5174 (1988).
[22] A. Yeung, J. Dinehart, and B. K. Jones, *Biochemistry* **27**, 6332 (1988).

minute (900 μW/cm^2) [as measured with a Blak Ray Model J255 UV Meter (Ultraviolet Products)]. The samples are placed into microcentrifuge tubes, and 2 μg glycogen and 6 μl of 3 M potassium acetate are added. The samples are precipitated by the addition of 200 μl (100%) ethanol and incubation at $-20°$ for at least 2 hr.

Alkali Reversal. Shi *et al.*[21] have reported that incubation of psoralen-cross-linked DNA under alkaline denaturing conditions (e.g., pH 10 at 60° for 2 hr in the presence of 3 M urea) results in the reversal of furan-side cyclobutane bonds, leaving exclusively pyrone-side monoadducts. Yeung *et al.* have utilized an incubation for 30 min at 90° in 0.1 M KOH to catalyze the reversal of cross-links.[22]

Assays for Psoralen Photobinding to DNA

Measurement of Total Psoralen Binding Using Radioactive Psoralens

Radioactive psoralens can be used to determine the total amount of psoralen photobound to DNA. The calculation of "psoralens per base pair" requires determination of the specific activity of the psoralens and the specific activity of the labeled DNA (see, e.g., Sinden *et al.*[5]). Radioactive psoralens are available from HRI Associates, Inc., or they can be made in a custom labeling reaction performed by major suppliers of radioactivity.

Monoadduct Assays

Monoadducts can be detected by two methods. First, monoadducts can be readily observed in short DNA fragments by the retardation in migration on acrylamide gels.[12,23] Second, monoadducts can be detected, and in some cases quantitated, by exonuclease assays that will be described in detail below.

Cross-link Assays

Psoralen cross-links can be detected by a variety of methods, as listed below. All involve the same basic premise that the cross-link prevents complete separation of DNA strands following denaturation of the DNA. Because the two strands are covalently held together, following rapid removal of the denaturing conditions the DNA is "reversibly renaturing." From the fraction of DNA that is cross-linked (or non-cross-linked) the average number of cross-links per DNA molecule can be calculated assum-

---

[23] H. Gamper, J. Piette, and J. E. Hearst, *Photochem. Photobiol.* **40**, 29 (1984).

ing a Poisson distribution of cross-links. M_{xl}, the average number of cross-links per fragment, can be calculated from Eq. (1):

$$M_{xl} = -\ln(1 - F_{xl}) \tag{1}$$

where F_{xl} is the fraction of the DNA that is cross-linked. From M_{xl} and the size of the DNA, cross-links per kilobase pair (or other unit) can be determined.

Cross-link Assay Utilizing Hydroxyapatite. Single-stranded DNA is eluted from hydroxyapatite at a lower phosphate concentration than double-stranded DNA. To quantitate cross-links using hydroxyapatite, DNA is denatured by heat or alkali and quickly renatured. Cross-linked DNA is "reversibly renaturing" and will reform double strands, whereas non-cross-linked DNA will remain single-stranded. The methods and procedures used to calculate cross-link yields have been described in detail previously.[1,24] The fraction of reversibly renaturing DNA (determined from hydroxyapatite chromatography) and the molecular weight of the DNA analyzed are used to calculate the cross-link yield. This method is useful for determining cross-link yields in "total" or "bulk" DNA purified from bacterial or eukaryotic cells which may be randomly sheared (or cut) to some average size.

Denaturing Agarose Gels. Cross-linked DNA migrates much slower than single-stranded DNA in alkaline agarose gels.[25] Following cross-linking, the quantitation of the fraction of defined restriction fragments migrating as cross-linked or single-stranded DNA can be used to calculate cross-link yields. This method is applicable for DNAs from several hundred base pairs to over 10 kilobases (kb) in size. Caution should be exercised in quantitating gels stained with ethidium bromide since the fluorescence yields from single- and double-stranded DNA are not comparable. Quantitation of radioactively labeled DNA can provide a reasonable estimate of cross-link yields.

Alkaline agarose gels are run in buffer consisting of 33 mM NaOH (made fresh) and 2 mM EDTA. The NaOH and EDTA should be added after the agarose is autoclaved (or otherwise dissolved). The buffer should be circulated during electrophoresis. Gels should be neutralized by incubation in several changes of 20 mM Tris, pH 7.6, prior to staining with ethidium bromide (0.5 μg/ml). Samples can be prepared for loading on alkaline agarose gels by incubation in 0.25 M NaOH for 30 min at 20°.

[24] R. R. Sinden and R. S. Cole, *in* "DNA Repair: A Laboratory Manual" (E. C. Friedberg and P. C. Hanawalt, eds.), Vol. 1A, p. 69. Dekker, New York, 1981.
[25] T. R. Cech, *Biochemistry* **20**, 1431 (1981).

Acrylamide Gels. We have found that cross-links can be easily quantitated in DNA from 10 bp to several hundred base pairs by analysis on denaturing (DNA sequencing) polyacrylamide gels.[12,26] The cross-linked DNA runs at a position distinct from single-stranded DNA in denaturing gels. For DNAs with lengths of 10–72 bp a retardation in the migration of cross-linked DNA is observed, as may be intuitively expected.[12,26] Interestingly, however, the migration of a cross-linked 122-bp fragment was faster than that of the 122-base single-stranded molecule.[26]

Standard denaturing DNA polyacrylamide (DNA sequencing) gels are used. Gels are made in TBE buffer (90 mM Tris, 90 mM borate, 2.5 mM EDTA, pH 8.3) containing 8.33 M urea. We have used acrylamide concentrations of 5–20% (with a 20:1 acrylamide–bisacrylamide ratio) depending on the size of the DNA to be examined. Samples are resuspended in formamide (containing bromphenol blue and xylene cyanol) and boiled for 3 min prior to electrophoresis.

Denaturation/Renaturation Assays Using Southern Hybridization. Vos and Hanawalt have developed an assay for measuring cross-links in defined genomic sequences.[27] Essentially, the DNA is cross-linked, alkali denatured (10-min incubation at 55° in 0.2 N NaOH), run on a neutral agarose gel, transferred to nitrocellulose, and then probed with the genomic sequence. The position of non-cross-linked and cross-linked DNAs will be different, allowing quantitation of each form of DNA. Non-cross-linked DNA will bind the hybridization probe efficiently. A potential complication with this approach is that DNA cross-links can prevent hybridization at very high doses of cross-linking if the DNA has not been sufficiently nicked during the transfer procedure. It would seem best to compare samples at similar, low levels of cross-linking (<1 cross-link per fragment of DNA). At high levels of cross-linking, quantitation could be affected by differential hybridization to the cross-linked and non-cross-linked species. In fact, Matsuo and Ross[28] have utilized the nonhybridization of cross-linked DNA as an assay for cross-linking, albeit under different conditions than described by Vos and Hanawalt.[27]

Denaturation/Renaturation Assays Utilizing Ethidium Bromide Fluorescence and S1 Nuclease Digestion. Matsuo and Ross have described a method for detecting cross-links that involves denaturation, renaturation, and detection of the "reversible renaturing" DNA by S1 nuclease digestion, ethidium bromide fluorescence, and hybridization to

[26] R. R. Sinden and T. J. Kochel, *Biochemistry* **26**, 1343 (1984).
[27] J. H. Vos and P.C. Hanawalt, *Cell* (*Cambridge, Mass.*) **50**, 789 (1987).
[28] N. Matsuo and P. M. Ross, *Arch. Biochem. Biophys.* **266**, 351 (1988).

the fraction of non-cross-linked DNA.[28] The three independent methods provide a check in that they should give similar values.

Exonuclease Assays to Detect Monoadducts and/or Cross-links

Exonuclease III/Photoreversal Assay. We have developed a procedure to map the site of psoralen monoadducts and cross-links in DNA with base pair resolution. This assay is extremely sensitive and can be utilized at levels of psoralen photobinding as low as 1 cross-link per 8000 bp.[29,30] For this we utilize exonuclease III (exoIII), which digests a single strand of double-stranded DNA from a recessed 3'-OH end. Exonuclease III stops quantitatively two bases 3' to the base containing a photobound psoralen molecule; exoIII stops at monoadducts in the digested strand but not at the positions of monoadducts in the opposite (nondigested) strand.[29] The site of photoaddition can be determined by digesting a cross-linked sample of DNA with exoIII and analyzing the digestion products on a DNA sequencing gel. Basically, the procedure can be outlined as follows: (1) Photobind psoralen to DNA. (2) Purify a DNA fragment of interest that is end-labeled at a unique 5' end. (3) Digest to completion with exoIII. (4) Photoreverse DNA cross-links. (5) Analyze the digestion products on a denaturing 5% (w/v) polyacrylamide gel, using sequenced DNA as a marker.

Step 1: Psoralen photobinding. Psoralen photobinding is described in detail above. One important consideration for this type of analysis is that, on average, less than one psoralen molecule should be introduced into the DNA fragment of interest. Consequently this type of assay allows analysis of psoralen binding at extremely low levels of total photobinding. This will minimize the topological and structural perturbation of DNA structure induced by psoralen photoaddition. When multiple psoralens are introduced, exoIII digestion will stop at the first adduct encountered. For analyzing 200- to 300-bp DNA fragments *in vitro,* irradiation times of less than 30 sec (<1 kJ m^{-2}) are sufficient using the irradiation conditions described above.

Step 2: Purification of DNA fragment uniquely labeled at 5' blunt end with recessed 3'-OH at other end. The procedures described below apply specifically to analysis of inserts cloned into plasmid pUC8. The procedure can be adapted to analysis of any restriction fragment with judicious selection of appropriate restriction enzymes. The rationale for using a labeled blunt end and an end with a recessed 3'-OH is that exoIII prefers digestion at a recessed 3'-OH end. This will direct digestion to the end of the DNA opposite to the label. It is best to have the region of DNA of

[29] T. J. Kochel and R. R. Sinden, *J. Mol. Biol.* **205**, 91 (1989).
[30] T. J. Kochel and R. R. Sinden, *BioTechniques* **6**, 532 (1988).

interest nearest the recessed 3'-OH end of the molecule. The disadvantage with this procedure is that the blunt ends are more difficult to label with [γ-^{32}P]ATP. For analysis of DNA sequences inserted into the polylinker region of plasmid pUC8, digestion with *Pvu*II works well. This is followed by digestion with an enzyme that cuts within the polylinker but to the 3' side of the insert of interest (we have used *Bam*HI or *Hin*dIII). One important consideration is that different enzymes have different sensitivites to psoralen adducts in DNA. A cross-link or monoadduct at or near a restriction site will likely prevent restriction. *Hin*dIII appears particularly sensitive to psoralen.[31] To ensure complete cutting we typically utilize at least a 3- to 10-fold excess of restriction enzyme.

Digestion. Following cross-linking, the DNA (5–15 μg) is precipitated and resuspended in 100 μl *Pvu*II buffer, and 30–100 units of *Pvu*II are added. The sample is incubated for a minimum of 3 hr at 37°. The sample is then extracted twice with phenol and twice with chloroform–isoamyl alcohol (24 : 1). The sample is adjusted to 0.3 M sodium acetate, 2–3 volumes of ethanol is added, and the sample is incubated at $-20°$ for at least 1 hr to precipitate the DNA.

Phosphatase treatment. We have had the most success labeling blunt-ended restriction fragments using a procedure based on that described by Sambrook *et al.*[32] Briefly, following ethanol precipitation and a 70% ethanol wash of the pellet, the DNA is resuspended in 50 μl BAP buffer (50 mM Tris, pH 9.0, 1 mM MgCl$_2$, 0.1 mM ZnCl$_2$, 1 mM spermidine) and 1 unit of calf intestine alkaline phosphatase (CIAP) (molecular biology grade, Boehringer-Mannheim, Indianapolis, IN) is added.[33] The DNA is incubated for 15 min at 37° and then 15 min at 56°. An additional 1 unit of CIAP is added, and the sample is incubated again for 15 min each at 37° and 56°. Following the incubations 40 μl of 10× STE buffer (10× STE is 100 mM Tris, pH 8.0, 1.0 M NaCl, 10 mM EDTA) and 5 μl of 10% sodium dodecyl sulfate (SDS) are added, and the sample is incubated at 75° for 15 min. The DNA is extracted twice with phenol and twice with chloroform–isoamyl alcohol (24 : 1), and the DNA is precipitated by the addition of 300 μl ethanol and incubation at $-20°$ for at least 2 hr.

End-labeling procedure. We use a procedure for end labeling blunt-ended molecules based on that described by Maniatis *et al.*[34] DNA is

[31] J. O. Carlson, O. Pfenninger, R. R. Sinden, J. M. Lehman, and D. E. Pettijohn, *Nucleic Acids Res.* **10,** 2043 (1982).

[32] J. Sambrook, E. F. Fritsch, and T. Maniatis, "Molecular Cloning: A Laboratory Manual," 2nd Ed. Cold Spring Harbor Laboratory, Cold Spring Harbor, New York, 1989.

[33] We typically dissolve DNA in water and add 0.1 volume of a 10× enzyme buffer.

[34] T. Maniatis, E. F. Fritsch, and J. Sambrook, "Molecular Cloning: A Laboratory Manual." Cold Spring Harbor Laboratory, Cold Spring Harbor, New York, 1982.

dissolved in 40 μl prekinase buffer (20 mM Tris, pH 9.5, 1 mM spermidine, 0.1 mM EDTA), incubated at 70° for 5 min, and quickly cooled on ice. Five microliters of 10× blunt end kinase buffer [500 mM Tris, pH 9.5, 100 mM MgCl$_2$, 50 mM dithiothreitol (DTT), 50% glycerol] and 1 μl (160 μCi) of [γ-^{32}P]ATP (>7000 Ci/mmol) are added. Six to ten units of T4 polynucleotide kinase are added (diluted to about 3–5 units /μl in 50 mM Tris, pH 8.0, if necessary), and the sample is incubated for 30 min at 37°. The sample is then incubated at 65° for 5 min, diluted to 100 μl with TEN buffer, and extracted with phenol and then chloroform–isoamyl alcohol, and then ethanol precipitated.

Secondary digestion and fragment purification. The labeled sample must now be cut with a restriction enzyme to produce a recessed 3'-OH. In the polylinker region of p$UC8$ any enzyme producing a recessed 3'-OH can be used for the secondary cutting (*Hin*dIII, *Sal*I, *Bam*HI, *Xma*I, or *Eco*RI) depending on the site used for cloning the insert to be analyzed. For inserts in the *Eco*RI and *Bam*HI sites we have utilized cutting with *Bam*HI and *Hin*dIII, respectively. Following ethanol precipitation, the sample is dissolved in 100 μl restriction enzyme buffer, 30–100 units enzyme is added, and the sample is incubated for 4–20 hr at 37°. The sample is then extracted successively with phenol and chloroform and then ethanol precipitated.

The sample is resuspended in 20 μl TB buffer (40 mM Tris–borate, pH 8.3), adjusted to 5% glycerol with a TB buffer containing 25% glycerol, bromphenol, and xylene cyanol, and run on a 5% (w/v) nondenaturing polyacrylamide gel (40 × 35 × 0.4 cm) in TB buffer containing 10% (v/v) glycerol. (If the reaction is done in a smaller volume, the sample can be mixed with dye mix containing 10 mM EDTA and 0.1% SDS and loaded directly on the polyacrylamide gel.) DNA is fractionated by electrophoresis at 300 V (15 mA) for sufficient time to resolve the fragment of interest. This fragment is identified by autoradiography, excised from the gel, placed in a 1.5-ml microcentrifuge tube, ground with a plastic pipette tip against the side of the tube, and extracted into 300 μl of extraction buffer (1 M sodium acetate, 2 mM EDTA) containing 2 μg glycogen for 3–20 hr at 37°. The solution is centrifuged through a siliconized glass wool pad in the bottom of a 1.5-ml microcentrifuge tube (with a small hole in the bottom) into a second microcentrifuge tube. This solution is centrifuged for 5 min at 12,000 g to pellet any acrylamide particles; the supernatant is removed and precipitated by the addition of 3–4 volumes of ethanol. The pellet is rinsed with 70% ethanol and dried.

Step 3: Digestion with exonuclease III. The DNA is resuspended in 100 μl of exoIII buffer (50 mM Tris-HCl, pH 8.1, 5 mM MgCl$_2$, 10 mM 2-mercaptoethanol), 200 units of exoIII (Bethesda Research Laboratories,

Gaithersburg, MD) are added, and the sample is incubated at 37° for at least 3 hr. The DNA is extracted twice with phenol, twice with chloroform, 1 μg glycogen is added, the sample is adjusted to 0.3 M potassium acetate, and the DNA is precipitated by the addition of 4 volumes ethanol. Four volumes of ethanol is used to ensure precipitation of small digestion products. To ensure complete digestion the entire exoIII digestion procedure is repeated. We have found that a single exoIII digestion often results in incomplete digestion, as indicated in part by multiple bands at exoIII stop sites and also exoIII stop sites in the control (non-cross-linked DNA) lane.

Step 4: UV photoreversal. The samples are resuspended in 50 μl water and photoreversed as described above. Following photoreversal the samples are adjusted to 0.3 M potassium acetate and precipitated with 4 volumes of ethanol. In actuality, if the fraction of cross-linked DNA is very low (<1%) photoreversal may be unnecessary. Nonphotoreversed cross-linked DNA should run approximately as a full-length strand on the DNA sequencing gel. This should not obscure identification of other sites of photoaddition. If a particular site contained exclusively cross-links it would not be observed, however. The photoreversal could be omitted to assay exclusively monoadducts since all cross-linked molecules should migrate approximately as full-length molecules.

Step 5: Analysis on denaturing polyacrylamide gel. DNA samples are applied to an 80 × 30 × 0.4 cm denaturing polyacrylamide gel in TBE buffer (described above). Several thousand counts per minute per sample are applied to the gel. As molecular weight markers, a Maxam–Gilbert or dideoxy-sequencing reaction should be run on the gel (using a primer that extends from the blunt end cut site). The band corresponding to the "exoIII stop" migrates three bases longer than a Maxam–Gilbert sequencing reaction and two bases longer than a dideoxy sequence marker corresponding to the psoralen photoadducted base. Taking mass and charge differences into account, it appears that exoIII stops two bases 3' to the base containing the monoadduct or cross-link.[29]

Other Nucleases. Depending on the experimental design, other exonucleases with different activities and characteristics can be used to map sites of photoaddition. *Bal*31 nuclease stops at only cross-links, as described by Zhen *et al.*[35] To digest the DNA in the 5' to 3' direction, λ-exonuclease can be used.[36] Digestion in the 5' to 3' direction would allow use of a primer extension assay rather than the end-labeling required with a 3' → 5'-exonuclease. λ-Exonuclease may suffice for identification purposes. However, quantitation with this enzyme is difficult, since it

[35] W. P. Zhen, O. Buchardt, H. Nielsen, and P. E. Nielsen, *Biochemistry* 25, 6598 (1988).
[36] E. A. Ostrander, R. A. Karty, and L. M. Hallick, *Nucleic Acids Res.* 16, 213 (1988).

pauses but does not completely stop at monoadducts. The $3' \rightarrow$ $5'$-exonuclease activity of T4 DNA polymerase is similar to exonuclease III and has been used to map psoralen photoproducts.[37]

Primer Extension Reactions

An alternative approach to mapping sites of psoralen photoaddition is to perform a polymerase fill-in or primer extension reaction which would pause or stop at the photoadducted base. Piette and Hearst have used *Escherichia coli* DNA polymerase I in a nick translation reaction in which the enzyme stopped only at cross-links.[38] T7 RNA polymerase stops at all psoralen photoproducts and can be used provided there is a promoter present.[39] Avian myeloblastosis virus (AMV) reverse transcriptase will stop at cross-links in RNA, and this may produce similar results with DNA.[40] It is expected that any DNA polymerase will be stopped by an interstrand cross-link. In principle any polymerase that pauses at monoadducts can be used to identify these sites. Enzymes that are stopped by monoadducts could be used to quantitate photobinding at individual sites.

Primer Extension Using Taq Polymerase to Map Psoralen Photoproducts with Nucleotide Resolution.

To investigate psoralen reactivity with eukaryotic genomes, an increase in the sensitivity would be required over the exoIII assay described above. Primer extension and polymerase chain reaction methodologies can be used to overcome some of the limitations of earlier procedures. The APEX technique developed by Cartwright and Kelly[40a] results in the linear amplification of the extension product. Axelrod and Majors[40b] have used a similar method to map UV induced photoproducts in yeast genes. Based on these procedures we have developed the following *Taq* polymerase primer extension assay to map psoralen photoproducts in plasmid DNA containing a region of potential Z-DNA (for example pCGTA-C[26]). This approach to mapping psoralen photoproducts is more sensitive and much faster than the ExoIII/Photoreversal assay described above.[40c]

Step 1: Psoralen photobinding. Psoralen photobinding is described in detail above and is used at doses similar to that for the ExoIII assay.

[37] V. Boyer, E. Moustacchi, and E. Sage, *Biochemistry* **27**, 3011 (1988).
[38] J. G. Piette and J. E. Hearst, *Proc. Natl. Acad. Sci. U.S.A.* **80**, 5540 (1983).
[39] Y.-B. Shi, H. Gamper, and J. E. Hearst, *J. Biol. Chem.* **263**, 527 (1988).
[40] G. Ericson and P. Wollenzien, *Anal. Biochem.* **174**, 215 (1988).
[40a] I. L. Cartwright and S. E. Kelly, *BioTechniques* **11**, 188 (1991).
[40b] J. A. Axelrod and J. Majors, *Nucl. Acids Res.* **17**, 171 (1989).
[40c] R. W. Hoepfner and R. R. Sinden, manuscript in preparation.

Step 2: Restriction digestion to generate a defined fragment for probing. This step may not be necessary; however, we have digested DNA into a defined restriction fragment for primer extension. This provides a defined termination site in the template strand which serves as an internal control to show that the polymerization in strands that do not contain a photoproduct have gone to completion. We have utilized fragments of about 400 bp (a *Pvu*II fragment from pUC8-based plasmids). After digestion, the DNA was extracted once with phenol, once with chloroform : isoamylalcohol, adjusted to 0.3 M NaOAc, and then precipitated by the addition of two volumes of ethanol.

Step 3: End labeling the primer. We have used 20mer primers that hybridize flush with the end of the template strand. These have been 70–150 bp from the site of interest (a Z-DNA sequence in our case). The primers are chemically synthesized which leaves a 5′ end containing a hydroxyl group. It is probably best to avoid primers containing 5′-TA dinucleotides as they are preferential photobinding sites for psoralen. An unfortunate "hot spot" for psoralen binding in the primer region would complicate the analysis. However, as discussed earlier, the number of photoproducts is low and only a small proportion of the fragments might contain a psoralen photoproduct in the primer region. A standard end labeling protocol[34] has been used to end label the primer with γ^{32}P-ATP.

Step 4: Thermal cycling primer extension using Taq polymerase. For the analysis of (GC)$_n$ Z-DNA forming regions we have used the Stoffel fragment of the *Taq* polymerase (Cetus) because of its high thermal stability. The Stoffel fragment of *Taq* polymerase (Cetus) stops at both monoadducts and interstrand cross-links. The reactions are done in 50 μl of 10 mM Tris, pH 8.3, 10 mM KCl, 4 mM MgCl$_2$, 5 mM of each dNTP containing 100 ng plasmid DNA and a >5-fold molar excess of labeled primer. Five units of Stoffel fragment polymerase are added and the sample is overlaid with mineral oil. The thermal cycling procedure will probably need to be optimized for each individual application. The specifics of our application of a G + C rich Z-DNA fragment in pUC8 are given below.

Denaturation. The fragments are denatured at 96° for 2 min.

Hybridization. Hybridization was for 30 sec at 5° above the calculated T_m of the primer (60° for a primer adjacent to a *Pvu*II site in pUC8). The stringent hybridization temperature was used to ensure specific hybridization but more importantly to try to reduce (or slow) the renaturation of any cross-linked DNA fragments. As discussed above, a cross-link provides a nucleation point for renaturation which would prevent the primer binding. Thus it is important to work with photobound samples containing few cross-links per fragment. A high concentration of primer may also help to overcome this potential problem.

Elongation. The reaction is elongated at 80° for 5 min. This high temperature is needed to destabilize DNA secondary structure formed by the palindromic $(GC)_n$ Z-DNA forming sequence. Secondary structures formed by the $(GC)_n$ Z-DNA region inhibit all different *Taq* polymerases we have examined. This high temperature may not be necessary when examining DNAs that do not contain regions of potential secondary structure.

Number of cycles. We have successfully used 10 cycles for the conditions described above. The number of cycles may need to be tailored for a particular template and primer.

Termination. After the cycling is complete as much oil as possible is removed. Fifty μl of water is added to increase the volume and the sample is extracted with 50 μl of phenol : chloroform (50 : 50). The sample is adjusted to 0.3 M NaOAc, 5 μg glycogen is added, and the DNA is precipitated by the addition of 3 volumes of ethanol.

Step 5: Analysis on a DNA sequencing gel. This step is described in detail above for the *Exo*III assay.

Psoralen as Probe for Alternate DNA Conformations

Psoralen photobinding is sensitive to subtle and not-so-subtle changes in DNA structure. Psoralen photobinding has been used as a probe of DNA supercoiling, cruciforms, Z-DNA, and DNA–protein interaction. The application of psoralen photobinding as a measure of DNA supercoiling is described elsewhere in this volume.[41]

Psoralen Cross-linking Assay for Cruciforms

We have described an assay for quantitating cruciforms in DNA that is applicable *in vitro* and *in vivo*.[42,43] Interstrand cross-links in DNA covalently bind the two strands of the double helix together. If a region of inverted repeat or palindromic symmetry exists in a cruciform structure, then *intra*strand cross-linking is possible. In this case, a cross-link will covalently lock the DNA into the cruciform configuration. By cross-linking supercoiled plasmid DNA containing an inverted repeat, it is possible to quantitate the fraction of the inverted repeat that exists as cruciforms. Although psoralen photobinding unwinds DNA and relaxes DNA su-

[41] R. R. Sinden and D. W. Ussery, this volume [18].

[42] R. R. Sinden, S. S. Broyles, and D. E. Pettijohn, *Proc. Natl. Acad. Sci. U.S.A.* **80**, 1797 (1983).

[43] R. R. Sinden, S. S. Broyles, and D. E. Pettijohn, *in* "Mechanisms of DNA Replication and Recombination" (N. R. Cozzarelli, ed.), Vol. 10, p. 19. Alan R. Liss, New York, 1983.

percoils, if photobinding is performed at 0–4°, there is no interconversion of cruciforms back to the linear configuration, even when the level of supercoiling is reduced below that required for the cruciform transition.[42,44] There are two ways to quantitate the fraction of DNA-containing cruciforms, which are described below. A limitation of this procedure is that the inverted repeat needs to contain at least one cross-link site (preferably a 5′-TA).

Analysis of Inverted Repeat Restriction Fragment. An inverted repeat restriction fragment can exist in one of two forms of DNA: (1) a linear double-stranded form which will be "full length" and (2) two hairpin arms (resulting from denaturation and snapping back of the individual palindromic strands) that will be half the length of the original DNA fragment. We have found that *Eco*RI will readily cut hairpin arms from supercoiled DNA when the *Eco*RI sites were near the base of the arms of the cruciform structure.[42] The cutting of hairpin arms provides an indication that cruciforms exist but does not allow quantitation. To quantitate cruciforms, the DNA must first be cross-linked, then relaxed (or linearized) to allow non-cross-linked cruciforms to return to the linear form, and finally cut with an enzyme that will liberate the palindromic fragment and/or cruciform arms. This is basically a cross-link assay in which the fraction of the inverted repeat cross-linked in the linear form, F_1, and the fraction of the inverted repeat cross-linked in cruciform form, F_c, must be determined. From these values the fraction of the inverted repeat existing as a cruciform, P_c, can be calculated from Eq. (2):

$$P_c = F_c/(F_c + F_1) \tag{2}$$

We have shown that P_c is independent of the total number of cross-links introduced per inverted repeat fragment under photobinding conditions where there is no interconversion between the cruciform form and the linear form.[42]

This method works well for cloned inverted repeats in plasmids both *in vitro* and *in vivo* in systems where the inverted repeat restriction fragments can be identified and quantitated. Theoretically, the method should work for inverted repeats inserted into chromosomal DNA. For this, however, it will be necessary to identify and quantitate the inverted repeat DNA. This may be accomplished by hybridization analysis if the cross-links can be reversed and the loop of the hairpin arm can be cut, or the tip of the hairpin arm cut off with a restriction enzyme.

A 50- to 100-bp inverted repeat restriction fragment (containing 1–4 5′-TA dinucleotides) is cross-linked by treatment with Me_3-psoralen and

[44] R. R. Sinden and D. E. Pettijohn, *J. Biol. Chem.* **259**, 6593 (1984).

a dose of 10–48 kJ m^{-2} of 360 nm light.[42-45] Following cross-linking, the DNA is cut (in excess) with one to three restriction enzymes to linearize the plasmid and relax all DNA supercoiling. Enzymes that cut the plasmid at one or two sites are used. An important consideration is to pick enzymes that will not generate fragments with sizes close to those of the inverted repeat and the hairpin arm. After this initial digestion the DNA is extracted with phenol and chloroform–isoamyl alcohol (24:1), then precipitated with ethanol. The DNA is digested with the enzyme that will cut out the linear inverted repeat and the cruciform arms. Typically, DNA is cut with a severalfold excess of restriction enzyme for 3–18 hr at 37°. For inverted repeats with low melting temperatures, it is advisable to cut DNA at temperatures lower than 37° (30–35°) to reduce the chance for denaturation of the non-cross-linked linear fragment.

Digestion products are analyzed on a 14 × 10 × 0.15 cm 5% polyacrylamide gel in 40 mM Tris–borate buffer containing 10% glycerol, as described above and previously.[42] Current is kept at 5–8 mA to avoid overheating and melting the inverted repeat. One-half of the DNA sample is applied directly to the gel, and the fraction of the total inverted repeat (full- and half-length molecules), migrating as half-size cruciforms arms, is determined. This corresponds to F_c, the fraction of the inverted repeat cross-linked as cruciforms. The other one-half of the DNA is denatured by incubation in a boiling water bath for 5 min followed by quick cooling in an ice water bath. The fraction of the inverted repeat migrating as full-length linear molecules corresponds to F_1, the fraction of the inverted repeat cross-linked in the linear form. F_1 and F_c can be quantitated by the densitometric analysis of negatives of ethidium bromide-stained gels or densitometric (or other) analysis of autoradiograms of DNA fragments end-labeled with [γ-^{32}P]ATP.[44]

We have recently improved the sensitivity of the cross-link assay by isolating hairpin arms from the polyacrylamide gel, cutting off the ends of the hairpin arms of a particular series of inverted repeats, and analyzing the products on a denaturing DNA sequencing gel.[46] This analysis demonstrates directly that the DNA migrating as hairpin arms contains intrastrand DNA psoralen cross-links.

Analysis of Inverted Repeat in Larger Fragment of DNA. It is also possible to determine F_c, the fraction of DNA cross-linked as cruciforms, on the basis of slower migration of a large DNA fragment containing a stable (cross-linked) cruciform in a polyacrylamide gel. We have used this

[45] G. Zheng and R. R. Sinden, *J. Biol. Chem.* **263**, 5356 (1988).
[46] G. Zheng, T. Kochel, R. W. Hoepfner, S. E. Timmons, and R. R. Sinden, *J. Mol. Biol.* **221**, 107 (1991).

approach to quantitate cross-linking of a 42-bp inverted repeat within a 339-bp DNA fragment.[45] Gough and Lilley have shown that cruciforms introduce a bend in linear DNA, resulting in the reduced mobility of cruciform containing DNA.[47]

This rationale should provide a general assay for the analysis of the conformation of an inverted repeat in chromosomal DNA. DNA could be treated with psoralen and light, restricted with an enzyme that generates a DNA fragment of several hundred base pairs (up to perhaps 1 kb or more), and separated on a polyacrylamide gel that will resolve the bending induced by the stable cruciform. Multiple bent bands may be observed corresponding to inverted repeats containing cross-links at different positions in the hairpin stem. The limitation of this procedure may be the selective analysis of the specific DNA bands. This might be done by DNA hybridization with a radioactive probe in a "genomic sequencing" fashion. A concern will be the removal of cross-links which may interfere with the hybridization. To date we have not attempted this procedure on chromosomal DNA, but, in principle, it should work.

Z-DNA Assays

4,5′,8-Trimethylpsoralen photobinding to B-form DNA is facilitated by the precise alignment of the photoreactive pyrone 3,4- and furan 4′,5′-double bonds of psoralen and the 5,6-double bond in pyrimidine bases in opposite strands.[1] From analysis of the structure of Z-DNA it appeared that even if psoralen intercalated into Z-DNA the bond angles and distances between the 5,6-double bonds of pyrimidines and the reactive bonds of psoralen would not be conducive to cross-link formation. This differential reactivity to photobinding and cross-linking provides an assay for the B or Z conformation of DNA. There is evidence that intercalating drugs do not bind by intercalation to Z-DNA.[48,49] However, addition of intercalating drugs will eventually drive the B–Z equilibrium back toward the B form by binding and stabilizing B-DNA (see Sinden and Kochel for discussion[26]). Thus, Me$_3$-psoralen photobinding provides an ideal probe for Z-DNA in that the assay will not drive the B–Z equilibrium toward the Z form. Both the relaxation of DNA from unwinding due to intercalation and the intercalative binding will favor B-DNA. Z-DNA, therefore, should not be introduced artifactually by the assay system. Based on this rationale we have described two psoralen-based assays for Z-DNA.[26,29,30]

[47] G. W. Gough and D. M. J. Lilley, *Nature (London)* **313**, 154 (1985).
[48] F. M. Pohl, T. M. Jovan, W. Baeht, and J. J. Holbrook, *Proc. Natl. Acad. Sci. U.S.A.* **69**, 3805 (1972).
[49] G. T. Walker, M. P. Stone, and T. R. Kough, *Biochemistry* **24**, 7462 (1985).

Cross-linking Assay for Z-DNA. Our initial psoralen-based Z-DNA assay relied on DNA cross-linking.[26] We demonstrated that when Z-DNA was stabilized by DNA supercoiling in plasmid DNA the rate of psoralen cross-linking to the Z-DNA region was reduced. In addition, psoralen did not form monoadducts appreciably with Z-DNA. The rate of cross-linking of a 30-bp *Eco*RI restriction fragment that can form Z-DNA, $AATT(CG)_6TA(CG)_6$, was compared to the rate of cross-linking of an adjacent 20-bp DNA fragment that could not form Z-DNA. The 20-bp fragment provides an "internal control" in this system. This relative rate of cross-linking provided an indication of the level of supercoiling in DNA and the extent of Z-DNA formation. The application of this assay *in vivo* is described elsewhere in this volume.[41]

For this assay, plasmid DNAs of various superhelical density are treated with psoralen and a dose of up to 24 kJ m^{-2} 360 nm light. Following cross-linking it is important to purify supercoiled DNAs away from any nicked molecules. This is an important consideration since the potential Z-DNA region will exist in the B-form in nicked DNA, and this population will readily photobind psoralen. Supercoiled DNA is separated from nicked DNA by purification from an ethidium bromide–CsCl gradient.[5] The DNA is then digested exhaustively with the appropriate restriction enzymes to liberate the Z-DNA-forming and "control" restriction fragments to be analyzed. The procedures for DNA cross-link analysis of small restriction fragments were described above (see Acrylamide Gels).

The rates of cross-linking vary as a function of superhelical density. As DNA becomes more negatively supercoiled and potential Z-DNA sequences form Z-DNA, the rate of cross-linking in the Z-DNA region decreases. The extent of Z-DNA existing in a population of topoisomers can be readily determined from two-dimensional agarose gel electrophoresis.[26] A "standard curve" of the relationship between the rate of cross-linking and superhelical density is determined. Analysis of the rate of DNA cross-linking of plasmids of unknown superhelical density can be used to estimate the fraction of plasmids containing Z-DNA and, consequently, the superhelical density of the DNA. Although this may not be the first method of choice for analyzing Z-DNA *in vitro* the advantage of the psoralen assay is its applicability *in vivo*.

This Z-DNA assay is applicable to DNA restriction fragments containing a crosslinkable site(s) only within (or perhaps directly adjacent to) the Z-DNA forming region. We have chemically synthesized a number of such sequences for use as "torsionally tuned probes" of DNA supercoiling. A limitation of this assay is that it requires the introduction of large numbers of psoralen adducts in DNA, which will significantly relax DNA supercoiling. In addition, the signal for Z-DNA is a reduction in DNA cross-linking. The assay described in the next section alleviates these limitations.

Exonuclease III/Photoreversal Assay for Z-DNA. The exonuclease III/photoreversal assay described above increases the sensitivity of analysis of psoralen photobinding and extends the applicability of the psoralen-based Z-DNA assay by allowing analysis of Z-DNA sequences that are not flanked by restriction sites. The exoIII/photoreversal assay allows quantitation of psoralen photobinding to a large number of reactive sites in nearly any DNA sequence. Significantly, the assay is applicable at levels of photobinding producing less than one cross-link per 8000 bp.[29]

The rationale behind the psoralen-based exoIII/photoreversal assay for Z-DNA is identical to that described above for the DNA cross-linking Z-DNA assay. Briefly, as Z-DNA forms in a potential Z-DNA region, the reactivity of this sequence to psoralen will decrease. Using this assay we have shown that the reactivity of a 5'-TA within the Z-DNA forming sequence $(CG)_6TA(CG)_6$ varies from strongly reactive to nonreactive as the negative superhelical density σ increases from 0 to -0.1.[29] The psoralen reactivity within the Z-DNA forming region is quantitated relative to the reactivity of a 5'-TA in a region of DNA that does not adopt an alternate DNA conformation. A "standard curve" can be obtained that relates the relative rate of photobinding to superhelical density.

The exoIII/photoreversal assay revealed another feature of this system that provides an important additional indication of the presence of Z-DNA and superhelical density. The $(CG)_6TA(CG)_6$ Z-DNA forming sequence was cloned as an *Eco*RI fragment. In relaxed DNA the 5'-AATT sequences within the *Eco*RI sites were not photoreactive to psoralen. When the $(CG)_6TA(CG)_6$ sequence existed as Z-DNA, however, the reactivity of the 5'-AATT sequences to psoralen photobinding increased several hundred-fold.[29] The 5'-AATT sequences are adjacent to the Z-DNA forming region and presumably represent B–Z junctions. The reason for the strong hyperreactivity is not understood, although it may reflect a preferential interacalation in the B–Z junction owing to a partially unwound state in the DNA. The reactivity of one junction provided a reasonable relationship between the reactivity relative to a "control 5'-TA" and superhelical density.

This assay has been applied to several Z-DNA forming sequences.[30] In all cases, as Z-DNA forms, the reactivity of psoralen photobinding to 5'-TA dinucleotides decreases. When Z-DNA sequences are cloned as *Eco*RI sites the 5'-AATT regions become hypersensitive to psoralen photobinding. The hyperreactivity of junctions flanking $(GC)_n$ Z-DNA forming sequences parallels the formation of Z-DNA. Junctions flanking $(GT)_n$ Z-DNA forming sequences become hyperreactive, but not until levels of supercoiling are higher than that required to form Z-DNA. This different reactivity may reflect a different structure of the B–Z junctions flanking (GC)$_n$ and (GT)$_n$ Z-DNA regions.

The exoIII/photoreversal assay for Z-DNA provides a dual signal for the presence of Z-DNA.[29] First, the relative reactivity of the 5'-TA within the Z-DNA region is analyzed. The reactivity of this site decreases as Z-DNA is formed. Second, the reactivity of the B–Z junctions provides a positive signal for the presence of Z-DNA. Analysis of both these reactivities provides a double signal for the presence of Z-DNA. The application of these "standard curves" to determining the existence of Z-DNA *in vivo* and application of this approach as a "torsionally tuned probe" for supercoiling are described elsewhere in this volume.[41]

[14] Photofootprinting DNA *in Vitro*

By MICHAEL M. BECKER and GREGORY GROSSMANN

Introduction

Since their introduction in 1976,[1,2] footprinting techniques have proved invaluable in elucidating the structural details of nucleic acids and their interactions with regulatory proteins. The use of light to footprint nucleic acids was first introduced in 1984 and is termed photofootprinting.[3] Two different photofootprinting methodologies exist. The first, termed ultraviolet (UV) footprinting, uses UV photons as the sole footprinting agent.[3] A second methodology utilizes visible light to photoactivate small inorganic or organic molecules which subsequently damage or cleave nucleic acid backbones.[3–6] In this chapter we concentrate exclusively on the use of UV light to footprint nucleic acids.

In contrast to other footprinting techniques, the UV footprinting technique appears to detect changes in the torsional flexibility of individual bases along the DNA helix.[7–9] The most prevalent form of UV damage appears to be dimerization between adjacent bases of the DNA helix.

[1] W. Gilbert, A. Maxam, and A. Mirzabekov, *in* "Control of Ribosome Synthesis" (N. C. Kjelgaard and O. Maaloe, eds.), p.139. Munksgaard, Copenhagen, 1976.

[2] D. Galas and A. Schmitz, *Nucleic Acids Res.* **5**, 3157 (1978).

[3] M. M. Becker and J. C. Wang, *Nature (London)* **309**, 682 (1984).

[4] P. E. Nielsen, C. Jeppesen, and O. Buchardt, *FEBS Lett.* **235**, 122 (1988).

[5] C. Jepperson, O. Buchardt, U. Henriksen, and P. E. Nielsen, *Nucleic Acids Res.* **16**, 5755 (1988).

[6] K. Uchida, A. M. Pyle, and J. K. Barton, *Nucleic Acids Res.* **17**, 10259 (1989).

[7] M. M. Becker and Z. Wang, *J. Mol. Biol.* **210**, 429 (1989).

Dimerization is not restricted to pyrimidines but also occurs at purines.[7] Before adjacent bases can photodimerize, their photoreactive bonds must overlap. Because adjacent bases along the DNA helix are rotated approximately 34° relative to one another, the DNA must unwind to allow photoreactive bonds to dimerize.[3] Although it is theoretically possible for a UV photon to supply the necessary energy for unwinding, it appears that unwinding is achieved by random thermal motions within the DNA helix.[7] On the absorption of a UV photon, only those thermally excited bases that are in a geometry capable of easily forming a photodimer during excitation photoreact.[7] Thus, agents which inhibit the torsional flexibility of DNA (e.g., the B → A transition,[10] sequence-specific protein–DNA contacts,[3,11–14] triple-strand formation[15–17]) inhibit the UV photoreactivity of DNA. In contrast, agents which enhance the torsional flexibility of DNA (e.g., melting of the double helix,[7] kinking of the double helix,[12] premelting of rigid dA · dT tracts[7]) enhance the photoreactivity of DNA. Finally, agents which do not alter the torsional flexibility of DNA [e.g., salt ($\leq$0.25 M),[8] organic solvents,[10] temperature,[7] chromosomal protein–DNA interactions[11]] do not alter the UV photoreactivity of DNA.

The insensitivity of UV damage rates to temperature, salt, or organic solvents makes UV light an ideal footprinting probe of nucleic acid structure since structural changes can be induced by a wide range of environmental perturbants and subsequently footprinted with UV light. Only when the environmental perturbant alters the structure of DNA will the UV photoreactivity of the DNA also be altered.[7,10]

When UV light is used to footprint protein–DNA complexes, only intimate sequence-specific contacts to the DNA helix are detected by UV

[8] M. M. Becker and Z. Wang, *in* "A Laboratory Guide for *in Vitro* Studies of Protein–DNA Interactions" (H. P. Saluz and J. P. Jost, eds.), in press. Birkhauser Verlag, Basel (1991).

[9] V. Lyamichev, *Nucleic Acids Res.* **19**, 4491 (1991).

[10] M. M. Becker and Z. Wang, *J. Biol. Chem.* **264**, 4163 (1989).

[11] Z. Wang and M. M. Becker, *Proc. Natl. Acad. Sci. U.S.A.* **85**, 654 (1988).

[12] M. M. Becker, D. Lesser, M. Kurpiewski, A. Barranger, and L. Jen-Jacobson, *Proc. Natl. Acad. Sci. U.S.A.* **85**, 6247 (1988).

[13] M. M. Becker, Z. Wang, G. Grossmann, and K. A. Becherer, *Proc. Natl. Acad. Sci. U.S.A.* **86**, 5315 (1989).

[14] S. B. Selleck and J. E. Majors, *Mol. Cell. Biol.* **7**, 3260 (1987).

[15] V. I. Lyamichev, M. D. Frank-Kamenetskii, and U. N. Seyfer, *Nature (London)* **344**, 568 (1989).

[16] V. I. Lyamichev, O. N. Voloshin, M. D. Frank-Kamenetski, and V. N. Soyfer, *Nucleic Acids Res.* **19**, 1633 (1991).

[17] M. Tang, H. Htun, Y. Cheng, and J. E. Dahlberg, *Biochemistry* **30**, 7021 (1991).

light. Less intimate contacts, which do not significantly alter the torsional flexibility of DNA, are only weakly detected by UV light.[8,11]

As described previously, UV footprinting patterns can be generated using chemical reactions to break the DNA backbone at the sites of UV damage[3] or by the use of DNA polymerases whose synthesis of complementary DNA is terminated by UV photoproducts.[11,13,18] Because detailed protocols describing the use of chemical reactions to reveal UV footprinting patterns have been given elsewhere,[8] we shall limit our discussion to the use of a polymerase to reveal UV footprinting patterns.

The use of DNA polymerases has several advantages over the use of chemical reactions to reveal UV photoproducts. First, the polymerase method is extremely simple since no chemical reactions are required to break the DNA backbone at the sites of photodamage. Instead, UV photoproducts are detected by their ability to terminate DNA synthesis of complementary DNA directly. Second, the polymerase method exhibits better signal-to-noise ratios than the chemical method since signal arising from unirradiated DNA is very low. Third, because of the excellent signal-to-noise ratio of the polymerase method, very low levels of UV irradiation can be used to generate UV footprinting patterns. We have previously shown that transilluminators commonly found in molecular biology laboratories can be used to irradiate DNA for UV footprinting analysis. At UV dosages of 150 mJ/cm^2 or less, protein–DNA complexes are not dissociated or cross-linked by UV light,[11-13] and subtle structural alterations along the DNA helix can readily be footprinted.[7,10] Below we describe the use of the Klenow fragment of *Escherichia coli* DNA polymerase to detect UV footprinting patterns in DNA.

Isolation of Viral DNA

Native SV40 minichromosomes are isolated according to the procedure of Varshavsky *et al.*[19] COS cells are infected with SV40 viral strain (1-3),[20] which contains one copy of the 72-base pair (bp) repeat enhancer element and bears an 8-bp *Bam*HI linker inserted at the *Hpa*II site at map position 346. Viral minichromosomes are extracted 40 hr after infection and purified by sucrose sedimentation. Peak fractions from the gradient are pooled, and the DNA is purified by phenol extraction, ethanol precipitation, and extensive dialysis.

[18] J. D. Axelrod and J. Majors, *Nucleic Acids Res.* **17,** 171 (1989).
[19] A. J. Varshavsky, O. H. Sundin, and M. J. Bohn, *Nucleic Acids Res.* **5,** 3469 (1978).
[20] H. Weiher, M. Konig, and P. Gruss, *Science* **219,** 626 (1983).

In Vitro Ultraviolet Irradiation of Viral DNA

Materials

Transilluminator, fitted with six germicidal UV bulbs (Ultraviolet Products, San Gabriel, CA, Model TL-33): Remove the face plate (filter) so that the bulbs will shine directly on the sample. Construct two supports, 15–20 cm high, on which to invert the device such that the bulbs face downward. Devise a shutter with which to control exposure times. An opaque sheet of cardboard should suffice.

UV-transparent quartz cuvette (Hellma Cells, Jamaica, NY, 165-Qs)

Radiometer (Ultraviolet Products, Model UVX)

Buffers: TE [10 mM Tris-HCl, pH 8, 1 mM ethylenediaminetetraacetic acid (EDTA)]; 3 M sodium acetate; 95% ethanol

Procedure

1. Dilute 200–500 ng of viral DNA into 250 μl TE and place the sample into the cuvette.

2. Remove the filter from the transilluminator, and invert it onto its supports. Warm up the bulbs for 2 min before use to ensure a steady output of light. Adjust the height such that the UV flux measured at the center of the irradiated area is 5–6 mW/cm^2. Mark this position as the central spot. With the shutter in position, place the filled cuvette on the central spot.

3. Remove the shutter, and irradiate the sample for the desired time. Transfer the irradiated DNA to a microcentrifuge tube, and rinse the cuvette.

4. Precipitate the irradiated sample by adding 0.1 volume of 3 M sodium acetate and 2.5 volumes of ethanol. Resuspend the pellet in 70 μl TE.

Linearization and Purification of Viral Template DNA

The template DNA is linearized at a unique restriction site located outside the region that will be analyzed by primer extension. Linearization of the template permits more efficient hybridization with the labeled primer. The template DNA is also digested with RNase A to eliminate RNA species that copurify with the minichromosomes. RNA interferes with the hybridization and extension reactions.

Materials

Enzymes: *Eco*RI (Boehringer-Mannheim, Indianapolis, IN, 10 U/μl); RNase A (United States Biochemicals, Cleveland, OH, 6 μg/μl)

Buffers: SEVAG [50% (v/v) phenol, 48% chloroform, 2% isoamyl alcohol]; 7.5 M ammonium acetate; 3 M sodium acetate; TE; 0.5 M EDTA; 1 mg/ml yeast tRNA (Sigma, St. Louis, MO)

Procedure

1. Digest the DNA sample with *Eco*RI, following the manufacturer's directions. Use no more than 1 U enzyme per 500 ng DNA.
2. Add RNase A to a final concentration of 10 ng/μl to the above solution and incubate for an additional 30 min.
3. Stop the reaction by adding EDTA to a final concentration of 20 mM. Adjust the volume to 200 μl with TE, and extract twice with SEVAG. After the second extraction, transfer the aqueous layer to a new tube and add 10 μg tRNA as carrier. (This RNA, necessary for precipitating the DNA, will be removed with RNase A immediately before the hybridization and extension reaction.)
4. Precipitate with ammonium acetate–ethanol as follows: add 100 μl of 7.5 M ammonium acetate, mix well, then add 750 μl ethanol. Mix, let stand at room temperature for 15 min, then spin in a microcentrifuge for 15 min. Rinse the pellet twice in ethanol, dry thoroughly, and resuspend in 300 μl TE.
5. Precipitate with sodium acetate–ethanol (0.1 volume 3 M sodium acetate, 2.5 volumes ethanol) and resuspend the pellet at a concentration of 2.5 ng/μl.

Synthesis of 5'-End-Labeled Primer

In a typical hybridization reaction, 8 ng of 5'-end-labeled primer is annealed to 25–50 ng viral template DNA, representing a 100- to 200-fold molar ratio of primer to template. For a sample set of 10 to 20 different template DNAs, 100 to 300 ng primer is labeled at once.

Materials

Label: [γ-^{32}P]ATP, $\geq$6000 Ci/mmol (New England Nuclear, Boston, MA, NEG-035C; 27.3 pmol/μl)

Primer: synthetic 16-mer oligonucleotide; 100 ng/μl, 18 pmol/μl

Enzyme: T4 polynucleotide kinase (PNK) freshly diluted to 2.5 U/μl (Pharmacia, Piscataway, NJ, 10 U/μl)

Buffers: 5$\times$ Kination Buffer (KB) [330 mM Tris-HCl, pH 7.6, 50 mM MgCl$_2$, 50 mM 2-mercaptoethanol (2-ME), 50 μg/ml gelatin]; TE; SEVAG

NENsorb-20 column (New England Nuclear), and buffers for its use; this column removes unincorporated label and salt from the end-labeled primer

Procedure

1. The kination reaction is performed in 1.5-ml screw-capped microcentrifuge tubes. A 50% molar excess of label is provided in the reaction. The 20 μl reaction mix is composed as follows: 2 μl primer (36 pmol), 2 μl label (54 pmol), 4 μl of 5× KB, and 12 μl TE. Add 1 μl (2.5 U) PNK and incubate at 37° for 20 min. Add another 1-μl aliquot of PNK and continue the incubation for 20 min.

2. Stop the reaction by adjusting the volume to 400 μl with TE, and extract once with SEVAG. Spin in a microcentrifuge for 5 min to separate the phases, and remove the aqueous layer to a new tube. Spin once more for 2 min.

3. Carefully load the aqueous phase onto a NENsorb-20 column that has been wetted and primed according to the product instructions. Wash the adsorbed primer free of unincorporated label and rinse with 2 volumes of sterile deionized water to desalt the column. Elute the primer into 1 ml of 50% ethanol and lyophilize to dryness.

4. Resuspend the primer at a concentration of 2 ng/μl.

Hybridization and Primer Extension with Klenow Enzyme

The hybridization and extension reactions are performed in screw-capped microcentrifuge tubes in a reaction volume of 30 μl. Twenty-five to 100 ng of template annealed with 8 ng primer should yield ample signal strength on the sequencing gel.

Materials

Water baths: 100°; 55°

Enzymes: DNA polymerase, Klenow fragment freshly diluted to 0.5 U/μl (Bethesda Research Laboratories, Gaithersburg, MD, 6U/μl); RNase A diluted to 0.6 ng/μl (United States Biochemicals, 6 mg/ml)

Deoxynucleotide triphosphate (dNTP) mix: 2.5 mM each of dA, dC, dG, and dT (Pharmacia, 100 mM solutions)

Chain-termination nucleotide mixes: 1 mM dideoxy-CTP (ddCTP); 0.1 mM dCTP; 2.5 mM each of dA, dG, and dT (Pharmacia, 100 mM solutions)

Buffers: 10× Hybridization buffer (HB = 500 mM Tris-HCl, pH 7.6, 100 mM MgCl$_2$); 4 mM 2-ME; 1 mg/ml gelatin; 1 mg/ml tRNA; 0.5 M EDTA; 7.5 M ammonium acetate; 3 M sodium acetate; sequencing gel loading buffer (95% deionized formamide, 0.1% xylene cyanol, 0.1% bromphenol blue)

Procedure

1. Per reaction mix, 20 μl (50 ng) viral DNA template, 4 μl (8 ng) 5′-end-labeled primer, 3 μl of 10× HB, 3 μl of 1 mg/ml gelatin, and 1.8 μl of 0.6 ng/μl RNase A. Preincubate the reaction mix for 15 min at 37° to allow the RNase A to degrade the carrier tRNA.

2. Boil the sample for 4 min, then transfer to 55° and incubate for 45 min.

3. Transfer the tube to 37° and allow the temperature to equilibrate for 2 min before adding the Klenow enzyme and dNTP mix as follows: for each reaction, mix 3 μl dNTP mix (2.5 mM each), 1.5 μl of 4 mM 2-ME, and 1 μl (0.5 U) Klenow fragment. Incubate at 37° for 30 min.

4. For ddC chain-termination reactions, add the following to the tube: 2 μl of 1 mM ddC, 0.8 μl of 0.1 mM dC, 2.4 μl of 2.5 mM each dA, dG, and dT, 1 μl of 4 mM 2-ME, and 2 μl (2 U) Klenow enzyme. Incubate at 37° for 30 min.

5. Bring the volume to 200 μl with ice-cold TE, including 2 μl 0.5 M EDTA and 5–10 μg tRNA carrier. Precipitate with ammonium acetate–ethanol at room temperature for 15 min, then spin for 15 min in the microcentrifuge. Rinse the pellet twice in ethanol and dry thoroughly.

6. Resuspend the pellet in 1 μl TE and 5 μl sequencing gel loading buffer and transfer the sample to a new tube.

7. Boil the samples for 4 min and load onto the sequencing gel.

Sequencing Gel Electrophoresis

The extension products are separated on a 50 cm long 10% (20 : 1) polyacrylamide sequencing gel run in 0.8× TBE (40 mM Tris, 40 mM borate, 0.8 mM EDTA, pH 8.3). The gel is cast with wedge side spacers, 0.8–1.8 mm from top to bottom. The gel will run at or below 50° at 80 W (1900 V, 50 mA). The xylene dye front will reach 10 to 15 cm from the bottom of the gel in about 5 hr; at this point the primer, and extension products 10 bases beyond it, will have been run off the gel. Autoradiography of the gel is performed by standard procedures.

Photoreactivity of DNA

A dose–response experiment in which purified SV40 DNA was irradiated with UV light for increasing amounts of time is shown in Fig. 1. The autoradiograph shows labeled primer extension products resolved on a sequencing gel. When unirradiated viral DNA is used as a template in a primer extension reaction (lane marked − UV in Fig. 1), a small number of

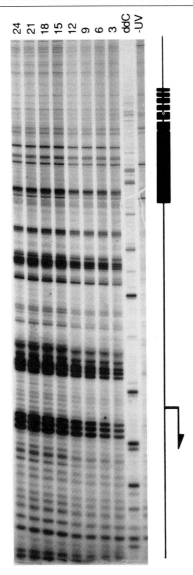

FIG. 1. Detection of UV photoproducts by primer extension analysis. SV40 DNA was UV irradiated *in vitro* at 5.5 mW/cm² for the indicated times (in seconds), and the Klenow-mediated primer extension products were resolved on a 10% polyacrylamide sequencing gel. The lane marked − UV contained unirradiated SV40 DNA. Lane ddC contained unirradiated SV40 DNA extended in the presence of ddC to introduce chain terminations at G residues in the template. Locations of landmark SV40 regulatory sequence elements are drawn to the side of the autoradiograph. The arrow marks the late transcriptional start site at map position 325. Rectangles mark the enhancer element and GC motifs of the viral promoter.

primer extension bands appear. These bands represent extension products resulting from both random and sequence-specific dissociations of Klenow enzyme from the unreacted template.

In the lanes from irradiated templates (Fig. 1), several observations may be made. First, increasing the UV dose enhances UV photoproduct formation at all reactive bases, resulting in enhanced band intensities. Second, some weak bands appear only at higher UV doses, most noticeably in the bottom one-third of the autoradiograph (Fig. 1). To some extent, therefore, new bases will become reactive with increasing UV dose. Third, within this range of UV dosage, the length of the longest extension products (evident at the top one-quarter of the gel) does not diminish. These long extension products diminish noticeably if longer or more intense UV light exposures are applied, which is an indication that the template has been overirradiated (results not shown).

Frequency of Ultraviolet Photoproduct Formation

Bases in the SV40 regulatory sequences that form UV photoproducts are indicated in Fig. 2. Of 728 bases analyzed, 343 (47%) are UV reactive. Sixty-seven percent of all pyrimidines and 28% of all purines are photoreactive. Where both strands of DNA were analyzed, 74% (157 out of 212) of the base pairs are photoreactive. The density of UV photoproduct formation under the conditions described here approaches three-fourths of all base pairs, which is sufficiently high to be useful as a probe for footprinting regulatory protein–DNA contacts and altered structures of DNA.

Predicting Ultraviolet Reactivity of Specific DNA Sequences

The sensitivity of double-stranded DNA to UV damage has been previously examined in experiments in which end-labeled DNA was UV irradiated, and the DNA chemically cleaved at the sites of photoreaction.[7] As summarized in Fig. 2, the UV photoproducts detected by the primer extension method agree well with those detected by the chemical cleavage method. Both methods reveal that whenever two or more pyrimidines are adjacent to one another, photoreactions take place at each pyrimidine. In contrast, nonadjacent pyrimidines exhibit little or no UV photoreactivity.

The finding that adjacent pyrimidines are photoreactive while nonadjacent ones are not is readily explained by the well-known ability of adjacent pyrimidines to covalently dimerize on absorption of UV light. For a covalent dimer one would expect that the 5′ residue would rarely terminate primer-directed DNA synthesis since the 3′ residue is encountered first by the DNA polymerase. Indeed, as summarized in Fig. 2, for two or

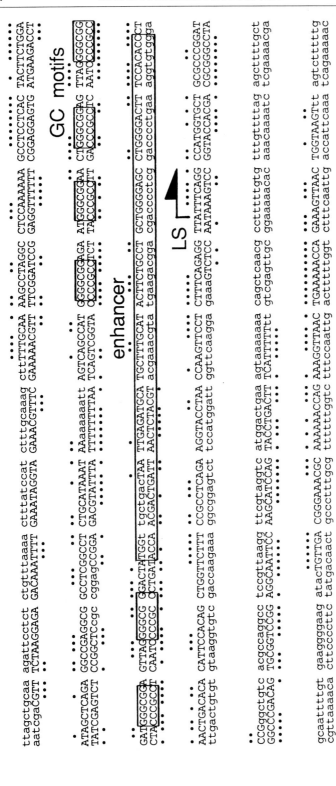

FIG. 2. DNA sequence of the regulatory region of the SV40 (1-3) mutant strain used in this study. Uppercase letters denote those bases which have been analyzed for photoproduct formation with the primer extension method. Lowercase letters denote those bases which have not been analyzed. Bases that form UV photoproducts in these experiments are denoted by black dots.

more contiguous pyrimidines, the 5′-most residue rarely terminates DNA synthesis.

In contrast to pyrimidines, the rules for predicting purine photoreactions are more complicated. Previous experiments employing chemical cleavage of UV photoproducts revealed that purines flanked on their 5′ side by two or more contiguous pyrimidines are the only purines in double-stranded DNA that are damaged by low levels of UV light.[7] As seen in Fig. 2, the primer extension method also detects UV photoreactions at these purines; however, only 40% of these purines efficiently terminate DNA synthesis.

In addition to photoreactions at these purines, the primer extension method also detects UV photoreactions at other purines. These purine photoreactions are rare (15% of all purines) and exhibit no obvious common characteristics.

[15] Mapping Adducts of DNA Structural Probes Using Transcription and Primer Extension Approaches

By HAN HTUN and BRIAN H. JOHNSTON

Introduction

Chemical probing techniques have proven to be powerful tools for analyzing the structure,[1-5] base pairing status,[6-10] and modification state[11-13] of DNA (also reviewed elsewhere in this volume). Potentially interesting segments of DNA are reacted with a chemical probe, and the

[1] B. H. Johnston and A. Rich, *Cell (Cambridge, Mass.)* **42**, 713 (1985).

[2] W. Herr, *Proc. Natl. Acad. Sci. U.S.A.* **82**, 8009 (1985).

[3] T. Kohwi-Shigematsu, R. Gelinas, and H. Weintraub, *Proc. Natl. Acad. Sci. U.S.A.* **80**, 4389 (1983).

[4] D. M. J. Lilley and E. Paleček, *EMBO J.* **3**, 1187 (1984).

[5] D. E. Pulleyblank, D. B. Haniford, and A. R. Morgan, *Cell (Cambridge, Mass.)* **42**, 271 (1985).

[6] W. W. Dean and J. Lebowitz, *Nature (London) New Biol.* **231**, 5 (1971).

[7] D. E. Pulleyblank and A. R. Morgan, *J. Mol. Biol.* **91**, 1 (1975).

[8] A. Melnikova, R. Beakealashvilla, and A. D. Mirzabekov, *Eur. J. Biochem.* **84**, 301 (1978).

[9] U. Siebenlist, *Nature (London)* **279**, 651 (1979).

[10] K. Kirkegaard, H. Buc, A. Spassky, and J. C. Wang, *Proc. Natl. Acad. Sci. U.S.A.* **80**, 2544 (1983).

[11] H. Ohmori, J. Tomizawa, and A. M. Maxam, *Nucleic Acids Res.* **5**, 1479 (1978).

[12] J. R. Miller, E. M. Cartwright, G. G. Brownlee, N. V. Fedoroff, and D. D. Brown, *Cell (Cambridge, Mass.)* **13**, 717 (1978).

METHODS IN ENZYMOLOGY, VOL. 212

positions of the resulting adducts are made visible on a sequencing gel by a technique that generates fragments ending at the adducts. In most cases, this has been accomplished by end-labeling the DNA and cleaving the backbone at the positions of the adducts by treatment with an agent such as piperidine, following the method of Maxam and Gilbert[14] in their procedure for sequencing DNA. An alternative approach, more akin to the polymerase termination sequencing method of Sanger et al.,[15] involves using the modified DNA as a template for an RNA or DNA polymerase which will stop at the adducts. The resulting nested set of newly synthesized fragments, made radioactive by incorporating labeled nucleoside triphosphates or a labeled primer, are analyzed on a sequencing gel.

The polymerase approach has several advantages over the chemical cleavage approach. Most importantly, it is rapid and simple, allowing the processing and analysis of many modified samples simultaneously by eliminating the time required to purify and prepare a singly end-labeled fragment, break the DNA backbone at the modified site by hot piperidine treatment, and remove the residual piperidine. Second, certain modifications at Watson–Crick base-paired positions that do not render the base alkali-labile, such as methylation at the N-1 position of adenines, can readily be detected as a block to the progression of polymerases. Third, little sample is needed for the analysis, since the signal can be amplified by incorporating multiple, radioactive labels in the newly synthesized strand and by multiple rounds of synthesis. Finally, because scission of the template strand at the modified site effectively prevents further elongation of the nascent chain, the polymerase approach can ultimately detect all adducts identified by the chemical cleavage approach; this is particularly useful when the initially formed adduct does not block the progression of polymerases, such as 7-methylguanine. Thus, the polymerase approach saves material and time, allows amplification of the original signal and, in certain instances, complements the chemical cleavage approach.

Balanced against these advantages are two drawbacks. First, owing to slight heterogeneity in termination at the modified site, attributing radioactive bands on the gel to certain nucleotides in the template strand is more difficult. Nevertheless, assignment can be made partly on the basis of the known specificity of the damaging reagent toward particular nucleotides. Second, premature termination in a small but significant fraction of newly synthesized strands at positions other than the end of the template or the

[13] E. Fritzsche, H. Hayatsu, G. L. Igloi, S. Iida, and H. Kossel, Nucleic Acids Res. **15**, 5517 (1987).

[14] A. M. Maxam and W. Gilbert, this series, Vol. 65, p. 499.

[15] F. Sanger, S. Nicklen, and A. R. Coulson, Proc. Natl. Acad. Sci. U.S.A. **74**, 5463 (1977).

modified site contributes to higher background, which is exacerbated by incorporating radioactivity at more than one position. Despite these drawbacks, the rapidity and ease of assay and amplification of the signal make the polymerase approach clearly the method of choice when optimizing conditions[16] for DNA modification or alternate structure formation or when analyzing single-copy genomic DNA.[17-21]

In this chapter we present protocols for the use of both DNA and RNA polymerases to detect adducts. DNA polymerases are used to extend an annealed primer, an approach applied to the analysis of altered DNA structures by Gralla[22] and akin to work in which reverse transcriptases were used to map adducts on RNA.[23] We have recently developed procedures for using RNA polymerases from phages to map adducts by transcribing modified DNA. A brief description of this approach has been given previously.[16] We shall refer to the RNA polymerase approach as transcription and the DNA polymerase approach as primer extension.

Materials

Because both the primer extension and transcription approaches will produce gel bands at any obstruction to the polymerase, whether covalently or noncovalently bound to the DNA (such as contaminating protein), it is especially important that the template DNA be free of contaminants. This is critical when examining DNA structure in the presence of proteins, namely, in footprinting experiments. In these cases, the DNA should be thoroughly extracted with ultrapure or redistilled phenol saturated with a 100 mM Tris-HCl (pH 8) buffer prior to the primer extension reaction. All reagents should be fresh and of the highest purity to avoid inadvertently introducing DNA damage.

DNA Modification

Protocols for modifying DNA using the various chemical probes described are given elsewhere in this volume[24-27] and therefore are not re-

[16] H. Htun and J. E. Dahlberg, *Science* **241**, 1791 (1988).
[17] P. R. Mueller and B. Wold, *Science* **246**, 780 (1989).
[18] G. P. Pfeifer, S. D. Steigerwald, P. R. Mueller, B. Wold, and A. D. Riggs, *Science* **246**, 810 (1989).
[19] M. M. Becker, Z. Wang, G. Grossmann, and K. A. Becherer, *Proc. Natl. Acad. Sci. U.S.A.* **86**, 5315 (1989).
[20] H. Saluz and J.-P. Jost, *Proc. Natl. Acad. Sci. U.S.A.* **86**, 2602 (1989).
[21] J. D. Axelrod and J. Majors, *Nucleic Acids Res.* **17**, 171 (1989).
[22] J. D. Gralla, *Proc. Natl. Acad. Sci. U.S.A.* **82**, 3078 (1985).
[23] T. Inoue and T. R. Cech, *Proc. Natl. Acad. Sci. U.S.A.* **82**, 648 (1985).
[24] E. Paleček, this volume [17].
[25] T. Kohwi-Shigematsu and Y. Kohwi, this volume [9].

peated here. However, for the samples shown in Figs. 1–6, pertinent information is provided in the figure legends to allow reproducibility under the modification conditions previously used.[16] Other chemical probes as well as ultraviolet light, described in this volume[28] and elsewhere[29,30] should also be applicable to the polymerase mapping approach.

Transcription of Modified DNA

The normal precautions for preventing ribonuclease cleavage should be observed in these protocols as with all RNA work. This includes the use of disposable plasticware or baked (RNase-free) glassware, preparation of reagents using water treated with diethyl pyrocarbonate, and use of an RNase inhibitor (e.g., RNasin; Promega Corp., Madison, WI) in the transcription reaction. Because of the limited duration and number of manipulations, degradation of RNA transcripts from nuclease contamination is normally not observed. Nevertheless, in those instances where such contaminations do occur, degradation of the RNA transcripts will be evident as a dark lane with numerous bands from the top to the bottom of the autoradiogram. Detailed description of phage polymerases and protocols for transcription can be found elsewhere[31] and in this series.[32–35]

Template

The principal requirement for the transcription approach is that an appropriate promoter should exist near the sequence to be analyzed. Several vectors containing convergent RNA phage promoters flanking a multiple cloning sequence are commercially available, into which the DNA of interest is inserted. Commonly used promoters are those derived from SP6, T3, and T7 phages. Although templates modified with the reagents

[26] B. H. Johnston, this volume [10].

[27] B. H. Johnston, this series, Vol. 211, [7].

[28] Reviewed by D. M. J. Lilley, this volume [7].

[29] C. Ehresmann, F. Baudin, M. Mougel, P. Romby, J.-P. Ebel, and B. Ehresmann, *Nucleic Acids Res.* **15**, 9109 (1987).

[30] R. D. Wells, S. Amirhaeri, J. A. Blaho, D. A. Collier, J. C. Hanvey, W.-T. Hsieh, A. Jaworski, J. Klysik, J. E. Larson, M. J. McLean, F. Wohlrab, and W. Zacharias, *in* "Unusual DNA Structures" (R. D. Wells and S. C. Harvey, eds.), p. 1. Springer-Verlag, New York, 1988.

[31] M. Chamberlin and T. Ryan, *in* "The Enzymes" (P. D. Boyer, ed.), 3rd Ed., Vol. 15, p. 85. Academic Press, New York, 1982.

[32] P. A. Krieg and D. A. Melton, this series, Vol. 155, p. 397.

[33] J. K. Yisraeli and D. A. Melton, this series, Vol. 180, p. 42.

[34] J. F. Milligan and O. C. Uhlenbeck, this series, Vol. 180, p. 51.

[35] V. M. Reyes and J. N. Abelson, this series, Vol. 180, p. 63.

described here can be analyzed directly by transcription without prior linearization of the DNA, the latter procedure is recommended, especially when studying supercoil-dependent structures. To avoid initiation of phage polymerases from the end of the linearized DNA, restriction enzymes that generate 5' overhangs or blunt ends should be used.[36]

RNA Polymerases

The transcription termination procedure has been successfully tested using SP6, T3, and T7 RNA polymerases (Promega Corp.), each of which has a high specificity for its cognate promoter.[31,37] The use of SP6 and T7 RNA polymerases, in separate reactions for each modified sample in a pGEM vector (Promega Corp.), allows analysis of each strand of the inserted DNA, as illustrated in Fig. 3.

Transcription Reaction

Transcription reactions are carried out on modified DNA as well as unmodified control DNA and also unmodified DNA in the presence of 3'-deoxyribonucleoside 5'-triphosphate (3'-dNTP) RNA chain terminators to provide a reference RNA ladder.

1. Place 0.2 μg of each sample in a 1.5-ml tube. Bring the volume to 4 μl with distilled water. Leave at room temperature.

2. Make the transcription cocktail in a separate 1.5-ml tube using the volumes given below after multiplying by the number of samples to be analyzed. Add each component, which is kept on ice, to a tube at room temperature in the following order:

> 2.0 μl of 5× transcription buffer [5× is 200 mM Tris-HCl (pH 7.5 at 37°), 30 mM MgCl$_2$, 10 mM spermidine, 50 mM NaCl]
> 1.4 μl distilled water
> 0.6 μl of 200 mM dithiothreitol (DTT)
> 0.5 μl of a mixture containing ATP, GTP, CTP, and UTP at 2.4 mM each[38]
> 1.0 μl [α-^{32}P]GTP or [α-^{32}P]CTP or [α-^{32}P]UTP (10–20 mCi/ml;

---

[36] E. T. Schenborn and R. C. Mierendorf, Jr., *Nucleic Acids Res.* **13**, 6223 (1985).

[37] E. D. Jorgensen, R. K. Durbin, S. S. Risman, and W. T. McAllister, *J. Biol. Chem.* **266**, 645 (1991).

[38] Although the optimal concentration of each dNTP in the final transcription reaction is determined to be 500 μM (see Krieg and Melton[32] and Milligan and Uhlenbeck[34]), we routinely perform our transcription assay with 120 μM of each unlabeled nucleotide without any adverse effect.

400–800 Ci/mmol), depending on the availability of the label and base composition of the template strand[39]

0.3 μl RNasin (40 units/μl; Promega Corp.) or equivalent RNase inhibitor[40]

0.2 μl RNA polymerase (SP6, T7, or T3; 18 units/μl)

3. Distribute 6.0 μl of the transcription cocktail into the tubes with the DNAs; mix well.

4. Incubate at 37–39° for 60 min.

5. Bring to 50 μl with 10 μg carrier tRNA in 40 μl distilled water and immediately extract with phenol–chloroform–isoamyl alcohol (50 : 50 : 1) previously equilibrated with 100 mM Tris-HCl (pH 8) buffer.

6. Transfer the aqueous (top) layer to a fresh tube and extract the aqueous phase a second time.

7. Precipitate RNA with 8 μl of 2 M sodium acetate (pH 5.5) and 136 μl cold ethanol or alternatively, to remove unincorporated radioactive nucleotides, with 50 μl 10 M ammonium acetate and 300 μl cold ethanol.

8. Place in dry ice–ethanol bath for 5–10 min.

9. Thaw briefly and spin down in microcentrifuge at 4° for 15 min.

10. Wash pellet with 1 ml cold 67% (v/v) ethanol.

11. Spin down in microcentrifuge for 3 min.

12. Remove radioactive supernatant, leaving behind the RNA pellet.

13. Repeat the 67% (v/v) ethanol wash in Steps 10 to 12.

14. Dry under reduced pressure.

15. Resuspend in 12 μl of 10 mM Tris-HCl (pH 8), 1 mM ethylenediaminetetraacetic acid (EDTA) buffer. The samples are analyzed by denaturing polyacrylamide gel electrophoresis immediately or within 12 hr after the completion of transcription.

16. Remove 2 μl, add 2 μl formamide loading dye [95% (v/v) deionized formamide containing 20 mM Tris-HCl (pH 8), 5 mM EDTA, 0.03% (w/v) bromphenol blue, and 0.03% (w/v) xylene cyanol], heat to 90° for 3 min, and load on denaturing polyacrylamide gels, prerun for about 1 hr. Note that proper precaution should be exercised in applying the samples to the well and in handling the highly radioactive gel, especially near the wells. Typically, overnight exposure of the gel without the aid of an intensifying screen is sufficient to visualize the radioactive transcripts. A detailed guide to preparation and running of polyacrylamide gels can be

---

[39] The amount of radioactivity can be decreased 5- to 10-fold, requiring longer gel exposure time or use of an intensifying screen during autoradiography.

[40] A minimum of 1 mM DTT is required for RNasin to be effective.

[41] J. Sambrook, E. F. Fritsch, and T. Maniatis, "Molecular Cloning: A Laboratory Manual," 2nd Ed. Cold Spring Harbor Laboratory, Cold Spring Harbor, New York, 1989.

found in Ref. 41. To examine transcripts 15 to 150 nucleotides long, use a 15% polyacrylamide [29 : 1 (w/w) acrylamide–N,N'-methylenebisacryl-amide] gel containing 7 M urea in 1× TBE (90 mM Tris–borate, 1 mM EDTA) buffer; the tracking dyes, bromphenol blue and xylene cyanol, comigrate with transcripts containing approximately 10 and 39 nucleotides, respectively. For transcripts longer than 100 nucleotides, use a lower percentage gel such as 8% polyacrylamide. In the latter case, bromphenol blue and xylene cyanol comigrate with transcripts approximately 17 and 83 nucleotides long, respectively.

Reference RNA Ladders

To determine the location of termination sites in the template, reference RNA ladders are loaded in lanes adjacent to the sample. These ladders are generated by supplementing the basic transcription reaction of the unmodified DNA template with 3′-dNTPs as RNA chain terminators.[42,43] Four separate tubes, one for each terminator, are set up at room temperature as follows:

   6.0 μl unmodified DNA template (0.1 μg/μl)
   1.9 μl of 500 μM 3′-dATP, 500 μM 3′-dGTP, 600 μM 3′-dCTP, or 480 μM 3′-dUTP (Pharmacia, Piscataway, NJ)[44]
   2.1 μl distilled water
   2.0 μl [α-^{32}P]GTP or [α-^{32}P]CTP or [α-^{32}P]UTP (10–20 mCi/ml; 400–800 Ci/mmol) depending on the availability of the label and the base composition of the template strand

The total volume in each tube is 12.0 μl. To each of the above tubes, add 18.0 μl of the following cocktail mix (which makes 73.9 μl total) at room temperature prepared immediately before the reaction:

   24.6 μl of 5× transcription buffer
   30.8 μl distilled water
   6.6 μl of 200 mM DTT
   6.2 μl of a mixture of ATP, GTP, CTP, and UTP at 2 mM each
   4.1 μl RNasin (40 units/μl; Promega Corp.) or equivalent RNase inhibitor
   1.6 μl RNA polymerase used to transcribe the modified template (SP6, T3, or T7; 18 units/μl)

[42] V. D. Axelrod and F. R. Kramer, *Biochemistry* **24**, 5716 (1985).
[43] J. F. Klement, M.-L. Ling, and W. T. McAllister, *Gene Anal. Tech.* **3**, 59 (1986).
[44] The actual concentration of 3′-dNTP chain terminators may have to be adjusted, depending on the base composition of the template and the region of interest.

Mix and incubate at 37–39°. Follow Steps 5 to 15 outlined above, except in Step 5 add 30 μl distilled water containing 10 μg of carrier tRNA (instead of 40 μl) prior to phenol extraction.

Because the reference RNA ladders generated using 3′-dNTP terminators end with a 3′-deoxyribose (i.e., 3′-H), whereas transcription products terminated by an adduct end with a ribose (i.e., 3′-OH), it is necessary to know the effect on mobility of the type of 3′ terminus in order to determine the position at which transcription terminates owing to the presence of an adduct in the template strand. RNAs having defined end points were obtained by limiting the amount of guanosine 5′-triphosphates in a transcription reaction (Fig. 1). Separation of these RNAs on a sequencing gel shows the expected series of bands for RNA transcripts ending before cytosines in the template strand (lane 1, Fig. 1). From the position of these bands relative to those for the products of 3′-dNTP RNA chain terminators (lanes 2 and 3, Fig. 1), it can be concluded that replacing the ribose moiety in the last nucleotide of a RNA chain with a 3′-deoxyribose (e.g., 3′-OH with 3′-H) increases the mobility of the RNA such that it migrates as if it were one-half nucleotide shorter (compare lane 1 with lanes 2 and 3, Fig. 1).

Transcription Blockage at Chemically Modified Bases

Osmium tetroxide (OsO$_4$ or OT) reacts at the 5,6-double bond of exposed thymines producing, in the presence of pyridine, bis(pyridine) osmate ester adduct that slowly hydrolyzes to yield a thymine glycol.[4,45] Figure 2 shows the result of transcribing a test DNA that had previously been exposed to osmium tetroxide and pyridine as double or single strands. As seen in lanes 4–6 (Fig. 2), brief exposure of single-stranded DNA to osmium tetroxide and pyridine results in the appearance of prematurely terminated products, whose average size decreases with increasing time of incubation. In contrast, DNA that has been denatured and renatured (lane 3, Fig. 2) or double-stranded DNA that has been exposed to the reagents for an extended period (lane 1, Fig. 2) yields a pattern of RNA transcripts not significantly different from untreated DNA (lane 2, Fig. 2). These results are consistent with bases in single-stranded DNA being exposed and hence reactive to osmium tetroxide, whereas those in native, double-helical DNA remain inaccessible to osmium tetroxide.

Transcription does not terminate at every position in the template strand (lanes 4–6, Fig. 2). Comparison with reference RNA ladders, obtained by including 3′-dNTPs in the transcription mixture (lanes 7 and 8,

[45] M. Schroder, *Chem. Rev.* **80,** 187 (1980).

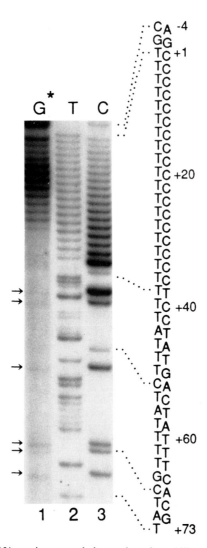

FIG. 1. Influence of 3′ terminus on gel electrophoretic mobility of RNAs. Products of T7 RNA polymerase transcription of the DNA described in Fig. 3 were subjected to electrophoresis through a 8% polyacrylamide (29 : 1)–7 M urea gel in 1× TBE buffer. The standard transcription reaction was modified by lowering the concentration of GTP to 2 μM, resulting in transcripts having 3′-OH ends that terminate before template cytosines (lane 1). Alternatively, the reaction was supplemented with 3′-deoxyadenosine 5′-triphosphate (lane 2) or 3′-deoxyguanosine 5′-triphosphate (lane 3), resulting in transcripts having 3′-H ends that terminate at template thymines (T) or cytosines (C), respectively. Because the bands in lane 1 contain RNAs one nucleotide shorter than the corresponding bands in lane 3, the RNAs ending with a ribose moiety (3′-OH) migrate one-half nucleotide step slower than RNAs ending with a deoxyribose moiety (3′-H). Arrows at left point to faint bands in lane 1 clearly visible in the original autoradiogram. The sequence of the template strand is given at right, and the nucleotides are numbered as in Fig. 3.

Fig. 2), shows that for every band in the "T" lane, there are two bands in the OT-treated single-stranded DNA lanes (lanes 4–6, Fig. 2) displaced upward by one-half and one-and-one-half nucleotide steps. Taking into account the fact that transcripts in lanes 7 and 8 (Fig. 2) terminate with a deoxyribonucleotide (i.e., 3'-H group) and therefore migrate one-half nucleotide faster than those in lanes 4 through 6 (Fig. 2), we see that termination occurs such that the final base of the transcript in most cases is opposite the osmium tetroxide–pyridine-modified thymines, whereas in other cases it extends one base beyond the lesion. Similar results are seen with T4 DNA polymerases and Klenow fragment of *Escherichia coli* DNA polymerase I; however, for these polymerases, termination occurs almost exclusively after inserting a base opposite a thymine glycol residue.[46,47]

A number of other adducts also block the progression of RNA polymerases. This was demonstrated by analyzing a $(dT-dC)_{18} \cdot (dA-dG)_{18}$ sequence, abbreviated as $(TC)_{18} \cdot (AG)_{18}$ (large open box in Fig. 3), inserted between the SP6 and T7 RNA promoters of pGEM-2 DNA (Promega Corp.). Under negative torsional stress or low pH, this sequence has been shown to form H-DNA, in which either the 5' or 3' half of the $(dT-dC)_{18}$ sequence folds back into the major groove of the other half of the sequence to form a triple helix and a displaced single-stranded polypurine loop. When present in highly negatively supercoiled molecules, the predominant H-DNA conformer is the H-y3(16) DNA structure, shown in Fig. 4A, whose triplex–duplex junction extends outward to include nucleotides in the two adjacent short direct repeats (small open boxes in Fig. 3), shown in Fig. 4B.[16,48,49]

Treatment with a variety of chemical probes and subsequent transcription, using T7 RNA polymerase to visualize the $(TC)_{18}$ strand (Fig. 5A) and SP6 RNA polymerase to visualize the $(AG)_{18}$ strand (Fig. 5B), show a pattern of transcription termination end points consistent with the structure shown in Fig. 4B. Transcription stops in the single-stranded regions of the $(TC)_{18}$ strand (center and to the 3' side) are seen for the pyrimidine reagents: osmium tetroxide (OT; lane 1, Fig. 5A, exposed thymines), methoxylamine (MA; lane 2, Fig. 5A, exposed cytosines and some exposed thymines), and hydroxylamine (HA; lane 3, Fig. 5A, exposed cytosines and, to a much lesser extent, some exposed thymines). The blockage of transcription in DNAs reacted with diethyl pyrocarbonate (DEPC; lane 5, Fig. 5A, exposed adenines and to a lesser extent guanines) or dimethyl

[46] H. Ide, Y. W. Kow, and S. S. Wallace, *Nucleic Acids Res.* **13**, 8035 (1985).
[47] R. C. Hayes and J. E. LeClerc, *Nucleic Acids Res.* **14**, 1045 (1986).
[48] B. H. Johnston, *Science* **241**, 1800 (1988).
[49] H. Htun and J. E. Dahlberg, *Science* **243**, 1571 (1989).

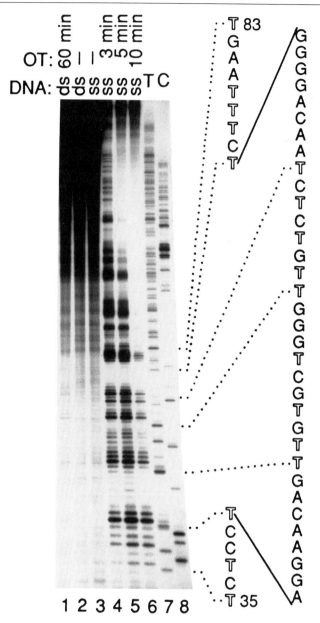

FIG. 2. Transcription termination at osmium tetroxide-modified thymines. Native linearized DNAs (lanes 1 and 2) or denatured DNAs (lanes 3–6) were treated with 2.5 m*M* osmium tetroxide, 2% pyridine (v/v) at pH 8 and room temperature for the indicated amount of time (lanes 1 and 4–6) or left untreated (lanes 2 and 3). Prior to transcription with T7 RNA

sulfate (DMS; lane 6, Fig. 5A, unpaired adenines and much less frequently unpaired cytosines) can be seen in the 3' side of the $(TC)_{18}$ sequence, which forms a single-stranded region linking the triplex and duplex DNA. On the opposite strand, transcription stops are seen in the unpaired 5' half of the $(AG)_{18}$ sequence with the purine reagents and in the 5' flanking sequences with all reagents (Fig. 5B).

Comparison with the reference RNA ladders shows that transcripts terminate primarily at the nucleotide (1) immediately 3' to (and hence before) the adducts for methoxylamine-modified pyrimidines; (2) immediately before or at the adducts for hydroxylamine-modified cytosines; (3) immediately before or to a lesser extent at the adducts for diethyl pyrocarbonate-modified adenines; or (4) immediately before the adducts for dimethyl sulfate-modified adenines. Taking these observations into consideration, the pattern of transcription stops seen with osmium tetroxide, methoxylamine, hydroxylamine, and diethyl pyrocarbonate reflects the distribution of alkali-labile sites in the DNA, as revealed by the chemical cleavage approach (see Introduction and Fig. 6). However, with dimethyl sulfate, modification at the heterocyclic N-1 does not render the methylated base alkali-labile and thus is not detectable by treatment with hot piperidine (data not shown).

Although the modified bases in the above example were detected by transcription without additional treatment, not all modified bases can be detected in such a manner. Notably, 7-methylguanine, the major dimethyl sulfate product, does not generally block the progression of polymerases, hence the lack of transcripts terminating at guanines in lane 5 of Fig. 5B. Additional treatment, such as with alkali or piperidine, is required to open the modified guanine ring and/or to cleave the backbone to allow detection of the lesion by transcription or primer extension.[50] Thus, the use of any reagent creating adducts that either directly block polymerases or ultimately lead to strand scission allows mapping of reactive sites by the polymerase approach. It should be noted that in terminating transcription

---

[50] T. R. O'Connor, S. Boiteux, and J. Laval, *Nucleic Acids Res.* **16,** 5879 (1988).

---

polymerase, the denatured DNAs were renatured at room temperature by resuspending the DNA pellet (3 μg) in 97% deionized formamide (82 μl) and then incubating for several hours in 50% formamide, 100 mM Tris-HCl (pH 8), 10 mM EDTA (160 μl total). The renatured DNA was concentrated and analyzed by transcription. Lanes marked by T (lane 7) and C (lane 8) refer to the base in the template that directed the incorporation of 3'-dATP and 3'-dGTP RNA chain terminators, respectively. Template DNA used was the same as in Fig. 1 except that it contained 5 (dT-dC) repeats instead of 18; the sequence of the template strand examined is given at right and numbered in this case with the first nucleotide indicating the start of transcription.

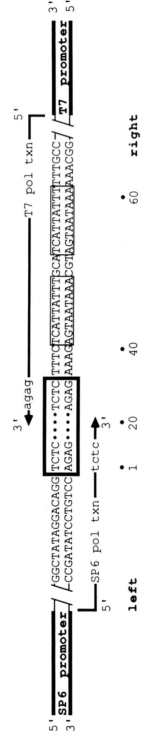

Fig. 3. (TC · AG)$_{18}$ repeat and flanking sequences examined by modification and transcription. A 180-base pair *Sac*I–*Bam*HI restriction fragment from a region 1.8 kilobases downstream of a human U1 gene was cloned between the SP6 and T7 RNA phage promoters of the pGEM-2 vector (Promega Corp). This fragment contains a (dT-dC)$_{18}$ · (dA-dG)$_{18}$ sequence, denoted by a heavy box, and two short (dA + dT)-rich direct repeats, enclosed by light boxes. The lowercase letters above or below the DNA sequence represent the RNA products when transcribed with T7 or SP6 RNA polymerases, respectively. Transcription that used the TC$_{18}$ strand as template (from the T7 promoter) resulted in the shortest products coming from the 3' side of the TC$_{18}$ strand, or the right-hand side of the structure as drawn here; likewise, transcripts that probed the AG$_{18}$ strand (from the SP6 promoter) had their shortest products at the left-hand side of the structure. The solid dots denote every twentieth nucleotide starting at the 5' end of the TC$_{18}$ repeat. [Reprinted with permission from H. Htun and J. E. Dahlberg, *Science* **241**, 1791 (1988). Copyright 1989 by the American Association for the Advancement of Science.]

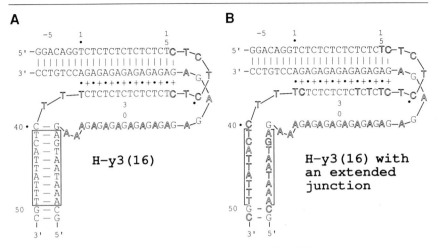

FIG. 4. Predominant H-DNA structures formed at a (TC · AG)$_{18}$ sequence in native and highly negatively supercoiled molecules at acid pH. In both cases, the H-y3(16) DNA conformers are shown, where the 3' half of the (TC)$_{18}$ sequence folds back to donate pyrimidines participating in Hoogsteen base pairs (dots and pluses) with the acceptor helix, maintained by Watson–Crick base pairs (long lines); the resulting triple-stranded structure contains 16 base triplets. The displaced polypurine strand is single stranded but may be structured [H. Htun and J. E. Dahlberg, in "Structure and Methods, Volume 3: DNA and RNA" (R. H. Sarma and M. H. Sarma, eds.), p. 185. Adenine Press, Schenectady, New York, 1990]. Two disrupted Watson–Crick base pairs link the donated strand and the surrounding double helix at native bacterial levels of negative supercoiling and pH 5 (A). With increasing levels of negative supercoiling, the number of disrupted base pairs increases [H. Htun and J. E. Dahlberg, Science **241**, 1791 (1988); B. H. Johnston, Science **241**, 1800 (1988)] (B) to permit greater relaxation of negative supercoils [H. Htun and J. E. Dahlberg, Science **243**, 1571 (1989)]. Nucleotides reactive in the H-y3(16) DNA conformations are indicated by large, open letters. [Adapted with permission from Htun and Dahlberg (1988, 1989). Copyright 1988 and 1989 by the American Association for the Advancement of Science.]

at the point of strand scission, phage RNA polymerases, like DNA polymerases (see next section), can add an extra nucleotide to its 3' end not encoded in the template strand.[34]

Primer Extension

Many of the generalizations made about transcription also apply to primer extension. The main differences are that the copying enzymes used are different, and have somewhat different sensitivities to the presence of adducts, and that an annealed primer is required for primer extension. Because primers can be directed at any segment of DNA, the analysis of modified DNA is no longer limited to a region contained between two phage promoters. Some genomic sequences can readily be examined through the use of a highly radioactive primer or by multiple rounds of DNA synthesis.[19-21] Although cleavage of the phosphodiester backbone is often not necessary, except in cases like the 7-methylguanine adducts of dimethyl sulfate noted above, cleavage may improve the signal-to-background ratio. Cleavage of modified sites also allows any genomic DNA to be analyzed by the ligation-mediated polymerase chain reaction.[17,18] Detailed protocols for primer extension with DNA are given by Sasse-Dwight and Gralla[51] elsewhere in this series. The brief account given here draws heavily on that source in addition to our own experience. Accounts of primer extension using RNA and reverse transcriptase[23,52-55] may also be profitably consulted.

[51] S. Sasse-Dwight and J. D. Gralla, this series, vol. 208, p. 146.
[52] G. S. Shelness and D. L. Williams, *J. Biol. Chem.* **260,** 8637 (1985).
[53] S. Stern, D. Moazed, and H. F. Noller, this series, Vol. 164, p. 481.
[54] A. Krol and P. Carbon, this series, Vol. 180, p. 212.
[55] W. R. Boorstein and E. A. Craig, this series, Vol. 180, p. 347.

FIG. 5. Blocks to transcription at the H-y3(16) DNA forming sequences reacted with several chemical probes. The H-y3(16) DNA structure formed in highly negatively supercoiled DNA (shown in Fig. 4B) was not reacted (buffer) or was reacted with several base-specific reagents: osmium tetroxide in the presence of pyridine (OT), methoxylamine (MA), hydroxylamine (HA), diethyl pyrocarbonate (DEPC), and dimethyl sulfate (DMS). Polarity of the template strand is indicated in the line drawing at the side, where the large box represents either the $(TC)_{18}$ or $(AG)_{18}$ sequence and the two smaller boxes the adjacent short direct repeats; the heavy dots denote every twentieth nucleotide, as in Fig. 3. For the reference RNA ladders, the nucleotide on the template strand that directed the incorporation of 3'-dUTP, 3'-dCTP, 3'-dGTP, or 3'-dATP is indicated as A, G, C, or T, respectively; the small dots to the left of each lane indicate the position of expected bands in places where reading of the sequence might be or is ambiguous. DNAs were linearized with SinI prior to transcription with T7 RNA polymerase to analyze the $(TC)_{18}$ strand (A) or SP6 RNA polymerase to analyze the complementary $(AG)_{18}$ strand (B). *Reaction conditions:* The DNA described in Fig. 3 was treated with topoisomerase I in the presence of ethidium bromide (25 μg/ml) to generate highly negatively supercoiled molecules, which were reacted separately under the following conditions: (lanes OT) 2.0 mM osmium tetroxide, 2% pyridine (pH 6), 100 mM sodium cacodylate (pH 6) for 5 min at room temperature; (MA) 3 M methoxylamine (adjusted with diethylamine to pH 5), 25 mM sodium acetate (pH 5) for 30 min at 37°; (HA) 3 M hydroxylamine (adjusted with diethylamine to pH 5), 25 mM sodium acetate (pH 5) for 60 min at 37°; (DEPC) 2.5% (v/v) diethyl pyrocarbonate, 100 mM sodium MES [2-(N-morpholino)ethanesulfonic acid)] (pH 6) for 40 min at room temperature; and (DMS) 85 mM dimethyl sulfate, 100 mM sodium cacodylate (pH 6) for 3 min at room temperature.

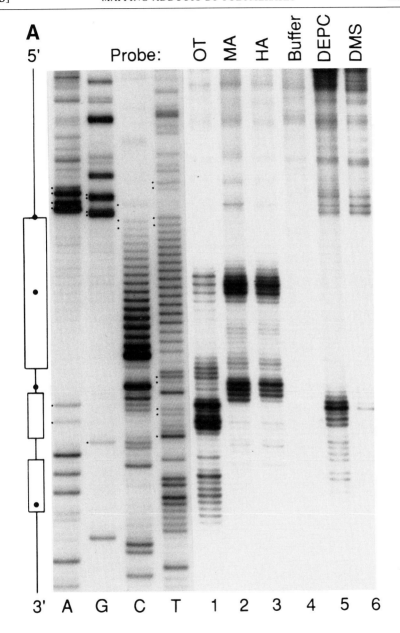

B

Probe:

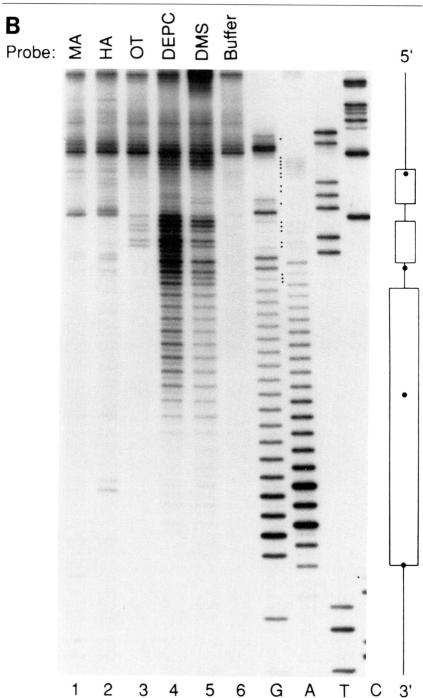

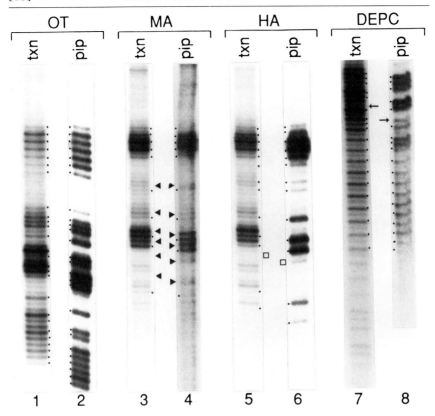

FIG. 6. Distribution of products from transcription and hot piperidine treatment. Each of the DNA samples analyzed in Fig. 5 by transcription (odd-numbered lanes) was also analyzed by the chemical cleavage approach (even-numbered lanes) as described in the Introduction, i.e., labeled with radioactive ^{32}P at a unique DNA end and treated with hot piperidine to induce strand scission at alkali-labile sites before separating the resultant fragments on a DNA sequencing gel. The symbols beside the piperidine (pip) lanes indicate DNA bands generated by cleavage at osmium tetroxide-modified thymines (dots) in lane 2, methoxylamine-modified cytosines (dots) and thymines (arrowheads) in lane 4, hydroxylamine-modified cytosines (dots) and thymine (open square) in lane 6, and diethyl pyrocarbonate-modified adenines (dots) and guanine (arrow) in lane 8. Corresponding bands in the transcription (txn) lanes are indicated for RNA transcripts terminating at the modified site (lane 1) or one nucleotide short of the modified site (lanes 3, 5, and 7). See text for explanation of additional bands in the transcription lanes.

Template

Freedom from contaminants as described above is essential, and a highly recommended step is to pass the DNA through a spin column (e.g., Sephadex G-50; Pharmacia, Piscataway, NJ) equilibrated with distilled water prior to primer extension. Note again that to detect the

7-methylguanine adducts in dimethyl sulfate-modified DNA, treatment with 1 M piperidine at 90° for 30 min, as described by Maxam and Gilbert,[14] should precede the primer annealing step; however, to detect the 1-methyladenine and 3-methylcytosine adducts, the piperidine treatment should be avoided.[23,29]

Primer

The preferred labeling technique is to use a 5'-^{32}P-end-labeled primer. [α-^{32}P]dNTPs can also be incorporated during the extension reaction but may yield a higher background. Obviously, good results require that there be only one annealing site for the primer, so sequences chosen for primers should be unique, and there should be no similar sequences elsewhere on the template DNA. They should also anneal with a melting temperature, T_m, of about 50–57° (17–20 nucleotides in length is good), preferably have G or C residues on the 3' end, and be situated at least 15–20 base pairs (bp) from the sequence to be analyzed. The melting temperature can be estimated from the formula

$$T_m = 81.5 + 16.6(\log M) + 0.41(\% \text{ GC}) - 500/n$$

where M is the molarity of the monovalent cation, % GC is the percent GC content, and n is the length of the primer.[56] Primers should be purified on a polyacrylamide gel after synthesis, and care should be taken to filter off acrylamide debris and to avoid precipitation with ammonium salts which will inhibit the 5'-labeling reaction. A total of 20–30 pmol of primer is labeled using polynucleotide kinase and [α-^{32}P]ATP, and unincorporated ATP is removed by passing through a spin column equilibrated with 10 mM Tris-HCl (pH 8), 1 mM EDTA, 100 mM NaCl. For an effluent volume of 50 µl, a specific activity of 0.7–5 × 10[6] counts/min (cpm)/µl is expected; when this decays to a level of 0.3 × 10[6] cpm/µl it is discarded, as decay products may interfere with the reaction.

Polymerases

A number of polymerases may be used to extend a primer from an annealed site. The Klenow fragment of *E. coli* polymerase I, T4 and T7 DNA polymerases, *Taq* and *Vent* polymerase, and reverse transcriptase are commonly used. Minor differences exist between the various polymerases in their ability to move through the site of the lesion[57–59] or add a

[56] G. Wahl, S. L. Berger, and A. R. Kimmel, this series, Vol. 152, p. 399.
[57] J. M. Clark and G. P. Beardsley, *Biochemistry* **26**, 5398 (1987).
[58] D. Sagher and B. Strauss, *Nucleic Acids Res.* **13**, 4285 (1985).
[59] M. Takeshita, C.-N. Chang, F. Johnson, S. Will, and A. P. Grollman, *J. Biol. Chem.* **262**, 10171 (1987).

single nucleotide to the 3' end of a double-stranded DNA, yielding a one base overhang.[60-62] DNA synthesis should always be performed under conditions of high polymerase fidelity to avoid significant "readthrough" and undesired incorporation of additional nucleotides at the modified site.[46]

Primer Extension by Klenow Enzyme

Precautions regarding the preparation of the template have been discussed above.

1. To 0.5 μg of each modified plasmid DNA sample in 35 μl distilled water, add 1 μl ^{32}P-end-labeled primer (0.3–0.5 × 10^6 cpm/μl).

2. Add 4 μl of 10 mM NaOH and mix well.

3. Heat for 2 min at 80° to denature template DNA; immediately place on ice for at least 5 min.

4. Add 5 μl freshly made 10× TMD buffer (0.5 M Tris-HCl, pH 7.2, 0.1 M MgSO$_4$, 2 mM DTT) and mix well.

5. Incubate at 45–50° (or T_m − 5°) for 3 min to allow hybridization of primer to the modified DNA template; by keeping the hybridization temperature close to the T_m, hybridization at unwanted sites will be minimized. Immediately, place each sample on ice.

6. Add 5 μl of dNTP mix (containing 5 mM dATP, 5 mM dCTP, 5 mM dGTP, 5 mM dTTP); mix and return to ice.

7. Keeping the enzyme cold, dilute the Klenow fragment of DNA polymerase I to 0.5–1.0 unit/ml using Klenow dilution buffer [50% (v/v) glycerol, 50 mM KH$_2$PO$_4$, 1 mM DTT, 100 μg/ml bovine serum albumin (BSA); stored at − 20°]. The enzyme should be placed at the bottom of a cold microcentrifuge tube without creating bubbles and the cold diluent added carefully. Mix gently.

8. Add 1 μl diluted Klenow fragment to each sample and mix gently. Incubate at 50° for exactly 10 min.

9. Immediately add $\frac{1}{3}$ volume of ammonium acetate, 20 mM EDTA to the sample and mix thoroughly.

10. Add 2.5–3.0 volumes cold 95% (v/v) ethanol and chill at − 70° for 20 min or overnight at − 20°. Spin in a microcentrifuge for 10 min at 4°, rinse the DNA pellet gently with 70% (v/v) ethanol, and dry the pellet. Proceed with gel electrophoresis as described below.

[60] J. M. Clark, *Nucleic Acids Res.* **16**, 9677 (1988).
[61] P. R. Mueller and B. Wold, *Methods* **2**, 20 (1991); P. R. Mueller and B. Wold, *Current Protocols in Molecular Biology* (*New York*), vol. 2, chaps. 15.5.1–15.5.16 (1991).
[62] P. A. Garrity and B. J. Wold, *Proc. Natl. Acad. Sci. U.S.A.* **89**, 1021 (1992).

Extension by Single-Primer Polymerase Chain Reaction

The following protocol, for the analysis of bacterial chromosomal DNA modified *in vivo*, is taken from Sasse-Dwight and Gralla.[51] This technique is often necessary when analyzing chromosomal DNA because it magnifies the normally low signal by allowing multiple rounds of extension and permits the extension to occur at a higher temperature, thereby minimizing nonspecific hybridization. It is equivalent to the normal polymerase chain reaction (PCR) except that only one primer is added to the reaction mix, and therefore the amplification is linear. Alternatively, the use of a highly radioactive primer with labeled phosphates at numerous positions and a greater mass of chromosomal DNA allows analysis of single-copy genes in more complex genomes by one[19] or multiple rounds[20,21] of primer extension.

1. Prepare the samples by adding the following solutions in order:

    49.5 μl distilled water

    10.0 μl of 10× *Taq* reaction buffer (as provided with enzyme; e.g, Perkin Elmer Cetus, Norwalk, CT)

    4.0 μl dNTP mix (contains 5 mM dATP, 5 mM dCTP, 5 mM dGTP, 5 mM dTTP)

    1.0 μl of ^{32}P-end-labeled primer (diluted to 0.5 × 10^6 cpm/μl with distilled water)

    35.0 μl DNA (0.5 μg) in distilled water after passing through a spin column; equivalent to isolated chromosomal DNA from 3 ml of *E. coli* cells at mid log phase

    0.5 μl *Taq* polymerase (2.5 units/μl)

2. Overlay samples with 100 μl mineral oil.

3. Treat samples as follows: (a) 1.5 min at 94°; (b) 2 min at 57° (or T_m of primer); (c) 3 min at 72°; (d) 1 min at 94°; (e) repeat steps (b) through (d) 15–20 times; (f) heat for 2 min at 57° (or T_m of primer); (g) heat for 10 min at 72°; (h) let samples cool to room temperature.

4. Separate the sample from the mineral oil by pipetting the aqueous phase up from the bottom of the tube and place it in a fresh tube.

5. Extract sample once with chloroform–isoamyl alcohol (24 : 1).

6. Precipitate the DNA samples with $\frac{1}{3}$ volume of 4 M ammonium acetate, 20 mM EDTA and 2.5 volumes cold 95% (v/v) ethanol for 20 min at −70° or overnight at −20°.

7. Pellet the DNA by spinning for 10 min at 4°. Wash the pellet with 70% (v/v) ethanol by gently rolling the tube, respin for 2 min, remove the ethanol, and dry the pellet. The samples are now ready for polyacrylamide gel electrophoresis.

Extension by Ligation-Mediated Polymerase Chain Reaction

Single-copy genes in more complex organisms can also be analyzed using a primer labeled at one position. However, in this case, the primary signal has to be amplified exponentially by the use of nonradioactive primers before analyzing the amplified products with the radioactive primer. See Wold *et al.*[61,62] for details of this procedure.

Polyacrylamide Gel Electrophoresis

The DNA pellets resulting from the primer extension procedure are resuspended in 4–6 μl of alkaline formamide–urea loading dye (75 mg ultrapure urea, 100 μl deionized formamide, 8 μl of 50 mM NaOH, 1 mM EDTA, and 8 μl 0.5% (w/v) xylene cyanol, 0.5% (w/v) bromphenol blue). For consistent results, it is best to treat each sample in an identical fashion when resuspending. This is best accomplished by placing the samples with dye at 45–50° and removing each to tap forcefully for 30 sec. Following this, the samples should be vortexed in pairs for 30 sec. This technique ensures that most of the sample is resuspended in the dye mix. This can be checked by making sure that most of the counts (over 90%) are now in the solution.

Following a brief spin to collect the liquid, the samples are ready for loading. DNA fragments are resolved in a standard denaturing polyacrylamide gel containing 19 : 1 (w/w) acrylamide–N,N'-methylenebisacrylamide; note that RNA transcripts were previously resolved in a denaturing polyacrylamide gel with a lower number of cross-links, that is, 29 : 1 (w/w) acrylamide–N,n'-methylenebisacrylamide. Following electrophoresis, the gel is transferred to a used film, covered with Saran Wrap, and placed on X-ray film, using an intensifying screen if necessary. Exposure time is usually 1 or 2 days.

Concluding Remarks

Although transcription or primer extension approaches may be used interchangeably in most cases, situations will arise when one approach should be used over the other. The transcription approach was developed because the Klenow fragment of *E. coli* DNA polymerase I could not proceed unimpeded through long $(dT-dC)_n \cdot (dA-dG)_n$ sequences associated with human U1 RNA genes.[63,64] A significant fraction of DNA syntheses failed within the copolymeric region when the template contained as

[63] H. Htun, E. Lund, and J. E. Dahlberg, *Proc. Natl. Acad. Sci. U.S.A.* **81**, 7288 (1984).
[64] H. Htun and J. E. Dahlberg, unpublished result (1984).

few as 15 (dA-dG) repeats in the polypurine strand or 30 (dT-dC) repeats in the polypyrimidine strand.[64] A more detailed analysis of this phenomenon along with ways to overcome this problem has been carefully studied by Manor and co-workers.[65,66] In contrast, RNA polymerases transcribed through the same copolymers with little difficulty, allowing the mapping of DNA adducts within these copolymers as demonstrated here and previously.[16] On the other hand, the presence of natural RNA polymerase pause or termination sites in the template strand may dictate the use of DNA over RNA polymerases.[31,33,34,67]

Acknowledgments

We thank Dr. Vladimir D. Axelrod for a generous supply of 3'deoxyribonucleoside triphosphates making the analysis by transcription possible, Dr. James E. Dahlberg for stimulating discussion leading to the development of the transcription mapping approach, Drs. James E. Dahlberg, Alexander Rich, and Carl S. Parker for support, Dr. Jay D. Gralla and Paul R. Mueller for communicating manuscripts prior to publication, and Dr. Paul R. Mueller, William R. Boorstein, and Michael Harrington for useful comments on the manuscript. This work was partly supported by a National Institutes of Health postdoctoral fellowship (1F32GM14220) to H.H. and a National Institutes of Health postdoctoral fellowship (3F32GM09031) and grant (GM41423) to B.H.J.

[65] A. Lapidot, N. Baran, and H. Manor, *Nucleic Acids Res.* **17**, 883 (1989).
[66] N. Baran, A. Lapidot, and H. Manor, *Proc. Natl. Acad. Sci. U.S.A.* **88**, 507 (1991).
[67] S. T. Jeng, J. F. Gardner, and R. I. Gumport, *J. Biol. Chem.* **265**, 3823 (1990).

[16] Enzyme Probes *in Vitro*

By Franz Wohlrab

Introduction

The interaction of proteins with DNA is greatly affected by secondary structure changes, a fact that has been exploited in the use of enzymes as probes for variations in DNA conformation *in vitro*. For many site-specific DNA binding proteins, structural alterations on the level of the nucleic acid lead to a reduction or abolishment of substrate recognition. Conversely, several enzymes recognize structural features not present on canonical B-DNA, such as single strandedness or unstacked bases. Several important enzymatic processes, such as homologous recombination, DNA repair, and the initiation of DNA replication among others, are known to involve structural transitions in the DNA. However, for practical purposes, the use of enzymes as structural probes has been essentially limited to two classes of reagents: restriction–modification systems and endonucleases. In addition, several other enzymes have been employed in special cases, such as resolvases for the detection of Holliday junctions.

Restriction Endonucleases

Restriction endonucleases and DNA methylases are generally unable to recognize their target sites when they are in a noncanonical conformation, except for certain enzymes that can work on single-stranded substrates. Thus, the *Hha*I and *Bss*HII restriction enzymes will not cleave their respective recognition sites (GCGC and GCGCGC) when they are in a left-handed conformation.[1] Other enzymes, notably *Eco*RI and *Bam*HI, have been shown to be inhibited if their recognition sequence is located adjacent to a region of left-handed Z-DNA.[2,3] This has been used to determine the extent of the structural aberrations (junctions) at the interface of B- and Z-DNA stretches, and it has been shown that, for alternating $(dG-dC)_n$ blocks, this junction extends not more than 7 base pairs (bp) into the flanking sequence.

Similarly, a restriction site in the center of an inverted repeat will be single stranded on cruciform extrusion. Most, but not all, restriction endonucleases cleave single strands at very slow rates, so that these enzymes have been used to measure the rate of formation or disappearance of cruciform structures in supercoiled plasmids.[4] The same procedure can, of course, also be employed for the study of hairpin-containing oligodeoxynucleotides. Similarly, when restriction sites are at least partly involved in an intermolecular triple-stranded structure (formed by the addition of complementary oligonucleotides to duplex DNA), nucleolytic cleavage is also expected to be strongly inhibited. This has been experimentally demonstrated for several restriction enzymes, among them *Ava*I, *Bam*HI, and *Sst*I.[5–7] Likewise, *Bam*HI, *Pst*I, *Hin*dIII, and *Ava*I have been shown to be incapable of cutting their recognition sequences in topologically constrained form V DNA.[8]

Site-specific methyltransferases are also affected by the conformation

[1] W. Zacharias, J. E. Larson, M. W. Kilpatrick, and R. D. Wells, *Nucleic Acids Res.* **12**, 7677 (1984).
[2] C. K. Singleton, J. Klysik, and R. D. Wells, *Proc. Natl. Acad. Sci. U.S.A.* **80**, 2447 (1983).
[3] F. Azorin, A. Nordheim, and A. Rich, *EMBO J.* **2**, 649 (1983).
[4] M. Gellert, M. H. O'Dea, and K. Mizuuchi, *Proc. Natl. Acad. Sci. U.S.A.* **80**, 5545 (1983).
[5] J.-C. Francois, T. Saison-Behmoaras, N. T. Thuong, and C. Helene, *Biochemistry* **28**, 9617 (1989).
[6] J. C. Hanvey, M. Shimizu, and R. D. Wells, *Nucleic Acids Res.* **18**, 157 (1990).
[7] L. J. Maher III, B. Wold, and P. B. Dervan, *Science* **245**, 725 (1989).
[8] Y. S. Shouche, N. Ramesh, and S.K. Brahmachari, *Nucleic Acids Res.* **18**, 267 (1990).

of the substrate DNA. This is of special interest because DNA methylases, when expressed in bacterial cells, provide a genetic method to detect structural alterations *in vivo*.[9] This is covered in detail elsewhere in this volume [19].

Nucleases

The most widely used enzymatic probes for DNA structure *in vitro* are endonucleases with little or no sequence specificity, such as the family of zinc-dependent single-strand-specific endonucleases that comprise S1 nuclease, P1 nuclease, mung bean nuclease, and others. In general, double-stranded DNA is cleaved by these enzymes at low rates, whereas single strands are digested rapidly.[10] If structural transitions in the DNA lead to unpairing of complementary bases, the nuclease then cuts at this site. Because this provides a positive signal for a structural transition, nucleases have been extensively used to monitor conformational parameters in nucleic acids. Their mechanism of action and the precise substrate requirements have been a matter of debate, and it has been argued that phosphodiester backbone conformations rather than single strandedness are the determining factor in cutting site selection by these enzymes.[11] Whatever the molecular mechanisms involved, these enzymes have been shown to be remarkably sensitive to changes in secondary structure.

S1 Nuclease

S1 nuclease from *Aspergillus oryzae* has been widely used to probe structural transitions in DNA. It has a pH optimum in the acidic range (pH 5 and below), which makes it a good probe for the investigation of protonated structures but restricts its use under physiological conditions. In certain cases, however, S1 nuclease has sufficient residual activity at neutral pH values to be useful as a probing tool.[12,13] It cleaves single-stranded DNA at 3'-phosphate phosphodiester bonds and does not show any base-specific recognition pattern.[14] It can cleave DNA strands oppo-

---

[9] A. Jaworski, W.-T. Hsieh, J. A. Blaho, J. E. Larson, and R. D. Wells, *Science* **238**, 773 (1987).

[10] K. Shishido and K. Ando, *in* "Nucleases" (S. M. Linn and R. J. Roberts, eds.), p. 155. Cold Spring Harbor Laboratory, Cold Spring Harbor, New York, 1985.

[11] D. E. Pulleyblank, M. Glover, C. Farah, and D. B. Haniford, *in* "Unusual DNA Structures: Proceedings of the First Gulf Shores Symposium" (R. D. Wells and S. C. Harvey, eds.), p. 23. Springer-Verlag, New York, 1988.

[12] R. F. Fowler and D. M. Skinner, *J. Biol. Chem.* **261**, 8994 (1986).

[13] J. Bernués, R. Beltrán, J. M. Casasnovas, and F. Azorin, *EMBO J.* **8**, 2087 (1989).

[14] H. R. Drew, *J. Mol. Biol.* **176**, 535 (1984).

site nicks and gaps and is stable at temperatures up to 65°. Citrate and phosphate buffers have been reported to inhibit its activity.

Because of its specificity for single-stranded nucleic acids, S1 nuclease has been employed to map sites on DNA that had been locally denatured by chemical reagents. Such reagents include bromoacetaldehyde, chloroacetaldehyde, and osmium tetroxide, since these modifications cause the unpairing of the two strands of the double helix.[15,16] However, this technique has been largely replaced by the use of hot alkali such as piperidine to achieve strand scission at the site of modification. S1 nuclease also recognizes junctions between right-handed B- and left-handed Z-DNA sequences.[17-19] The location of the cleavage sites differs between different junctions. In the case of left-handed $(dG-dC)_n$ blocks, the major S1 nuclease nicks are located within 3–20 bp from the first alternating residue, with minor nicks extending into the $(dG-dC)_n$ region.[20] In contrast, alternating $(dC-dA)_n \cdot (dT-dG)_n$ stretches embedded in right-handed DNA produce different S1 nuclease susceptibility patterns. Here the nicks extend over a wider region, and, in the case of longer repeats, the entire alternating region becomes nuclease sensitive.[21] This probably reflects the existence of smaller Z-DNA segments within the insert as well as conformational heterogeneity of the junctions, a finding corroborated by studies on the thermodynamics of the B–Z transitions.[22] The increase of $\Delta G_{junction}$ with increasing NaCl concentrations and the theoretical unwinding of the adjacent right-handed helix by 0.4 turns/junction supports the single-stranded character of the B–Z interfact.[3,23] However, other data have argued against single strandedness of the junctions, so that the precise nature of the structure recognized by S1 nuclease remains to be elucidated.

In cruciform structures or hairpins, S1 nucleases recognizes the unpaired loop regions of the inverted repeats.[24,25] In the case of supercoil-

[15] T. Kohwi-Shigematsu, R. Gelinas, and H. Weintraub, *Proc. Natl. Acad. Sci. U.S.A.* **80**, 4389 (1983).

[16] J. C. Hanvey, J. Klysik, and R. D. Wells, *J. Biol. Chem.* **263**, 7386 (1988).

[17] C. K. Singleton, J. Klysik, S. M. Stirdivant, and R. D. Wells, *Nature (London)* **299**, 312 (1982).

[18] C. K. Singleton, M. W. Kilpatrick, and R. D. Wells, *J. Biol. Chem.* **259**, 1963 (1984).

[19] M. W. Kilpatrick, J. Klysik, C. K. Singleton, D. A. Zarling, T. M. Jovin, L. H. Hanau, B. F. Erlanger, and R. D. Wells, *J. Biol. Chem.* **259**, 7268 (1984).

[20] F. Wohlrab and R. D. Wells, in "Gene Amplification and Analysis" (J. G. Chirikjian, ed.), Vol. 5, p. 247. Elsevier, New York (1987).

[21] T. E. Hayes and J. E. Dixon, *J. Biol. Chem.* **260**, 8145 (1985).

[22] B. H. Johnston, W. Ohara, and A. Rich, *J. Biol. Chem.* **263**, 4512 (1988).

[23] T. R. O'Connor, D. S. Kang, and R. D. Wells, *J. Biol. Chem.* **261**, 13302 (1986).

[24] C. K. Singleton and R. D. Wells, *J. Biol. Chem.* **257**, 6292 (1982).

[25] D. Lilley and L. R. Hallam, *J. Biomol. Struct. Dyn.* **1**, 169 (1983).

stabilized Z-DNA, a single nick by the nuclease will lead to relaxation of the plasmid and reversion to the right-handed B form, so that under limiting enzyme conditions only one cleavage site per molecule is expected. As cruciform-to-B-DNA transitions have large kinetic barriers, more than one hit on the same molecule is possible. The specificity of S1 nuclease depends on the type of cruciform and the conditions employed, but it is not clear if this is due to properties of the cruciforms or to a substrate preference of the nuclease.

S1 nuclease also cleaves within regions of homopurine–homopyrimidine (pur–pyr) character in supercoiled plasmids.[16,26–28] This has been attributed to the formation of triple-stranded DNA in which one DNA strand folds back and Hoogsteen pairs with half of the pur–pyr sequence, leaving a single strand as target for the nuclease.[29] Depending on the salt conditions, the presence of divalent metal ions, and the supercoil density of the plasmid, either a pyr–pur–pyr triplex (H-DNA) or a pur–pur–pyr (H*-DNA) structure can be formed, as well as different isomers of each type. For oligo(dG · dC) sequences four-stranded structures have been postulated.[13,30–32] In addition to the exposed single-stranded region, the loop of an intramolecular triple-strand as well as one junction with neighboring B-DNA are also unpaired and therefore susceptible to nuclease attack.

Other Nucleases

P1 nuclease from *Penicillium citrium* acts in many ways similarly to S1 nuclease. An important difference is its ability to cut DNA well at neutral pH values, although its pH optimum is acidic.[33,34] The products of P1 nuclease action are oligonucleotides with 5′-phosphates. BAL31 nuclease (from *Alteromonas espejiana*) is also active at neutral pH, requires Ca^{2+} for activity, and can work at extremely high NaCl concen-

[26] C. C. Hentschel, *Nature (London)* **295**, 714 (1982).

[27] D. E. Pulleyblank, D. B. Haniford, and A. R. Morgan, *Cell (Cambridge, Mass.)* **42**, 271 (1985).

[28] F. Wohlrab, M. J. McLean, and R. D. Wells, *J. Biol. Chem.* **262**, 6407 (1987).

[29] R. D. Wells, D. A. Collier, J. C. Hanvey, M. Shimizu, and F. Wohlrab, *FASEB J.* **2**, 2939 (1988).

[30] V. I. Lyamichev, S. M. Mirkin, and M. D. Frank-Kamenetskii, *J. Biomol. Struct. Dyn.* **3**, 327 (1985).

[31] Y. Kohwi and T. Kohwi-Shigematsu, *Proc. Natl. Acad. Sci. U.S.A.* **85**, 3781 (1988).

[32] H. Htun and J. E. Dahlberg, *Science* **243**, 1571 (1989).

[33] D. B. Haniford and D. E. Pulleyblank, *Nucleic Acids Res.* **13**, 4343 (1985).

[34] J. A. Blaho, J. E. Larson, M. J. McLean, and R. D. Wells, *J. Biol. Chem.* **263**, 14446 (1988).

trations. It has therefore been used to investigate the salt-induced B-to Z-DNA transitions.[35] However, its associated potent 3'- and 5'-exonuclease activity has limited its use for fine mapping of junction regions at the base pair level.

Mung bean nuclease is related to the S1 and P1 nucleases.[36] The enzyme is highly sensitive to small variations in DNA structure[37] and converts single-stranded DNA to mono- or oligonucleotides with 5'-phosphates. Unlike S1 nuclease, it does not cleave opposite a nick, which can be an advantage for fine mapping of structural aberrations. Mung bean nuclease works under a variety of conditions including relatively high temperatures (up to 70°) and 40% formamide.

DNase I is a Mg^{2+}-dependent enzyme which produces oligodeoxynucleotides with a 5'-phosphate.[38] It has been proposed that the width of the DNA minor groove and the twist angle at a base pair step can affect the rate of hydrolysis catalyzed by DNase I.[39,40] The enzyme is widely used for footprinting of proteins on the DNA as well as for the detection of nuclease-hypersensitive sites on chromatin.[41-43] It has also been used for studies on purified DNA, such as the determination of double helix parameters on DNA fragments fixed on a solid support, and for structural studies on d(A,T)-rich oligomers.[44] The activity of DNase II is anticorrelated with that of DNase I, so that regions reactive to DNase I treatment are protected from DNase II attack and vice versa.[39]

Materials and Methods

S1 nuclease, P1 nuclease, and mung bean nuclease are obtained from Bethesda Research Laboratories (Gaithersburg, MD) or Boehringer-Mannheim (Indianapolis, IN). The reaction conditions are as follows:

---

[35] M. W. Kilpatrick, C.-F. Wei, H. B. Gray, and R. D. Wells, *Nucleic Acids Res.* **11**, 3811 (1983).
[36] D. Kowalski, *Nucleic Acids Res.* **12**, 7071 (1984).
[37] L. G. Sheflin and D. Kowalski, *Nucleic Acids Res.* **12**, 7087 (1984).
[38] S. Moore, *in* "The Enzymes" (P. D. Boyer, ed.), Vol. 14A, p. 281. Academic Press, New York, 1981.
[39] H. R. Drew and A. Travers, *Cell (Cambridge, Mass.)* **37**, 491 (1984).
[40] G. P. Lomonossoff, P. J. G. Butler, and A. Klug, *J. Mol. Biol.* **149**, 745 (1981).
[41] H. Weintraub, *Cell (Cambridge, Mass.)* **32**, 1191 (1983).
[42] S. C. H. Elgin, *J. Biol. Chem.* **263**, 19259 (1988).
[43] D. S. Gross and W. T. Garrard, *Annu. Rev. Biochem.* **57**, 159 (1988).
[44] J. W. Suggs and R. R. Wagner, *Nucleic Acids Res.* **14**, 3703 (1986).

S1 Nuclease

Supercoiled or relaxed plasmid DNA (1–2 µg) is incubated in a total volume of 50–100 µl with 40 mM sodium acetate (pH 4.6), 1 mM ZnSO$_4$, and 50 mM NaCl. After addition of 1.4 units of S1 nuclease per microgram of DNA, the reaction is incubated for 10–30 min at 37°. The reaction is terminated by cooling the mixture on ice, followed by addition of 0.5 M ethylenediaminetetraacetic acid (EDTA) (pH 8.0) to a final concentration of 37 mM. The samples can then either be microdialyzed against 10 mM Tris-HCl (pH 8.0), 0.1 mM EDTA (TE buffer) or be ethanol precipitated, dried, and redissolved in TE buffer. For mapping purposes, the resulting DNA is digested with a restriction endonuclease (best is a single cutter) and analyzed by agarose electrophoresis alongside marker DNA. For fine mapping of S1 nuclease-induced nicks, supercoiled DNA is digested as described above except that 14 units of enzyme per microgram DNA is used and the reaction time is reduced to 1 min.

After isolation of the reacted plasmid DNA as described above, aliquots of the samples are digested with appropriate restriction endonucleases and dephosphorylated using calf intestinal phosphatase (Boehringer-Mannheim). This is followed by phenol–chloroform extraction and ethanol precipitation, and the DNA is 5′-end-labeled using T4 polynucleotide kinase and [γ-^{32}P]ATP (Amersham Corp., Arlington Heights, IL). This protocol does not result in labeling of the nicks. Following a second restriction endonuclease digest, the labeled fragments are separated by electrophoresis on a polyacrylamide gel. After elution of the appropriate band from the gel, the location of S1 nuclease-specific nicks is determined by electrophoresis of the samples on denaturing polyacrylamide gels in parallel with Maxam–Gilbert sequencing ladders of the same labeled fragments. The bands are then visualized by autoradiography and can be quantitated either by densitometry or by excision from the gel and subsequent scintillation counting.

P1 Nuclease

Reactions are performed in a total volume of 50–100 µl on 1–1.5 µg DNA in 50 mM NaCl, 10 mM MgCl$_2$, and 50 mM Tris-HCl (pH 7.5). Enzyme, 0.7–1.0 units of P1 nuclease per microgram of DNA, is added, and the digestion is allowed to proceed for 10–30 min at 37°. Termination of the reaction and mapping of cleavage sites are as described above for S1 nuclease. For fine mapping of the single-stranded nicks, 10 units of P1 nuclease is used, and the reaction time is reduced to 1 min at 37°. Labeling and analyis of the nicked DNA are performed as described above for S1 nuclease reactions.

Mung Bean Nuclease

The reaction mixture contains 10 mM Tris-HCl (pH 7.0), 1 mM EDTA (pH 8.0) and 1.5–2.0 μg supercoiled DNA in the presence of 0.05–0.1 units of mung bean nuclease in a total volume of 30 μl. The reaction is terminated by phenol extraction and subsequent ethanol precipitation. Because the nuclease does not efficiently cut across nicks, the conversion of supercoiled to nicked DNA can be monitored on agarose gels. The DNA can then be labeled at a restriction site, with T4 polynucleotide kinase and [γ-^{32}P]ATP. After denaturation with glyoxal, the individual strands of the plasmid DNA can be separated by electrophoresis through agarose gels.The position of the nicks is then ascertained by autoradiography.

Section III

Analysis of DNA Structure inside Cells

[17] Probing of DNA Structure in Cells with Osmium Tetroxide–2,2'-Bipyridine

By EMIL PALEČEK

Introduction

In contrast to our knowledge about DNA structure in crystals, fibers, and solution, information about DNA structure in its natural environment, namely, in cells, is very limited. Only recently have techniques capable of providing data concerning the details of local DNA structures in the cell been developed. We may thus expect substantial progress in the study of DNA structure *in vivo* as well as in our understanding of DNA structure–function relationships within the next few years.

At the present time two approaches of direct probing of DNA structure *in situ* are available: (1) molecular genetic methods based on the interaction of an enzyme (induced in the cell) with a local DNA structure[1,2] and (2) chemical methods based on a single-strand-selective chemical probe that can penetrate into the cell interior and react with the target DNA.[3,4] In 1987 we proposed the first single-strand-selective chemical probe capable of providing information about local DNA structures in the cell.[3] This probe was osmium tetroxide–2,2'-bipyridine (Os–bipy).[3–7]

Single-Strand-Selective Chemical Probes of DNA Structure *in Situ*

Local supercoil-stabilized DNA structures contain bases accessible to single-strand-selective probes. Such bases form a part of cruciform loops and of some four-way junctions, B–Z junctions, triplex loops, and displaced strands. These structures are sometimes called "open" DNA structures. Several osmium tetroxide complexes have the ability to site-specifi-

[1] N. Panayotatos and A. Fontaine, *J. Biol. Chem.* **262**, 11364 (1987).
[2] A. Jaworski, W.-T. Hsieh, J. A. Blaho, J. E. Larson, and R. D. Wells, *Science* **238**, 773 (1987).
[3] E. Paleček, P. Boublikova, and P. Karlovsky, *Gen. Physiol. Biophys.* **6**, 593 (1987).
[4] E. Paleček, E. Rasovska, and P. Boublikova, *Biochem. Biophys. Res. Commun.* **150**, 731 (1988).
[5] P. Boublikova and E. Paleček, *Gen. Physiol. Biophys.* **8**, 475 (1989).
[6] J. A. McClellan, P. Boublikova, E. Paleček, and D. M. J. Lilley, *Proc. Natl. Acad. Sci. U.S.A.* **87**, 8373 (1990).
[7] P. Karlovsky, P. Pecinka, M. Vojtiskova, E. Makaturova, and E. Paleček, *FEBS Lett.* **274**, 39 (1990).

cally modify open supercoil-stabilized DNA structures *in vitro*.[8-10] In addition the osmium tetroxide–pyridine reagent (Os–py) has been successfully applied to study "open complexes" of RNA polymerase as well as structural changes induced in DNA by drugs *in vitro*.[10] We were interested in whether the osmium tetroxide complexes that were successfully applied to study DNA structures *in vitro* would also be applicable *in situ*.

We tested osmium tetroxide complexes with the following ligands: pyridine (Os–py), 2,2'-bipyridine (Os–bipy), tetramethylethylenediamine (Os–TEMED), 1,10-phenanthroline (Os–phe), and bathophenanthroline-disulfonic acid (Os–bpds). We have found that Os–py, Os–bipy, Os–TEMED, and Os–phe penetrate easily into *Escherichia coli* cells[5,9,11] without any pretreatment to permeabilize the cells. Of these four complexes only Os–bipy and Os–TEMED appeared suitable as single-strand-selective probes of the DNA structure in *E. coli* cells. Os–py induced cell lysis (which started earlier at higher pyridine concentrations), and Os–phe displayed low specificity toward local DNA structures and also reacted at other DNA sites. Our preliminary results[12] suggest that Os–phe may become useful in DNA footprinting *in situ*. Treatment of *E. coli* cells with Os–bipy[7] or Os–TEMED[11] did not result in any disturbance of cell integrity or in other changes indicating damage to the cell surface observable in a microscope (using phase contrast). Recently Sasse-Dwight and Gralla[13] have shown that potassium permanganate can also be applied to studies of DNA structure *in situ*.

Site-Specific Modification of Open Local DNA Structures in Intracellular
    Plasmids by Osmium–Bipyridine

Principle

Hypersensitivity of local open DNA structures toward Os–bipy is exploited for analysis in bacterial cells. Cells are treated with a millimolar concentration of Os–bipy for a short time, the excess reagent is removed, plasmid DNA is isolated by the boiling method, and osmium binding sites are determined.

[8] E. Paleček, *in* "Highlights of Modern Biochemistry" (A. Kotyk, J. Skoda, V. Paces, and V. Kostka, eds.), p. 53. VSP International Science Publ., Zeist, The Netherlands, 1989.
[9] E. Paleček, P. Boublikova, F. Jelen, A. Krejcova, E. Makaturova, K. Nejedly, P. Pecinka, and M. Vojtiskova, *in* "Structure and Methods, Volume 3: DNA and RNA" (R. H. Sarma and M. H. Sarma, eds.), p. 237. Adenine Press, Schenectady, New York, 1990.
[10] E. Paleček, *Crit. Rev. Biochem. Mol. Biol.*, **26**, 151 (1991).
[11] P. Boublikova and E. Paleček, *FEBS Lett.* **263**, 281 (1990).
[12] P. Boublikova, M. Vojtiskova, J. Stokrova, and E. Paleček, unpublished results, 1991.
[13] S. Sasse-Dwight and J. D. Gralla, *J. Biol. Chem.* **264**, 8074 (1989).

Reagents and Solutions

8 m*M* Osmium tetroxide in 0.2 *M* sodium phosphate, pH 7.4
8 m*M* 2,2'-Bipyridine in 0.2 *M* sodium phosphate, pH 7.4
0.2 *M* Sodium phosphate, pH 7.4
Chloramphenicol at a concentration of 10 mg/ml in ethanol

Procedure. The following procedure can be used if only a small amount of DNA is required, for example, if a single technique such as S1 nuclease cleavage is used for the osmium binding site analysis. A larger-scale procedure has been described elsewhere.[3,6]

Prepare *E. coli* carrying the plasmid of interest by growing cells in 10 ml LB medium in a 50-ml flask with shaking at 37° to an OD_{550} of 0.8. Add 150 μl chloramphenicol solution and shake the culture for 15 hr at 37° (to amplify the plasmid), collect the cells by centrifugation, and wash with 10 ml of 0.2 *M* sodium phosphate, pH 7.4. To prepare the Os–bipy reagent combine 400 μl of 8 m*M* osmium tetroxide with 400 μl of 8 m*M* 2,2'-bipyridine and mix. Resuspend the cells in 800 μl of 0.2 *M* sodium phosphate, pH 7.4, in a 2 ml Eppendorf tube. (The amount of bacteria in the suspension should not exceed 2 mg dry weight cells per milliliter. An OD_{550} reading of 0.5 corresponds roughly to 200 μg dry weight of *E. coli* cells per milliliter.[3]) Add the prepared Os–bipy reagent to the cell suspension and mix. Incubate for 15 min at 37°. (Time of incubation with Os–bipy can be reduced to 5 min.[3,7] The temperature at which cells are treated with Os–bipy can be adjusted to the purpose of the given study; site-specific modification of a B–Z junction in an intracellular plasmid was observed even at 0°.[3]) Stop the reaction by cooling the tube in an ice bath and immediately centrifuging in an Eppendorf centrifuge (14,000 rpm, 1 min, 4°). Rinse the cells 3 times with 1.4 ml of an ice-cold phosphate buffer. At this stage cells can be frozen and kept at −70° for several days (e.g., over the weekend).

Isolation of DNA by Modified Boiling Method[14]

Solutions

STET buffer: 8% sucrose, 5% Triton X-100, 50 m*M* ethylenediaminetetraacetic acid (EDTA), 50 m*M* Tris-HCl, pH 8.0
Lysozyme at a concentration of 20 mg/ml in 10 m*M* Tris-HCl, pH 8.0
2-Propanol
2.5 *M* Sodium acetate
Chloroform

Procedure. Resuspend the cell pellet in 350 μl of a STET buffer in a screw-capped Eppendorf tube. Add 25 μl lysozyme solution, close the

---

[14] D. S. Holmes and M. Quigley, *Anal. Biochem.* **114**, 193 (1981).

tube, and mix briefly by vortexing. Place the tube in a boiling water bath for 60 sec. Centrifuge immediately in an Eppendorf centrifuge for 10 min (at room temperature). Remove the pellet with a pipette tip (or with a toothpick). To the supernatant, add 40 μl of 2.5 M sodium acetate and 420 μl 2-propanol. Mix by vortexing and store in a dry ice–ethanol bath for 15 min. Centrifuge for 10 min at 4° in an Eppendorf centrifuge. Redissolve the pellet in 400 μl water. Add 600 μl chloroform. Mix by vortexing and shake for 15 min. Centrifuge briefly, remove the aqueous layers with a Gilson pipette, and place the sample into a new tube (discard the precipitate and the chloroform layer). Repeat this deproteination procedure one more time. (If the presence of RNA interferes with the following steps, RNA should be removed by precipitation with LiCl[15] and/or by treatment with DNA-free RNase.) Precipitate DNA with ethanol: add 80 μl of 2.5 M ammonium acetate and 800 μl of 95% ethanol. Close the cap and mix. Chill at $-70°$ for 15 min (dry ice–ethanol bath). Centrifuge for 15 min at 4°. Remove the supernatant and dry under reduced pressure. Add 20 μl distilled water, close the cap, and redissolve the DNA (Vortexer). Osmium binding sites are then determined in the same way as in experiments *in vitro*.[16]

Using this approach it has been shown that left-handed DNA [formed by (C-G)$_n$ sequences], cruciforms, and protonated H triplexes can exist in *E. coli* cells.

Local Open DNA Structures in *Escherichia coli* Cells

Left-Handed DNA and Junctions with B-DNA

After the initial excitement evoked by the detection of anti-Z antibody binding sites in eukaryotic chromosomes, it has been recognized that immunochemistry does not yield evidence for the presence of Z-DNA in the cell, but shows only the presence of potential Z-DNA sequences.[10,17] Attempts to demonstrate the existence of left-handed DNA in (C-G)$_n$ segments of plasmid DNA in *E. coli* by the linking number assay did not solve the problem because the results of this method are not free from ambiguity.[10,17]

In 1987 we applied the direct chemical DNA probing method to search for left-handed DNA in (C-G)$_n$ segments of an intracellular plasmid.[3] We showed that Os–bipy site-specifically modified the B–Z junction inside

[15] Z. Lev, *Anal. Biochem.* **160**, 332 (1987).
[16] E. Paleček, this volume [8].
[17] A. Rich, A. Nordheim, and A. H. J. Wang, *Annu. Rev. Biochem.* **53**, 791 (1984).

E. coli cells. At the same time Jaworski *et al.*[2] showed independently that Z-DNA exists in the plasmid (C-G)$_n$ segments and elicits a biological response in *E. coli* cells. These authors used a special molecular genetic technique based on the inhibition of methylation of the *Eco*RI recognition site when it is near or in Z-DNA. The results of the genetic[2] and chemical[3,4] studies provided evidence suggesting that left-handed DNA can exist in bacterial cells. Further results obtained with Os–bipy[5] and Os–TEMED,[11] as well as with other methods,[10] are in good agreement with this conclusion. Recently Rahmouni and Wells,[18] using the Os–bipy probing method, showed that (dC-dG) stretches as short as 12 base pairs (bp) could adopt a left-handed DNA structure when cloned upstream from the *tet* gene, whereas no left-handed DNA was detected when the (dC-dG) segments (up to 74 bp in length) were cloned downstream. These results supported the notion of domains with different degrees of supercoiling related to transcriptional activity in *E. coli* cells.[10]

Cruciforms

Evidence of cruciform existence in *E. coli* cells has recently been obtained similarly to left-handed DNA by the molecular genetic method and direct chemical probing of the DNA structure. Panayotatos and Fontaine[1] demonstrated cleavage of the ColE1 inverted repeat (which was actively transcribed from its natural promoter) in the intracellular plasmid by T7 endonuclease induced in *E. coli* cells. Using an Os–bipy probe the presence of cruciforms in intracellular plasmids of *E. coli* was demonstrated[4,6] in (dA-dT) inserts of various lengths. In the system biased in favor of cruciform formation by topoisomerase mutation or by salt shock (to shift the intracellular superhelical density to more negative values), cruciform modification patterns characteristic of cruciform extrusion were observed in (dA-dT)$_n$ when n was 34, 22 and 15, but not when n was 12.[6] There was no transcription through the (dA-dT)$_n$ tracts in the plasmids studied. These results made it possible to calculate the effective intracellular DNA superhelical density which responded directly to the environmental and genetic influences.

Protonated Triplexes

Attempts to identify protonated H triplexes in cells have only a very short history, as this structure was proposed[10,16] and demonstrated *in*

---

[18] A. R. Rahmouni and R. D. Wells, *Science* **246,** 358 (1989).

vitro[10,19-23] only a few years ago. In 1987 binding of a monoclonal antibody specific for triplex DNA to metaphase chromosomes and interphase nuclei was reported.[24,25] The antibody does not differentiate between the various types of triplexes, and the results obtained *in situ* suffer from drawbacks similar to those of Z-DNA immunofluorescent staining.[10]

We have attempted to search for triplex DNA in cells with the Os–bipy probe. The presence of triplex H-DNA *in vitro* is manifested by a strong modification in the center of the $(C-T)_n$ sequence and a weaker modification at the triplex boundary (Fig.1) (which was observed at acidic pH values up to pH 6.0). To probe the homopurine strand in the triplex structure *in vitro*, diethyl pyrocarbonate or glyoxal was applied.[19-23] So far no reports on the application of these or other single-strand-selective probes reacting with purines to studies of DNA structure in cells have been published. Thus information about triplex structure in cells obtained by direct chemical probing is limited to the pyrimidine strand. Fortunately, the Os–bipy modification pattern of the triplex homopyrimidine strand *in vitro* differs from those observed in any of the local supercoil-stabilized structures so far studied and is thus sufficiently characteristic to be used for triplex identification in cells.

Direct probing of DNA in cells with Os–bipy at neutral pH showed no triplex formation in the homopurine–homopyrimidine segment of the intracellular pEJ4 plasmid.[9] If *E. coli* cells were preincubated and treated with Os–bipy at pH 4.5 or 5.0, a modification pattern characteristic of H triplexes was obtained. Shifting the intracellular superhelical density to more negative values (by cultivating the cells at a higher NaCl concentration) resulted in a stronger site-specific modification.

Probing of the intracellular pL153 plasmid (containing the pEJ4 homopurine–homopyrimidine sequence not undergoing transcription) at acid extracellular pH values resulted in an Os–bipy modification pattern corresponding to the triplex structure (Fig. 1), but detailed analysis revealed some differences between the patterns obtained *in vitro* and *in situ*.[7] In the latter pattern more bases were modified in the triplex loop, and modifi-

[19] V. I. Lyamichev, S. M. Mirkin, and M. D. Frank-Kamenetskii, *J. Biomol. Struct. Dyn.* **3**, 667 (1986).

[20] M. Vojtiskova, S. M. Mirkin, V. I. Lyamichev, and M. D. Frank-Kamenetskii, and E. Paleček, *FEBS Lett.* **234**, 295 (1988).

[21] J. C. Hanvey, J. Klysik, and R. D. Wells, *J. Biol. Chem.* **263**, 7386 (1988).

[22] H. Htun and J. E. Dahlberg, *Science* **241**, 1791 (1988).

[23] B. H. Johnston, *Science* **241**, 1800 (1988).

[24] J. S. Lee, G. D. Burkholder, L. J. P. Latimer, B. L. Haug, and R. P. Braun, *Nucleic Acids Res.* **15**, 1047 (1987).

[25] J. S. Lee, J. L. P. Latimer, B. L. Haug, D. E. Pulleyblank, D. M. Skinner, and G. D. Burkholder, *Gene* **82**, 191 (1989).

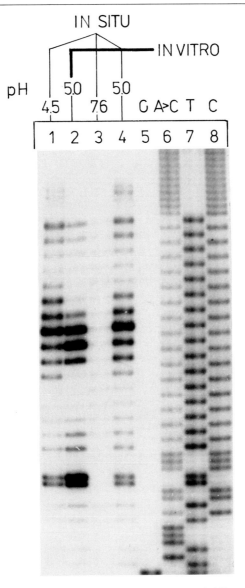

FIG. 1. Osmium binding sites in the homopurine–homopyrimidine sequence of pL153. *Escherichia coli* JM109 containing intracellular plasmid pL153 was treated with 2 m*M* Os–bipy (30 min at 37°) at pH 4.5 (lane 1), pH 5.0 (lane 4), and pH 7.6 (lane 3). Supercoiled pL153 modified *in vitro* at pH 5.0 (lane 2) is also shown.

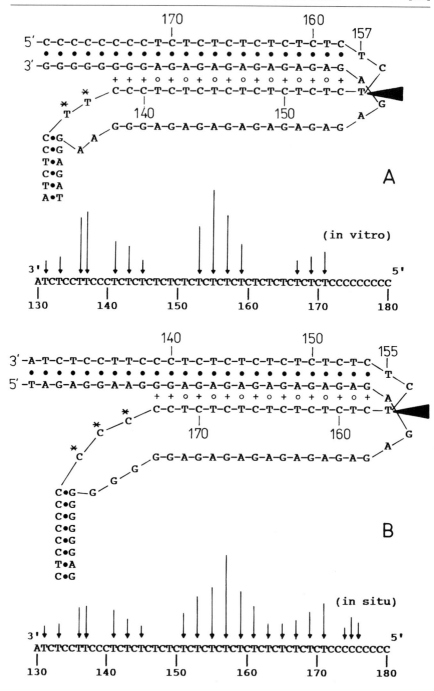

cation of two thymine residues (T136 and T137) at the 3' end of the (C–T)$_{16}$ sequence was weaker (Fig. 1). On the other hand, modification of three cytosine residues at the 5' end of the sequence was observed; these cytosines, which were unmodified *in vitro* (Figs. 1–3), form the B–H junction in the H-y5 conformer (Figs. 1 and 2), whereas T136 and T137 are contained in the B–H junction of the H-y3 conformer.

It has been tentatively suggested that the H-y5 conformer (which is formed *in vitro* at lower superhelical density than H-y3 observed mainly in isolated plasmids) prevails in *E. coli* cells.[7] The differences in the triplex loop modifications observed *in situ* and *in vitro* might be due to some interactions occurring *in situ* but not in solution. Practically the same modification patterns were observed over a pH range of 4.5–5.2 (Fig. 3); at pH 5.4 no modification characteristic of the triplex was found. The intracellular pH values were higher by about 0.5 as compared to extracellular pH values. The pH values at which cells were modified are not fully physiological for *E. coli;* but *E. coli* cells can grow at an extracellular pH of 5.2 and 5.0, and these pH values (unlike pH 4.3) do not induce acid-shock proteins.[26] It has been concluded[7,9] that protonated triplex DNA can exist in *E. coli* cells under some conditions, but its structure may differ in some details from that observed *in vitro*.

Immunochemical Detection of DNA–Osmium–Ligand Adducts

To extend the possibilities of detecting the DNA–Os–ligand adducts we elicited polyclonal antibodies in rabbits.[27] For immunization we used thermally denatured calf thymus DNA modified with Os–py or Os–bipy (60 μg) mixed with methylated bovine serum albumin (BSA, 90 μg) and Freund's complete adjuvant for a total volume of 1 ml. In subsequent

[26] M. Heyde and R. Portalier, *FEMS Microbiol. Lett.* **69**, 19 (1990).
[27] E. Paleček, A. Krejcova, M. Vojtiskova, V. Podgorodnichenko, T. Ilyina, and A. Poverennyi, *Gen. Physiol. Biophys.* **8**, 491 (1989).

FIG. 2. (A) H-y3 and (B) H-y5 conformers of the H-DNA triplex and the Os–bipy modification of bases in the polypyrimidine tract of the pL153 insert (A) *in vitro* and (B) *in situ*. The lengths of the vertical arrows in the nucleotide sequence represent the relative intensities of the bands on the sequencing gel (Fig. 1) obtained by densitometric tracing (A) after Os–bipy treatment of the supercoiled plasmid pL153 DNA *in vitro* (Fig. 1, lane 2) and (B) after treatment of cells with Os–bipy at an external pH of 5.0 (Fig. 1, lane 4). The main element of both H conformers is a triple helix which includes the Watson–Crick duplex (●) associated with the homopyrimidine strand by Hoogsteen base pairing (○, +) where cytosines are protonated. The triangle shows the strongest modified base in the triplex, and asterisks denote modification at the B–H junctions.

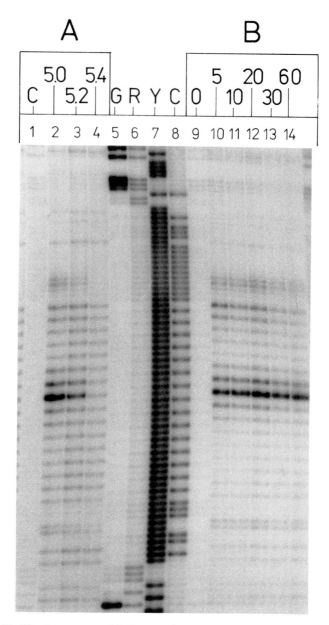

Fig. 3. Modification patterns of the homopurine–homopyrimidine sequence of intracellular pL153. (A) pH dependence: cells were treated with Os–bipy at pH 5.0 (lane 2), pH 5.2 (lane 3), and pH 5.4 (lane 4). Lane 1, Unmodified control. (B) Dependence on modification time: cells were treated with Os–bipy at pH 5.0 for 0 (lane 9), 5 (lane 10), 10 (lane 11), 20 (lane 12), 30 (lane 13), and 60 min (lane 14). R, Purine; Y, pyrimidine. [Adapted from P. Karlovsky, P. Pecinka, M. Vojtiskova, E. Makaturova, and E. Paleček, *FEBS Lett.* **274,** 39 (1990), with permission.]

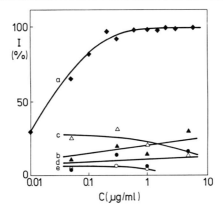

FIG. 4. Competitive binding assay of anti-DNA–Os–py measured by ELISA. Dependence of binding inhibition (*I*) on the competitor concentration (*C*). Calf thymus DNA–Os–py (0.5 μg) was the antigen on the plate; and anti-DNA–Os–py (3 μg) was used in the second layer. Competitors: (a) thermally denatured calf thymus DNA–Os–py, (b) RNA–Os–py, (c) unmodified RNA, (d) BSA–Os–py, (e) unmodified BSA. The mixture of antibody and a competitor was preincubated (60 min at 37°) before application to the plate. [From E. Paleček, A. Krejcova, M. Vojtiskova, V. Podgorodnichenko, T. Ilyina, and A. Poverennyi, *Gen. Physiol. Biophys.* **8,** 491 (1989), with permission.]

injections an incomplete adjuvant was used. Rabbits were bled via the marginal ear vein 7 days after immunization.

The antibodies obtained against DNA modified with Os–py (anti-DNA–Os–py) or Os–bipy (anti-DNA–Os–bipy) showed a remarkable specificity *in vitro*. They did not react with unmodified DNA, RNA, and proteins or with Os–py- or Os–bipy-modified proteins (Fig. 4). They showed, however, a small cross-reaction with Os–py- or Os–bipy-modified RNA. Cross-reaction of anti-DNA–Os–bipy was reduced by antibody affinity column fractionation.[28] Recently a monoclonal antibody has been obtained in which cross-reaction with RNA–Os–bipy is virtually absent.[29]

The antibodies significantly extend the possibilities of detecting DNA–Os–ligand adducts *in vitro* by various immunological methods including enzyme-linked immunosorbent assay (ELISA), DNA gel retardation, and immunoblotting.[27,28] Their application *in situ* seems to be even more important. We have found that Os–bipy is able to penetrate easily into isolated nuclei,[30] eukaryotic cells, and even salivary glands of *Chironomus thummi* and *Drosophila melanogaster* and react with DNA

---

[28] A. Kuderová-Krejčová, A. Poverennyi, and E. Paleček, *Nucleic Acids Res.* **19,** 6811 (1991).
[29] A. Kabakov and A. Poverennyi, unpublished results, 1991.
[30] M. Robert-Nicoud, A. Poverennyi, and E. Paleček, unpublished results, 1991.

in chromosomes. The reaction sites can be detected by means of anti-DNA–Os–bipy using indirect immunofluorescence. Competition experiments demonstrated a very good specificity of the antibodies *in situ*.

Unlike the approach based on the analysis of isolated DNA (see above, Site-Specific Modification of Open Local DNA Structures), histoimmunochemical techniques cannot provide modification patterns specific to the given DNA structure; they yield, however, information about the location of open DNA structures in the cells, thus complementing already established techniques of chemical probing of DNA structure *in situ*. An important point in the application of the chemical probe in histoimmunofluorescence is that Os–bipy is applied to living cells and not to cells drastically changed by fixation procedures which produce serious difficulties in the immunofluorescence studies of unusual DNA structures (such as Z-DNA) in eukaryotic cells.[10,17] The DNA–Os–bipy adducts resist fixation based on acid or organic solvent treatment. Removal of proteins (induced by fixation) and subsequent changes in the DNA superhelical density[10] (which can induce extrusion of a local DNA structure) cannot influence the results because the anti-DNA–Os–bipy applied as the first antibody does not recognize the specific DNA structure but only osmium-labeled base residues in DNA. In other words, if bases (e.g., in the cruciform loop) are selectively modified in the cell by Os–bipy and the cruciform disappears during the fixation, then the adducts formed in the original cruciform structure will be detected by the antibody; vice versa, if a triplex structure is secondarily induced in the cell as a result of acid fixation, it will not be recognized by anti-DNA–Os–bipy because the chemical probe is applied prior to fixation (when the triplex is absent). Thus, the experimenter need not worry about the influence of fixation on DNA structure and can choose an optimum fixation procedure.

Reactivity of Osmium–Bipyridine Probe in Cells

Osmium tetroxide, widely used as a tissue fixative and staining reagent, attacks unsaturated entities present in tissues.[31] It has been shown that addition of ligands to osmium tetroxide markedly affects its staining properties (see Ref. 32 and references therein). As a probe of DNA structure in cells Os–bipy is applied at concentrations substantially lower than those used for staining and fixation. It can thus be expected that the steric arrangement of the cell will not be destroyed by a short interaction with

[31] E. D. Korn, *J. Cell. Biol.* **34**, 627 (1967).
[32] E. J. Behrman, *in* "Science of Biological Specimen Preparation" (J.-P. Revel, T. Barnard, and G. H. Haggis, eds.), p. 1. SEM, Chicago, 1984.

Os–bipy. On the other hand, under the conditions used in probing experiments, the ability of bacterial cells to multiply is lost.

So far little is known about the reactions of Os–bipy with cell components other than DNA. It has been shown *in vitro* that osmium tetroxide complexes do not react at a significant rate with polysaccharides.[32] They do react with RNA bases, but the reaction of Os–bipy with thymine in single-stranded DNA is faster than that with bases in single-stranded RNA.[33] Osmium tetroxide reacts with unsaturated fatty acids, amino acids, and proteins; detailed information about these reactions is, however, lacking. Studies of the relative reactivity of amino acid residues in proteins showed that the most rapidly reacting side chains are those of methionine, cystine, tryptophan, histidine, and lysine.[32,34] Osmium tetroxide-induced protein cleavage was markedly slowed by ligands such as bipy and TEMED. Binding of anti-Z-DNA IgG in DNA was not substantially decreased in the presence of 0.025% osmium tetroxide (15 min at 37°).[35] Data about reactions of Os–bipy with RNA, proteins, polysaccharides, and lipids inside the cell are completely absent.

At present three single-strand-selective probes are available for probing DNA structure in the cell, namely, Os–bipy,[3-10,18] Os–TEMED,[11] and $KMnO_4$.[13] The shortest incubation times and lowest concentrations of the probes used to produce a modification pattern of a local DNA structure in cells have been about the same for Os–bipy[3,7] and $KMnO_4$.[13] Compared to $KMnO_4$ the osmium tetroxide complexes are more selective, which might represent an important advantage in applications *in situ*. For instance, Os–bipy reacts with DNA and RNA bases but not with *cis*-diols at the terminal ribose in UpA and ApU. Changing the oxidation state of the osmium reagent from $+8$ to $+6$ results in the formation of an osmate ester of the ribose *cis*-diol,[16] but produces no reaction with the nucleic acid bases.

Further advantages of osmium tetroxide complexes include (1) the strong antigenicity of their adducts with DNA and high specificity of the anti-DNA–Os–ligand antibodies and (2) the possibility of making a choice (by changing the ligand) between complexes with different properties (size, electric charge, etc.) which may come into play in the selectivity of the probes for specific DNA structures as well as in penetration of the probe into the cell interior and its interactions with cellular structures. Further work is needed to obtain information about changes in the cell induced by

[33] E. Paleček and P. Pecinka, unpublished results, 1991.

[34] J. S. Deetz and E. J. Behrman, *Int. J. Pept. Protein Res.* **17**, 495 (1981).

[35] M.-O. Soyer-Odile, M.-L. Geraud, D. Coulaud, M. Barray, B. Theveny, B. Revet, and E. J. Delain, *J. Cell Biol.* **111**, 293 (1990).

the probe and about the fate of the probe after its penetration into the cell. Experiments aimed at these goals are in progress.

Perspectives

Direct chemical probing of DNA structure in cells is a new technique which offers a number of possibilities in the study of structural details of DNA and their functional relations. The number of single-strand-selective probes available for this purpose is limited; development of further probes can, however, be expected. The three chemical probes available (Os–bipy, Os–TEMED, and $KMnO_4$) react preferentially with thymine. It would be valuable to develop probes reacting with other bases and especially with purines.

If the principles of Maxam and Gilbert sequencing are utilized to trace the probe reaction in a local DNA *in situ* the lowest concentration of Os–bipy capable of producing a specific modification pattern is close to 0.5 mM. Application of more sensitive detection systems such as the polymerase chain reaction (PCR) may allow the use of smaller probe concentrations (which would be less harmful to cell multiplication) and/ or the possibility of working with substantially smaller cell amounts.

The results obtained so far by means of chemical probes have mainly been concerned with local DNA structures of *E. coli* intracellular plasmids. Studies of DNA structures in chromosomal DNAs will require further development of immunological and other methods. Our preliminary results suggest that application of pulse-field gel electrophoresis may be of particular interest.[33]

To obtain unequivocal results DNA structures should be probed *in situ* by different techniques. They may include various chemical and radiation probes, as well as enzymes induced in the cell. It may be expected that combining these approaches will help to obtain important information about the details of DNA structure *in vivo* and about their relations to cell functions.

Acknowledgments

I am indebted to Dr. J. Soška for critical reading of this chapter and to Dr. F. Jelen for help with the preparation of the manuscript.

[18] Analysis of DNA Structure *in Vivo* Using Psoralen Photobinding: Measurement of Supercoiling, Topological Domains, and DNA–Protein Interactions

By RICHARD R. SINDEN and DAVID W. USSERY

Introduction: Rationale and Advantages of Using Psoralen as an *in Vivo* Probe

The characteristics of psoralen and its photoreaction with DNA make it an exceptional *in vivo* probe of DNA structure. Bacterial cells are permeable to psoralen although the permeability of different strains varies considerably.[1,2] Eukaryotic cells in tissue culture are readily permeable to psoralens.[2,3] The addition of psoralen has little, if any, effect on cells in the absence of long-wavelength UV light. Psoralen enters cells and binds to DNA, RNA, and proteins. Psoralen binds to DNA by intercalation, and the level of bound psoralen varies for different derivatives. In general, levels of this dark binding are sufficiently low that the topology of the DNA is not perturbed. At saturation *in vitro* about 15 4,5′,8-trimethylpsoralen (Me$_3$-psoralen) molecules bind per 1000 base pairs (bp) of DNA.[4] This value may be much less inside cells, even in a medium that is saturated with Me$_3$-psoralen (about 0.6 μg/ml). Use of lower concentrations will result in less psoralen binding to DNA. The binding of psoralen is sufficiently weak that neither dark binding nor photobinding will displace nucleosomes.[5,6] On irradiation with 360 nm light, psoralen forms monoadducts and cross-links as described elsewhere in this volume.[7] The rate of photoaddition is controlled by varying the incident light intensity; the rate is also dependent on psoralen concentration.

Cells can survive high levels of exposure to psoralen and light that will

[1] R. R. Sinden and R. S. Cole, "DNA Repair: A Laboratory Manual" (E. C. Friedberg and P.C. Hanawalt, eds.), Vol. 1A, p. 69. Dekker, New York, 1981.

[2] R. R. Sinden, J. O. Carlson, and D. E. Pettijohn, *Cell (Cambridge, Mass.)* **21,** 773 (1980).

[3] G. D. Cimino, H. B. Gamper, S. T. Isaacs, and J. E. Hearst, *Annu. Rev. Biochem.* **54,** 1154 (1985).

[4] J. E. Hearst, S. T. Isaacs, D. Kanne, H. Rapoport, and K. Straub, *Q. Rev. Biophys.* **17,** 1 (1984).

[5] T. Cech and M. L. Pardue, *Cell (Cambridge, Mass.)* **11,** 613 (1977).

[6] W. De. Bernardin, T. Koller, and J. M. Sogo, *J. Mol. Biol.* **191,** 469 (1986).

[7] D. W. Ussery, R. H. Hoepfner, and R. R. Sinden, this volume [13].

introduce many cross-linking damages into DNA.[8,9] In bacteria, the repair of psoralen cross-links requires recombination; a single cross-link is lethal to a *recA* cell. This suggests that any effect other than direct damage to the DNA by treatment with psoralen and irradiation with long-wavelength UV light may be relatively innocuous to cells. In many cases, the measurement of supercoiling or alternate conformations can be made at low levels of cross-linking that would minimally affect the survival of cells in culture.

Psoralen Photobinding as Probe for Supercoiling *in Vivo*

Rationale for Application of Trimethylpsoralen Photobinding as Probe of DNA Supercoiling: In Vitro Photobinding Analysis

Intercalating drugs bind supercoiled DNA better than relaxed DNA.[10] The binding of intercalating drugs to supercoiled DNA is thermodynamically favored, owing to the relaxation of supercoils that results from the unwinding of the double helix. We have demonstrated that Me_3-psoralen photobinds to naturally supercoiled DNA ($\sigma = -0.07$) at a rate that is 1.9 times faster than the binding to relaxed DNA. Moreover, photobinding of Me_3-psoralen is linearly proportional to the level of supercoiling in DNA.[2] The preferential binding to supercoiled DNA is independent of Me_3-psoralen concentration and temperature, and moderately dependent on ionic strength.[2] Measurement of the rate of Me_3-psoralen photobinding to DNA can provide a reliable estimate of the level of unrestrained supercoiling (or torsional tension) in the DNA.

This analysis is done experimentally by photobinding radioactive Me_3-psoralen to DNA and measuring the rate of binding, which is determined by the specific activity of the DNA. To ensure that the photobinding conditions are identical, supercoiled DNA is mixed with relaxed (nicked) DNA and treated with Me_3-psoralen and 360 nm light. The nicked and supercoiled DNAs are separated on CsCl–propidium iodide (or CsCl–ethidium bromide) gradients, and the specific activities of Me_3-psoralen labeling are determined.[2] From these experiments, a near linear relationship between negative superhelical density and the rate of Me_3-psoralen photobinding is observed. This careful analysis provides a ''standard curve'' for estimating the level of torsional tension in DNA.

[8] R. R. Sinden and R. S. Cole, *J. Bacteriol.* **136**, 538 (1978).
[9] M. E. Zolan, C. A. Smith, and P. C. Hanawalt, *Biochemistry* **23**, 63 (1984).
[10] W. R. Bauer, *Annu. Rev. Biophys. Bioeng.* **7**, 287 (1978).

In Vivo Approaches to Measuring Unrestrained Supercoiling

Measurement of Rate of Trimethylpsoralen Photobinding: Global Measurement. The basic approach for detecting unrestrained supercoiling is to measure the rate of photobinding to DNA in cells under conditions where the DNA should be supercoiled and under conditions where the unrestrained supercoiling should be relaxed. The relative rate of Me_3-psoralen binding should be proportional to the level of supercoiling as defined by the standard curve. Clearly, the standard curve results from *in vitro* measurements in a defined buffer which cannot be duplicated *in vivo*. Therefore, the *in vivo* superhelical density can be related to an "effective superhelical density" as defined by the *in vitro* conditions. Because the change in the relative rate of Me_3-psoralen photobinding is only a factor of 1.9 for "naturally supercoiled" DNA purified from *Escherichia coli* with σ equal to -0.07, measurements of supercoiling or changes in supercoiling require careful analysis. Small changes in the permeability of cells or other conditions which could change the rate or extent of reactivity of DNA with Me_3-psoralen would influence the results. Therefore, we feel it important to measure rates of Me_3-psoralen photobinding relative to an internal control or standard for *in vivo* experiments. In bacterial cells we have utilized the reactivity of Me_3-psoralen with RNA in cells.[2] In eukaryotic cells (*Drosophila* and human) the rate of Me_3-psoralen photobinding to bulk chromosomal DNA does not change as nicks are introduced into DNA.[2] Therefore, measurement of the rate of photobinding relative to bulk DNA has been used to analyze torsional tension in herpes simplex virus.[11] This approach could also be used as an internal standard for measurement to specific regions of the chromosome.

There are two ways to relax supercoils in *E. coli* cells: treatment with drugs that inhibit DNA gyrase or introduction of nicks into DNA. For *E. coli* we have utilized treatment with coumermycin to relax supercoils and treatment with ^{60}Co (γ-irradiation) to introduce nicks.[2] Following either of these treatments, a decrease in the rate of Me_3-psoralen photobinding was observed. In both cases a decrease of 1.7 was observed which suggested an "effective superhelical density" of about -0.05 for the bacterial chromosome. (This is determined by extrapolation from the standard curve.) It is important to ensure that treatment with conditions that relax DNA or introduce nicks into the DNA go to completion. Otherwise the ratios determined may not measure all unrestrained tension in the DNA. Without prior knowledge of the topological domain size, measurements of the rate of Me_3-psoralen photobinding should be made at

---

[11] R. R. Sinden, D. E. Pettijohn, and B. Francke, *Biochemistry* **21**, 4484 (1982).

several different doses of DNA nicking (or relaxation). The relative rate of Me$_3$-psoralen photobinding should decrease and reach a minimum. Once that minimum is reached the rate of photobinding should remain constant. If this important criterion is met, then one can assume that all torsional tension has been lost from the DNA *in vivo* and that the treatment with γ-irradiation or gyrase inhibitors are responsible for the reduction in the relative rate of photobinding.

We have used this approach to provide the first direct evidence that DNA in living *E. coli* cells is wound with unrestrained torsional tension.[2] In addition, this approach was applied to bacteriophage T4 to demonstrate that a linear DNA molecule could become wound with unrestrained supercoils in cells.[12] This method has also been used to demonstrate changes in the level of DNA supercoiling in *Bacillus brevis* during sporulation.[13]

Photobinding trimethylpsoralen to DNA in bacterial cells. Bacterial cells grown to a density of 4×10^8 cells/ml in K medium are centrifuged and resuspended at a density of about 2×10^9 cells/ml in cold M9 buffer (1 g NH$_4$Cl, 5.8 g Na$_2$HPO$_4$, and 3.0 g KH$_2$PO$_4$ per liter water). For cells grown in Luria broth we frequently wash cells once in cold M9 buffer before resuspending for the Me$_3$-psoralen addition. (The incubation in M9 buffer at 4° prevents DNA repair enzymes from introducing *uvrABC*-dependent nicks at DNA monoadducts and cross-links.) Nonradioactive Me$_3$-psoralen is added to saturation by adding 10 μl of a saturated Me$_3$-psoralen solution in ethanol per milliliter cell solution (a 1% solution, v/v). Radioactive Me$_3$-psoralen is usually added near or below the saturation point (0.6 μg/ml for 4,5′,8-trimethylpsoralen). The following steps are performed in a 4° cold room. Cells are incubated on ice for 2–5 min to allow equilibration with the Me$_3$-psoralen. Cells are transferred to a plastic petri dish (30–150 mm). The layer of cells should be as thin as possible and not more than 1 cm deep. Cells are irradiated wtih 360 nm light to introduce the desired level of photoaddition of Me$_3$-psoralen. The apparatus used to deliver about 1.2 kJ/m^2 360 nm light per minute is described elsewhere in this volume.[7] To obtain higher doses we have utilized a Blak-Ray Model B-100A ultraviolet lamp with either a 100-W spot or flood ultraviolet bulb. This lamp can produce much more intense light (although the light beam is not as uniform as that produced by the fluorescent bulbs). For the incorporation of radioactive Me$_3$-psoralen for the supercoiling measurements, 5–10 kJ/m^2 is used. To ensure uniform irradiation, the cell suspension should be mixed frequently during the irradiation.

Following irradiation, cells are centrifuged, and, following an optional

[12] R. R. Sinden and D. E. Pettijohn, *J. Mol. Biol.* **162**, 659 (1982).
[13] A. Bohg and H. Ristow, *Eur. J. Biochem.* **170**, 253 (1987).

wash in M9 buffer, DNA can be purified using any number of standard procedures.[1,2,11,12] We have used the following procedure to determine the specific activity of [^{3}H]Me$_3$-psoralen-labeled DNA and RNA. Cells (4 × 10^9) are washed with M9 buffer, resuspended in 1.5 ml TEN buffer, 0.3 ml of 30% sucrose in 0.6 M Tris, pH 8.1, 0.2 ml of 10% lysozyme (in water), and 0.5 ml of 32 mM ethylenediaminetetraacetic acid (EDTA) are added, and the suspension is incubated 10 min at 0–4°. The cells are lysed by addition of sodium dodecyl sulfate (SDS) to 0.5%. The sample is adjusted to 3.5 ml with TEN, and the DNA is sheared by vortex mixing at high speed for 30 sec. The sample is then incubated at 60° for 10 min, extracted 3 times with phenol, 3 times with chloroform–isoamyl alcohol (24 : 1), and adjusted to 0.3 M sodium acetate. The nucleic acids are precipitated by the addition of 2 volumes of ethanol and incubation at −20°. The nucleic acids are redissolved in 0.3 ml TEN, 10 μg pancreatic RNase and 5 units (U) T1 RNase are added, and the sample is incubated at least 2 hr at 37°. To separate the DNA and digested RNA, the samples are applied to a BioGel A-15 (Bio-Rad, Richmond, CA) column (1 × 27 cm) equilibrated with TEN and eluted. The specific activities of the 1-ml fractions are determined by measuring A_{260} and the radioactivity. (The DNA elutes with the void volume.)

Depending on the application, the dose of 360 nm light can vary. To cross-link a large fraction of 40- to 100-bp inverted repeats in cells, doses as high as 48 kJ/m^2 have been used. For *in vivo* analysis of Z-DNA torsionally tuned probes, doses of 1–5 kJ/m^2 are sufficient. Because Me$_3$-psoralen photodestructs after about 20 kJ/m^2 of irradiation, additional Me$_3$-psoralen is added periodically to ensure a linear rate of photoaddition.

Most *E. coli* strains are reasonably permeable to psoralens. Some derivatives, such as strains selected for sensitivity to antibiotics, are much more permeable to psoralen than wild-type cells.[2] In addition, *E. coli* can be permeabilized by treatment with EDTA to increase the intracellular concentration of psoralen.[14] For this, log phase cells are washed twice with 0.12 M Tris, pH 8.0, and then resuspended at 4–6 × 10^9 cells/ml in 0.12 M Tris, pH 8.0, at 37°. Then 0.3 μmol of EDTA are added per 10^{10} cells, and the cells are incubated for 2 min at 37° with gentle shaking. The permeabilization is terminated by the addition of 1.5 μmol of MgCl$_2$ per 10^{10} cells and chilling to 0–4°. Me$_3$-psoralen is added and the photobinding is performed.

Photobinding trimethylpsoralen to DNA in eukaryotic cells. Treatment of eukaryotic cells with Me$_3$-psoralen and 360 nm light is essentially identical to that described for bacterial cells with the exception that cells are

[14] L. Leive, *J. Biol. Chem.* **243**, 2373 (1968).

incubated in phosphate-buffered saline (PBS) for the addition of Me$_3$-psoralen and 360 nm irradiation. Cells are washed once in PBS, then suspended in PBS at about 5 × 10^6 cells/ml, and 1% (v/v) of a saturated Me$_3$-psoralen solution in ethanol is added.[2] For tissue culture cells attached to a plate, the medium is removed, the cells are washed with PBS, and 1–2 ml of PBS containing Me$_3$-psoralen is added. The cells are irradiated directly on the open plate. Eukaryotic cells seem to be "freely permeable" to Me$_3$-psoralen in that the rate of photobinding *in vivo* is not dissimilar to the rate observed for DNA *in vitro*. The procedure for determining the specific activities of [^{3}H]Me$_3$-psoralen labeling to DNA and RNA in eukaryotic cells is essentially identical to that described above with a few exceptions. Following photobinding cells are resuspended in NTE buffer (10 mM Tris, pH 7.6, 100 mM NaCl, 5 mM EDTA), lysed by addition of 0.5% SDS, 1 mg/ml proteinase K is added, and the sample is incubated 60 min at 60°. Subsequent steps are identical to those described above for bacterial cells.

Relaxation of unrestrained DNA supercoils in vivo. In *E. coli* supercoils can be relaxed by antibiotics that inhibit DNA gyrase. We have used growth of cells in K medium at 37° for 20 min in the presence of 50 μg/ml coumermycin to relax all unrestrained supercoils.[2] To introduce nicks into DNA we have utilized two methods; γ-irradiation and 5-bromodeoxyuridine (BrdUrd) photolysis. About 40 krad of γ-irradiation is required to relax all supercoiling in *E. coli* cells.[2,15] The dose required to relax all supercoils will likely be dependent on the geometry of the source employed. Cells should be resuspended in a small volume of cold M9 buffer and incubated on ice during the irradiation. The time between the initial centrifugation and the treatment with Me$_3$-psoralen and light should be minimized as much as possible. Immediately after γ-irradiation, Me$_3$-psoralen is added, and after a 1-min incubation cells are exposed to 360 nm light. We have calculated that 8.5 single-strand DNA breaks were introduced per kilorad per 2.7 × 10^9 daltons (Da) of DNA [or 1 break per 480 kilobase pairs (kbp) per krad].[15]

Photolysis of BrdUrd-labeled DNA can be used to introduce nicks into DNA. We have used this method for introducing nicks into bacteriophage T4 in *Escherichia coli*[12] and have also used it for nicking plasmid *in vivo*.[16] Bacteriophage T4 are labeled with BrdUrd by the addition of 50 μg BrdUrd/ml and 1 μg thymidine/ml to K medium 5 min after infection of cells at a multiplicity of infection of 0.01–0.1.[12] *Escherichia coli* can be labeled by growth in the presence of BrdUrd. We have grown wild-type

[15] R. R. Sinden and D. E. Pettijohn, *Proc. Natl. Acad. Sci. U.S.A.* **78**, 224 (1981).
[16] T. J. Kochel and R. R. Sinden, unpublished results (1987).

cells in K media in the presence of 20 μg uridine, 125 μg adenosine, and 50 μg/ml BrdUrd and obtained nicking of plasmid *in vivo*.[16]

Procedures used to label eukaryotic cells with BrdUrd have been described.[17,18] We have also labeled *Drosophila* tissue culture cells with BrdUrd. For this Schneider line 3 cells are grown in 10 ml of Schneider's complete medium [Schneider's medium supplemented with 7% fetal calf serum and 0.5% Bacto-peptone (Difco, Detroit, MI)] in a T-75 flask for 3–4 days at 25°. This culture is then diluted to 100 ml in complete medium and grown in a spinner flask to a density of 3–5 × 10⁶ cells/ml. Fifty milliliters of cells is then used to inoculate a spinner flask containing 500 ml of complete medium supplemented with 50 μM BrdUrd. These cells are grown 24–36 hr to a density of 3 × 10⁶ cells/ml before harvesting for analysis. Cells are irradiated as described below at a distance about 5 cm from the lamp source. Irradiation times between 5 and 20 min appear sufficient to nick selected domains of *Drosophila*.[19]

We have irradiated cells with 313 nm light at 5–10 cm from a GE 100-W high-pressure mercury lamp through a K_2CrO_4/NaOH filter. We use a Blak-Ray Model B-100A ultraviolet lamp without the 360 nm transmission filter. Light is filtered through a 1 cm deep solution of 0.1 mM K_2CrO_4 in 0.1 M NaOH in a sealed plastic petri dish. The two layers of plastic filter out light below 280 nm, and the K_2CrO_4–NaOH solution transmits a band of light between 290 and 335 nm with a peak at 313 nm. For the bacteriophage T4 application, the rate of nicking (at a 10 cm light-to-sample distance) was 3.2 breaks per T4 DNA molecule in *E. coli*. It is difficult to give a general nicking rate for this system. This can be calibrated by analyzing the introduction of nicks into chromosomal DNA using alkaline sucrose gradients[20] or alkaline agarose gels. Nicks in plasmid DNA can be measured on agarose gels by monitoring the conversion of supercoiled to nicked DNA.

Measurement of Rate of DNA Cross-linking: Localized Measurement. The rate of total Me_3-psoralen photobinding to DNA is nearly linearly proportional to superhelical density.[2] It is expected that the rate of cross-linking to DNA would also be dependent on superhelical density. We have shown that the rate of cross-linking of short fragments of DNA cloned into a plasmid is dependent on superhelical density.[21] In terms of a more general application, the cross-link assay of Vos and Hanawalt can be used to measure the rate of cross-linking to a DNA restriction fragment in eukary-

[17] E. R. Kaufman and R. L. Davidson, *Proc. Natl. Acad. Sci. U.S.A.* **74**, 4982 (1978).
[18] E. R. Kaufman, *Mol. Cell. Biol.* **4**, 2449 (1984).
[19] E. R. Jupe, R. R. Sinden, and I. L. Cartwright, manuscript in preparation.
[20] W. D. Rupp and P. Howard-Flanders, *J. Mol. Biol.* **31**, 291 (1968).
[21] R. R. Sinden and T. J. Kochel, *Biochemistry* **27**, 1343 (1987).

otic chromosomal DNA.[22] Briefly, this method involves the introduction of cross-links, alkali denaturation and rapid renaturation, and separation and analysis of cross-linked and non-cross-linked DNA on a neutral agarose gel.[23] Two groups of investigators have applied this approach of measuring the rate of cross-linking (under supercoiled and relaxed conditions) as an assay for torsional tension *in vivo* in bacteria and algae.[24,25] Differential *in vivo* cross-linking of regions of λ DNA replicative intermediates (θ structures) has shown that the parental sections of replicating λ are supercoiled whereas daughter regions are not covalently closed. This suggests a differential level of supercoiling on either side of the replication forks in intracellular bacteriophage λ.[26]

Psoralen Photobinding as Probe for Measuring Size of Topological Domains *in Vivo*

DNA in bacterial and eukaryotic cells is organized into independent topological domains. A domain of supercoiling is defined as a region of DNA bounded by topological constraints on the rotation of the double helix. If DNA is organized into a single topological domain (consider a plasmid molecule *in vitro* for instance), then a single nick will provide a swivel and all supercoils will be lost. If, on the other hand, a DNA molecule is organized into independent topological domains, a nick must be introduced into each domain before all supercoiling will be relaxed. We have shown that about 160 nicks per genome equivalent must be introduced before all supercoiling is relaxed in living *E. coli*.[15] This suggests that there are about 43 ± 10 domains of supercoiling per genome equivalent of DNA in living *E. coli*. We have also demonstrated using this approach that the linear T4 genome becomes organized into a single supercoiled topological domain shortly after infection into *E. coli*.[12]

Samples are treated with various doses of γ-irradiation or BrdUrd photolysis as described above. Each time point is treated with radioactive Me₃-psoralen and the same two different doses of 360 nm light that are in the "linear range" of photobinding. The relative rate of Me₃-psoralen photobinding, R, is determined. R is expressed as the rate of photobinding to DNA normalized to the rate of photobinding to total cellular RNA. To calculate the domain size, the average R values from the two doses of

[22] J. H. Vos and P. C. Hanawalt, *Cell (Cambridge, Mass.)* **50**, 789 (1987).
[23] This method is briefly discussed by D. W. Ussery, R. H. Hoepfner, and R. R. Sinden, this volume [13].
[24] D. N. Cook, G. A. Armstrong, and J. E. Hearst, *J. Bacteriol.* **171**, 4838 (1989).
[25] R. J. Thompson and G. Mosig, *Nucleic Acids Res.* **18**, 2625 (1990).
[26] R. B. Inman and M. Schnös, *J. Mol. Biol.* **193**, 377 (1987).

Me$_3$-psoralen and light are used to calculate F_r, which is plotted as a function of the number of nicks in the DNA. F_r is the fractional change in R relative to the change between R values in nonnicked or supercoiled DNA in unirradiated cells and R values in cells treated extensively with γ-irradiation to fully relax all DNA supercoiling.

To determine the average number of domains of supercoiling (m) per DNA molecule we assume the simple model that the domains are of equal size and that each is wound with the same level of supercoiling (a model that may not be correct, but which on average probably gives a reasonable assessment of domain sizes). We also assume that the distribution of nicks is near random and is described by a Poisson distribution. Because a single nick is sufficient to relax all supercoiling within a topological domain, the fraction of the original level of tension remaining in the DNA after the introduction of a defined number of nicks will be described by the zero-order term of the Poisson distribution. The zero-order term of the Poisson, $p(0)$, describes the fraction of the population of domains with no nicks:

$$p(0) = e^{-(x/m)} \tag{1}$$

where x is the number of nicks introduced per genome equivalent of DNA and m is defined above as the number of domains per genome equivalent of DNA. The rate of decay of supercoiling as nicks are introduced is close to first order. The experimental data are fitted to theoretical curves generated by varying the values of m. The best fit of a theoretical curve to the experimental results defines the appropriate value of m.[12,15]

Torsionally Tuned Probes for Measuring Supercoiling at Specific Sites in DNA *in Vivo*

Cruciform and Z-DNA-forming sequences used with the psoralen-based assays described previously provide torsionally tuned probes for measuring the level of torsional tension in DNA at specific sites in a plasmid, bacterial chromosome, or eukaryotic chromosome.[21,27,28] The formation of cruciforms and Z-DNA is dependent on superhelical density. The superhelical density dependence of the formation of the cruciforms is a function of the base composition of the inverted repeat at the center of symmetry.[29] Similarly, the superhelical density dependence of Z-DNA formation is dependent on base composition and length of the Z-DNA-

[27] R. R. Sinden, S. S. Broyles, and D. E. Pettijohn, *Proc. Natl. Acad. Sci. U.S.A.* **80**, 1797 (1983).
[28] T. J. Kochel and R. R. Sinden, *J. Mol. Biol.* **205**, 91 (1989).
[29] G. Zheng and R. R. Sinden, *J. Biol. Chem.* **263**, 5356 (1988).

forming sequence.[30] Therefore, the fraction of an inverted repeat existing as cruciforms or the fraction of a Z-DNA-forming sequence existing as Z-DNA *in vivo* provides an indication of the level of unrestrained tension in the DNA double helix.

We have used the torsionally tuned probes shown in Table I to demonstrate the existence of Z-DNA and cruciforms *in vivo* and to estimate the level of supercoiling in wild-type and *topA10* cells.[21,31] In addition, we have analyzed the level of cruciforms in different locations of plasmid pBR322 and demonstrated that different levels of supercoiling can exist in localized regions of DNA[31] as suggested by Liu and Wang.[32] A similar application using Z-DNA sequences has been described.[33]

Cruciform Torsionally Tuned Probes

Determination of Level of Supercoiling from Fraction of Inverted Repeats Existing as Cruciforms in Vivo. We have described an approach for estimating the width and center of a distribution of supercoils in DNA *in vivo* from analysis of the level of existence of several torsionally tuned cruciform probes.[31] Description of this analysis is beyond the scope of this chapter. To explain briefly the rationale, a particular level of supercoiling in cells should support formation of a defined level of cruciforms for a particular inverted repeat (F10S, for example). For an inverted repeat that forms cruciform at a lower (F14C) or higher (F14S) level of negative supercoiling, more or fewer cruciforms, respectively, should form *in vivo*. From analysis of the levels of existence of three or more cruciforms *in vivo*, it is possible to estimate the mean level of supercoiling and the width of the topoisomer distribution.

Cruciform Assay. The assay for cruciforms is described elsewhere in this volume.[7] Basically, to detect and to quantitate cruciforms *in vivo*, *E. coli* cells harboring a plasmid containing an inverted repeat are grown to stationary phase in K medium or Luria broth. Cells are treated with Me$_3$-psoralen and 360 nm light as described above. For analysis of the *Eco*RI fragment npF series inverted repeats inserted into plasmid pKTAC-O (Table I), doses of 12 and 24 kJ/m^2 were used to cross-link about 20–50% of the inverted repeat *in vivo* (at the highest dose). We have purified sufficient plasmid DNA for the quantitation of cruciforms from 40–100 ml of cells per cross-link analysis. For cruciforms existing at levels of 2–50% the assay originally described was used.[27] Briefly, DNA was

[30] M. J. McLean and R. D. Wells, *Biochim. Biophys. Acta* **950**, 243 (1988).
[31] G. Zheng, T. J. Kochel, R. W. Hoepfner, S. E. Timmons, and R. R. Sinden, *J. Mol. Biol.* **221**, 107 (1991).
[32] L. F. Liu and J. C. Wang, *Proc. Natl. Acad. Sci. U.S.A.* **84**, 7024 (1987).
[33] A. H. Rahmouni and R. D. Wells, *Science* **246**, 358 (1989).

TABLE I
TORSIONALLY TUNED CRUCIFORM AND Z-DNA PROBES

Probe	$\sigma_c{}^b$

Cruciform: npF series inverted repeats[a]

F14C   −0.038

1	11	21	31	41	
GAATTCCCAA	TTGATAGTGG	TAAAACTACA	TTAGCAGATG	GGCCCGATAT	
51	61	71	81	91	101
TTATAAATAT	CGGGCCCATC	TGCTAATGTA	GTTTTACCAC	TATCAATTGG	GAATTC

F10S   −0.049

1	11	21	31	41	
GAATTCCCAA	TTGATAGTGG	TAAAACTACA	TTAGCAGATG	GGCCCGATAT	
51	61	71	81	91	101
TTAATTTAAT	CGGGCCCATC	TGCTAATGTA	GTTTTACCAC	TATCAATTGG	GAATTC

F14S   −0.065

1	11	21	31	41	
GAATTCCCAA	TTGATAGTGG	TAAAACTACA	TTAGCAGATG	GGCCCGATAT	
51	61	71	81	91	101
TTAATTTATA	CGGGCCCATC	TGCTAATGTA	GTTTTACCAC	TATCAATTGG	GAATTC

Z-DNA[c]
pCGTA-C   −0.040

1	11	21	31
GAATTCGCGC	GCGCGCGTAC	GCGCGCGCGC	GAATTC

pTGTA-C   −0.057

1	11	21	31
GAATTCTGTG	TGTGTGTGTA	TGTGTGTGTG	TGAATTC

pUCTA-1   −0.050

1	11	21	31	41
GAATTCCCGT	GTGTGTGTGT	GTGTGTGTGT	GTATGTGGGG	ATCC

[a] In the npF series inverted repeats the central 14-bp AT-rich centers are underlined. In F10S and F14S the central 10 and 14 bp, respectively, are nonpalindromic. This is indicated by double underlining. The doubly underlined center sequences have mirror repeat symmetry. The nonpalindromic nature of the center affects σ_c by requiring a higher level of torsional energy to drive the cruciform transition. These inverted repeats have been described by Zheng *et al.*[31]

[b] σ_c is the critical superhelical density required for the transition to a cruciform or Z-DNA.

[c] For the Z-DNA sequences the *Eco*RI sites are underlined. The *Bam*HI site is double underlined in pUCTA-1. pCGTA-C has been described by Sinden and Kochel.[21] pTGTA-C and pUCTA-1 have been described by T. J. Kochel and R. R. Sinden [*BioTechniques* 6, 532 (1988)].

purified using a cleared lysate procedure without the CsCl–ethidium bromide gradient.[2] Plasmid is then cut to linearize the DNA. For the npF series EcoRI inverted repeat fragments in pUC8 (Table I), we digest 2 μg DNA with 4 U AvaII in 20 μl AvaII buffer for 1 hr at 37°. (In pMB9 we utilized digestion with HindIII, SalI, and BamHI to ensure complete linearization.[27]) The inverted repeat is then cut into linear form and hairpin arms by digestion with 50 U EcoRI in 25 μl EcoRI buffer for 3 hr at 35°. The DNA fragments are labeled with 10 μCi of [α-^{32}P]dATP in a Klenow fill-in reaction.[34] We separate the 100- and 50-bp forms of the npF inverted repeats, in native and denatured/rapidly renatured samples, on a 10% (w/v) polyacrylamide gel (containing 10% glycerol) in TB buffer (40 mM Tris–borate, pH 8.3). The fraction of the inverted repeats migrating as full-length linear fragments or hairpin arms are quantitated, and P_c, the fraction of the inverted repeat existing as cruciforms, is calculated as described.[7]

When cruciforms exist at levels below 2% we have employed a more sensitive assay for cruciforms.[31] For this, 2 μg of DNA is digested with AvaII as described. The linear plasmid band is then purified from a 0.8% (w/v) agarose gel run at 2 V/cm in TAE buffer (40 mM Tris, pH 8.3, 25 mM sodium acetate, 1 mM EDTA). This is to eliminate the possibility of analyzing supercoiled DNA on digestion with EcoRI. The gel slice is masticated and vortex mixed with 0.5 phenol. The mixture is frozen to −70° for 10 min and centrifuged at 16,000 g for 10 min at room temperature. The mixing, freezing, and centrifugation are repeated for two cycles. The DNA is then purified from the aqueous phase by three extractions each with phenol and chloroform–isoamyl alcohol (24 : 1), followed by two ethanol precipitations. Purified linear plasmid DNA is then cut with EcoRI, labeled, and separated on a 10% (w/v) polyacrylamide gel as described above. The full-length and hairpin arm-sized bands are purified from this gel. To ensure that cross-links are indeed introduced into the hairpin arms, the tip of the npF hairpin arm is cut off by digestion with 20 U HaeIII in 20 μl HaeIII buffer for 3 hr at 37°. This DNA is then run on an 8% (w/v) polyacrylamide–urea DNA sequencing gel. It is possible on this gel to distinguish non-cross-linked hairpin molecules from cross-linked molecules (which migrate slower in the gel). Stepwise quantitation allows calculation of F_c, the fraction of the inverted repeat cross-linked as cruciforms. F_1, the fraction of the inverted repeat cross-linked as linear molecules, can be determined from cross-link analysis (on the 8% sequencing gel) of

[34] J. Sambrook, E. F. Fritsch, and T. Maniatis, "Molecular Cloning: A Laboratory Manual," 2nd Ed. Cold Spring Harbor Laboratory, Cold Spring Harbor, New York, 1989.

*Hae*III digestion products from the full-length (linear) inverted repeat molecules purified from the 10% polyacrylamide gel.

Z-DNA Torsionally Tuned Probes

Determination of Level of Supercoiling from Level of Z-DNA Existing in Vivo. We have described a sensitive, psoralen-based exonuclease III (exoIII)/photoreversal assay for detecting Z-DNA that is applicable in living cells.[7,28] This assay requires determination of the relationship between the relative rate of Me$_3$-psoralen photobinding to a 5'-TA within a Z-DNA-forming sequence and the superhelical density of the DNA. In addition, the relationship between the relative rate of hypersensitive Me$_3$-psoralen photobinding to B–Z junctions and superhelical density is also determined. The rates of photobinding to these sites are measured relative to a 5'-TA outside the Z-DNA-forming region which does not change as a function of superhelical density. The relative rates of photobinding are calculated as the Z-TA/C-TA ratio, which is the intensity of the exoIII stop at the 5'-TA within the Z-DNA region divided by the intensity of the exoIII stop at the 5'TA within a "control" region of DNA. Likewise, the B-ZJII/C-TA ratio represents the intensity of the exoIII stop at the 5'-AATT at the B–Z junction divided by the intensity of the exoIII stop at the 5'-TA within a "control" region of DNA. Analysis of topoisomer populations on two-dimensional agarose gels establishes the relationship between the relative reactivities of the Z-DNA region to Me$_3$-psoralen photobinding and the fraction of individual topoisomers containing Z-DNA in the DNA population. These relationships are shown in Figs. 1 and 2. They provide standard curves which can be used to determine the superhelical densities from the relative rate of photobinding for the Z-DNA torsionally tuned probes.[35]

This basic method is applicable for any Z-DNA sequence that demonstrates a differential reactivity to Me$_3$-psoralen photobinding as a function of superhelical density. The reactivity to Me$_3$-psoralen should be characterized *in vitro* as a function of superhelical density to establish a standard curve. Analysis of the reactivity of Me$_3$-psoralen *in vivo* then provides an estimate of the effective superhelical density of the DNA *in vivo*.

Exonuclease III Z-DNA Assay. The exoIII assay for quantitating the reactivity of Me$_3$-psoralen photoproducts in DNA at base pair resolution is described elsewhere in this volume.[7] To apply the assay in living cells, we have grown, per sample, 50–100 ml of *E. coli* containing plasmid pCGTA-C [containing the torsionally tuned AATT(CG)$_6$TA(CG)$_6$AATT

---

[35] T. J. Kochel and R. R. Sinden, *BioTechniques* **6**, 532 (1988).

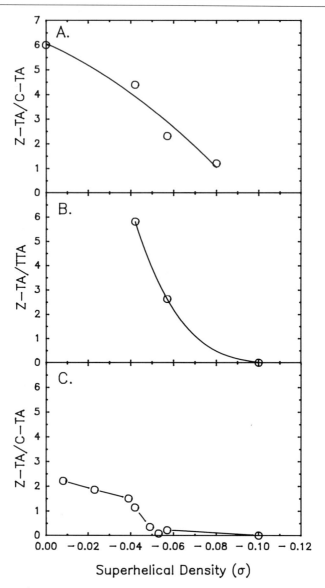

Fig. 1. Superhelical density dependence of the Z-TA/C-TA ratios. Zero-dose Z-TA/C-TA ratios are plotted as a function of superhelical density of the DNA. (A) Z-DNA sequence in pTGTA-C. The Z-TA/C-TA ratio is determined from the 5'-TA in the Z-DNA region $(TG)_6Ta(TG)_6$ and the 5'-TA within the flanking 20-bp control sequence. (B) Z-DNA sequence in pUCTA-1. The integrated area of the Z-TA exoIII stop was divided by the integrated area of the TTA exoIII stop 5' to the $(GT)_n$ sequence. (C) Z-DNA sequence in pUCTA-C. The Z-TA/C-TA ratio was determined as in (A). (From Kochel and Sinden.[35] Used with permission.)

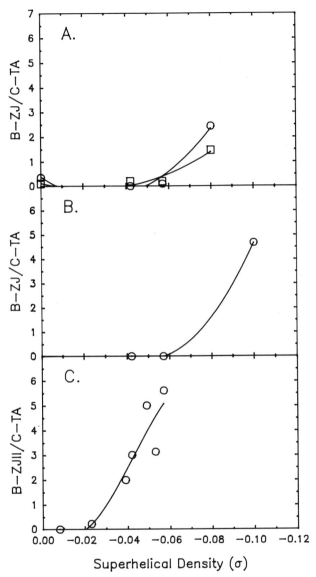

FIG. 2. Superhelical density dependence of the hyperreactivity of B–Z junctions on Me₃-psoralen photobinding. Zero-dose B–Z junction/C-TA ratios are plotted as a function of superhelical density of the DNA. (A) pTGTA-C. The integrated area of the exoIII stop of the B–Z junction was divided by the integrated area of the C-TA exoIII stop in the 20-bp control sequence. Junction 1 (○), junction 2 (□). (B) pUCTA-1. The integrated area of the B–Z junction exoIII stop was divided by the integrated area of the TTA stop. (C) pCGTA-C. The B–Z junction II was divided by the 5′-TA in the control DNA as in (A). (From Kochel and Sinden.[35] Used with permission.)

probe]. Analysis has been performed in chloramphenicol-treated cells in K medium (to amplify the amount of plasmid DNA to facilitate analysis) as well as in cells growing in log phase in Luria browth. Cells are chilled to 4° and then harvested by centrifugation and resuspended in M9 buffer at 0–4°. Me₃-psoralen is added, and after a 2-min incubation cells are irradiated up to 10 kJ/m² with 360 nm light. It is necessary to obtain a "zero-dose" ratio of Me₃-psoralen photobinding, which requires analysis of four or five different irradiation times for each sample of cells to be analyzed.[28,35] Following irradiation the supercoiled DNA is purified using a cleared lysate procedure including a CsCl–ethidium bromide gradient step.[2] It is important to remove any nicked DNA from further analysis since psoralen will photobind readily to the 5'-TA in the Z-DNA-forming sequence in nicked DNA, and this will affect the analysis of the fraction of Z-DNA. Details of the exoIII/photoreversal assay have been described earlier.[7]

Other Applications of Psoralen as DNA Structure Probe in Vivo

Psoralens have been applied in many ways both in vitro and in vivo to the analysis of DNA and RNA structure. Some of these applications have been reviewed by Cimino et al.[3] A few in vivo applications of psoralen photobinding will be mentioned here.

Psoralen as Probe of Nucleosome Organization in Vivo

Psoralen has been shown to cross-link linker DNA regions between nucleosomes, while the DNA wrapped around the nucleosome is protected from cross-linking.[5,6] When the DNA is denatured and examined using electron microscopy, bubbles can be seen where the nucleosomes once were.[36] This technique has been used to map nucleosomes in vivo.[37] It appears that the extent of psoralen cross-linking is representative of the chromatin structure.[38] From analysis of incorporation of radioactive psoralens, it has been shown that the chromatin organization in purified SV40 virus is different from that observed in vivo.[39] A 400-bp region at the origin of DNA replication is preferentially labeled, consistent with independent evidence for a nucleosome-free region in some fraction of intracellular SV40 minichromosomes.[39]

More recently, Conconi et al.[40] have extended the psoralen cross-link

[36] G. P. Wiesehahn, J. E. Hyde, and J. E. Hearst, Biochemistry 16, 925 (1977).
[37] L. M. Hallick, H. A. Yokota, J. C. Bartholomew, and J. E. Hearst, J. Virol. 27, 127 (1978).
[38] J. M. Sogo, P. J. Ness, R. M. Widmer, R. W. Parish, and T. Koller, J. Mol. Biol. 178, 897 (1984).
[39] G. K. Kondoleon, G. W. Robinson, and L. M. Hallick, Virology 129, 261 (1983).
[40] A. Conconi, R. M. Widmer, T. Koller, and J. M. Sogo, Cell (Cambridge, Mass.) 57, 753 (1989).

analysis to look at chromatin organization of ribosomal RNA genes *in vivo*, at various points in the cell cycle. Their approach was based on the observation that psoralen-cross-linked DNA will migrate slower than DNA not cross-linked and that digestion with various exonucleases can be used to produce a "nucleosome ladder."[41] It was observed that ribosomal RNA chromatin that was cross-linked *in vivo* migrated as two distinct *Eco*RI bands; both bands migrated slower than non-cross-linked DNA but faster than cross-linked naked DNA. The slower migrating chromatin band was shown to correspond to transcriptionally active chromatin, whereas the faster chromatin band represented inactive chromatin. (Inactive chromatin would be expected to contain less psoralen because of nucleosome protection from photobinding.) By lightly cutting with micrococcal nuclease and digesting with an exonuclease (λ-exonuclease or exonuclease III) and S1 nuclease, a nucleosome-type ladder was produced for the nontranscribed chromatin, whereas a smear was obtained for the transcriptionally active gene. This result suggests that the active conformation was nucleosome free (or at least did not contain positioned nucleosomes), whereas the inactive chromatin contained positioned nucleosomes. This approach could be easily extended to study a variety of systems.

Psoralen as Probe of Other DNA/Protein Interactions

As described above for nucleosomes, the binding of psoralen to DNA is sensitive to protein association. The rate of psoralen photobinding can change as the general organization of DNA with proteins changes. A change that is independent of torsional tension can reflect a change in the general level of DNA/protein organization as demonstrated for herpes simplex virus.[11] Zhen *et al.*, using a *Bal*31 exonuclease assay, have shown that cross-linking to DNA binding sites for λ repressor or RNA polymerase is prevented when these proteins are bound to the DNA.[42]

Bifunctional Psoralen Derivatives

Bifunctional psoralens or bispsoralens have been used to study the three-dimensional organization of DNA in viruses.[43] For example, using this approach, coupled with electron microscopy, the linear adenovirus genome has been shown to be organized into supercoiled loops in the virion.[44]

[41] R. M. Widmer, T. Koller, and J. M. Sogo, *Nucleic Acids Res.* **16**, 7013 (1988).
[42] W. Zhen, C. Jeppesen, and P. E. Nielsen, *FEBS Lett.* **229**, 73 (1988).
[43] J. Welsh and C. R. Cantor, *J. Mol. Biol.* **198**, 63 (1987).
[44] M.-L. Wong and M.-T. Hsu, *Nucleic Acids Res.* **17**, 3535 (1989).

[19] DNA Methylation *in Vivo*

By W. Zacharias

Introduction

Enormous progress has been made in the *in vitro* characterization of DNA structures which deviate from the classic right-handed B-form double helix.[1,2] The development of chemical and enzymatic probes which can focus in on a tiny segment of structural alteration in a vast background of unperturbed DNA helix has been a prerequisite for this progress. The specificities and applications of a variety of these probes have been reviewed recently.[3,4]

It has become increasingly important to identify biological functions of such non-B-form DNA structures (left-handed Z-DNA, cruciforms, bent regions, triple-helical segments, and others). Of course, only a combination of *in vivo* functional analysis and structural determination inside a cell can reveal the biological roles of these structures. In the past, this has created a severe problem. Physicochemical and optical techniques could not be used owing to the large excess of nontarget chromosomal or plasmid DNA contained in the cell. Several of the chemical probes (OsO_4,[5,6] $KMnO_4$[7]) have been applied to bacterial cell cultures, aiming at the analysis of non-B-DNA structures contained inside the cell. However, the cells usually do not survive this treatment with reactive chemicals; thus, there is doubt that the reaction conditions reflect a true *in vivo* situation.

Recently, a new concept has emerged that uses DNA-modifying enzymes, induced and/or expressed inside the cell to be analyzed, as probes for the detection of non-B-form DNA segments occurring in growing cells. Bacterial DNA methyltransferases accept only a right-handed, double-

---

[1] R. D. Wells, *J. Biol. Chem.* **263**, 1095 (1988).

[2] R. D. Wells, S. Amirhaeri, J. A. Blaho, D. A. Collier, J. C. Hanvey, A. Jaworski, J. E. Larson, A. Rahmouni, M. Rajagopalan, M. Shimizu, F. Wohlrab, and W. Zacharias, *in* "Structure and Methods, Volume 2: DNA Protein Complexes and Proteins" (R. H. Sarma and M. H. Sarma, eds.), p. 25. Adenine Press, Schenectady, New York, 1990.

[3] J. K. Barton, *Chem. Eng. News* **66**, 30 (1988).

[4] P. E. Nielsen, *J. Mol. Recognit.* **3**, 1 (1990).

[5] A. Rahmouni and R. D. Wells, *Science* **246**, 358 (1989).

[6] E. Palecek, E. Rasovska, and P. Boublikova, *Biochem. Biophys. Res. Commun.* **150**, 731 (1988).

[7] S. Sasse-Dwight and J. D. Gralla, *J. Biol. Chem.* **264**, 8074 (1989).

stranded DNA helix as substrate for the methylation reaction.[8,9] Thus, they can be used to detect DNA structural alterations that include or neighbor the recognition site of the enzyme, even if the helix perturbation comprises only a few base pairs in an environment of many kilobase pairs of B-type DNA structure.[10-15] This chapter summarizes methods and applications of this new approach to analyze Z-DNA and triplex structures in recombinant plasmids *in vivo*.

In Vitro Experiments as Basis for *in Vivo* Studies

The basic idea for the detection of unusual DNA secondary structures with DNA methyltransferases is the same for Z-DNA,[8,9] triplexes,[15,16] cruciforms,[10] and form V DNA,[17] and can be outlined as follows. If the recognition site for a methyltransferase in a supercoiled plasmid is located within a structural alteration, or at the interphase between this structure and the unperturbed right-handed Watson–Crick helix, the enzyme is not able to accept this site as a substrate. Thus, the rate of methylation is drastically decreased at this site, whereas other recognition sites for the same enzyme, located in the same molecule but farther away, will be methylated with the expected uninhibited kinetics. After terminating the methylase reaction, the extent of methylation at each site can easily be determined by restriction digest with the corresponding restriction endonuclease and analysis of the digestion products on agarose or polyacrylamide gels.

However, it has to be kept in mind that all the above-mentioned DNA secondary structures are supercoil-dependent, and that the endonuclease cleavage is also inhibited by these structures.[8,18] Therefore, if the target

[8] W. Zacharias, J. E. Larson, M. W. Kilpatrick, and R. D. Wells, *Nucleic Acids Res.* **12,** 7677 (1984).

[9] L. Vardimon and A. Rich, *Proc. Natl. Acad. Sci. U.S.A.* **81,** 3268 (1984).

[10] A. Jaworski, W.-T. Hsieh, J. A. Blaho, J. E. Larson, and R. D. Wells, *Science* **238,** 773 (1987).

[11] W. Zacharias, A. Jaworski, J. E. Larson, and R. D. Wells, *Proc. Natl. Acad. Sci. U.S.A.* **85,** 7069 (1988).

[12] A. Jaworski, W. Zacharias, W.-T. Hsieh, J. A. Blaho, J. E. Larson, and R. D. Wells, *Gene* **74,** 215 (1988).

[13] W. Zacharias, M. Caserta, T. R. O'Connor, J. E. Larson, and R. D. Wells, *Gene* **74,** 221 (1988).

[14] W. Zacharias, A. Jaworski, and R. D. Wells, *J. Bacteriol.* **172,** 3278 (1990).

[15] P. Parniewski, M. Kwinkowski, A. Wilk, and J. Klysik, *Nucleic Acids Res.* **18,** 605 (1990).

[16] J. C. Hanvey, M. Shimizu, and R. D. Wells, *Nucleic Acids Res.* **18,** 157 (1990).

[17] S. K. Brahmachari, Y. S. Shouche, C. R. Cantor, and M. McClelland, *J. Mol. Biol.* **193,** 201 (1987).

[18] F. Azorin, R. Hahn, and A. Rich, *Proc. Natl. Acad. Sci. U.S.A.* **81,** 5714 (1984).

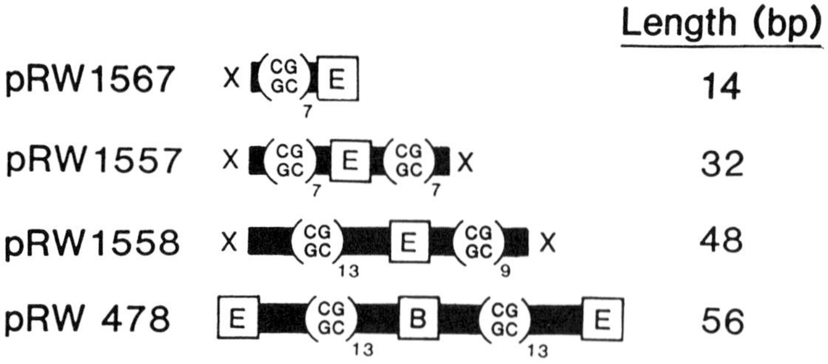

Length (bp)

pRW1567   X (CG/GC)7 [E]   14

pRW1557   X (CG/GC)7 [E] (CG/GC)7 X   32

pRW1558   X ■(CG/GC)13 [E] (CG/GC)9 ■ X   48

pRW 478   [E] ■(CG/GC)13 [B] ■(CG/GC)13 [E]   56

FIG. 1. Structures of Z-DNA inserts used for methylation inhibition assays. Four typical inserts with the potential to form Z-DNA *in vitro* are shown. E, *Eco*RI site, GAATTC; B, *Bam*HI site, GGATCC; X, *Xho*II site, RGATCY (R, purine; Y, pyrimidine). When E or B is embedded in the alternating dC-dG region, 4 bp is added to the total length of the insert.

site is a single site in the molecule, the plasmid has to be linearized with a different enzyme first, in order to remove supercoiling and reverse the structure back to a right-handed B-type helix. If there are several target sites in the molecule, and only some of them under the influence of the altered structure, then a restriction digest with the corresponding endonuclease can be performed directly, since cleavage at the remaining unperturbed sites will also remove supercoiling.

Left-Handed Z-DNA in Vitro

The formation of left-handed Z-DNA in recombinant plasmids containing alternating dC-dG inserts was analyzed *in vitro* by using *Hha*I methylase (M*Hha*I) and *Hha*I restriction endonuclease (*Hha*I).[8] Quantitation of C^3H_3 incorporation into the dC-dG insert as a function of superhelical density of the plasmid (by using ^{3}H-labeled S-adenosylmethionine as methyl donor) revealed that the inhibition of methylation in the insert was indeed dependent on supercoiling and went in parallel with the supercoil-induced B → Z transition, as determined by two-dimensional gel electrophoresis.[8] Also, chemical sequencing of the insert region after the methylation reaction demonstrated the same supercoil dependence of methyl incorporation into the insert.[9]

With a similar approach, M*Eco*RI and *Eco*RI were successfully used to monitor Z-DNA formation in dC-dG inserts that contain an *Eco*RI recognition site in the center of the insert or at a B–Z junction region (Fig. 1).[10,11] In addition, a mammalian DNA methyltransferase (isolated from murine erythroleukemia cells), which methylates cytosines in CpG dinucleotide units, was able to distinguish between a right-handed and a Z-helix structure in a 32-base pair (bp) dC-dG insert.[19] Interestingly, this enzyme could also sense other types of DNA structural deviations in pBR322 and in a *lac*UV5 promoter fragment which actually led to an increase in methylation at CpG sites, although this type of structural activation remains unclear at present.

Other systems described the lack of excision of formamidopyrimidine from left-handed poly(dG-5MedC) · poly(dG-5MedC) containing methylformamidopyrimidine residues by *Escherichia coli* DNA glycosylase[20] or the lack of repair by *E. coli* O^6-methylguanine methyltransferase of O^6-methylguanine-containing Z-DNA polymer.[21]

Two reports in the literature described that dC-dG regions in left-handed helix forms were apparently methylated by M*Hha*I in the Z form of poly(dG-dC) · poly(dG-dC)[22] or by M*Bsu*E in supercoiled ϕX174 RF.[23] These seemingly contradicting results may be explained by the equilibrium nature of the B → Z transition. Because the transition is reversible, a minor fraction of molecules may transiently exist in a B form, thus being substrates for the methylase. During prolonged incubation, this methylated population may increase significantly, leading to the conclusion that the enzyme seemingly methylated the Z form of the substrate. This shows that care must be taken in controlling the reaction conditions as well as the stability of the Z form of the substrate. Preferentially, methylation should be measured relative to unperturbed substrate sites as reference, or on supercoiled molecules relative to the linear form of the same plasmid.[8-11]

DNA Triple Helices in Vitro

A DNA triplex structure in polypurine–polypyrimidine sequences has three main components: a triple-helical stem, an unpaired half of the purine strand, and a central single-stranded loop region.[24] Each of these motifs should not serve as substrate for enzymes that require a double-stranded helix for recognition and reaction. This assumption was recently confirmed by experiments that used M*Eco*RI to detect intermolecular triple helix formation between a double-stranded $(GAA)_8$ region and a single-stranded $(CTT)_8$ oligonucleotide.[16] When the annealed $(CTT)_8$ oligomer overlapped

[19] T. Bestor, *Nucleic Acids Res.* **15**, 3835 (1987).
[20] C. Lagravere, B. Malfoy, M. Leng, and J. Laval, *Nature (London)* **310**, 798 (1984).
[21] S. Boiteux, R. Costa de Oliveira, and J. Laval, *J. Biol. Chem.* **260**, 8711 (1985).
[22] G. Soslau, J. Parker, and J. W. Nelson, *Nucleic Acids Res.* **14**, 7237 (1986).
[23] M. L. Gaido, C. R. Prostko, and J. S. Strobl, *J. Biol. Chem.* **263**, 4832 (1988).
[24] R. D. Wells, D. A. Collier, J. C. Hanvey, M. Shimizu, and F. Wohlrab, *FASEB J.* **2**, 2939 (1988).

a suitably located *Eco*RI site in the center of the GAA repeat region, methylation by M*Eco*RI was inhibited, thus showing that triplex formation can abolish recognition by a site-specific methylase.

On the other hand, *E. coli* dam methylase (DNA adenine methylase) was able to modify a dam site in the central loop region of an intramolecular triplex to the same extent as other dam sites in the plasmid molecule (as detected by restriction digest with *Mbo*I).[15] The reasons for this unexpected behavior of dam methylase are not understood.

Form V DNA in Vitro

Form V DNA is a topological form obtained by annealing two complementary single-stranded circles, thus necessitating the formation of right-handed as well as left-handed regions in the double-stranded product (including some amounts of single-stranded regions).[17] It is obvious that form V DNA is a useful object for studying the effects of structural alterations of the DNA helix on the activities of methyltransferases. Form V obtained from pBR322 was tested for reactivities with various methylases (M*Alu*I, M*Hha*I, M*Mbo*II, M*Msp*I, and M*Hpa*II).[17] The methylation patterns obtained with each of these methylases, correlated with the DNA sequence of pBR322, enabled classifying qualitatively different types of structural alterations in the product. However, a defined interpretation, in terms of what kind of structures were formed, could not be reached for many of the unmethylated or partially methylated restriction sites.

Experimental Procedures

Plasmid DNA at native supercoil density (approximately -0.05) is isolated from *E. coli* HB101 cells and purified by two CsCl density gradient centrifugations.[25] For the analysis of the supercoil dependence of methylation, plasmid topoisomer populations with defined average negative supercoil densities are prepared by the topoisomerase I/ethidium bromide method described previously.[26] Restriction enzymes and DNA methyltransferases are obtained from New England Biolabs (Beverly, MA) and used in the buffers recommended by the manufacturer.

Plasmid DNA (2–5 μg) is incubated at 37° in 50 μl of M*Eco*RI buffer containing 100 μM of *S*-adenosylmethionine and 20 to 40 units of methylase. At varying time points, 5-μl aliquots are withdrawn and heated at 65° for 15 min to inactivate the enzyme. Each aliquot is then digested with 10

[25] T. Maniatis, E. F. Fritsch, and J. Sambrook, "Molecular Cloning: A Laboratory Manual." Cold Spring Harbor Laboratory, Cold Spring Harbor, New York, 1982.
[26] C. K. Singleton and R. D. Wells, *Anal. Biochem.* **122**, 253 (1982).

TABLE I
DNA METHYLTRANSFERASES USED TO PROBE DNA STRUCTURES *in Vitro* AND *in Vivo*

DNA structure	Enzyme	Target sequence	Ref.
In vitro			
Z-DNA in plasmid inserts	M*Hha*I	GCGC	8, 9
	M*Eco*RI	GAATTC	10, 11
	M*Bsu*E	CGCG	23
	MEL MTase III	CG	19
Z-DNA in polymers	M*Hha*I	GCGC	22
	E. coli O^6-MeG-methyltransferase	O^6MeG	21
Cruciform	M*Eco*RI	GAATTC	10
Triplex, intramolecular	Mdam	GATC	15
Triplex, intermolecular	M*Eco*RI	GAATTC	16
Form V DNA	M*Alu*I	AGCT	17
	M*Hha*I	GCGC	17
	M*Mbo*II	GAAGA	17
	M*Msp*I	CCGG	17
	M*Hpa*II	CCGG	17
In vivo			
Z-DNA	M*Eco*RI	GAATTC	10, 11
Cruciform	M*Eco*RI	GAATTC	10
Triplex, intramolecular	Mdam	GATC	15

units of *Pst*I and 10 units of *Eco*RI in 20 μl of 50 mM Tris-HCl, 100 mM NaCl, 20 mM MgCl$_2$, and 1 mM dithiothreitol (DTT) for 2 hr at 37°. After ethanol precipitation, the digestion products are analyzed on 0.8% or 1.5% agarose gels and quantitated by densitometric tracing of Polaroid (Cambridge, MA) type 55 negative film with a Bio-Rad (Richmond, CA) Model 260 videodensitometer. An alternative procedure to quantitate the extent of methylation by measuring the incorporation of ^{3}H-labeled methyl groups into the DNA has been published.[8]

In Vivo Structural Analysis

The detection of a supercoil-dependent DNA secondary structure in a living cell relies on the same principles as the *in vitro* analyses described above (Table I). However, the system inside a cell is, naturally, much more complex and dynamic than the reaction with a minimum of purified components and a well-defined DNA substrate. Consequently, several preconditions have to be met in order to perform a successful *in vivo* analysis.

The probing methylase must be available in the form of a cloned gene that can use the host components (RNA polymerase, translation machinery, codon usage) for a productive expression of the protein product. Because the structure to be analyzed is expected to be induced by the *in vivo* supercoiling level, it is desirable to have the methylase gene in an inducible form (by temperature shift, chemical induction, or other means). If the gene is permanently expressed, methylase activity would be present in the cell while the plasmid is replicating and not yet supercoiled.[27] Thus, the target sites would be methylated, even if subsequently these sites would be included in the altered structure once the *in vivo* supercoiling level has been established. The controlled expression of an inducible methylase, on the other hand, will allow the measurement of time intervals of methylase activity, or the induction of enzyme activity only at certain growth stages. This will greatly diminish the problem of overmethylation by a constitutively expressed enzyme.

When cytosine methylases are used, the host strain for the experiment must be selected properly. Many *E. coli* strains cannot tolerate methylcytosine-containing DNA owing to the intrinsic methylcytosine-restricting (mcr) functions of the strain.[28,29] However, a variety of mcr⁻ strains is available in which these deleting functions are inactivated.[29] Finally, it is desirable to have at least one additional restriction target site, away from the unusual structure region, as a reference site for the rate of *in vivo* methylation by the enzyme.

Left-Handed Z-DNA in Vivo

A procedure was described that utilized a temperature-sensitive *Eco*RI methylase (M*Eco*RI[ts]) to detect Z-DNA formation in *E. coli* in a series of dC-dG inserts of varying lengths.[10,11] The gene for the probing methylase was provided in a pACYC184 vector, cotransformed into host cells that contained a pBR322 derivative into which each of the inserts was cloned. These inserts had varying capabilities *in vitro* to form supercoil-dependent left-handed helices, as determined by two-dimensional gel electrophoresis.[10,11] Each had an *Eco*RI target site either in the center (forming a Z–Z junction) or at the border with the vector sequence (forming a B–Z junction) (Fig. 1). The cotransformants were grown at the nonpermissive temperature of the methylase (42°), then shifted to the permissive temperature (22°) and grown for various periods of time. The extent of *in vivo* methylation at *Eco*RI target sites was determined by restriction double

[27] S. M. Lyons and P. F. Schendel, *J. Bacteriol.* **159**, 421 (1984).
[28] E. A. Raleigh, N. E. Murray, H. Revel, R. M. Blumenthal, D. Westaway, A. D. Reith, P. W. J. Rigby, J. Elhai, and D. Hanahan, *Nucleic Acids Res.* **16**, 1563 (1988).
[29] D. M. Woodcock, P. J. Crowther, J. Doherty, S. Jefferson, E. DeCruz, M. Noyer-Weidner, S. S. Smith, M. Z. Michael, and M. W. Graham, *Nucleic Acids Res.* **17**, 3469 (1989).

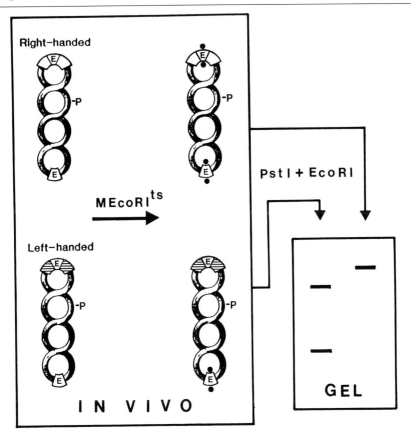

FIG. 2. Scheme for the *in vivo* analysis of Z-DNA. A recombinant plasmid is shown with an insert that can be either right-handed or left-handed in the cell. E, *Eco*RI site; P, *Pst*I site; filled circle, methyl group; M*Eco*RI[ts], temperature-sensitive *Eco*RI DNA methyltransferase. Other details are described in the text.

digests with *Pst*I and *Eco*RI, followed by agarose gel electrophoresis and densitometric quantitation of the digestion products (Fig. 2).

Substantial inhibition of methylation was found for longer dC-dG inserts (42 bp and longer), whereas the *Eco*RI sites near the very short inserts behaved quite similarly to the control site on the pACYC184 derivative. This result demonstrated the existence of left-handed helix structures in *E. coli* for some of the inserts. The assay employed living cells that continued to grow; no poisonous drugs, inhibitors, or disruptive procedures were used during the methylation step. Thus, the true *in vivo* milieu of the cell was unperturbed.

Cruciforms in Vivo

The *in vivo* assay for Z-DNA described above was used in an identical way to probe an inverted repeat sequence with a central *Eco*RI site in the loop region of the potential cruciform structure, cloned into the same location in place of the Z-DNA inserts.[10] No inhibition of methylation at the central *Eco*RI site was observed, however. This indicated that, at least for this sequence and during the exponential growth phase, no cruciform formation had occurred in *E. coli*.

Triplexes in Vivo

An approach similar to the M*Eco*RI *in vivo* assay for Z-DNA was used to analyze the behavior of triplex-forming polypurine–polypyrimidine sequences inside *E. coli* cells.[15] The dam methylase (Mdam) of *E. coli* was used to probe dam recognition sites, GATC, in the vector DNA and an isolated dam site in the center of an $(AG)_7ATCGATCG(AG)_7$ insert cloned into pUC18. However, two important aspects were different from the M*Eco*RI approach.

First, Mdam was the intrinsic (and not an induced) methylase constitutively expressed from the *E. coli* chromosome. Second, *in vitro* methylation with purified Mdam showed that the central dam target site in the insert was methylated as easily as the numerous vector dam sites, although a triplex structure clearly was present. It was therefore surprising that *in vivo* the triplex dam site was strongly undermethylated compared to the vector sites. Although this behavior is not quite understood, it may indicate that in the cell a triplex structure was actually formed and further stabilized by intracellular factors which were not present in the *in vitro* methylation reaction.[15]

Experimental Procedures

The *in vivo* assay for Z-DNA in recombinant plasmids is performed in *E. coli* HB101 ($R^- M^-$, $mcrA^+ mcrB^-$). This strain allows the use of M*Eco*RI as probe without interference of any host-owned chromosomal M*Eco*RI. The gene for M*Eco*RI[ts] is cloned into pACYC184, whereas Z-DNA inserts are cloned into a modified pBR322 vector.[10] Both plasmids are maintained in cotransformed *E. coli* HB101 by selection with ampicillin (50 μg/ml) and chloramphenicol (34 μg/ml). Cultures of 200 ml are grown in L broth at 42° to an OD_{600} of 0.8, then shifted to 22° and cultivated for varying periods of time. At each time point, 20 ml of the culture is immediately brought back to 42° and lysed at this temperature by the phenol–lysozyme method.[25] After purification, plasmids from different time points are analyzed for the extents of methylation at the *Eco*RI sites by restriction digests with *Pst*I and *Eco*RI, as described above for the *in vitro* experiments.

Summary and Outlook

The approaches described above have been applied successfully to analyze the formation of left-handed Z-DNA or intramolecular triplex structures in *E. coli* cells. The methods can be considered as true *in vivo* experiments, since the probing enzyme is provided in the cell by a nondisruptive technique, namely, by its expression inside the host cell using host cellular transcription and translation components. The experiments can be performed easily without any specialized or expensive equipment. Analysis of the probing results is done *in vitro* with commercially available enzymes, simple gel electrophoresis, and densitometry. The only requirement is the availability of the gene and DNA sequence of the DNA methyltransferase in order to construct a recombinant molecule with inducible expression of the enzyme. A list of all currently known bacterial restriction and modification enzymes and their recognition sequences has been published recently.[30]

Despite this obvious advantage, there are certain drawbacks that have to be kept in mind. Inhibition of methylation at a certain site per se does not allow the identification of the type of structure that was formed in the cell. Only a combination of the *in vivo* results with a characterization of the *in vitro* behavior of the inserts and suitably designed control sequences allows a clear identification. Furthermore, lack of methylation at a target site could, in principle, also indicate blocking of that site by a bound protein which prevents access for the methylase. Thus, additional control experiments should be performed (protein synthesis inhibition, *in vivo* UV cross-linking, or footprinting) in order to obtain unambiguous results.

At present, methylases have been used for DNA structural analyses only in bacteria. However, there is little doubt that the basic approach can also be applied to eukaryotic cells, although more complex procedures will be necessary for an inducible methylase expression and targeting of the protein to the nucleus, as well as for the analysis of the methylation products in nucleosomal DNA by Southern transfer and hybridization.

Finally, it should be pointed out that the concept of using enzymatic probes for *in vivo* structural analyses is not limited to DNA as target. It is conceivable that RNA-modifying enzymes, protein phosphorylases, or other enzymes whose target recognition depends on the conformational

[30] C. Kessler and V. Manta, *Gene* **92,** 1 (1990).

state of the target site in a macromolecule could also serve very well as tools for structural investigations inside living cells.

Acknowledgments

The assistance and key contributions of Dr. R. D. Wells and the members of his research group in some of the experiments described above are greatly appreciated.

[20] Topological Approaches to Studies of Protein-Mediated Looping of DNA *in Vivo*

By Hai-Young Wu and Leroy F. Liu

Introduction

Looping of DNA due to protein–protein interaction has been demonstrated in a number of systems. However, there are only a limited number of procedures for studying this type of interaction *in vivo* (for a review, see Ref. 1). We describe two new methods for studying protein-mediated looping of DNA *in vivo* using the *lac* repressor–operator complex as a model system.

Transcription-Mediated Template Supercoiling: Multimer Effect

The tetrameric *lac* repressor has been shown to bind simultaneously to two operators both *in vitro* and *in vivo*.[2–4] The simultaneous binding of the *lac* repressor to two operators on the same DNA molecule causes looping of the intervening DNA. Such *lac* repressor-mediated looping of DNA interferes with template supercoiling during RNA transcription and can cause changes in the supercoiled state of plasmid DNAs.[5]

Looping of DNA due to *lac* repressor binding requires the presence of at least two *lac* repressor binding sites on the same molecule. For plasmid DNAs containing only a single *lac* repressor binding site, only the multimeric forms of the plasmid DNAs can be efficiently looped by the

[1] M. Ptashne and A. A. F. Gann, *Nature (London)* **346**, 329 (1990).
[2] M. Besse, B. von Wilcken-Bergmann, and B. Müller-Hill, *EMBO J.* **5**, 1377 (1986).
[3] W.-T. Hsieh, P. A. Whitson, K. S. Matthews, and R. D. Wells, *J. Biol. Chem.* **262**, 14583 (1987).
[4] P. A. Whitson and K. S. Matthews, *Biochemistry* **25**, 3845 (1986).
[5] H.-Y. Wu and L. F. Liu, *J. Mol. Biol.* **219**, 615 (1991).

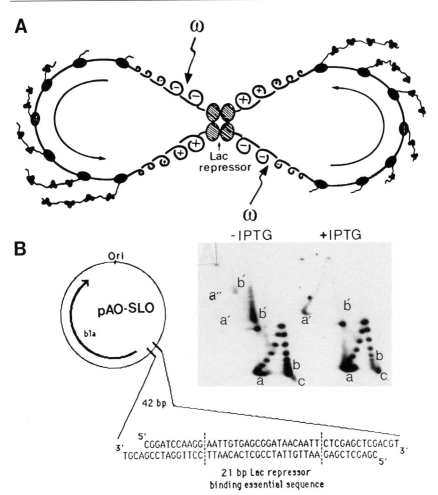

FIG. 1. *lac* repressor-mediated looping of DNAs: preferential positive supercoiling of the multimeric populations of plasmid DNAs. (A) pAO-SLO contains a single copy of a synthetic 42-bp oligomer containing a 21-bp *lac* repressor binding sequence and one major transcription unit, the *bla* gene [see (B), plasmid map]. pAO-SLO was transformed into *E. coli* AS19, a novobiocin permeable strain [M. Sekiguchi and S. Iida, *Proc. Natl. Acad. Sci. U.S.A.* **58,** 2315 (1967)]. Plasmid DNAs were isolated from a log phase culture grown in L broth in the presence of 50 μg/ml ampicillin. Thirty minutes prior to DNA isolation, novobiocin was added to the culture (100 μg/ml) with or without IPTG (1 m*M*). pAO-SLO DNA was isolated by the alkaline lysis method and analyzed by two-dimensional gel electrophoresis containing chloroquine.[6] a, a', and a" indicate the positions of negatively supercoiled monomeric, dimeric, and trimeric pAO-SLO DNAs, respectively. b, b', and b" mark the positions of positively supercoiled monomeric, dimeric, and trimeric pAO-SLO DNAs, respectively.

tetrameric *lac* repressor complex. The preferential supercoiling of the multimeric forms of the plasmid DNAs can therefore be a good indication of protein-mediated looping of DNA *in vivo*.

The principle of the method is diagrammatically shown in Fig. 1A. pAO-SLO DNA contains a single 21-base pair (bp) *lac* repressor binding sequence. The major transcription unit is the *bla* gene. The dimeric form of the plasmid DNA contains two *lac* repressor binding sequences diametrically opposed to each other. The simultaneous binding of the tetrameric *lac* repressor complex to both *lac* repressor binding sequences results in the partitioning of the dimeric DNA molecules into two equally sized loops. The formation of the loops restricts the diffusion of positive and negative supercoils generated by transcription of the *bla* gene, and hence increases the degree of positive and negative supercoiling in the vicinity of the tetrameric repressor complex. To measure the degree of positive supercoiling generated by transcription, negative supercoils are selectively removed by *Escherichia coli* DNA topoisomerase I. This is achieved *in vivo* by inhibiting the other major topoisomerase, *E. coli* DNA gyrase, with novobiocin.

The results of such an experiment are shown in Fig. 1. pAO-SLO DNA was isolated from *E. coli* AS19 (a novobiocin permeable strain) treated with novobiocin. The topoisomer distributions of both monomeric and dimeric pAO-SLO DNA populations were revealed by two-dimensional gel electrophoresis in gels containing chloroquine.[6] In the absence of isopropyl-β-D-thiogalactopyranoside (IPTG), the topoisomer distribution of the monomeric pAO-SLO DNA was heterogeneous and ranged from highly positively supercoiled topoisomers (marked b in Fig. 1) to highly

[6] H.-Y. Wu, S. H. Shyy, J. C. Wang, and L. F. Liu, *Cell* (*Cambridge, Mass.*) **53**, 433 (1988).

FIG. 2. Catenation of plasmid DNAs due to simultaneous binding of the *lac* repressor complex to two plasmid DNA molecules. pTO was constructed by deleting the 561-bp *Ssp*I–*Pst*I fragment of pAT153. The deletion removes the *bla* gene promoter. The resulting plasmid contains the *tetA* gene and a possible anti-*tet* transcription unit which is transcribed from a promoter overlapping the promoter region of the *tetA* gene (see plasmid map). A 42-bp synthetic DNA oligomer containing the 21-bp *lac* repressor binding sequence (the *lac* operator) was inserted into pTO outside the 3' end of the *tetA* gene to create pTO3. These two plasmid DNAs were transformed into *E. coli* AS19 separately. A single colony from each plate (10 hr posttransformation) was inoculated into LB medium in the presence or absence of IPTG (1 m*M*), and growth was continued for another 3 hr at 37°. Plasmid DNAs were isolated and analyzed by agarose gel electrophoresis. Lane 1, pTO DNA without IPTG; lane 2, pTO DNA with IPTG; lane 3, pTO3 without IPTG; lane 4, pTO3 DNA with IPTG. The interlocked dimers are distributed between supercoiled (I') and nicked (II') dimeric pTO3 DNA. III' marks the position of the linear dimeric pTO3 DNA.

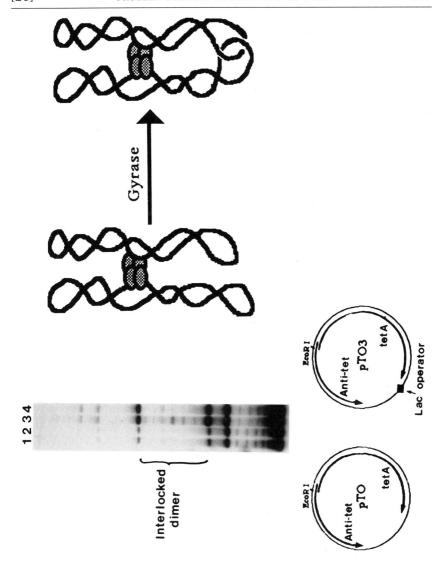

negatively supercoiled topoisomers (marked a in Fig. 1). The topoisomer distribution of the dimeric pAO-SLO DNA was quite different. Most of the topoisomers of the dimeric pAO-SLO DNA were highly positively supercoiled (marked by b'). On IPTG induction, the topoisomer distributions of the monomeric and dimeric populations were no longer different. The preferential positive supercoiling of the dimeric (and in fact multimeric) plasmid DNA populations is therefore indicative of *lac* repressor-mediated looping of DNA *in vivo*.

This type of analysis can probably be extended to studies of other DNA-binding proteins which involve either homologous or heterologous protein–protein interactions. It is interesting to point out that similar studies can also be performed in mutant *Saccharomyces cerevisiae* which express *E. coli topA* and are defective in both *top1* and *top2*.[7]

Formation of Catenated Dimer Plasmid DNAs Facilitated by lac Repressor-Mediated Protein–Protein Interaction in Vivo

The simultaneous binding of the tetrameric *lac* repressor complex to two plasmid DNA molecules, each of which contains a single *lac* repressor binding sequence, is expected to increase the population of interlocked dimeric plasmid DNA in cells. The principle for this method is diagrammatically shown in Fig. 2. The ability of gyrase to catenate the two plasmid DNA molecules brought together by the tetrameric repressor complex produces interlocked dimer DNA.

An experiment which demonstrates the formation of interlocked plasmid DNAs due to the simultaneous binding of the tetrameric *lac* repressor complex to two plasmid DNA molecules is shown in Fig. 2. pTO3 DNA contains a single *lac* repressor binding sequence near the 3' end of the *tetA* gene (see Fig. 2). In the absence of IPTG, interlocked dimers of pTO3 DNA were detectable (Fig. 2, lane 3), which migrated in a region between the positions of the nicked monomeric pTO3 DNA (marked II) and nicked dimeric pTO3 DNA (marked II'). On IPTG induction, no interlocked dimers could be readily detected (Fig. 2, lane 4). Similarly, the parent plasmid pTO which does not contain the *lac* repressor binding sequence did not produce any detectable interlocked dimer with or without IPTG induction (Fig. 2, lanes 1 and 2). The formation of the interlocked dimer is therefore indicative of the simultaneous binding of *lac* repressor to two plasmid DNA molecules.

This procedure in principle can be applied to other DNA-binding proteins which may interact homologously or heterologously, and it can be performed in yeast or higher eukaryotic cells. One caveat for this proce-

[7] G. N. Giaever and J. C. Wang, *Cell (Cambridge, Mass.)* **55**, 849 (1988).

dure is that the position of the *lac* repressor binding site may influence the results. In one instance, when the *lac* repressor binding sequence was inserted into the *Aat*II site of pTO DNA, the formation of interlocked dimers was not readily detectable (data not shown). Transcription from the anti-*tet* promoter may be responsible for destabilizing the *lac* repressor complex bound in this region.

Section IV

Interaction of DNA and Proteins

[21] Assay of Anti-DNA Antibodies

By DAVID G. SANFORD and B. DAVID STOLLAR

Introduction

Anti-DNA antibodies are valuable biochemical reagents. They identify specific bases or conformations among mixtures of DNA structures, measure transitions between differing DNA conformations, and detect several forms of chemically modified DNA.[1-3] Base-specific antibodies are induced by immunization with natural or modified purines, pyrimidines, nucleosides, or nucleotides covalently conjugated to protein carriers. Antibodies are also induced by immunization with noncovalent complexes formed with methylated bovine serum albumin (MBSA) and denatured DNA, helical duplex polynucleotides that differ from B-helical DNA (such as Z-DNA), triple-helical DNA, cruciform DNA, and chemically modified DNA (such as alkylated or UV-irradiated DNA). Although native B-helical DNA is not usually a strong immunogen, alone or in complexes formed with MBSA, antibodies to B-DNA occur in the sera of patients or mice with the autoimmune disease systemic lupus erythematosus (SLE). Both monoclonal and polyclonal autoantibodies and immunization-induced antibodies are sensitive probes for the presence of specific helical conformations.

Measured affinities of antinucleic acid antibodies are modest. Estimated by radioimmunoassay,[4] fluorescence,[5-7] gel filtration,[8] precipitation,[9] and filter binding radioimmunoassay,[10,11] they range from 10^6 to 3×10^8 (Table I). When bivalent antibody interacts with multivalent DNA, however, the apparent binding affinity is greatly enhanced, from 60-fold in some cases[7] to several orders of magnitude[12] in others.

Procedures for immunization and preparation of anti-DNA antibodies

[1] T.W. Munns and M. K. Liszewski, *Prog. Nucleic Acids. Res. Mol. Biol.* **24,** 109 (1980).
[2] B. D. Stollar, *Crit. Rev. Biochem.* **20,** 1 (1986).
[3] B. D. Stollar, *Int. Rev. Immunol.* **5,** 1 (1989).
[4] T. W. Munns and S. K. Freeman, *Biochemistry* **28,** 10048 (1989).
[5] J. S. Lee, D. F. Dombroski, and T. R. Mosmann, *Biochemistry* **21,** 4940 (1982).
[6] R. R. P. Kardost, P. A. Billing, and E. W. Voss, Jr., *Mol. Immunol.* **19,** 913 (1982).
[7] E. Sage and M. Leng, *Biochemistry* **16,** 4283 (1977).
[8] R. P. Braun and J. S. Lee, *J. Immunol.* **139,** 175 (1987).
[9] R. Ali, H. DerSimonian, and B. D. Stollar, *Mol. Immunol.* **22,** 1415 (1985).
[10] S. Iswari and T. M. Jacob, *FEBS Lett.* **176,** 43 (1984).
[11] S. C. Yuhasz, D. F. Senear, J. Adamkiewicz, M. F. Rajewsky, P. O. P. Ts'o, and L. Kan, *Biochemistry* **26,** 2334 (1987).
[12] R. P. Taylor, D. Weber, A. V. Broccoli, and J. B. Winfield, *J. Immunol.* **122,** 115 (1979).

TABLE I
AFFINITIES OF ANTI-DNA ANTIBODIES

Antibody	Antigen	K_a (M^{-1})	Method	Ref.
Autoantibodies				
MRL anti-GMP	$(dG)_{10}$	1×10^7	Radioimmunoassay	4
NZB MAb anti-ssDNA	Poly(dT)	1.2×10^7	Fluorescence	5
MRL MAb anti-ssDNA	d(GpG)	1.7×10^6	Fluorescence	6
NZB MAb anti-dsDNA	Poly(dA-dT)	4.26×10^6	Gel chromatography	8
Human anti-dsDNA	DNA fragment	1×10^7	Precipitation	9
Experimentally induced antibodies				
Rabbit anti-poly(I)	$(IpI)_6I$	6×10^7	Fluorescence	7
Rabbit anti-Guo[a]	Guo(ox–red)[b]	1.78×10^7	Filter binding	10
MAb anti-O^6-EtdGuo[c]	O^6-EtdGuo	3×10^8	Filter binding	11

[a] Guanosine.
[b] Guanosine oxidized with periodate and then reduced with borohydride.
[c] O^6-Ethyldeoxyguanosine.

have been described in a previous volume of this series.[13] This chapter describes the affinity purification of anti-DNA antibodies and measurement of the antibody–DNA interactions. Methods used in earlier studies of DNA immunochemistry included precipitation from solution,[14] immunodiffusion in agarose gel medium,[4,15] passive agglutination,[16] and quantitative microcomplement fixation.[17] This chapter describes assays that are more widely used at present: enzyme-linked immunosorbent assay (ELISA), radioimmunoassay (RIA), and gel electrophoresis shift assay. Interaction with intracellular DNA is also measured by immunofluorescence. More specialized techniques include electron microscopy of antibody–DNA complexes, chemical or UV-induced cross-linking, and NMR spectroscopy. Application of some of these techniques to the characterization of antibody–DNA interactions and the detection of conformational transitions in the DNA will be discussed.

Standard Procedures

Affinity Purification

Affinity columns with DNA bound to a solid support provide a matrix for the purification of anti-DNA antibodies from sera or tissue culture

[13] B. D. Stollar, this series, Vol. 70, p. 70.
[14] M. Seligmann, A. Cannat, and M. Hamard, Ann. N.Y. Acad. Sci. 124, 817 (1965).
[15] J. Oudin, this series, Vol. 70, p. 166.
[16] D. Koffler, R. Carr, V. Agnello, R. Thoburn, and H. G. Kunkel, J. Exp. Med. 134, 294 (1971).
[17] B. D. Stollar, Ann. Rheum. Dis. 36 (Suppl. 1), 102 (1977).

media. The procedure described here is that of Kubota *et al.*,[18] in which poly(L-lysine) is bound to cyanogen bromide-activated Sepharose 4B (Pharmacia, Piscataway, NJ) and DNA is cross-linked to the poly(L-lysine). We have applied this method to purification of antibodies specific for denatured DNA, native B-DNA, and Z-DNA. Alternative choices for immobilized antigens are commercially available poly(dT)-cellulose and DNA-agarose for purification of antibodies to denatured or native DNA.

Poly(L-lysine) (PLL)-Sepharose is prepared by incubation of 80 mg of PLL (MW 4000, Sigma, St. Louis, MO) with 15 ml of CNBr-activated Sepharose 4B in 0.2 M sodium bicarbonate (pH 8.3), 1.9 M NaCl, for 2 hr at room temperature. Residual sites on the CNBr-Sepharose are blocked by incubation of the gel in 0.2 M glycine, pH 8.0. DNA is linked to the PLL-Sepharose by incubation of 8 mg of DNA with 5 ml of PLL-Sepharose in 25 mM sodium phosphate buffer (pH 7.0) containing 2% (v/v) formaldehyde. The mixture is stirred overnight at 4°. Before use, the affinity column is washed with 20 mM sodium carbonate (pH 10.5) containing 5% (v/v) dimethylsulfoxide (DMSO), to remove any DNA that is not cross-linked to PLL. The column is then equilibrated with TBS (10 mM Tris-HCl, 150 mM NaCl, pH 7.4).

Antibody samples are applied to the column in TBS buffer. The antibody binding capacity is about 0.5 mg/ml packed bed volume. The column is washed with TBS–EDTA [TBS buffer containing 5 mM ethylenediaminetetraacetic acid (EDTA)], and anti-DNA antibodies are then eluted in 20 mM sodium carbonate, pH 10.5 (or pH 10.8 if necessary). Fractions containing antibody are neutralized with an equal volume of 1 M Tris (pH 7.0) to minimize denaturation of the protein. The neutralized antibody solution is dialyzed against TBS or PBS (phosphate-buffered saline: 10 mM sodium phosphate, 150 mM NaCl, pH 7.4) and centrifuged at 10,000 rpm to remove aggregated protein. Measurement of the UV absorbance spectrum indicates whether the product is purely protein, for which the A_{260}/A_{280} ratio is approximately 0.6. Purified antibody yields two distinct bands (heavy and light chain) on sodium dodecyl sulfate (SDS)–polyacrylamide gel electrophoresis (PAGE) in the presence of 2-mercaptoethanol.

Monoclonal antibodies in tissue culture media can be purified with an affinity column containing immobilized anti-immunoglobulin (Ig) antibodies (anti-mouse or anti-human Ig, depending on the source of the hybridoma). From 2 to 5 mg of the anti-Ig is immobilized per milliliter of CNBr-activated Sepharose, and excess activated sites on the gel are blocked with 0.2 M glycine, pH 8, or 0.1 M ethanolamine, pH 8. The column is washed with neutral buffer, then with 0.1 M glycine, pH 3, and again with

[18] T. Kubota, T. Akatsuka, and Y. Kanai, *Clin. Exp. Immunol.* **62**, 321 (1985).

neutral buffer. Tissue culture fluid is applied, and the column is washed with TBS or PBS until the A_{280} is less than 0.01. Antibody is eluted with 0.1 M glycine, pH 3, neutralized, dialyzed, and centrifuged. Purified antibody is characterized by measurement of its UV absorbance spectrum and by SDS–PAGE.

Enzyme-Linked Immunosorbent Assay

ELISA measures antibody binding to antigen immobilized on a solid phase, using an enzyme-linked second antibody as a detecting reagent. General features of the ELISA have been described in this series.[19] In a direct binding assay, DNA is immobilized on the surface of wells of a polystyrene microtiter plate. Anti-DNA antibody is incubated with the immobilized DNA, and bound antibody is detected by an anti-Ig linked to an enzyme which reacts with a chromogenic substrate (Fig. 1a). Alkaline phosphatase–Ig and horseradish peroxidase–Ig are commonly used conjugates.

Direct binding ELISA is a convenient method for analysis of many antibody and DNA samples, but it is not reliable for determining the relative affinities of an antibody for various antigens. Competitive assays, in which nucleic acid competitors are preincubated with the antibody in solution before incubation with the solid-phase antigen, provide a more reliable measurement of the relative affinities for various polynucleotides (Fig. 1b). Reaction of the competitors and antibody in solution avoids artifacts that may occur in the direct binding assay. Artifacts include variable binding of different polynucleotides to the solid phase and conformational changes of nucleic acids that may occur when they bind to the solid phase. It is difficult to determine precise affinities with DNA even in solution, because DNA presents many binding sites and it interacts with bivalent IgG or multivalent IgM.[12]

Direct Binding ELISA. Denatured DNA binds readily to the surface of polystyrene microtiter plates, but native DNA does not.[20] A common method used for enhancing the immobilization of native DNA is coating the plate with a positively charged polypeptide such as poly(L-lysine)[21] or protamine,[22] to which negatively charged DNA then binds tightly. Recently, streptavidin was used as a precoating, and DNA containing biotinylated nucleotides was bound to the streptavidin-coated surface.[23] Pre-

[19] E. Engvall, this series, Vol. 70, p.419.
[20] F. Fish and M. Ziff, *Arthritis Rheum.* **24,** 534 (1981).
[21] R. B. Eaton, G. Schneider, and P. H. Schur, *Arthritis Rheum.* **26,** 52 (1983).
[22] J. L. Klotz, R. M. Minami, and R. L. Teplitz, *J. Immunol. Methods* **29,** 155 (1979).
[23] W. Emlen, P. Jarusiripipat, and G. Burdick, *J. Immunol.Methods* **132,** 91 (1900).

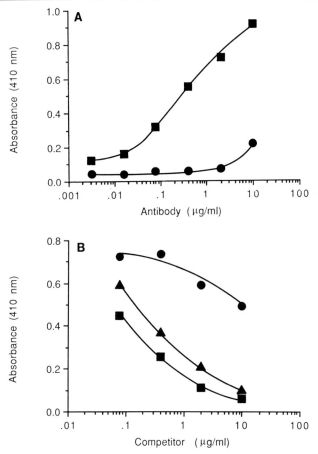

FIG. 1. Enzyme-linked immunosorbent assay with anti-B-DNA antibodies. (a) Direct binding ELISA. Various amounts of purified monoclonal autoantibody 2C10 were added to wells of UV-irradiated microtiter plates coated with DNA. Wells were coated with either native calf DNA (■) or poly(dG-md5C) (Z-DNA) (●). One hundred microliters of a 2.5 μg/ ml solution of DNA in PBS with 20 mM Mg^{2+} was added to each well. Bound antibody was detected with alkaline phosphatase–anti-mouse IgG. (b) Competitive ELISA. Relative affinities of polynucleotides were evaluated by their binding to autoantibody 2C10 in solution. Various concentrations of native calf DNA (▲), poly(dA-dT) (■), or poly(dG-dC) (●) were incubated with a constant amount of antibody in solution before the mixture was added to wells of a native DNA-coated microtiter plate.

treatment of polystyrene plates with UV light also enhances binding of nucleic acids.[24] It avoids the use of DNA–protein complexes, which may alter DNA conformation or measure antibodies that bind to the positively charged protein rather than to DNA. UV treatment also eliminates the step of precoating the plate and is at least as sensitive as the poly(L-lysine) procedure.

The UV lamp of a laminar flow biosafety hood can be used to irradiate microtiter plates. Polystyrene plates (Immulon type 1, Dynatech Labs., Chantilly, VA) placed at a distance of 65 cm from the UV lamp for at least 12 hr give maximal binding of DNA. Treated plates have enhanced native DNA binding for at least 1 week following irradiation. The UV-treated plates also bind denatured DNA and synthetic polynucleotides.

The wells of the plate are coated by incubation of 100 μl of a DNA solution (1–5 μg/ml in TBS) per well for 2 hr at room temperature. Negative controls (wells with no antigen) are required, so for each lane that is coated with DNA a corresponding lane on the same plate is left uncoated. To prevent nonspecific binding of antibodies to the plate, the surfaces of both the coated and the uncoated wells are blocked with 150 μl of 1% bovine serum albumin (BSA, RIA grade, Sigma) in PBS for 2 hr at room temperature.

The typical concentration range to be tested is 0.01 to 10 μg/ml of purified antibody or serum dilutions between 1/100 and 1/10,000. Antibody dilutions are made in PBS with 1% BSA, and 100 μl of each dilution is incubated in both an antigen-coated well and an uncoated (but BSA-blocked) well for 1 hr. The wells are then washed 3 times with PBS containing 0.1% Tween 20 (Fisher, Pittsburgh, PA).

An appropriate antibody–enzyme conjugate (i.e., goat antibody to mouse, human, or rabbit Ig) is used to detect bound antibodies. A wide variety of such conjugates is available commercially. The detecting antibody is diluted, typically between 1 : 1000 and 1 : 5000 in TBS, and 100 μl of the diluted reagent is incubated in each well for 1 hr at room temperature. The plate is washed 3 times with PBS–Tween. Substrate is added, and the absorbance is read at a time between 15 min and 1 hr (or, for more precise kinetic mesurement of enzyme, at several time points). For alkaline phosphatase conjugates, the substrate p-nitrophenyl phosphate is prepared as a 0.1% (w/v) solution in 50 mM carbonate buffer, pH 9.5, containing 2 mM MgCl$_2$, and the absorbance of the product is read at 410 nm. Horseradish peroxidase is assayed with o-phenylenediamine as substrate, yielding a product measured at 492 nm. All absorbance readings are adjusted by

[24] M. Zouali and B. D. Stollar, *J. Immunol. Methods* **78**, 1173 (1986).

subtraction of the absorbance of a negative control well to which no anti-DNA antibody was added.

Competitive ELISA. The competitive assay is most efficient at the concentration of anti-DNA antibody that gives 50% of maximal binding in a direct binding ELISA. The plate for a competitive assay is coated with antigen and blocked in the same manner as that used for the direct binding ELISA. Before addition of the antibody to the wells, the competitor is incubated with a standard concentration of antibody. A range of competitor concentrations is used, typically between 0.01 and 100 μg/ml. For the control measurement of uninhibited binding, one sample of antibody is mixed with buffer in place of competitor. After incubation for 1 hr, 100 μl of each mixture is added to a well on the plate and kept at room temperature for 1 hr. As in the direct binding assay, a background control is a well that is not coated with antigen but receives all other reagents. Addition of the antibody–enzyme conjugate and the substrate is performed as in the direct binding procedure described above.

A precaution to be noted in competitive ELISA or other competitive assays described below is that certain polynucleotides could inhibit antibody binding by annealing to the test antigen and blocking its sites rather than by binding to the antibody. This is unlikely to occur with very heterogeneous DNA but should be kept in mind when antibodies to homopolymers are studied.

Radioimmunoassay

RIA measures the interaction of antibody with radioactively labeled antigen. Although it is more tedious then ELISA, RIA is a solution-phase assay in which the total concentration of antigen available for reaction is known. This allows a more quantitative analysis of the data. Many methods have been used for the separation of bound from free antigen, including relatively nonspecific precipitation of antibody–DNA complexes with polyethylene glycol or ammonium sulfate,[25] specific precipitation of complexes with an anti-Ig antibody,[26] or immobilization of antibody–DNA complexes on nitrocellulose filters.[27] Bound DNA can then be determined by measuring the radioactivity in the precipitate or bound to the filter. Competitive RIA is frequently used to determine the relative affinities of different DNA competitors for an antibody.

Direct Binding Immunoprecipitation RIA. DNA can be labeled by incorporation of [3H]thymidine during bacterial growth and purification of

[25] R. Smeenk, G. van de Lelij, and L. Aarden, *J. Immunol.* **128,** 73 (1982).
[26] M. Papalian, E. Lafer, R. Wong, and B. D. Stollar, *J. Clin. Invest.* **65,** 469 (1980).
[27] R. M. Lewis, B. D. Stollar, and E. B. Goldberg, *J. Immunol. Methods* **3,** 365 (1973).

the labeled DNA,[17,26] or by incorporation of ^{32}P or ^{35}S into purified DNA by nick translation.[28] Iodination with ^{125}I can also be used to label denatured DNA.[29] Oligonucleotides are end-labeled with T4 polynucleotide kinase.[30] For the immunoassay, a solution of 0.02 to 2 μg/ml of labeled DNA is used. In a typical experiment with DNA labeled *in vivo* by [^{3}H]thymidine, 1500–2000 counts/min (cpm) in 50 μl of DNA is added to antibody. DNA labeled with ^{32}P or ^{32}S provides much higher counts.

Serially diluted antibody (100 μl) in phosphate–EDTA buffer (60 mM Na$_2$HPO$_4$, 30 mM EDTA, 100 mM NaCl, pH 8.0) is mixed with 50 μl of labeled DNA, in 1.5-ml microcentrifuge tubes. After incubation at room temperature for 1 hr, antibody and antibody–antigen complexes are precipitated with 50 μl of an appropriate anti-Ig antiserum. Precipitating sera are available commercially. It is important that the precipitating anti-Ig be added in excess to ensure precipitation of all of the antigen–antibody complex. The amount required can be determined with a preliminary titration of a standard anti-DNA serum with a known total Ig content or from a quantitative precipitation analysis.[31] The precipitate is pelleted by centrifugation, washed twice with TBS, redissolved in 0.2 ml of 0.1 N NaOH, and counted in 3 ml of scintillation fluid. A portion of the total amount of DNA added to each sample is also counted in 0.2 ml of 0.1 N NaOH plus scintillation fluid. As a negative control, a nonimmune serum or IgG is used in place of the test serum or antibody.

RIA does not measure mass units of antibody. Activity units can be expressed by comparison with dilutions of a known standard serum required for binding of a given percentage (usually 50% or 33%) of antigen.[32] The World Health Organization maintains a standard serum for anti-DNA autoantibody measurement.[33] Various dilutions of the standard serum are included in each experiment for internal calibration of activity units.

Ammonium Sulfate Precipitation RIA. Labeled DNA and antibody are prepared and incubated as described for the immunoprecipitation RIA. In place of anti-Ig, 100 μl of a saturated ammonium sulfate solution, adjusted

[28] J. Sambrook, E. F. Fritsch, and T. Maniatis, "Molecular Cloning: A Laboratory Manual," 2nd Ed., p. 10.6. Cold Spring Harbor Laboratory, Cold Spring Harbor, New York, 1989.
[29] S. L. Commerford, this series, Vol. 70, p. 247.
[30] J. Sambrook, E. F. Fritsch, and T. Maniatis, "Molecular Cloning: A Laboratory Manual," 2nd Ed., P. 10.59. Cold Spring Harbor Laboratory, Cold Spring Harbor, New York, 1989.
[31] E. A. Kabat, this series, Vol. 70, p. 3.
[32] J. Holian, I. D. Griffiths, D. N. Glass, R. N. Maini, and J. T. Scott, *Ann. Rheum. Dis.* **34,** 438 (1975).
[33] T. E. W. Feltkamp, T. B. Kirkwood, R. N. Maini, and L. A. Aarden, *Ann. Rheum. Dis.* **47,** 740 (1988).

to pH 8, is added to the 150 μl antibody–DNA mixture to precipitate the immune complexes. The mixture is allowed to stand at 4° for 15 min and is then centrifuged in a microcentrifuge in the cold. The precipitate is washed twice with 40% saturated ammonium sulfate, dissolved in 0.2 ml of 0.1 N NaOH, and counted with a scintillation counter.

Filter-Binding RIA. Nitrocellulose, cellulose ester, or glass fiber filters bind protein and protein–DNA complexes and separate them from free native DNA. Nitrocellulose and cellulose ester filters are not suitable for assays with denatured DNA, however, because they bind free denatured DNA. Glass fiber filters (GF/C, Whatman, Clifton, NJ) bind neither free native DNA nor denatured DNA and can be used to measure antibodies to both forms.

Samples of 0.1 μg of radiolabeled DNA are incubated with varying amounts of antibody in 1 ml of 10 mM Tris, 1 mM EDTA, 200 mM NaCl for 1 hr at room temperature. In preparation for use, nitrocellulose or cellulose ester filters (Millipore, Bedford, MA, type HA, 0.45 μm) are soaked in 0.3 M NaOH for 10 min and washed with distilled water. After incubation, the DNA–antibody solutions are filtered through the prepared membranes and washed 3 times with 1 ml of the Tris–EDTA buffer. The filters are dried under a heating lamp, placed in scintillation vials, and covered with scintillation fluid for counting bound radioactivity. A toluene-based fluid is used for nitrocellulose filters and a xylene-based fluid for glass fibers. Background counts are determined from a control mixture containing an antibody that does not bind DNA.

Competitive RIA. An antibody concentration that binds 50% of the labeled DNA in the absence of competitor is used for the competitive RIA. Samples of this amount of antibody are incubated with various concentrations of competing unlabeled DNA for 30 min before addition of the radiolabeled DNA. The mixtures of antibody, copmpetitor, and labeled DNA are incubated for 1 hr, and the antibody and complexes are then precipitated with anti-Ig reagent or ammonium sulfate. Like competitive ELISA, the competitive RIA is useful for comparing relative affinities of a given antibody with different polynucleotides. Under suitable conditions, competitive RIA can be used to determine the actual affinity.[34] If one type of DNA is being measured, the competitive assay is also useful for quantifying the DNA in the unlabeled sample. For this purpose, the amount of inhibition given by a sample of unknown concentration is conpared with that caused by a standard of known concentration.

---

[34] R. Müller and M. Rajewsky, this series, Vol. 92, p. 589.

Gel Shift Assay

Procedures for electrophoretic assays of DNA–protein interactions are described elsewhere in this volume.[35] They have been applied to measurement of antibodies binding DNA in the form of plasmids,[36] minicircles,[37] or oligonucleotides.[38] This chapter describes the binding of oligonucleotides by antibodies.

If the antigen adopts multiple conformations that can be resolved on the gel, the electrophoresis shift method can demonstrate the preferential binding of antibody to one of the forms. For example, a mouse monoclonal anti-DNA autoantibody, H241, binds to the duplex form of the oligonucleotide $rA_{12}d(GC)_6dT_{12}$ but not to the single-stranded loop form present in the same preparation.[38] Bound and free oligonucleotides can be detected by ethidium bromide staining of the gel or, if the DNA is radiolabeled, by autoradiography. With autoradiography, bands on the photographic film can be scanned and quantified with a densitometer. Whole antiserum should not be used with ethidium bromide staining, because serum albumin binds the dye and gives a fluorescent band unrelated to antibody binding. Ethidium bromide is not efficient for the staining of single-stranded oligonucleotides or oligonucleotides in the Z-DNA conformation.

The gels used for measuring the binding of oligonucleotides (of 20–50 bp) by antibody contain a 20% acrylamide "running gel" and a 3.5% acrylamide "stacking gel." The gels contain TBM buffer (90 mM Tris–borate, 5 mM $MgCl_2$, pH 8.3), and the ratio of acrylamide to bisacrylamide is 29 : 1 (w/w). Samples of oligonucleotide (20 μl, in PBS) are incubated with varying amounts of antibody (5–20 μl) for 1 hr. For analysis with ethidium bromide staining, 1 to 5 μg of duplex oligonucleotide is used; for autoradiographic analysis, 20 to 200 ng of oligonucleotide, usually end-labeled with ^{32}P, is sufficient. A loading solution containing 30% glycerol and 0.01% bromphenol blue is added immediately before application of the sample to the gel. Electrophoresis is carried out at 4° with low voltage (50–100 V) to prevent heating of the gel. Electrophoresis is continued until the dye front approaches the bottom of the gel (usually 2–4 hr).

Immunofluorescence Assay with Crithidia

SLE anti-native DNA antibodies are often assayed by their immunofluorescent staining of giant mitochondria (kinetoplasts) of the trypano-

[35] A. Nordheim, this volume [22].

[36] D. A. Zarling, D. J. Arndt-Jovin, M. Robert-Nicoud, L. P. McIntosh, R. Thomae, and T. M. Jovin, *J. Mol. Biol.* **176**, 369 (1984).

[37] A. Nordheim and K. Meese, *Nucleic Acids Res.* **16**, 21 (1988).

[38] D. G. Sanford, K. J. Kotkow, and B. D. Stollar, *Nucleic Acids Res.* **16**, 10643 (1988).

some *Crithidia luciliae*. The kinetoplast contains a large network of catenated circular native DNA free of histone or other nuclear proteins. The method described here is that of Aarden *et al.*[39]

Crithidia luciliae are grown in Bacto-tryptone medium, pH 7.4, at 24°. They are washed with PBS, pH 7.4, with low-speed centrifugation at 3000 *g* for 10 min at room temperature. The washed organisms are suspended in distilled water at a density of 2×10^7/ml. Ten-microliter drops are placed on glass slides, dried, and fixed in 96% (v/v) ethanol for 10 min at room temperature. The slides may be used immediately or stored at $-20°$. When the slides are stained with ethidium bromide (1 mg/ml in PBS) for 1 min, washed and viewed by fluorescence microscopy, the kinetoplast and the nucleus are seen as distinct fluorescent structures.

Fixed organisms on the slides are covered at room temperature for 30 min with 2 drops of serially diluted serum (starting from 1 : 10), then washed 3 times with PBS at 37° for 10 min each time. Two drops of an appropriate fluorescein-labeled anti-Ig are placed on the slides, which are incubated for 30 min at room temperature and washed again with PBS. One drop of buffered glycerol (50% PBS and 50% glycerol) is added, and the slides are covered with coverslips and sealed with Parafilm. They are examined under a fluorescence microscope with settings for fluorescein.

Immunofluorescence with Polytene Chromosomes

Polytene chromosomes of *Drosophila* or *Trichosia* have served as useful substrates for antinucleic acid antibodies, including anti-Z-DNA[40,41] and anti-RNA–DNA hybrid[42] antibodies. It is necessary to note that fixation with acid or acid–ethanol extracts proteins[43] and affects the exposure and conformation of nucleic acids in the chromosomes. Less protein extraction occurs if the chromosomes are prefixed with dilute formaldehyde.[44] As described by Lancillotti *et al.*,[41] larval salivary glands are excised and fixed in 50% acetic acid and then squashed on glass slides in the same acetic acid, frozen, transferred to 100% ethanol for 30 min, and rehydrated through a series of steps of decreasing ethanol concentration

[39] L. A. Aarden, E. R. de Groot, and T. E. W. Feltkamp, *Ann. N.Y. Acad. Sci.* **254**, 505 (1975).

[40] A. Nordheim, M. L. Pardue, E. M. Lafer, A. Möller, B. D. Stollar, and A. Rich, *Nature (London)* **294**, 417 (1981).

[41] F. Lancillotti, M. C. Lopez, C. Alonso, and B. D. Stollar, *J. Cell Biol.* **100**, 1759 (1985).

[42] W. Büsen, J. M. Amabis, O. Leoncini, B. D. Stollar, and F. J. S. Lara, *Chromosoma* **87**, 247 (1982).

[43] R. J. Hill, F. Watt, and B. D. Stollar, *Exp. Cell Res.* **153**, 469 (1984).

[44] F. Lancillotti, M. C. Lopez, P. Arias, and C. Alonso, *Proc. Natl. Acad. Sci. U.S.A.* **84**, 1560 (1987).

in PBS. The prepared chromosomes on the slides are incubated for 1 hr with various dilutions of anti-DNA antibody, washed with PBS, incubated for 1 hr with fluorescein-labeled anti-Ig, and washed again with PBS. The glands are mounted in glycerol–1 M Tris (9 : 1), pH 8, and examined with a fluorescence microscope with settings for fluorescein. After the chromosomes are photographed under fluorescent microscopy, the total DNA can be stained with acetoorcein. A stock solution, which contains 2 g of orcein in 50 ml acetic acid and 50 ml lactic acid, is diluted 1 : 3 in acetic acid–lactic acid. Chromosomes on the glass slides are washed with tap water, stained with the diluted acetoorcein for 5 min, washed with 3 changes of 45% acetic acid (5 min each), mounted, and viewed by direct light microscopy.

Specialized Techniques

Cross-linking

Covalent cross-linking preserves an antigen–antibody complex for subsequent analysis under conditions in which the noncovalent antibody–DNA bond is not stable. For example, anti-Z-DNA antibodies bind to supercoiled plasmids containing alternating purine–pyrimidine sequences,[45] but when the plasmid is cut with restriction enzymes for analysis of the antibody-linked sites, the Z-DNA conformation is lost. The antibody binding sites of such plasmids were located by cross-linking antibodies to the plasmids with glutaraldehyde before the DNA was cleaved.[46] The restriction fragments were passed through a nitrocellulose filter, which removed antibody–antigen complexes. The free DNA fragments, which passed through the filter, were analyzed by gel electrophoresis to determine which restriction fragments had been removed or reduced in amount by the membrane binding.

UV irradiation, which also yields covalent bonds between the protein and DNA, provides an affinity labeling of antibody.[47] The UV treatment is effectively a "zero–length" cross-linker that identifies very close contacts between the antibody and the DNA. All nucleic acid bases have the potential to form cross-links with protein on UV irradiation.[48] Fifteen of the common amino acids can be cross-linked to native or denatured

[45] A. Nordheim, E. M. Lafer, L. J. Peck, J. C. Wang, B. D. Stollar, and A. Rich, *Cell* (*Cambridge, Mass.*) **31**, 309 (1982).

[46] A. Nordheim, R. E. Herrera, and A. Rich, *Nucleic Acids Res.* **15**, 1661 (1987).

[47] Y. J. Jang and B. D. Stollar, *J. Immunol.* **145**, 3353 (1990).

[48] E. Livneh (Noy), S. Tel-Or, J. Sperling, and D. Elad, *Biochemistry* **21**, 3698 (1982).

DNA.[49] The major sites of addition are the C-5 and C-6 positions of pyrimidines and the C-8 of purines, which reside in the major groove of helical DNA.

The following procedure was used to cross-link anti-DNA antibodies to duplex oligonucleotides,[47] to identify close contacts with the antibody heavy and light chains. It was designed to yield a small number of antibody–DNA cross-links so that a single antibody would be linked at no more than one position. Similar procedures have been applied to protein interactions with high molecular weight nucleic acids.[50]

Antibody (20 μg) is mixed with 0.5 μg of ^{32}P-end-labeled duplex oligonucleotide in a total volume of 50 μl, and the mixture is incubated, on ice, for 1 hr. Samples of the antibody–oligonucleotide mixture are placed into the wells of a polystyrene microtiter plate, which rests on an ice bath. A short-wavelength UV light (Mineral Lamp Model UVGL-25, UVP Inc., San Gabriel, CA) is placed 4 cm above the plate, and the samples are irradiated for 10 min. This exposure results in cross-linking of approximately 2–5% of the DNA to antibody and no indication of multiple cross-links. Exposures of 1 hr lead to protein–protein cross-links and aggregation. SDS-PAGE of the sample under reducing conditions is used to separate the heavy chain and light chains of antibodies cross-linked to ^{32}P-labeled oligonucleotides. Autoradiography reveals a small percentage of the DNA migrating slightly more slowly than the Coomassie blue-stained heavy and/or light chain bands; the DNA–protein complex is larger than the free protein. The relative amount of radioactivity linked to the heavy and light chains varies with different monoclonal antibodies. No cross-linking is observed with normal mouse IgG. The radioactive oligonucleotide remains associated with peptide fragments obtained by proteolytic digestion of the labeled chains and reanalyzed by gel electrophoresis under denaturing and reducing conditions.

Electron Microscopy

Antibodies bound to DNA (e.g., anti-Z-DNA antibodies bound to supercoiled plasmids) can be seen by electron microscopy.[51,52] The locations of bound antibodies can be determined by cutting the plasmid with a restriction enzyme and measuring the distance from the restriction site to

---

[49] M. D. Shetlar, J. Christensen, and K. Hom, *Photochem. Photobiol.* **39**, 125 (1984).

[50] M. D. Shetlar, *Photochem. Photobiol. Rev.* **5**, 105 (1976).

[51] F. D. Miller, K. F. Jorgenson, R. J. Winkfein, J. H. van de Sande, D. A. Zarling, J. Stockton, and J. B. Rattner, *J. Biomol. Struct. Dyn.* **1**, 611 (1983).

[52] H. Castleman, L. H. Hanau, W. Zacharias, and B. F. Erlanger, *Nucleic Acids Res.* **16**, 3977 (1988).

the bound antibodies. By analysis of many plasmids, a number of potential sites for Z-DNA formation were located, even though only a few were in the Z-DNA conformation in any one plasmid.[52] In many cases, an antibody cross-links different sites in the same plasmid, resulting in the formation of loop structures.[52]

For one study,[51] 2 μg of plasmid DNA containing a Z-DNA-forming insert was incubated with 3 μg of purified antibody in a total volume of 40 μl of membrane-filtered 150 mM 4-morpholineethane sulfonate (MES), pH 7.4, for 30 min at room temperature. The mixture was chilled, and 100 μl of MES was added, followed by 16 μl of 10% formaldehyde. The 37% formaldehyde stock solution had been heated at 90° for 5 min before dilution in MES. The antigen–antibody–formaldehyde mixture was incubated at 4° for 15 min. Then 26 μl of fresh 4% glutaraldehyde was added, and the mixture was kept at 4° for 15 min more. The mixture was then passed through a 1 × 15 cm column of Sepharose 2B with 150 mM MES, pH 7.4. DNA–antibody complexes were obtained in the void volume.

For microscopy, mixtures were prepared with 50 μl of 99% formamide, 10 μl of 1 M Tris-HCl, 0.1 M EDTA, pH 8.5, 35 μl of a sample containing DNA at a concentration of 1–1.5 μg/ml, and 5 μl of a 1% aqueous solution of cytochrome c (Sigma type IV). A 25-μl sample of this mixture was spread on a water surface at the interface formed by a glass slide placed at a 45° angle to the surface. After 45 sec, the film that formed was picked up with a flat grid coated with Parlodion (Electron Microscopy Sciences, Fort Washington, PA) and carbon and immediately transferred into 95% ethanol for 30 sec. It was stained for 1 min with uranyl acetate (2.5 ml of a 1% aqueous solution of uranyl acetate diluted in 22.5 ml of 95% ethanol) and then washed again for 10 sec in 95% ethanol. The grid was dried and shadowed with platinum and palladium (80 : 20) from an angle of 10°.

NMR Spectroscopy of Antibody–Oligonucleotide Complexes

NMR spectroscopy can identify specific DNA sites that interact with antibodies.[53] Because the antibody is a large molecule the relaxation times of the nuclei in the protein and in a protein-bound oligonucleotide are short, with the result that resonances of bound antigen are broadened. The size of the antibody–DNA complex can be reduced by use of the monovalent Fab fragment of antibody. The Fab fragment is prepared by digestion of the antibody with papain,[54] and it is purified by affinity chromatography with a DNA column (described above) or by removal of intact antibody and Fc fragments with a protein A-Sepharose column.

[53] D. Sanford and B. D. Stollar, *J. Biol. Chem.* **265**, 18608 (1990).
[54] M. G. Mage, this series, Vol. 70, p. 142.

The line width of resonances in the Fab–oligonucleotide complex can be reduced further by incorporation of deuterium in place of protons in the antibody. This decreases the relaxation rate and simplifies the ^{1}H NMR spectrum by eliminating some of the proton resonances. Many exchangeable protons are replaced with deuterium by lyophilizing the samples several times from 99.9% D_2O and redissolving them in D_2O. Specifically deuterated amino acids can be incorporated into monoclonal antibody by growing the hybridoma cells in medium containing the desired deuterated amino acids.[55] We have prepared monoclonal anti-Z-DNA anti-bodies with fully deuterated aromatic amino acids, thereby eliminating the aromatic protein resonances from the ^{1}H NMR spectrum.[53]

Saturation transfer experiments were particularly useful for identifying resonances of DNA bound to antibody. When excess antigen is added to antibody, the resonances of the free antigen are narrow relative to those of the bound form and are more easily observed. Saturation of a proton of the bound form can result in transfer of saturation to the free form by chemical exchange. This reduces the intensity of the free form resonance, and the decrease can be observed in difference spectra. In early experiments, designed to develop techniques for defining antibody–DNA interactions by NMR spectroscopy, a monoclonal anti-GMP antibody was studied.[56] In a stoichiometric mixture of the antibody with GMP the guanine H8 resonance was markedly broadened. Addition of excess antigen resulted in a line corresponding to free GMP but with some exchange broadening. Systematic irradiation of the spectrum at various frequencies near the guanine H8 resonance resulted in the location of the bound resonance by saturation transfer to the free resonance.

In more recent experiments the binding of anti-Z-DNA antibodies to Z-form oligonucleotides was studied.[53] Irradiation at a frequency within the protein spectrum resulted in spin diffusion, which rapidly saturated the entire protein spectrum. This prepared a saturated antigen binding site. When oligonucleotide d(CG)$_3$ was bound to this site, cross-relaxation occurred between the saturated protein and protons on the DNA at points of close contact. Subsequent chemical exchange between the free and bound forms of d(CG)$_3$ resulted in saturation transfer to the free form, and the effect of cross-relaxation could be observed in the free DNA resonances. This experiment provided direct evidence that the anti-Z-DNA antibody, Z44, binds to the convex surface (corresponding to the

[55] J. Anglister, T. Frey, and H. M. McConnell, *Nature* (*London*) **315**, 65 (1985).
[56] B. D. Stollar, B.-S. Huang, and M. Blumenstein, *in* "Fourth Conversation in Biomolecular Stereodynamics" (R. H. Sarma, ed.), p. 69. Adenine Press, Guildersland, New York, 1985.

major groove) of Z-DNA, whereas a different antibody, Z22, does not interact with the convex surface.[53]

Applications

Characterization of Binding Sites on DNA

Synthetic oligonucleotides are useful for determining the structural features of certain antibody–DNA interactions, allowing a more precise description than can generally be obtained with polynucleotides or native DNA. The binding site in native DNA for an anti-native DNA antibody, H241 (a murine monoclonal autoantibody), was mapped by its binding to a series of oligonucleotides.[57] Various oligonucleotides were used as competitive inhibitors of the interaction with ^{3}H-labeled E. coli DNA in immunoprecipitation RIA. H241 shows a preference for alternating GC sequences in polynucleotides.[56] By competitive assays of a series of duplex oligonucleotides with the sequence d[ATATA(GC)$_n$TATAT], in which the (GC)$_n$ core varied in length, the minimum length required to form the binding site was determined to be 5–6 base pairs.

Points of H241–DNA interaction within the GCGCGCGC core were identified by comparison of reactions of chemically modified forms of d(ATATAGCGCGCGCTATAT). The modifications included replacement of one or all phosphate(s) with phosphorothiolate linkages and replacement of one or all guanine(s) by hypoxanthine, of cytosine(s) by 5-methylcytosine, or of the central guanine by 7-deazaguanine. Computer display of the sites where substitution reduced binding suggested that the binding site straddles the phosphate backbone, has its main interactions in the major groove, and has some interaction with the backbone and a guanine amino group in the minor groove.

Measuring Conformational Transitions

Antibodies that bind specifically to Z-DNA but not to B-DNA or single-stranded DNA[58] are useful for measuring the B–Z transition induced by negative supercoiling in closed circular DNA[45] or by increasing ionic strength.[59] An important consideration in this application is that the antibodies themselves can affect the transition, especially when high concentrations of high affinity antibody are used. The salt-induced B–Z transition

[57] B. D. Stollar, G. Zon, and R. W. Pastor, Proc. Natl. Acad. Sci. U.S.A. 83, 4469 (1986).
[58] E. M. Lafer, A. Möller, A. Nordheim, B. D. Stollar, and A. Rich, Proc. Natl. Acad. Sci. U.S.A. 78, 3546 (1981).
[59] E. M. Lafer, R. Sousa, A. Rich, and B. D. Stollar, J. Biol. Chem. 261, 6438 (1986).

for poly(dG-dC) occurs at 2.25 M NaCl. In a filter-binding RIA with radiolabeled poly(dG-dC), anti-Z-DNA antibody binding was essentially zero between 0.5 and 1.5 M NaCl, increased between 1.5 and 2.5 M, and leveled off above 2.5 M NaCl.[60] For several antibodies, the binding occurred at nearly the same salt concentration as the B–Z transition observed by circular dichroism (in the absence of antibody). In one case, however, the transition occurred at 2.0 M NaCl, significantly below the transition midpoint in the absence of antibody (2.25 M NaCl). Antibody binding shifted the B–Z equilibrium in favor of the Z conformation. The polyclonal antibody that produced this shift contained a larger population of high affinity antibodies than the other antibodies tested. Anti-Z-DNA antibodies also induced a shift of the B–Z equilibrium at very low salt concentrations and at modest degrees of plasmid supercoiling. In these studies, antibodies served as useful models of the stabilizing effects of Z-DNA-binding proteins. When antibodies are to be used as reporters of preexisting structures, however, they should be used at concentrations that do not affect the equilibrium between conformational forms.

Acknowledgments

Work in the authors' laboratory was supported by grants from the National Institutes of Health, currently GM32375, AI19794, and AI26450.

[60] E. M. Lafer, R. Sousa, and A. Rich, *EMBO J.* **4**, 3655 (1985).

[22] Topoisomer Gel Retardation: Protein Recognition of Torsional Stress-Induced DNA Conformations

By PETER DRÖGE and ALFRED NORDHEIM

Introduction

Negative torsional strain in a topologically underwound DNA can lead to the induction and stabilization of a variety of DNA conformations that deviate significantly from classic right-handed B-DNA,[1,2] which may include Z-DNA, H-DNA, DNA cruciforms, or locally denatured DNA

[1] A. Nordheim, R. E. Herrera, K. Meese, L. Runkel, P. M. Scholten, H. Schröter, and P. E. Shaw, *in* "Architecture of Eukaryotic Genes" (G. Kahl, ed.), p. 3. VCH, Weinheim, Germany, 1988.
[2] R. D. Wells, *J. Biol. Chem.* **263**, 1095 (1988).

segments.[3] Such local alterations in DNA conformation can represent recognition sites leading to specific protein–DNA interactions. To facilitate the study of such protein–DNA interactions, we developed an electrophoretic band shift assay, which employs the separation of small topoisomeric DNA circles from each other and from their respective complexes with proteins.[4] An equivalent approach was developed independently by Prunell and co-workers[5] for the study of nucleosome reconstitution.

Our technique, termed topoisomer gel retardation, has since proved useful for the study of interactions of anti-Z-DNA antibodies with left-handed Z-DNA,[4] reevaluation of binding specificities displayed by presumptive Z-DNA-binding proteins,[6] study of repressor-mediated DNA looping in the *lac* and *araBAD* operons,[7,8] as well as for study of the influence of DNA supercoiling on such looping phenomena.[8] In this chapter we describe the basic features of the gel retardation technique and demonstrate some applications, namely, the binding of anti-Z-DNA antibodies [both intact immunoglobulin (Ig) molecules and Fab fragments] to Z-DNA, the identification of HeLa nuclear extract activities with an apparent affinity for structural discontinuities associated with B–Z junctions, and the association of *Escherichia coli* single-strand-specific DNA-binding protein with negatively $(-)$ supercoiled minicircles containing stretches of Z-DNA.

Materials and Methods

Proteins

Monoclonal anti-Z-DNA antibodies Z22[9] and Z-D11[10] are prepared according to standard procedures using protein A affinity chromatography. For the generation of Z22 Fab fragments, papain cleavage is employed as described followed by protein A chromatography.[11] Extracts from HeLa

---

[3] M. D. Frank-Kamenetskii, *in* "DNA Topology and Its Biological Effects" (N. R. Cozzarelli and J. C. Wang, eds.), p. 185. Cold Spring Harbor Laboratory, Cold Spring Harbor, New York, 1990.

[4] A. Nordheim and K. Meese, *Nucleic Acids Res.* **16**, 21 (1988).

[5] I. Goulet, Y. Zivanovic, A. Prunell, and B. Revet, *J. Mol. Biol.* **200**, 253 (1988).

[6] K. J. Rohner, R. Hobi, and C. C. Kuenzle, *J. Biol. Chem.* **265**, 19112 (1990).

[7] R. B. Lobell and R. F. Schleif, *Science* **250**, 528 (1990).

[8] H. Krämer, M. Amouyal, A. Nordheim, and B. Müller-Hill, *EMBO J.* **7**, 547 (1988).

[9] A. Möller, J. E. Gabriels, E. M. Lafer, A. Nordheim, A. Rich, and B. D. Stollar, *J. Biol. Chem.* **257**, 12081 (1982).

[10] R. Thomae, Ph.D Thesis, University of Konstanz, Germany.

[11] J. W. Goding, *in* "Monoclonal Antibodies: Principles and Practice." Academic Press, London, 1983.

nuclei are prepared using the chloroquine extraction procedure.[12] The extract is passed over a heparin column, and bound proteins are eluted with a KCl gradient. The identified binding activity to supercoiled minicircles elutes at a salt concentration of 300 to 500 mM KCl ($E. coli$ single-strand-specific DNA-binding protein was a gift from K. Geider, Max-Planck-Institute, Heidelberg, FRG.) Restriction enzymes and other DNA modifying enzymes were purchased from Boehringer Mannheim.

Available Plasmid Constructs for Generation of Minicircles

A schematic presentation of the available substrates to test for Z-DNA binding of proteins along with the respective control DNAs is shown in Fig. 1. Briefly, plasmid pBR322 and its d(CG)$_n$-containing derivatives pLP14, pLP24, and pLP32 have been described elsewhere[4] (see Fig. 1A). pAN700 is a derivative of pUC7, which has inserted at its polylinker $Hinc$II site a pUC18–PvuII fragment extending from pUC18 coordinates 302 to 628.[4] Plasmids pAN701 to 704 (Fig. 1B) are all derived from pUC7 by inserting into its $Hinc$II site the d(CG)$_n$-containing PvuII fragments of plasmids pAN014 [d(CG)$_7$], pFP316 [d(CG)$_8$], pAN022 [d(CG)$_{11}$], and pUC32 [d(CG)$_{16}$], respectively. pAN500 was obtained by placing the Taq fragment of pBR322 into the blunt-ended SmaI site of pUC18M. pUC18M was constructed by insertion of an EcoRI linker into the blunt-ended HindIII site of pUC18, thereby generating symmetrical EcoRI sites at both ends of the polylinker (a gift from M. Caserta, "La Sapienza," Rome). pAN501, pAN502, and pAN503 (Fig. 1C) were obtained by inserting the d(CG)$_7$-, d(CG)$_{12}$-, and d(CG)$_{16}$-containing TaqI fragment of pLP14, pLP24 and pLP36, respectively, into the blunt-ended SmaI site of pUC18M. Plasmid constructions and purifications are performed according to standard cloning procedures.

Generation of DNA Minicircles

TaqI as well as EcoRI restriction fragments are dephosphorylated with calf intestinal alkaline phosphatase, phenol extracted twice, and purified by gel electrophoresis and electroelution out of the gel (both agarose as well as polyacrylamide gels may be used and give comparable results). Purified restriction fragments are labeled with [γ-^{32}P]ATP (Amersham), using polynucleotide kinase. Labeled fragments are ligated at low DNA concentrations (not exceeding 1 μg/ml in a total reaction volume of 250 μl) into unit length minicircles in a buffer consisting of 30 mM Tris-HCl (pH 7.5), 10 mM dithiothreitol (DTT), 7 mM MgCl$_2$, and 1 mM ATP.

[12] H. Schröter, P. E. Shaw, and A. Nordheim, *Nucleic Acids Res.* **14**, 10145 (1987).

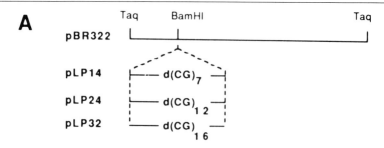

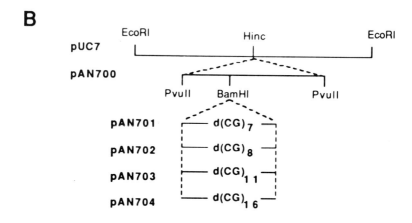

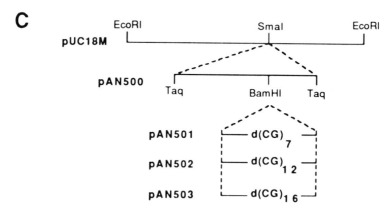

FIG. 1. Plasmid constructs for the generation of minicircles. A description of the available plasmid constructs is given in the text. The fragments used for minicircle ligation are approximately 360 base pairs in length.

Ligation reactions containing various amounts of ethidium bromide (as indicated in the figure legends) are usually performed at 4°. It should be mentioned that large amounts of ethidium bromide can significantly reduce the ligation efficiency. Reactions are stopped by phenol–chloroform extractions, and the DNA is precipitated with ethanol and subsequently resuspended in TE buffer [10 mM Tris-HCl, pH 7.5; 1 mM ethylenediaminetetraacetic acid (EDTA)].

Protein Binding to Topoisomers

Binding reactions are performed in 10- to 20-μl reaction volumes, usually containing between 0.1 and 1.0 pg of radiolabeled minicircle DNA and protein in the concentration range given in the respective figure legends. The reaction buffer contains 20 mM Tris-HCl (pH 7.5) and NaCl in the range of 50 to 100 mM. Reactions are performed both on ice or at room temperature as indicated for 15 to 30 min. Prior to the loading of the samples, 3 μl of a dye solution containing 15% (w/v) Ficoll and 0.5% (w/v) bromphenol blue is added to the reaction mixtures.

Gel Electrophoresis

Electrophoretic separation of topoisomers is achieved in 4% polyacrylamide (29 : 1) gels, buffered with 0.5-fold TBE (89 mM Tris–borate, pH 8.3; 2.5 mM EDTA), that are run in the cold room at 4° for about 4 to 6 hr at 30 mA. The gels are exhaustively electrophoresed prior to the loading of the samples. We also use polyacrylamide–agarose (2%/0.5%) (w/v) composite gels (Figs. 2 to 4) that are run under identical conditions. These gels are easier to handle but do not give as good a separation of topoisomers as obtained with pure polyacrylamide gels.

Results

Electrophoretic Separation of Negatively Supercoiled DNA Minicircles

As an example for the generation of negatively supercoiled minicircles, we show in Fig. 2A the result of a ligation reaction using the *Eco*RI fragment of pAN500, which does not contain a d(CG)$_n$ insert (see Fig. 1C). The DNA was ligated in the presence of increasing amounts of ethidium bromide, and, after removal of the intercalating drug by phenol extraction, the ligated minicircles show different electrophoretic mobilities. This is due to the negative superhelical tension introduced into the DNA in dependency of the amount of drug added. Ligation in the absence of ethidium bromide (lane 1, Fig. 2A) results in the formation of form IV DNA,

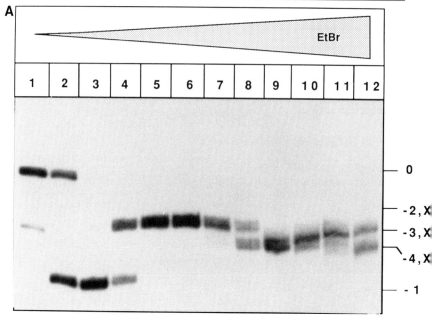

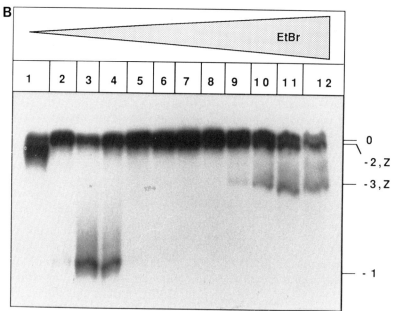

that is, covalently closed unconstrained DNA (0). We did not attempt to identify the species in lane 1 (Fig. 2A) which migrates faster than form IV DNA close to a position marked as (-2, X), but this form is likely to represent positively supercoiled DNA containing one superhelical twist.

With increasing amounts of drug, the relaxed form IV DNA is converted to a faster migrating species, most likely characterized by one negative superhelical turn (-1) (lanes 2 and 3, Fig. 2A). Further increase in the amount of drug added results in a reduction of electrophoretic mobility indicative of a structural transition driven by an additional ($-$) superhelical twist (-2, X) (lanes 4 to 7, Fig. 2A). This was unexpected because AN500 does not contain long stretches of alternating purines–pyrimidines, nor does it contain long palindromic sequences that would indicate a potential for the extrusion of a cruciform. We shall show below by antibody binding that the observed shift in electrophoretic mobility is most likely due to a B $\rightarrow$ Z transition. Further increase in negative supercoiling of AN500 finally results in two additional species with increasing electrophoretic mobilities, namely, (-3, X) and (-4, X) (lanes 8 to 12, Fig. 2A). (For a titration of ethidium bromide in the ligation reaction see also figures in Nordheim and Meese.[4])

An example of the construction of negatively supercoiled minicircles containing a $d(CG)_{16}$ insert placed within the pUC18M polylinker (pAN503; Fig. 1C) is given in Fig. 2B. As observed with AN500 in the absence or presence of low concentrations of ethidium bromide, the DNA migrates as form IV DNA (0) or as negatively supercoiled topoisomer

---

FIG. 2. DNA minicircles generated by circularization in the presence of increasing amounts of ethidium bromide. (A) Minicircles derived from *Taq* fragments of pAN500, which lacks a stretch of alternating $d(CG)_n$, are generated by ligation in the presence of 0 to 25 μM ethidium bromide as indicated by the shaded arrow above the gel lanes. The resulting minicircles are topologically underwound proportionally to the amount of ethidium bromide present in the reaction mixture, whereby in lane 1 the DNA is ligated in the absence of the drug and appears to be mainly unconstrained (0). The underwound state becomes apparent by a change in the electrophoretic mobility, which is likely to be due to the compensating changes in negative writhe. Thus, we mark these changes as -1, -2, and so on, to indicate the introduction of superhelical twists into the minicircles. The index X to the numbers indicates an unidentified DNA conformational change within the minicircles that alters the electrophoretic mobility. Topoisomers are displayed by electrophoresis using an agarose–polyacrylamide composite gel (0.5%/2%) and are visualized by autoradiography. (B) Minicircles derived from *Taq* fragments of pAN503 containing a $d(CG)_{16}$ insert. Conditions for the generation of minicircles are identical to those given for (A). The index Z indicates a change in the $d(CG)_{16}$ stretch from the right-handed B conformation to the left-handed Z conformation. The sample analyzed in lane 1 was ligated in the absence of ethidium bromide, and electrophoresis was as in (A).

containing one superhelical twist (-1), respectively (lanes 1 to 4, Fig. 2B). With increasing amounts of drug, the minicircles show a reduced electrophoretic mobility, now comigrating at a position identical to that of form IV DNA (lanes 5 to 9, Fig. 2B). We mark this species as (-2, Z) indicative for topoisomer -2 in which a structural transition in the d(CG)$_{16}$ stretch from the right-handed B to the left-handed Z conformation has occurred (see below). The end point of the ethidium bromide titration in this experiment leads to a topoisomer species with an increased mobility, marked as (-3, Z), which most likely contains one negative superhelical twist in addition to the stretch of left-handed DNA (lanes 10 to 12, Fig. 2B).

Differential Binding of Anti-Z-DNA Antibodies or Fab Fragments to Negatively Supercoiled Minicircles

To test whether the observed structural transition in AN500 minicircles (Fig. 2A) represents a B → Z conformational change, we generated a mixture of different topoisomers and incubated it with increasing amounts of a well-characterized monoclonal antibody, designated Z-D11.[10] This antibody is specifically directed against stretches of alternating d(CG)$_n$ in the left-handed conformation and exhibits an unusually high binding affinity for this alternate DNA conformation.[10,13] It can be seen in Fig. 3 that the antibody at high concentrations preferentially binds to and thus retards the electrophoretic mobility of topoisomer species from (-2, Z) to (-4, Z). This result indicates that in pAN500 minicircles left-handed DNA is indeed induced by ($-$) torsional strain.

When the same range of antibody concentration was tested with topoisomers (-2, Z) and (-3, Z) of AN503 minicircles, it becomes evident (Fig. 4) that the antibody exhibits at least a 40-fold greater affinity to the d(CG)$_{16}$ stretch in the left-handed conformation than to the unspecified left-handed conformation detected in AN500. The species which is not recognized and bound by the antibody is most likely unconstrained circular DNA with at least one broken phosphodiester bound in one of the two polynucleotide chains (0) (Fig. 4, lanes 1 to 7).

The high specificity of monoclonal antibody (MAb) Z-D11 for left-handed d(CG)$_n$ stretches is further demonstrated in Fig. 5. When both form IV DNA (0) and topoisomer (-1) from AN503 is incubated with MAb Z-D11. Binding to topoisomer (-1) can only be detected at high antibody concentrations, whereas form IV DNA is not bound at all, except at very high protein concentrations when nonspecific aggregation is ob-

[13] F. M. Pohl, *Proc. Natl. Acad. Sci. U.S.A.* **83**, 4983 (1986).

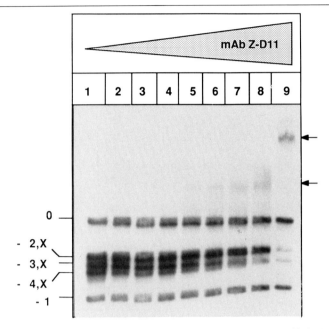

FIG. 3. Binding of monoclonal antibody Z-D11 to negatively supercoiled AN500 minicircles. Increasing amounts (0.003 to 1 μg) of monoclonal antibody (MAb) Z-D11 was incubated with a mixture of AN500 topoisomers for 15 min at room temperature, whereby the sample in lane 1 served as a control and was incubated in the absence of Z-D11. Protein binding is analyzed by electrophoresis through an agarose–polyacrylamide composite gel as before. The positions of the resulting nucleoprotein complexes are indicated by the arrows to the right of the autoradiogram. Note that the antibody binds first to the topoisomers with the highest degree of (−) supercoiling (−4, X). Abbreviations and symbols used are as in Fig. 2A.

served (lane 8). We interpret this result to reveal the antibody shifting the B–Z equilibrium of the d(CG)$_{16}$ stretch in topoisomer (− 1) in favor of the left-handed form, which is likely to be due to the intrinsic Z-DNA-stabilizing activity of the antibody.[13,14]

A qualitatively similar result presented in Fig. 6 was obtained with Z22 Fab fragments. In this case, protein binding was assayed with minicircles containing a stretch of d(CG)$_7$ (derived from pAN701). As observed before with monoclonal Z-D11, preferential binding to topoisomers (− 2, Z) (lanes 7 to 14, Fig. 6) occurs at much lower protein concentrations than an interaction with topoisomers (− 1) (lanes 12 and 13, Fig. 6). Again, in the

[14] F. M. Pohl, *Biophys. Chem.* **26**, 385 (1987).

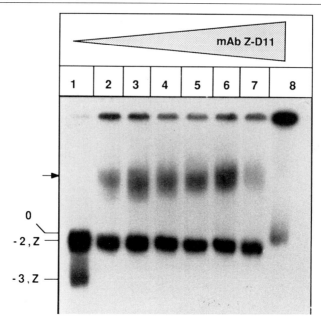

FIG. 4. Binding of monoclonal Z-D11 to Z-DNA-containing minicircles. A mixture of topoisomers (-2, Z) and (-3, Z) derived from pAN503 was incubated with increasing amounts of monoclonal antibody Z-D11 under conditions identical to those stated in Fig. 3 (lane 1, no antibody). After binding, the samples were analyzed using polyacrylamide gels (4%). The arrow at left points to the position of the specific nucleoprotein complexes formed between the antibody and topoisomers (-2, Z) and (-3,Z). Note, however, that at high protein concentrations (100 μg/ml), the DNA tends to aggregate and is prevented from entering the gel matrix (lane 8). Abbreviations and symbols used are as in Fig. 3.

latter case the equilibrium between the two DNA conformations appears to be shifted toward the left-handed conformation on protein binding.

Detection of Topoisomer-Binding Activity in Extracts from HeLa Cells

The results presented above demonstrate that the topoisomer gel retardation assay can be successfully used to test for the presence of Z-DNA-binding proteins. This encouraged us to search for such an activity in HeLa cell nuclear extracts, fractionated over a heparin column. As a substrate, we again used minicircles derived from pAN701 in which the d(CG)$_7$ stretch flips from the right-handed to the left-handed conformation when more than one ($-$) superhelical twist is introduced. It is evident in Fig. 7 that binding activities are present in these extracts that specifically interact with topoisomers (-2, Z) and (-3, Z) (lanes 9 to 20, Fig. 7). This

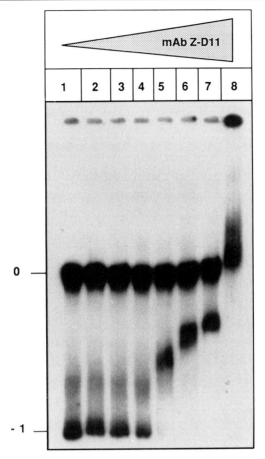

Fig. 5. The presence of monoclonal antibody Z-D11 shifts the B–Z equilibrium in topoisomer (-1) toward the left-handed conformation. A mixture of topoisomers (0) and (-1) derived from pAN503 was incubated with increasing amounts of monoclonal Z-D11 (lane 1, no antibody) as described in Fig. 3. Protein binding was monitored by polyacrylamide (4%) gel electrophoresis. Note again that at high antibody concentrations nonspecific DNA aggregation occurs (lane 8).

activity, on the other hand, neither binds to form IV (0) nor topoisomer (-1), at least in the concentration range tested here (Fig. 7, A, B, and E). These results indicate that a factor(s) exists in the extract exhibiting an increased affinity for a DNA structural discontinuity associated with the B $\rightarrow$ Z transition.

To test whether this factor specifically interacts with the stretch of left-handed DNA in AN701, binding to topoisomer (-3, Z) was competed by

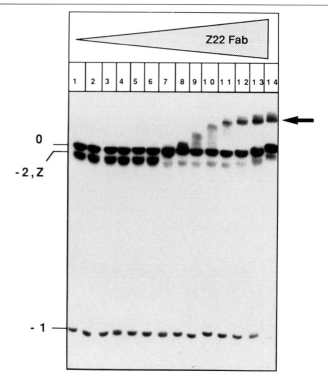

Fig. 6. Binding of Fab fragments to Z-DNA. Increasing amounts (0 to 20 μg/ml) of Fab fragments derived from monoclonal anti-Z-DNA antibody Z22 were incubated with a mixture of different topoisomers [(0), (-1), and (-2, Z)] which was obtained by ligation of the EcoRI fragment of pAN701. After a 15-min incubation on ice, protein binding was analyzed on 4% polyacrylamide gels (lane 1, no protein added). The arrow at right points to the position of specific nucleoprotein complexes between topoisomer (-2, Z) and the Fab fragment. Also detectable is an interaction of Fab fragments with topoisomer (-1) at high protein concentrations (lanes 13 and 14).

the addition of increasing amounts of either single-stranded or double-stranded DNA. Our preliminary results indicate that the presence of single-stranded DNA prevents the factor(s) from binding to topoisomer (-3, Z) more efficiently than negatively supercoiled B-DNA, or stretches of Z-DNA present in a ($-$) supercoiled plasmid (data not shown). We conclude that the HeLa factor(s), which preferentially binds to topoisomers containing the stretch of $d(CG)_7$ in the left-handed conformation, most likely recognizes Z-DNA-associated regions with single-stranded character, possibly B–Z junctions.

To investigate directly whether regions with single-stranded character

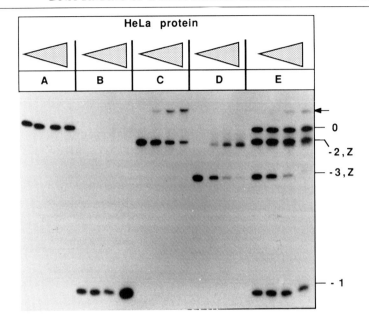

FIG. 7. Binding of HeLa protein(s) to Z-DNA-containing pAN701-derived minicircles. Increasing amounts of heparin-fractionated HeLa cell nuclear extracts were incubated with either (A) unconstrained topoisomer (0); (B) topoisomer (-1); (C) topoisomer (-2, Z); (D) topoisomer (-3, Z); or (E) an equimolar mixture of all four topoisomers. After 20 min on ice, protein binding was monitored by polyacrylamide gel electrophoresis and autoradiography. The arrow at right points to the position of the (-2, Z) topoisomer–protein complex within the gel. Under the conditions used for electrophoresis, the (-3, Z) topoisomer–protein complex migrates at a position identical to that of topoisomer (-2, Z), and thus the two species are not resolved in (E).

are present in higher negatively supercoiled minicircles of AN701, we tested a well-characterized single-strand-specific DNA-binding protein (SSB) from *E. coli*. Whereas neither form IV DNA (0) nor topoisomer (-1) are bound by the *E. coli* protein (Fig. 8), an interaction is indeed detectable with topoisomer (-3, Z). It is interesting to note that the apparent binding of this SSB protein results in an increased electrophoretic mobility of the nucleoprotein complexes (indicated by the arrow on the left-hand side of the autoradiogram in Fig. 8), rather than in an electrophoretic retardation as observed with the HeLa factor in Fig. 7. We have presently no explanation for this effect, but it is tempting to speculate that the protein–DNA interaction induces conformational changes in the minicircles which in turn alter the electrophoretic mobility. However, the result supports our previous conclusion that the identified nuclear HeLa

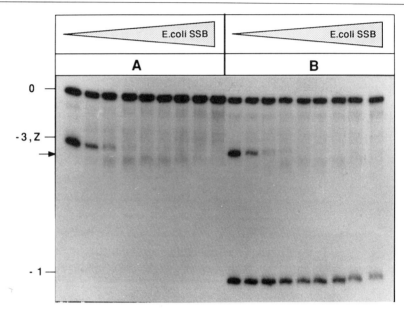

Fig. 8. Binding of *E. coli* single-strand DNA-binding protein to negatively supercoiled minicircles. Increasing amounts of *E. coli* SSB (0 to 100 μg/ml) were incubated with a mixture of AN701 topoisomers (0) and (−3, Z) (A), or (0), (−1), and (−3, Z) (B) for 20 min on ice. Binding was analyzed as described for Fig. 7. The arrow at the left-hand side of the autoradiogram points to the position of the (−3, Z) protein complex, which migrates slightly faster than the protein-free topoisomer. The DNA displayed in the left-hand lanes of (A) and (B) was incubated in the absence of protein.

factor(s) is indeed a DNA-binding protein that shows an affinity for DNA conformations with single-stranded character.

Discussion and Comments

Protein Binding to Stress-Induced DNA Structures

We describe above the use of the topoisomer gel retardation technique to analyze protein–Z-DNA interactions. We demonstrate recognition of Z-DNA by the specific binding of monoclonal antibodies or their Fab fragments to stretches of d(CG)$_n$ in the left-handed conformation.

In principle, our test system can be employed to identify the specific binding of proteins to any stress-induced alternate DNA conformation. The system is particularly useful because two different DNA conformations of identical nucleotide sequences, namely, B- and Z-DNA, can be analyzed in direct competition by offering different topoisomers simultaneously in one binding reaction. Thus, preferential binding of a protein to topoisomers that contain the stretch of d(CG)$_n$ in a left-handed conformation is a good first indication that the protein may indeed interact with the alternate DNA conformation. However, it is strongly recommended to test further for the specificity of protein binding by competition experiments as described earlier for both the monoclonal antibody Z-D11[10] and the Z22 Fab fragment[9] examined in this chapter. In so doing, a nuclear HeLa protein fraction which interacts preferentially with a subset of topoisomers [(-2, Z) and (-3, Z); Fig. 7] was found to be efficiently competed by single-stranded DNA. Intriguingly, *E. coli* SSB was also found to interact specifically with Z-DNA containing topoisomers (Fig. 8). Thus, it is very likely that the HeLa factor recognizes and binds to stretches of DNA with single-stranded character which have been identified previously to be located at the B–Z junctions.[15]

We want to point out that high protein concentrations in the binding reactions can lead to nonspecific protein–DNA aggregation (Fig. 4, lane 8; Fig. 5, lane 8). This could falsely be taken as evidence for a specific protein–Z-DNA interaction. Thus, careful titration of the protein under investigation and the inclusion of different competitor DNA types appear to be important. Furthermore, comparison to a well-characterized standard protein (e.g., monoclonal anti-Z-DNA antibodies) is strongly recommended in order to obtain conclusive results.

Finally, another cautionary note should be added to ensure correct interpretation of binding data. From the results presented in Fig. 2A, we strongly advise checking by ethidium bromide titration for structural transitions in the respective parent minicircles which lack the DNA segment with a potential for undergoing a supercoiled-driven structural transition as present in the "test" minicircles. Our observation that MAb Z-D11 specifically interacts with AN500 [topoisomers (-2, X) to (-4, Z); Fig. 3] indicates that the observed conformational change has indeed created a left-handed helical structure. By comparing the relative antibody affinities to either a stretch of left-handed d(CG)$_{16}$ or the left-handed structure composed of an as yet unidentified sequence, it becomes evident that the antibody exhibits at least a 40-fold higher affinity for the d(CG)$_{16}$ stretch (compare Figs. 3 and 4). Because the antibody was specifically raised and selected for high affinity binding to d(CG)$_n$ in the left-handed conformation,[10] this result is not unexpected. From previous comparative experiments,[4] we can localize the sequence which is likely to be responsible for

[15] T. Kohwi-Shigematsu, T. Manes, and Y. Kohwi, *Proc. Natl. Acad. Sci. U.S.A.* **84**, 2223 (1987).

the observed transition in the pUC18M polylinker region of pAN500. The sequence that appears most likely to have the potential to undergo a B → Z transition at a superhelical density of about -0.06 [topoisomer (-2, Z); Figs. 2A and 4] is a stretch of seven alternating purine and pyrimidine residues, which make up the recognition sequence for the restriction enzyme *Sph*I.

DNA Structural Transitions in Supercoiled Minicircles

The introduction of negative supercoiling into DNA minicircles was used to induce B → Z transitions in stretches of alternating purine and pyrimidine residues.We showed that the structural transition in both d(CG)$_7$ and d(CG)$_{16}$ occurs when the minicircles become topologically underwound by more than one helical twist, that is, when the superhelical density of the minicircles is increased from about -0.03 to -0.06. This is in agreement with previous observations where the minimal energetic requirements for B → Z transitions as a function of length of the d(CG)$_n$ inserts were examined, although plasmids several kilobase pairs in length were examined in these studies.[16]

We noticed that, after the supercoiling-driven conformational change has occurred in both the d(CG)$_7$ and d(CG)$_{16}$ stretches, the minicircles migrate at a position identical or close to that of form IV DNA, indicating that no torsional strain remains in the minicircles (Figs. 2B and 6). It appears, therefore, that although the d(CG)$_n$ stretches differ in length, two negative superhelical twists have been released due to the conformational change in both cases. Because of the length difference, however, one would expect that about two ($-$) supercoils are released when the entire d(CG)$_7$ stretch flips into the left-handed conformation, whereas about five ($-$) supercoils are expected to be relaxed concomitant with the transition of the entire d(CG)$_{16}$ stretch. Thus, in the latter case, introduction of three ($+$) supercoils into the minicircles would be expected when the entire purine–pyrimidine stretch has flipped, leading to an increased electrophoretic mobility [note, e.g., that the mobility of topoisomer (-3, Z) is increased in comparison to that of topoisomer (-2, Z); Figs. 2B and 7]. As the introduction of three ($+$) supercoils into the minicircles is likely to be prohibited on energetic grounds, the most reasonable explanation for the two comigrating Z-DNA-containing topoisomers, AN701 (-2, Z) and AN503 (-2, Z), is that only a portion (about one-half) of the alternating purine–pyrimidine stretch in d(CG)$_{16}$ is changing its conformation to the left-handed form at the expense of two ($-$) superhelical turns. Such a

[16] L. J. Peck and J. C. Wang, *Proc. Natl. Acad. Sci. U.S.A.* **80,** 6206 (1983).

partial flipping in longer stretches of alternating purine–pyrimidine residues has been previously observed at threshold negative superhelical densities.[17] In our case with AN500, a second transition step is therefore expected when the energy content of the minicircles is increased by the introduction of additional (−) supercoils, that is, with topoisomer (− 4, Z). However, in order to understand the described phenomena in more detail, chemical footprinting of the Z-stretch in different topoisomers, for example, is necessary.

Acknowledgments

We thank H. Schröter and K. Geider for protein samples and K. Meese for technical assistance. We are grateful to F. M. Pohl, University of Konstanz, for an ample supply of monoclonal Z-D11. The gift of plasmid pUC18M by M. Caserta is acknowledged. This work was funded by grants from the BMFT (BCT-0381-5 and Forschungsschwerpunkt "Grundlagen der Bioprozesstechnik") and by the Fonds der Chemischen Industrie.

[17] B. H. Johnston, W. Ohara, and A. Rich, *J. Biol. Chem.* **263**, 4512 (1988).

[23] Algorithms for Prediction of Histone Octamer Binding Sites

By WILLIAM G. TURNELL and ANDREW A. TRAVERS

Nucleosome Positioning

The complex of the histone octamer with 145 base pairs (bp) of DNA, the core nucleosome particle, can be precisely positioned with respect to a given DNA sequence, yet, paradoxically, the octamer can associate with an immense variety of DNA sequences, both *in vivo* and *in vitro*. The resolution of this apparent paradox was first proposed by Trifonov and Sussman,[1] who suggested that the major determinant of nucleosome positioning was the bendability, or anisotropic flexibility, of DNA. Thus a DNA of particular sequence would bend most easily in a certain direction and so assume a preferred configuration. Subsequently it has been demonstrated that the direction of curvature of both a natural octamer binding site and a bacterial DNA sequence that has not been subject to the selective constraints of nucleosome formation is the same when the respective DNA molecules are bent in the absence of any protein and also when they are

[1] E. N. Trifonov and J. L. Sussman, *Proc. Natl. Acad. Sci. U.S.A.* **77**, 3816 (1980).

reconstituted into core nucleosome particles.[2] The available evidence is thus wholly consistent with the notion that bendability is a determinant of nucleosome positioning.

The anisotropic flexibility of a DNA molecule is a sequence-dependent property that is determined by the physicochemical characteristics of individual base steps. It is thus an intrinsic property of the DNA. This means that although precise binding sites for the histone octamer can be determined, such sites cannot be described or defined in terms of a consensus sequence of particular nucleotides such as would characterize the binding site of a sequence-specific DNA-binding protein. Instead, the positioning of a histone octamer on a particular DNA sequence can be described in terms of two parameters, rotational, which determines the orientation of the DNA molecule relative to the protein surface, or more accurately relative to the direction of curvature, and translational, which determines the preferred precise binding site of the histone octamer.[2]

The average sequence organization of nucleosome core DNA has been derived from a collation of 177 sequences of cloned DNA molecules isolated from chicken erythrocyte core nucleosomes.[3] These sequences were aligned on and averaged about their midpoints. From this alignment it was shown that for these sequences the probability of occurrence of particular base steps at equivalent positions is nonrandom. Moreover, the positions of preferential occurrence of any such base steps occur periodically with an average period repeat over the entire octamer binding site of 10.2 bp. This number is close to the helical periodicity of DNA and arises because the wrapped DNA duplex tends to present the same side of the double helix toward the protein core of the nucleosome. It was consequentially inferred that these preferred occurrences represented the sequence dependence of the smooth anisotropic bending of DNA. In particular short A/T-rich sequences are preferred where the minor groove points in toward the histone octamer, and conversely, where the minor groove points out, short G/C-rich sequences are preferred.[4]

These sequence preferences and the periodicity of their occurrence constitute the essential determinants of rotational positioning. The sequence periodicity directly reflects the relative or local twist of the DNA on the surface of the protein, that is, it is the number of base pairs between, for example, successive outward facing minor grooves and is thus a property of the geometry of the complex. This parameter should be clearly distinguished from the intrinsic twist of the DNA molecule. The latter

[2] H. R. Drew and A. A. Travers, *J. Mol. Biol.* **186,** 773 (1985).
[3] S. C. Satchwell, H. R. Drew, and A. A. Travers, *J. Mol. Biol.* **191,** 659 (1986).
[4] A. A. Travers and A. Klug, *Philos. Trans. R. Soc. London B* **317,** 537 (1987).

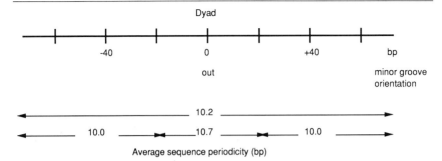

FIG. 1. Sequence organization of nucleosome core DNA. The rotational sequence period-icities are from Travers and Klug.[4]

quantity is a measure of the average rotation between successive base pairs measured about the local axis of the DNA double helix.[2,4,5]

An essential characteristic of rotational signals is their regular periodic nature. By contrast translational signals would be expected to identify unique structural features in the path of the double helix around the histone octamer. It has been demonstrated experimentally that within the octamer binding site the major translational determinants are confined to the four central double-helical turns.[6,7] Within this region distinct sequence features have been identified. The sequence "dyad" itself has a distinct pattern of sequence preferences,[8] whereas other features are asymmetrically arranged with respect to the midpoint of the binding site—the sequence dyad—and include an enhanced probability of occurrence of "flexi-DNA" sequences,[9] together with significant variations in the local sequence periodicity (Fig. 1). To date the significance of this sequence organization has not been tested experimentally.

It should be emphasized that the general utility of positioning algorithms, particularly with respect to the prediction of "natural" positions, is potentially dependent on the provenance of the basis set of sequences from which the general sequence organization has been deduced. For example, the core nucleosomal DNA analyzed by Satchwell et al.[3] was isolated from core nucleosomes that were positioned under conditions of relatively high ionic strength. Consequently it cannot be assumed a priori that the parameters of periodicity and sequence preference which are

[5] J. H. White and W. R. Bauer, Cell (Cambridge, Mass.) **55,** 9 (1989).
[6] P. C. Fitzgerald and R. T. Simpson, J. Biol. Chem. **260,** 15318 (1985).
[7] N. Ramsay, J. Mol. Biol. **189,** 179 (1986).
[8] W. G. Turnell, S. C. Satchwell, and A. A. Travers, FEBS Lett. **232,** 263 (1988).
[9] S. C. Satchwell and A. A. Travers, EMBO J. **8,** 229 (1989).

characteristic of this set of DNA molecules are also characteristic of nucleosomes assembled *in vivo,* although they should, and do, predict accurately the positions of nucleosomes reconstituted under similar conditions *in vitro.*[10] Nevertheless the dominant sequence periodicity (v = 1/10.2), and therefore the structural periodicity of the DNA contained in these particles, is identical to the periodicity of the probability of pyridimine dimer formation of nucleosomes in intact nuclei.[11,12] Because this latter parameter is also structure-dependent it is reasonable, at least to a first approximation, to assume that the rotational signals are conserved.

Algorithms for Nucleosome Positioning

Methods for the prediction of preferred binding sites for the histone octamer on a defined DNA sequence are of two types. In statistically based algorithms the tested sequences are compared against the sequence organization of experimentally determined binding sites.[13,14] By contrast, for structurally based algorithms the preferred configuration of a sequence is calculated on the basis of known and inferred structures of individual base steps.[15,16] This approach is essentially a calculation of preferred curvature.

The direction of the local axis of the double helix will change whenever the planes of adjacent base pairs are inclined relative to each other (Fig. 2). The assignment of this principal component of curvature—the "wedge" angle of a base step—depends heavily on the limited number of available crystal structures of short DNA oligomers. This wedge angle has two components, a relative roll, ρ, and a relative tilt, τ. Of these two components the relative roll is generally substantially larger and energetically more favorable.[17] Therefore in some algorithms[15] τ is justifiably ignored. Where these parameters are not available from crystal structures they are either estimated from conformational energy calculations[16] or by reference to rotational sequence preferences.[15] An alternative and less direct approach is to calculate the wedge angles from the anomalous gel

[10] H. R. Drew, *J. Mol. Biol.* **219**, 391 (1991).

[11] J. M. Gale, K. A. Nissen, and M. J. Smerdon, *Proc. Natl. Acad. Sci. U.S.A.* **84**, 6644 (1987).

[12] J. R. Pehrson, *Proc. Natl. Acad. Sci. U.S.A.* **86**, 9149 (1989).

[13] H. R. Drew and C. R. Calladine, *J. Mol. Biol.* **195**, 143 (1987).

[14] B. Pina, M. Truss, H. Ohlenbusch, J. Postma, and M. Beato, *Nucleic Acids Res.* **18**, 6981 (1990).

[15] C. R. Calladine and H. R. Drew, *J. Mol. Biol.* **192**, 907 (1986).

[16] D. Boffelli, P. De Santis, A. Palleschi, and M. Savino, *Biophys. Chem.* **39**, 127 (1991).

[17] R. E. Dickerson, M. L. Kopka, and P. Pjura, *in* "Biological Macromolecules and Assemblies" (F. A. Jurnak and A. McPherson, eds.), Vol. 2, p. 37. Wiley, New York, 1985.

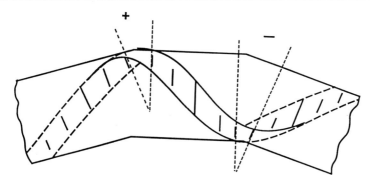

FIG. 2. Bending of the double-helical axis of DNA. The direction of the helical axis changes at positions where adjacent base pairs are inclined along their short axes relative to each other. This inclination or relative roll, ρ, can be toward either the major ($+$) or minor ($-$) groove. The path of the minor groove is outlined and shows the effect of positive and negative roll angles separated by half a double-helical turn separately producing a bend in the same direction.

mobilities of a variety of DNA fragments of different sequence.[18,19] These methods yield widely disparate values for particular individual base steps. However, it should be emphasized that for the calculation of the preferred configuration of a DNA molecule of mixed sequence the important parameter is not the absolute value for an individual base step but the relative differences between base steps. Put another way, the prefered direction of curvature of a particular sequence would be unchanged if the values of all the assigned roll angles were overestimated by, for example, a uniform $+ 10°$.

This approach necessarily assumes that the wedge angle is an intrinsic property of a base step and is largely independent of the sequence context. This assumption is clearly a simplification but is a useful approximation. A second general limitation of this approach is that only a single wedge value is assigned to each base step, whereas there is considerable evidence that certain base steps, notably YpR (where Y is a pyrimidine and R a purine) and also GpC, are bistable and can adopt alternative conformations with different wedge components.[20] However, because the rotational sequence preferences are redundant, in any 145-bp stretch of DNA of mixed sequence it is unlikely that there will be sufficient occurrences of a wrongly assigned wedge angle to influence the assignment of a rotational position

[18] C. R. Calladine, H. R. Drew, and M. J. McCall, *J. Mol. Biol.* **201,** 127 (1988).
[19] A. Bolshoy, P. McNamara, R. E. Harrington, and E. N. Trifonov, *Proc. Natl. Acad. Sci. U.S.A.* **88,** 2312 (1991).
[20] C. R. Calladine and H. R. Drew, *J. Mol. Biol.* **178,** 773 (1984).

TABLE I
BASES FOR ALGORITHMS

Algorithm	Basis	Comments
Calladine and Drew (1986)[a]	Vectorial addition of roll angles for each base step; constant periodicity, phase change at dyad	Roll angles derived from crystal structures
Boffelli et al. (1990)[b]	Vectorial addition of roll and tilt angles for each base step	Roll and tilt angles from energy
Drew and Calladine (1987)[c]	Probability of occurrence of base steps at given rotational orientation, constant periodicity; also translational parameters	Probability table calculated by reiterative procedure
Pina et al. (1990)[d]	Same probability table as Drew and Calladine (1987)[c]	Used to calculate rotational orientation for experimentally defined site

[a] C. R. Calladine and H. R. Drew, *J. Mol. Biol.* **192,** 907 (1986).
[b] D. Boffelli, P. De Santis, A. Palleschi, and M. Savino, *Biophys. Chem.* **39,** 127 (1991).
[c] H. R. Drew and C. R. Calladine, *J. Mol. Biol.* **195,** 143 (1987).
[d] B. Pina, M. Truss, H. Ohlenbusch, J. Postma, and M. Beato, *Nucleic Acids Res.* **18,** 6981 (1990).

significantly. In other words, in this respect the algorithms are relatively robust.

Statistical algorithms depend on the assignment of a rank order of the probability of occurrence of di- or trinucleotides at defined positions, usually corresponding to rotational orientations. In the simplest case the rank order for rotational signals is determined by the amplitude of the periodic modulation of occurrences in a set of aligned sequences.[3] Because the alignment of sequenced binding sites is inevitably imprecise, attempts have been made to refine the coefficients by reiterative procedures which realign individual sequences within the set by selecting for the strongest rotational signal with constant periodicity (e.g., Ref. 13). Such procedures may, however, obscure translational information. The bases of published algorithms are summarized in Table I.

Principles of Positioning Algorithms

If we describe the DNA as a sequence of steps between successive bases in one dimension, the predictions can be made either in real space or, via the Fourier transform, in reciprocal space. For example, Calladine and Drew[15] have shown that with a constant relative twist the corresponding roll angles, ρ, derived from structural models and from crystal structures determine the plane curve that a particular DNA sequence would adopt in space. If the left-handed supercoil of DNA around the histone octamer[21] is approximated to an idealized plane curve, then a fit between the curve of DNA generated by probable roll angles ρ adopted by successive base steps and the ideal curve may be calculated. The closeness of fit is then a measure of the likelihood of a particular segment of DNA adopting the idealized shape for wrapping around the histone core.

Calladine and Drew[15] calculated

$$fit_k = \sum_{j=1}^{W} [(\rho_I - \rho_S)^2]_j \qquad (1)$$

Here the measure of fit is determined for each position k along the sequence of a moving window of W base steps. The fit is derived from the differences between the ideal DNA curvature described by the roll angle ρ_I at each of the steps j and the curvature ρ_S accessible to the equivalent steps in the natural sequence. The squared differences are summed over W for each position k of W, and the minimal values of the sum give predicted core positions.

The predictions are thus calculated in real space. From this relatively straightforward derivation these predictions are largely consistent with known experimentally determined positions, but of limited accuracy. The inherent sequential and structural periodicity along the wrapped DNA of the nucleosome confers automatic advantages to a description of the sequence of base steps in one-dimensional reciprocal space.

Calladine and Drew[15] modeled their ideal plane curve for DNA wrapped around the histones as

$$[\rho_I]_j = (2 \times 4.5°) \cos(2\pi j/10.2) + 2° \qquad (2)$$

We may generalize this as

$$[\rho_I]_j = \$ \cos(2\pi j\nu) + [\overline{\rho_I}]_W \qquad (3)$$

Rearranging, we have

$$R \propto \frac{1}{\$} = ([\rho_I]_j - [\overline{\rho_I}]_W) \cos(2\pi j\nu) \qquad (4)$$

Equation (4) describes how the radius of curvature R of an ideally wrapped DNA segment is locally modulated by the helical periodicity of the duplex,

[21] T. J. Richmond, J. T. Finch, B. Rushton, D. Rhodes, and A. Klug, *Nature (London)* **311**, 532 (1984).

as shown in Fig. 2. The equation is the real part of a Fourier transform that describes the distribution of ρ_1 about a mean at intervals $1/\nu$ within a window of W base steps.

Expressing the path of DNA in this form enabled Calladine and Drew[15] to refine empirically their model set of roll angles ρ_1 against a basis set of 177 aligned sequences from chicken erythrocyte nucleosomes.[3] Incorporating this refined model into their real-space fitting algorithm, and adopting a window of 145 base pairs, the authors were able to successfully predict rotational core positions in 75% of the 177 sequences that made up the basis set. The results indicated how realistic was the idealized model generated by $[\rho_1]_j$ in Eqs. (2)–(4).

If the occurrence of a particular angle ρ_1 is strongly periodic with a frequency ν, then the intensity, I of the fluctuations in R will be relatively high:

$$I_k(\nu) = \left| \sum_{j=1}^{W} ([\rho_1]_j - \overline{[\rho_1]}_W \cos(2\pi j\nu)) \right|^2 \tag{5}$$

for each position k of the window W along the sequence. Equation (5) describes a power spectrum of which each component ν has an amplitude $[\rho_1]_j - [\rho_1]_W$ at positions j along the sequence, together with a phase $j\nu$. Structurally, $[I_k(\nu)]^{1/2}$ represents the magnitude of change in curvature of the DNA supercoil between successive positions $j \pm (1/\nu)$. The phase, $\phi = j\nu$, represents the relative direction of curvature at j for the Fourier component of the spectrum with frequency ν. Each component represents an idealized path through space of DNA whose sequence of bases has been rewritten as successive coefficients, $[\rho_1]_j$ assigned to base steps j having a periodicity $1/\nu$.

The idealized model for the path of DNA around the histone core can be made more realistic by allowing for the sigmoidal shape of the supercoil that occurs in the vicinity of the structural dyad of the nucleosome.[21] Only the first and last 60 base steps of the wrapped DNA are approximated by a plane curve generated by ρ_1 with $\nu = 1/10.2$ and a constant phase angle. When the sequence organization is symmetrized about the sequence dyad, approximately 10 base steps on either side of the dyad are about 180° out of phase with the remainder of the nucleosome supercoil.[15] Phase relationships between segments of the string of coefficients assigned to the DNA sequence can be incorporated into Eq. (5) as

$$I_k(\nu) = \left| \sum_{j=1}^{W} ([\rho_1]_j - \overline{[\rho_1]}_W \cos 2\pi([\phi + \Delta\phi]_j)) \right|^2 \tag{6}$$

where $\Delta\phi$ is a relative shift in phase $\phi = j\nu$ of component ν at position j.

The improvement in their model around the structural dyad enabled Drew and Calladine[13] to predict, with moderate success, the translational positions of nucleosome cores along a piece of frog DNA that had not formed part of the basis set used to refine ρ_1 empirically.

The one-dimensional Fourier transform provides a formulation for more generalized predications of DNA wraps.

For example, ρ_1 in Eq. (6) could be replaced by ρ_S and $[I_k(\nu)]_S$ compared with $[I_k(\nu)]_1$ to produce a calculation of fit in reciprocal space. Indeed, Eq. (6) may be generalized further:

$$I_k(\nu) = \left| \sum_{j=1}^{W} ([C_{M,S}]_j - \overline{[C_{M,S}]_W}) \exp\{2\pi i\,[(j\nu) + [\Delta\phi]_j]\} \right|^2 \qquad (7)$$

where $C_{M,S}$ are coefficients ascribed to base steps in either a model, M, or a sequence, S. The coefficients C_M need not bear a structural interpretation, such as roll angle, but may, in principle, be derived statistically by comparison of sequences known to wrap around cores.

Boffelli et al.[16] have built up a set of coefficients C_M by correlating structurally derived base step twists as well as roll angles with the statistical data provided by the 177 sequences from Satchwell et al.[3] With $\nu = 1/10.4$ and $W = 31$ base steps, these authors were able to predict the position of the most stable nucleosome in the regulatory region of the SV40 gene[22] to ± 2 rotational positions.

Future Developments

Predictions based on fits between spectra derived in reciprocal space by generalized expressions such as Eq. (7) offer scope for variation on and extension of the work by previous authors already described. We list here three main aspects.

1. The global shape described by a single component ν of Eq. (7) is unlikely to be sufficiently accurate to describe the path of DNA wrapped around the histone core, and especially the set of conformations that a particular sequence could adopt in practice. With the generalized form of Eq. (2), as in Eq. (7), simultaneous comparisons of several components of the power spectrum generated by C_M becomes trivial. Expectation of improvements in predications resulting from multicomponent fitting is tantamount to saying that the necessary assumption held in previous work that $[C_S]_j$ is independent of $j \pm \Delta j$ (i.e., that the assignment of coefficients is independent of the context of each base step) is unlikely to hold true

[22] C. Ambrose, A. Rajadhyaksha, H. Lowman, and M. Bina, J. Mol. Biol. 209, 255 (1989).

beyond a first approximation. In particular, information for translational positioning will come from Fourier components with $\nu \neq 1/10$. Rotational information predominates for components $\nu \approx 1/10$.

2. A prefixed phase relation, $\Delta\phi$, between different parts of the DNA wrap is also unlikely to be sufficiently accurate for reliable predictions. In Eq. (7) the phase term is not only a function of frequency ν but also of j, and therefore of k. In practice this does not introduce as much noise as might be expected provided that the moving window is long ($W \gg 1/\nu$), as adopted by previous authors. Furthermore, a certain amount of smoothing is introduced by ignoring the dependence of ν on W. This dependency is described as follows.

For $W \gg 1/\nu$, $[C_S]_j$ may be regarded as a smooth function of j. More realistically, the DNA sequence of base steps is better represented by a string of discrete coefficients $[C_S]_j$ in which successive values of j are separated by increments in phase of $2\pi/N$, where N equals $W - 1$. The Fourier transform of such a string of W coefficients has components given by

$$\nu = n/N \tag{8}$$

where n is an integer. Moreover, all the information contained in these components will be obtained by sampling ϕ by a set of discrete phases $[\phi]_n$ such that

$$\phi = j\nu = jn/N = [\phi]_n \tag{9}$$

where integer n ranges from zero to $N - 1$.

Thus, spectra $[I_k(\nu)]_n$ may be calculated. This is illustrated in Fig. 3. For each k and ν, the dependence of $I_k(\nu)$ on n is due to local deviations of the DNA wrap from a plane curve supercoil, and these discontinuities provide translational information.

Fourier transformation is often used to calculate sequence periodicities from an aligned set of sequenced binding sites.[3] The validity of the periodicities so obtained depends crucially on both the average amplitude and regularity of the periodic modulations in sequence as well as on the limitations inherent in the method itself.

We have shown how the sampling theorem limits the selection of components that contribute to the power spectrum calculated from a string of discrete coefficients. It follows from Eq. (8) that short window lengths permit few components, thus severely limiting the obtainable resolution of periodicities. This problem can be circumvented by application of pseudowindows, p times longer than N coefficients. A pseudowindow (Fig. 4) contains (pN) coefficients of magnitude $[fC_S]_j$, where

$$f_{j,k} = -1 \{[\overline{C_S}]_W + \cos 2\pi([j\nu]_p)\} \tag{10}$$

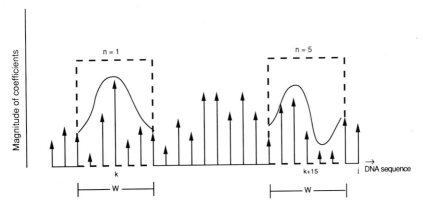

FIG. 3. A DNA sequence is depicted as a horizontal string of j coefficients. The relative magnitudes of the coefficients are represented by vertical arrows. Two positions k and $k + 15$ of a window W are shown as dashed boxes. At each position the contribution to a component ν of the power spectrum $[I_k(\nu)]_n$ is calculated from Eqs. (7) and (9) (see text). Here $W = 7$ coefficients, and $\nu = \frac{1}{6}$ [from Eq. (8)]. At both these positions of the window the contribution to $I_k(\nu)$ would be relatively high, whereas intervening coefficients would contribute less.

Here, $j_p = 1$ to (pN); $\nu_p = 1/2(pN)$.

The effect of this operation is to raise a positive cosine wave of periodicity ν_p to the mean value of $C_{S,j}$ for values of j within the window W, and thence to multiply $C_{S,j}$ by the part of the cosine wave f_j. This is the portion of the cosine function that corresponds to the real window as shown in

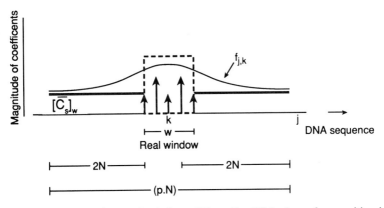

FIG. 4. The creation of a pseudowindow pN from Eq. (10) is shown for a position k of a real window W. The coefficients within the real window (vertical arrows) are to be multiplied by the curve f_{jk}. Here $W = 5$ and $p = 5$, so $pN = 20$ (see text). The resulting permitted components ν_p are listed in Table II.

TABLE II
PERIODICITIES PERMITTED WITH
PSEUDOWINDOW OF LENGTH pN^a

		p		
1	3	5	7	9
20.00^b	20.00	20.00	20.00	20.00
				18.00
			17.50	
		16.67		16.36
	15.00		15.56	15.00
		14.29	14.00	13.85
	12.00	12.50	12.73	12.00
		11.11	11.67	11.25
			10.77	10.59
10.00	10.00	10.00	10.00	10.00
			9.33	9.47
		9.09		9.00
	8.57		8.57	8.57
		8.33	8.24	8.18
		7.69	7.78	7.83
	7.50		7.37	7.50
		7.14	7.00	7.20
				6.92
6.67	6.67	6.67	6.67	6.67
		6.25	6.36	6.21
	6.00		6.09	6.00

[a] For a moving window of 22 base pairs: $W =$ 21 base steps, $W = 21$ coefficients; $N = 20$.
[b] Periodicities, $1/\nu$ are displayed in columns. Those values independent of p are boxed in rows.

Table II. Permitted values of $1/\nu$ resulting from the replacement of N by pN in Eq. (8) are shown in Table II.

3. Advantages of this approach are as follows: (a) Long-range features of the DNA wrap (i.e., the overall shape of the supercoil) need not be anticipated because, as Fig. 3 shows, the phase relationships between different segments of coefficients $[C_S]_j$ are determined empirically. (b) A single sequence, either from or independent of a basis set, but which is known from experiment to wrap around the histone core may be assigned

coefficients. Components of the resulting power spectrum can then be fitted empirically to the equivalent components from a sequence of unknown properties, and predictions made. The only assumption inherent in this operation is the nature of the coefficients assigned to both sequences. However, even these may be empirical, being derived statistically from a basis set of aligned experimental sequences. No physical or topological assumptions need underline the set of coefficients. (c) Crystallography has shown that the Fourier synthesis of an accurate image of an object depends more on the faithful reproduction of phase relationships between Fourier components than of their respective amplitudes. Analogously, accurate knowledge of the relative magnitudes of coefficients assigned to different base steps should prove to be of less importance than retention of the phase relations between different segments of the string of coefficients, as illustrated in Fig. 3.

General Applications

In principle, generalized algorithms for nucleosome positioning may also be applied to determine potential DNA binding sites for other protein complexes whose selectivity depends on the utilization of structural properties of the DNA double helix. The principal constraint for most algorithms is a requirement for a regular sequence periodicity. This condition is met by the *Escherichia coli* DNA gyrase but not by the bacterial RNA polymerase. In the latter case the rotational signals in promoter sites are not sufficiently regular to be accurately averaged by the assignment of a constant periodicity. Nevertheless, in such examples, the derivation of the power spectrum of the sequence is potentially of considerable utility. This is also particularly relevant where structural changes in a nucleoprotein complex may change the structural constraints on the bound DNA and hence the pattern of sequence preferences. In this context the transition from the nucleosome core particle to the chromatosome is a pertinent example.

We would also point out that structural periodicities reflected by sequence periodicities are also characteristic of certain proteins; the coiled coils present in myosin and keratins[23,24] are notable examples. Consequently the same algorithms that are used for nucleosome positioning could with appropriate modification be used for analysis of protein sequences.

[23] A. D. McLachlan and J. Karn, *J. Mol. Biol.* **164**, 695 (1983).
[24] A. D. McLachlan, *J. Mol. Biol.* **124**, 297 (1978).

[24] Thermodynamics of Ligand–Nucleic Acid Interactions

By TIMOTHY M. LOHMAN and DAVID P. MASCOTTI

I. Introduction

The noncovalent interactions of proteins and other ligands with DNA and RNA are of central importance to DNA metabolism and gene expression in all organisms. As a result, these interactions have been investigated by a variety of approaches in order to understand them at the thermodynamic, kinetic, and structural level. The structural details of the individual macromolecules and their complexes can provide information about some of the molecular contacts that can occur within the complex.[1,2] However, an understanding of the basis for their function and control requires knowledge of their stability, specificity, and mechanisms of interaction, which can only be achieved through thermodynamic and kinetic studies.[3–6]

Protein–nucleic acid interactions can be classified generally as either specific or nonspecific, depending on whether the proteins bind with high selectivity to a particular region of DNA, usually defined by its nucleotide sequence or conformation. However, specific nucleic acid-binding proteins, such as gene regulatory proteins, also bind to nonspecific regions with lower affinity. In fact, the competition of nonspecific DNA binding sites for the *Escherichia coli lac* repressor is believed to play an essential role in the regulation of expression of the *lac* operon *in vivo*,[7–9] and this is likely to be the case for other operons. There is also a general class of nucleic acid-binding proteins that bind to DNA exclusively in a nonspecific manner, with little dependence on the nucleotide sequence.[10] Examples of

[1] C. O. Pabo and R. T. Sauer, *Annu. Rev. Biochem.* **53**, 293 (1984).

[2] S. C. Harrison and A. K. Aggarwal, *Annu. Rev. Biochem.* **59**, 933 (1990).

[3] M. T. Record, Jr., J.-H. Ha, and M. A. Fisher, this series, Vol. 208, p. 291.

[4] D. R. Lesser, M. W. Kurpiewski, and L. Jen-Jacobson, *Science* **250**, 776 (1990).

[5] T. M. Lohman, *Crit. Rev. Biochem.* **19**, 191 (1986).

[6] P. H. von Hippel, D. G. Bear, W. D. Morgan, and J. A. McSwiggen, *Annu. Rev. Biochem.* **53**, 389 (1984).

[7] P. H. von Hippel, A. Revzin, C. A. Gross, and A. C. Wang, *Proc. Natl. Acad. Sci. U.S.A.* **71**, 4808 (1984).

[8] P. H. von Hippel, *in* "Biological Regulation and Development" (R. F. Goldberger, ed.), Vol. 1, p. 279. Plenum, New York, 1979.

[9] M. T. Record, Jr., and R. S. Spolar, *in* "Nonspecific DNA–Protein Interactions" (A. Revzin, ed.), p. 33. CRC Press, Boca Raton, Florida, 1990.

[10] A. Revzin, *in* "Nonspecific DNA–Protein Interactions" (A. Revzin, ed.), p. 5. CRC Press, Boca Raton, Florida, 1990.

METHODS IN ENZYMOLOGY, VOL. 212

these are the helix-destabilizing (single-strand DNA binding) proteins,[11–15] histones,[16] as well as a variety of enzymes involved in DNA replication, recombination, and repair.[17]

The stability of specific protein–nucleic acid complexes is dependent on the specific array of hydrogen bonds and other contacts that can form between the protein and the nucleic acid ("direct readout") as well as the conformation of the nucleic acid (single stranded and various duplex forms), which can also be dependent on the nucleotide sequence ("indirect readout").[2] The specificity of the interaction of a protein for a particular nucleic acid site is also dependent on the affinity of the protein for the large number of other nonspecific sites that exist within a linear nucleic acid.[3] However, the stabilities and specificities of protein–nucleic acid interactions, just as for all macromolecular interactions, are also extremely sensitive to the solution environment (salt concentration and type, pH), since these small molecules (e.g., ions) generally interact directly and preferentially with the free protein, free DNA, and their complexes.[9,18,19] Because the specific and nonspecific interactions of any single protein can display differential sensitivities to solution variables, the nucleic acid binding specificity of proteins can also be a function of solution variables.[4,20,21]

The binding of proteins and small positively charged ligands to nucleic acids is particularly sensitive to the ionic environment in solution. This is due to the polyelectrolyte nature of linear nucleic acids, which results in the local accumulation to high concentrations of cations (e.g., K^+, Mg^{2+}) in the vicinity of the nucleic acid,[22–25] some of which are released on

[11] S. C. Kowalczykowski, D. G. Bear, and P. H. von Hippel, in "The Enzymes" (P. D. Boyer, ed.), Vol. 14, p. 373. Academic Press, New York, 1981.

[12] T. M. Lohman and W. Bujalowski, in "The Biology of Nonspecific DNA–Protein Interactions" (A. Revzin, ed.), p. 131. CRC Press, Boca Raton, Florida, 1990.

[13] R. L. Karpel, in "Nonspecific DNA–Protein Interactions" (A. Revzin, ed.), p. 103. CRC Press, Boca Raton, Florida, 1990.

[14] K. R. Williams and J. W. Chase, in "Nonspecific DNA–Protein Interactions" (A. Revzin, ed.), p. 197. CRC Press, Boca Raton, Florida, 1990.

[15] J. W. Chase and K. R. Williams, Annu. Rev. Biochem. 55, 103 (1986).

[16] M. J. Behe, in "Nonspecific DNA–Protein Interactions" (A. Revzin, ed.), p. 229. CRC Press, Boca Raton, Florida, 1990.

[17] M. M. Cox, in "Nonspecific DNA–Protein Interactions" (A. Revzin, ed.), p. 171. CRC Press, Boca Raton, Florida, 1990.

[18] M. T. Record, Jr., T. M. Lohman, and P. H. deHaseth, J. Mol. Biol. 107, 145 (1976).

[19] M. T. Record, Jr., C. F. Anderson, and T. M. Lohman, Q. Rev. Biophys. 11, 103 (1978).

[20] M. T. Record, Jr., P. H. deHaseth, and T. M. Lohman, Biochemistry 16, 4791 (1977).

[21] M. C. Mossing and M. T. Record, Jr., J. Mol. Biol. 186, 295 (1985).

[22] G. S. Manning, J. Chem. Phys. 51, 924 (1969).

[23] G. S. Manning, Q. Rev. Biophys. 11, 179 (1978).

formation of the protein–nucleic acid complex.[5,18,19] In fact, much of the stability of complexes between gene regulatory proteins and nonspecific DNA is largely due to the large positive entropy changes that result from ion release on formation of the complex.[26] As a result of the dependence of protein–nucleic acid stability and specificity on solution conditions, it is impossible to understand the forces that drive these interactions based solely on structural considerations. Rather, as discussed in the first part of this chapter, the equilibrium binding properties (thermodynamics, energetics) of these interactions must be investigated as a function of solution conditions in order to understand the origins of stability and specificity.

Studies of the thermodynamics of protein– and ligand–nucleic acid interactions can in principle be examined by two approaches: calorimetry and the measurement of equilibrium binding constants as a function of solution variables. The combined use of both approaches has proved extremely useful in studies of small ligand–DNA interactions.[27,28] However, to date, there have been no calorimetric studies reported for any protein–nucleic acid interaction, and all of the thermodynamic information available on these interactions has been obtained from equilibrium binding studies. Therefore, in this discussion we focus on the use of equilibrium binding studies to obtain such information.

To obtain meaningful thermodynamic information from studies of the dependence of equilibrium binding constants on solution variables, techniques and methods of analysis are required that yield rigorous equilibrium binding isotherms, which can be analyzed to extract equilibrium binding parameters. The rigorous use of spectroscopic methods in this regard has been discussed[29,30] (see also [25] in this volume[31]).

II. Equilibrium Binding Parameters

A. Equilibrium Binding Constant (K_{obs})

Consider the interaction of a ligand, L, with a nucleic acid, D, to form a complex, LD, as in Eq. (1):

[24] C. F. Anderson and M. T. Record, Jr., *Annu. Rev. Phys. Chem.* **33**, 191 (1982).
[25] M. T. Record, Jr., M. Olmsted, and C. F. Anderson, *in* "Theoretical Biochemistry and Molecular Biophysics" (D. L. Beveridge and R. Lavery, eds.), p. 285. Adenine Press, Schenectady, New York, 1990.
[26] P. L. deHaseth, T. M. Lohman, and M. T. Record, Jr., *Biochemistry* **16**, 4783 (1977).
[27] L. A. Marky and K. J. Breslauer, *Proc. Natl. Acad. Sci. U.S.A.* **84**, 4359 (1987).
[28] W. Y. Chou, L. A. Marky, D. Zaunczkowski, and K. J. Breslauer, *J. Biomol. Struct. Dyn.* **5**, 345 (1987).
[29] W. Bujalowski and T. M. Lohman, *Biochemistry* **26**, 3099 (1987).
[30] T. M. Lohman and W. Bujalowski, this series, Vol. 208, p. 258.
[31] T. M. Lohman and D. P. Mascotti, this volume [25].

$$L + D \overset{K_{obs}}{\rightleftharpoons} LD \qquad (1)$$

This equilibrium can be described by the intrinsic (microscopic) equilibrium constant, K_{obs}, defined in Eq. (2):

$$K_{obs} = \frac{[LD]}{[L][D]} \qquad (2)$$

We emphasize that the equilibrium in Eq. (1), described by the intrinsic equilibrium binding constant K_{obs}, reflects the binding of only one form of the ligand, L, to yield one type of complex, LD. Therefore, if multiple forms of the ligand exist in equilibrium (e.g., a protein that exists in equilibrium between monomer, dimer, and tetramer forms), then a separate intrinsic binding constant is required to describe the interaction of each form of the ligand with the nucleic acid. Similarly, if a ligand can bind to the DNA in a number of different modes, then the binding of each mode is described by a separate intrinsic binding constant. Only if K_{obs} is such an intrinsic equilibrium constant can it be used to calculate the true thermodynamic quantities, $\Delta G°$, $\Delta H°$, and $\Delta S°$, for the equilibrium. The relationships among these quantities are given in Eq. (3), where R is the gas constant and T the absolute temperature:

$$\Delta G° = -RT \ln K_{obs} = \Delta H° - T\Delta S° \qquad (3)$$

If L, D, or LD are involved in equilibria in addition to that defined by Eq. (1), and these other equilibria are not explicitly considered in the analysis of the interaction, then only an "apparent" binding constant will be obtained from an analysis of the binding interaction. This "apparent" binding constant will be a composite parameter reflecting all of the multiple equilibria and hence will not be related in a simple (or necessarily known) manner to the free energy change for the reaction in Eq. (1). To determine whether such multiple equilibria exist, it is generally advisable to examine the ligand–nucleic acid interactions over a range of ligand and nucleic acid concentrations, at constant solution conditions. If L and D are involved only in the equilibrium defined in Eq. (1), then the value of K_{obs} will be independent of the ligand and nucleic acid concentration. If L and D are involved in other equilibria, however, then the experimental binding constant will be a function of the ligand and/or nucleic acid concentrations and therefore will not represent the intrinsic binding constant for the reaction in Eq. (1). In this context, it is essential to understand and characterize the solution behavior of the free ligand (quaternary structure, self-assembly equilibria) and the ligand–nucleic acid complex, quantitatively, before embarking on a quantitative investigation of its equilibrium binding and thermodynamic properties.

B. Ligand Binding Site Size

The ligand binding site size, n, represents the number of bases or base pairs that are occupied by the ligand on binding to the nucleic acid. The site size for ligand binding defines the region of the nucleic acid that is occluded by the ligand and hence not accessible for interaction with other ligands. This occluded site size is not necessarily the same as the number of nucleotides contacted by the ligand, although n will provide an upper estimate for the number of contacts. Knowledge of the ligand site size is necessary for an accurate description of the nonspecific equilibrium binding of large ligands to long, linear nucleic acids, since the number of free sites on the nonspecific nucleic acid lattice is dependent on n as well as the ligand binding density.[32] These aspects are discussed further in Section III,B.

C. Cooperativity of Ligand Binding

Cooperative binding of ligands to linear nucleic acids is possible when multiple ligands (proteins) can bind to the same nucleic acid molecule. Cooperativity is a thermodynamic quantity that reflects the influence of one bound ligand on the binding affinity of a second ligand. If the free energy changes on binding two or more ligands to the same nucleic acid are not independent, then cooperative interactions are indicated. Cooperativity can be either positive or negative; that is, the binding affinity of a second ligand can be either enhanced or reduced with respect to the affinity of the first ligand. True cooperativity reflects changes in the intrinsic binding parameters and should not be confused with statistical effects, such as the apparent negative cooperativity that results from the overlap of potential nonspecific ligand binding sites.[32]

Cooperative effects can be due to nearest-neighbor interactions, namely, interactions resulting from ligands bound to adjacent sites on a linear DNA molecule, or non-nearest-neighbor interactions. Nearest-neighbor cooperativities can result from ligand–ligand interactions and/or short-range conformational changes that are induced in the DNA on ligand binding. Non-nearest-neighbor cooperativity has been observed between DNA-binding proteins that are bound to specific sites that are well separated along the contour length of a linear nucleic acid molecule.[33,34] These cooperativities have generally been explained by models that invoke direct

[32] J. D. McGhee and P. H. von Hippel, *J. Mol. Biol.* **86**, 469 (1974).
[33] J. A. Borowiec, L. Zhang, S. Sasse-Dwight, and J. D. Gralla, *J. Mol. Biol.* **196**, 101 (1987).
[34] S. Oehler, E. R. Eismann, H. Kramer, and B. Müller-Hill, *EMBO J.* **9**, 973 (1990).

interactions between the proteins, resulting in "looping" of the intervening DNA.[35,36]

Positive cooperativity has been observed for a number of nonspecific nucleic acid-binding proteins, including the *E. coli* SSB (single-strand-specific binding) protein,[37–39] the bacteriophage T4 gene 32 protein,[40,41] and the bacteriophage gene 5 protein,[42–44] as well as for some sequence-specific DNA-binding proteins.[45,46] Both positive and negative cooperative effects have been observed for the binding of some small ligands, such as drugs and dyes, to duplex DNA,[47,48] although negative cooperativity in these systems is small and difficult to resolve from the statistical effects that result from overlap of potential nonspecific binding sites. Section III,B discusses different statistical thermodynamic models for the treatment of nearest-neighbor cooperativity between ligands bound nonspecifically to linear nucleic acids, as well as a quantitative definition of cooperativity.

III. Sequence Specific *versus* Nonspecific Binding

Protein (ligand)–nucleic acid interactions can generally be divided into two classes: (1) site-specific interactions, in which the protein binds with high affinity to a specific site(s), usually defined by the nucleotide sequence, and (2) nonspecific interactions in which the protein binds to nucleic acids with the same general affinity, nearly independent of nucleotide sequence. Of course, ligands that bind with sequence specificity can also bind to nonspecific regions of the nucleic acid, although with signifi-

[35] R. Schleif, *Science* **240**, 127 (1988).
[36] G. R. Bellomy and M. T. Record, Jr., *in* "Progress in Nucleic Acid Research and Molecular Biology" (W. Cohn and K. Moldave, eds.), Vol. 39, p. 81. Academic Press, New York, 1990.
[37] N. Sigal, H. Delius, T. Kornberg, M. L. Gefter, and B. Alberts, *Proc. Natl. Acad. Sci. U.S.A.* **69**, 3537 (1972).
[38] T. M. Lohman, L. B. Overman, and S. Datta, *J. Mol. Biol.* **187**, 603 (1986).
[39] W. Bujalowski and T. M. Lohman, *J. Mol. Biol.* **195**, 897 (1987).
[40] B. Alberts and L. Frey, *Nature (London)* **227**, 1313 (1970).
[41] S. C. Kowalczykowski, N. Lonberg, J. W. Newport, and P. H. von Hippel, *J. Mol. Biol.* **145**, 75 (1981).
[42] B. Alberts, L. Frey, and H. Delius, *J. Mol. Biol.* **68**, 139 (1972).
[43] S. J. Cavalieri, K. E. Neet, and D. A. Goldthwait, *J. Mol. Biol.* **102**, 697 (1976).
[44] D. Porschke and H. Rauh, *Biochemistry* **22**, 4737 (1983).
[45] A. D. Johnson, A. R. Poteete, G. Lauer, R. T. Sauer, G. K. Ackers, and M. Ptashne, *Nature (London)* **294**, 217 (1981).
[46] G. K. Ackers, A. D. Johnson, and M. A. Shea, *Proc. Natl. Acad. Sci. U.S.A.* **79**, 1129 (1982).
[47] W. D. Wilson and I. G. Lopp, *Biopolymers* **18**, 3025 (1979).
[48] L. S. Rosenberg, M. J. Carvlin, and T. R. Krugh, *Biochemistry* **25**, 1002 (1986).

cantly lower affinity. The nonspecific binding of otherwise specific DNA-binding proteins appears to play important roles in the regulation of gene expression[7,8] as well as in facilitating the location of the specific DNA site by the protein.[5,49-53]

A. Site-Specific Equilibrium Binding Isotherms

The intrinsic observed equilibrium constant, K_{PS}, for a site-specific protein–nucleic acid interaction is given in Eq. (4):

$$K_{PS} = \frac{[PS]}{[P][S]} \tag{4}$$

where P, S, and PS represent the free protein, specific nucleic acid site, and the complex, respectively. In the absence of cooperativity, the fractional occupancy of the specific site, $\Theta \equiv [PS]/S_{TOTAL}$, is described by the simple Langmuir isotherm given in Eq. (5):

$$\Theta = \frac{K_{PS}[P]}{(1 + K_{PS}[P])} \tag{5}$$

However, as the specific site is usually contained within a larger region of nonspecific DNA, one must also account for the competitive binding of protein to the nonspecific sites.[7] For the case of multiple specific binding sites, where cooperativity between proteins bound to the specific sites can occur, then the simple expression in Eq. (5) must be modified as in the case of the bacteriophage λ cI repressor, which binds to three specific sites in both the λO_R and λO_L operators.[45,46]

B. Nonspecific Equilibrium Binding Isotherms

The quantitative description of the equilibrium isotherm for a large ligand (protein) binding nonspecifically to linear nucleic acids is more complicated than Eq. (5), since potential nonspecific ligand binding sites overlap. Large ligands are defined as those with site sizes, n, greater than 1 residue (nucleotide or base pair). As a result, the number of available binding sites for the ligand decreases nonlinearly with increasing ligand binding density, ν (moles of ligand bound per residue). This statistical effect results in curvature in the equilibrium binding isotherm, even for a

[49] P. H. Richter and M. Eigen, *Biophys. Chem.* **2**, 255 (1974).
[50] O. G. Berg and C. Blomberg, *Biophys. Chem.* **4**, 367 (1976).
[51] D. R. Dowd and R. S. Lloyd, *J. Biol. Chem.* **265**, 3424 (1990).
[52] O. G. Berg and P. H. von Hippel, *Annu. Rev. Biophys. Biophys. Chem.* **14**, 131 (1985).
[53] O. G. Berg, R. B. Winter, and P. H. von Hippel, *Trends Biochem. Sci.* **7**, 52 (1982).

noncooperative binding ligand, and has been described quantitatively for both infinite nucleic acid lattices[32,54,55] as well as finite nucleic acid lattices.[56,57]

We briefly describe two models for the nonspecific binding of large ligands to homogeneous infinite nucleic acid lattices that consider cooperative ligand binding. Both of these treat cooperative interactions only between nearest-neighbor bound ligands. Although more complicated non-nearest-neighbor models have been described,[58–60] these models require more parameters than can usually be determined from an experimental binding isotherm. The two nearest-neighbor cooperativity models that we consider differ in that the ligands possess either "unlimited" or "limited" cooperativity, as defined below and depicted in Fig. 1.[61] In the unlimited cooperativity model, clusters of bound ligands are limited only by the length of the nucleic acid, whereas in the limited cooperativity model, the ligand cluster size is limited, even for high values of cooperativity.

1. Unlimited Cooperativity. A detailed description of the quantitative aspects of this model has been given[32,55] and Fig. 2 defines the binding parameters for this model. In the "unlimited" cooperativity model, a ligand with site size, n residues (nucleotides or base pairs), can bind to an infinite, linear, homogeneous nucleic acid lattice in any of three modes: an isolated mode, with intrinsic equilibrium constant, K; a singly contiguous mode (one nearest-neighbor ligand), with equilibrium constant $K\omega$; a doubly contiguous mode (two nearest-neighbor ligands), with equilibrium constant, $K\omega^2$. The cooperativity parameter, ω, is a unitless equilibrium constant for the process of moving two bound isolated ligands so that both are bound singly contiguously (see Fig. 2). The cooperativity parameter, ω, is related to the standard free energy change for this process by $\Delta G°_{coop} = -RT \ln \omega$. Positive cooperativity is reflected by $\omega > 1$ ($\Delta G°_{coop} < 0$), negative cooperativity results when $\omega < 1$ ($\Delta G°_{coop} > 0$), and noncooperative interactions are described by $\omega = 1$ ($\Delta G°_{coop} = 0$).

The model predicts that when positive cooperativity exists ($\omega > 1$), the ligands will form clusters of variable length along the nucleic acid,

[54] A. S. Zasedatelev, G. V. Gursky, and M. V. Volkenshtein, *Mol. Biol. (U.S.S.R.)* **5**, 245 (1971).

[55] J. A. Schellman, *Isr. J. Chem.* **12**, 219 (1974).

[56] I. R. Epstein, *Biophys. Chem.* **8**, 327 (1978).

[57] J. A. Schellman, *in* "Molecular Structure and Dynamics" (M. Balaban, ed.), p. 245. International Science Services, Philadelphia, 1980.

[58] Y. Chen, *Biophys. Chem.* **27**, 59 (1987).

[59] G. V. Gursky and A. S. Zasedatelev, *Sov. Sci. Rev. Physicochem. Biol.* **5**, 53 (1984).

[60] Y. D. Nechipurenko and G. V. Gursky, *Biophys. Chem.* **24**, 195 (1986).

[61] T. M. Lohman, W. Bujalowski, and L. B. Overman, *Trends Biochem. Sci.* **13**, 250 (1988).

A

"Unlimited" Cooperativity

B

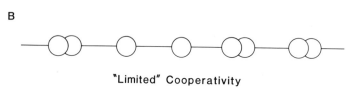

"Limited" Cooperativity

FIG. 1. Diagram illustrating two different types of nearest-neighbor cooperative binding of ligands to linear nucleic acid. (A) "Unlimited" cooperativity, typified by T4 gene 32 protein,[41] in which long clusters of bound protein can form owing to nearest-neighbor interactions occurring on both sides of a bound protein. As the cooperativity parameter, ω (see Fig. 2 for definition), increases to infinity, a single protein cluster will form along the nucleic acid lattice. (B) "Limited" cooperativity, typified by the $E.$ $coli$ SSB tetramer binding to single-stranded DNA in its $(SSB)_{65}$ binding mode.[39] In this case, each circle represents an SSB tetramer, and clustering is limited to the formation of dimers of tetramers (octamers), such that no doubly contiguously bound tetramers can form. With this type of cooperativity, it is difficult to saturate the DNA lattice. (From Lohman et $al.$[61])

depending on the binding density, ν (for given values of n, K, and ω), which are limited in size only by the length of the nucleic acid. At a given ligand binding density, the average ligand cluster length is determined by the values of ω and n.[32,55] McGhee and von Hippel[32] obtained separate closed-form expressions for the equilibrium binding isotherms for this model for $\omega = 1$ and $\omega \neq 1$; however, these can be combined into a single expression for the binding isotherm[62] that is valid for all values of $\omega > 0$, and this is given in its Scatchard form in Eq. (6):

$$\nu/L = K(1 - n\nu)\{[2\omega(1 - n\nu)]/[(2\omega - 1)(1 - n\nu) + \nu + R]\}^{(n-1)}\{[1 - (n + 1)\nu + R]/[2(1 - n\nu)]\}^2 \quad (6)$$

where L is the free ligand concentration and $R = \{[1 - (n + 1)\nu]^2 + 4\omega\nu(1 - n\nu)\}^{1/2}$. This nearest-neighbor unlimited cooperativity model provides a good description of the nonspecific, cooperative binding of the bacteriophage T4 gene 32 protein to single-stranded polynucleo-

[62] W. Bujalowski, T. M. Lohman, and C. F. Anderson, $Biopolymers$ **28**, 1637 (1989).

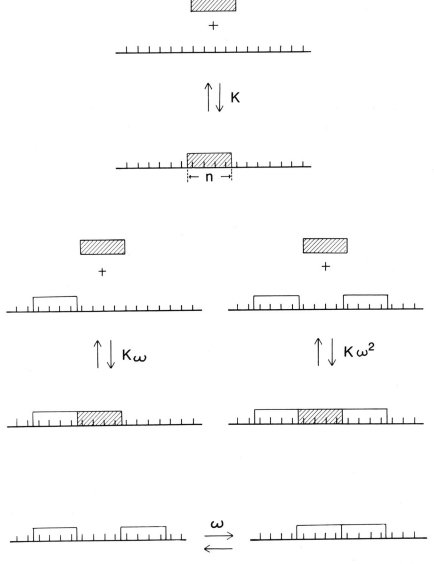

FIG. 2. Definition of the equilibrium binding parameters n (site size), K (intrinsic binding constant), and ω (cooperativity parameter) for the nearest-neighbor, "unlimited" cooperativity model[32] for binding of a ligand to an infinite, one-dimensional, homogeneous lattice.

tides.[11,41,63,64] However, a number of nonspecific cooperatively binding proteins do not seem to conform to this model, even though their equilibrium binding isotherms can be described well by this model. These include the phage fd gene 5 protein, which displays non-nearest-neighbor cooperative interactions,[65] and the *E. coli* SSB protein in its $(SSB)_{65}$ binding mode, which appears to bind with limited cooperativity,[12,39] although it may bind in an unlimited cooperativity mode in its $(SSB)_{35}$ binding mode.[12,38]

An alternative "free sliding ligand" model has been presented[66] to describe the nonspecific binding of large ligands to an infinite linear lattice, including nearest-neighbor cooperativity. In this model, the ligand is not constrained to bind to distinct sites on the nucleic acid lattice, rather it is assumed to translate freely along the nucleic acid lattice. Attempts have also been made to incorporate intrinsic negative cooperative effects that should arise for the binding of small positively charged ligands to a linear nucleic acid at high binding density.[67] However, this approach[67] overestimates the negative cooperativity, at least at low binding densities, for the binding of oligolysines to linear nucleic acids.[68]

2. Limited Cooperativity. In the "limited" cooperativity model, cooperative interactions also occur only between nearest-neighbor bound ligands; however, the interactions are such that clustering is limited to the formation of dimers of ligands; hence, cooperativity is "limited" (see Fig. 1B).[39] The closed-form expression for the binding isotherm (Scatchard form[69]) for this model is

$$v/L = K[q^2 - 2vq + (1 - \omega)v^2]^n/q^{(2n-1)} \qquad (7)$$

where $q = [1 - (n - 1)v] + \{[1 - (n - 1)v]^2 - v(1 - \omega)[2 - (2n - 1)v]\}^{1/2}$. This limited cooperativity model provides a better description of the nonspecific binding of the *E. coli* SSB tetramer to single-stranded nucleic acids in its $(SSB)_{65}$ binding mode than does the unlimited cooperativity model.[39]

3. Noncooperative Binding. The closed form expression for the equilibrium binding isotherm for noncooperative, nonspecific ligand binding to an infinite, linear nucleic acid can be obtained by substituting $\omega = 1$ into Eq. (6), with the resulting expression[32]

$$v/L = K(1 - nv)[(1 - nv)/(1 - (n - 1)v)]^{(n-1)} \qquad (8)$$

[63] T. M. Lohman, *Biochemistry* **23**, 4656 (1984).
[64] T. M. Lohman, *Biochemistry* **23**, 4665 (1984).
[65] J. E. Coleman and J. L. Oakley, *Crit. Rev. Biochem.* **98**, 247 (1980).
[66] C. P. Woodbury, *Biopolymers* **20**, 2225 (1981).
[67] R. A. G. Friedman and G. S. Manning, *Biopolymers* **23**, 2671 (1984).
[68] D. P. Mascotti and T. M. Lohman, unpublished, 1992.
[69] G. Scatchard, *Ann. N.Y. Acad. Sci.* **51**, 660 (1949).

We note that Eq. (8) can also be obtained from Eq. (7) on substituting $\omega = 1$; however, the equilibrium constants for the two expressions differ by a statistical factor of 2, since in the unlimited cooperativity model, the ligands are assumed to bind in only one orientation along the nucleic acid, whereas the limited cooperativity model assumes bidirectional ligand binding. However, if the same polarity of ligand binding is assumed for both models then the definitions of K for each model are identical.

IV. Dependence of K_{obs} on Salt Concentration

The intrinsic binding constant defined in Eqs. (1) and (2) is an "observed" binding constant, that is, K_{obs} is defined in terms of only the ligand and nucleic acid species, independent of their interactions with solvent or low molecular weight solutes (e.g., ions, protons).[19] The true thermodynamic equilibrium constant, K_T, is defined in terms of the activities of all species that participate in the reaction and hence is dependent only on temperature and pressure. Therefore, if preferential interactions of cations, anions, water, or protons occur with any of the species, L, D, or LD, then K_{obs}, which is the quantity that is measured experimentally, will be dependent on the bulk salt concentration and pH, as well as temperature and pressure.

A. Cation Effects on Ligand–Nucleic Acid Equilibria

1. Polyelectrolyte Effect: Cation Release from Nucleic Acid. The majority of protein–nucleic acid interactions, as well as positively charged small ligand–nucleic acid interactions, are extremely sensitive to the salt concentration in solution.[5,18,19] This stems from the fact that linear nucleic acids are highly charged polyanions and as a result sequester cations in their vicinity in order to reduce the net charge on the nucleic acid.[22–25,70] The interaction of counterions (M^+) with a linear polyelectrolyte such as DNA is equivalent, thermodynamically, to having a constant fraction of a counterion bound per nucleic acid phosphate.[18] In aqueous media, the fraction of a counterion thermodynamically bound per phosphate, ψ, is dependent only on the structural charge density along the nucleic acid and the counterion valance and is independent of the bulk salt concentration, as long as it is in excess over the phosphate charge.[19,22,23] This has been verified experimentally by ^{23}Na NMR studies of Na–DNA solutions[71] and by examining the salt dependence of a number of oligocations interacting with duplex and single-stranded nucleic acids. For double-stranded B-form

[70] C. F. Anderson and M. T. Record, Jr., *Annu. Rev. Biophys. Chem.* **19**, 423 (1990).
[71] C. F. Anderson, M. T. Record, Jr., and P. A. Hart, *Biophys. Chem.* **7**, 310 (1978).

DNA, ψ has a value of 0.88,[18] whereas for single-stranded nucleic acids, near neutral pH, ψ is approximately 0.70–0.74.[72,73] The lower value of ψ for single-stranded nucleic acids reflects its lower charge density, relative to the duplex form.[74,75]

As a result, the association of a simple oligocation (net charge $+z$) with a linear nucleic acid in the presence of excess univalent salt, MX, can be represented, thermodynamically, by the following equilibrium:

$$L^{z+} + D \rightleftharpoons LD + z\psi M^+ \qquad (9)$$

We note that the coefficient $z\psi$ includes contributions due to both the direct release of cations previously bound to the nucleic acid as well as perturbations of the electrostatic screening of the phosphates by the remaining bulk ions (activity coefficient effects). Application of Le Chatelier's principle to Eq. (9) indicates that K_{obs} will increase as the salt (MX) concentration is decreased. The predicted quantitative dependence of K_{obs} on $[M^+]$ is[18,19]

$$\frac{\partial \log K_{obs}}{\partial \log[M^+]} = -z\psi \qquad (10)$$

Because the coefficient, $z\psi$, is constant for a given ligand and nucleic acid, this predicts that the logarithm of K_{obs} will decrease as a linear function of the logarithm of the monovalent counterion (M^+) concentration. Furthermore, since $z\psi$ is generally large, the decrease of K_{obs} with increasing salt concentration is also quite dramatic.[18,19]

The salt dependence of K_{obs} for the binding to poly(U) of a series of oligolysines with varying charge, $+z$, is shown in Fig. 3. For each oligopeptide, log K_{obs} decreases linearly with log$[K^+]$, and the absolute value of the slope, $|\partial \log K_{obs}/\partial \log[K^+]|$, is proportional to the oligopeptide charge. The value of the proportionality constant determined from the data is $\psi = 0.71 \pm 0.03$, which is consistent with the lower charge density for a single-stranded nucleic acid.[73] Similar behavior has been observed for the interaction of positively charged oligopeptides and polyamines with duplex nucleic acids[18,73,76-79]; however, $\psi = 0.90 \pm 0.05$ due to the higher charge density of the duplex nucleic acids.[18]

[72] M. T. Record, Jr., C. P. Woodbury, and T. M. Lohman, *Biopolymers* **15**, 893 (1976).
[73] D. P. Mascotti and T. M. Lohman, *Proc. Natl. Acad. Sci. U.S.A.* **87**, 3142 (1990).
[74] M. T. Record, Jr., *Biopolymers* **14**, 2137 (1975).
[75] G. S. Manning, *Biopolymers* **11**, 937 (1972).
[76] S. A. Latt and H. A. Sober, *Biochemistry* **6**, 3293 (1967).
[77] S. A. Latt and H. A. Sober, *Biochemistry* **6**, 3307 (1967).
[78] T. M. Lohman, P. H. deHaseth, and M. T. Record, Jr., *Biochemistry* **19**, 3522 (1980).
[79] W. H. Braunlin, T. J. Strick, and M. T. Record, Jr., *Biopolymers* **21**, 1301 (1982).

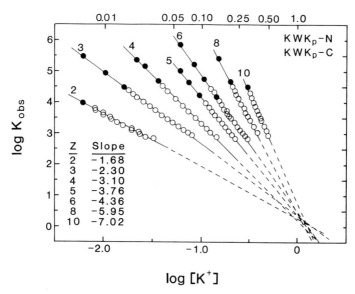

FIG. 3. Dependence of log K_{obs} on log[K⁺] for a series of oligolysines binding to poly(U) (25.0°, pH 6.0). KCH_3CO_2 is used to vary the K⁺ concentration. The oligopeptide sequences are KWK_p-NH_2, with $p = 1, 2, 4, 6,$ and 8 ($z = +3, 4, 6, 8,$ and 10) and KWK_p-CO_2, with $p = 1$ and 4 ($z = +2$ and 5). The net positive charge of each peptide is indicated for each line, and the slopes of each line are given. (●), Values of K_{obs} determined from an individual reverse titration at a constant salt concentration; (○), values of K_{obs} determined from salt back-titrations. (Data from Mascotti and Lohman.[73])

The release of thermodynamically bound counterions into the bulk salt solution provides a favorable entropic contribution ($\Delta S > 0$) to the free energy of binding (a free energy of dilution).[18] This free energy of dilution has been referred to as the polyelectrolyte effect[80] and is given by Eq. (11) for a ligand–nucleic acid equilibrium occurring in a univalent salt solution[18]:

$$\Delta G_{PE}^{\circ} = z\psi RT \ln[M^+] \tag{11}$$

Equation (11) indicates that the polyelectrolyte effect is eliminated at 1 M K⁺, which is the standard state for the reaction. This behavior has been verified for a number of simple oligocations binding to both duplex[18,78,79] and single-stranded nucleic acids, an example of which is shown in Fig. 3

[80] M. T. Record, Jr., in "Unusual DNA Structures" (R. D. Wells and S. C. Harvey, eds.), p. 237. Springer-Verlag, New York, 1988.

for the oligolysine-·poly(U) interaction.[73] This polyelectrolyte effect (counterion release from the nucleic acid) provides the major driving force for the interactions of simple oligocations with linear nucleic acids at low salt concentrations.[18,73,78] This also appears to be the case for a number of nonspecific ligand (protein)–nucleic acid interactions, including the *E. coli lac* repressor.[26]

Of course, preferential cation interactions with the nucleic acid will also result from any process that affects the charge spacing of the nucleic acid. The latter effect has been explicitly considered in the interpretation of the salt dependence of K_{obs} for small planar ligands that can intercalate between the base pairs of duplex nucleic acids.[47] In this case, cation release results from both the neutralization of phosphate charge as well as the increase in the axial charge spacing of the duplex nucleic acid due to intercalation, although the former effect dominates the salt dependence.

Various methods have detected small differences in the affinities of different monovalent cations for duplex DNA.[81–84] ^{23}Na NMR measurements[84] indicate the following hierarchy of affinities: $Na^+ < Li^+ < K^+ < Cs^+ < NH_4^+$. However, it has been observed generally that the effects of cations on the equilibrium binding of simple oligocations to linear DNA are not sensitive to the type of monovalent cation, M^+. For example, identical values of K_{obs} and $\partial \log K_{obs} / \partial \log[M^+]$ are observed for the interaction of oligolysines with poly(U), independent of whether the cation is Na^+, K^+, or $NH_3(n\text{-}C_4H_9)^+$.[68] A similar independence of K_{obs} on the cation type (K^+, Na^+, NH_4^+) has been observed for the specific binding of the *E. coli lac* repressor to the O^{sym} [85] and wild-type *lac* operator,[86] although, for the latter interaction, K_{obs} in Cs^+ is larger by a factor of 2 and K_{obs} in Li^+ is smaller by a factor of 10. The latter effects may reflect differences in the affinity of these ions for the DNA; however, preferential interactions of these cations with the protein have also not been ruled out.

In an alternative view, Manning[23] has suggested that the coefficient ψ equals unity, which predicts that one monovalent cation should be released per oligocation positive charge on complex formation, independent of the nucleic acid charge density. However, recent studies of the equilibrium binding of a series of oligolysines to the single-stranded polynucleotide

[81] B. Wolf and S. Hanlon, *Biochemistry* **14,** 1661 (1975).
[82] A. Chan, R. Kilkuskie, and S. Hanlon, *Biochemistry* **18,** 84 (1979).
[83] P. Anderson and W. Bauer, *Biochemistry* **17,** 594 (1978).
[84] M. L. Bleam, C. F. Anderson, and M. T. Record, Jr., *Proc. Natl. Acad. Sci. U.S.A.* **77,** 3085 (1980).
[85] J.-H. Ha, Ph.D. Thesis, University of Wisconsin, Madison, Wisconsin (1990).
[86] M. D. Barkley, P. A. Lewis, and G. E. Sullivan, *Biochemistry* **20,** 3842 (1981).

poly(U) indicate that for this interaction ψ is significantly less than unity (see Fig. 3).[73]

  2. Electrostatic versus Nonelectrostatic Contributions to ΔG°_{obs}. The standard free energy change for a ligand–nucleic acid interaction, ΔG°_{obs}, can be viewed as having contributions arising from electrostatic and non-electrostatic (noncoulombic) interactions. In the absence of preferential ion interactions with the ligand, Record *et al.*[18] have described a method for estimating the relative contributions from these interactions. This method is based on the fact that the polyelectrolyte effect [see Eq. (11)] is eliminated at $1 \, M \, M^+$. Therefore, the free energy change on extrapolation to $1 \, M \, M^+$ [$\Delta G^{\circ}_{obs} (1 \, M \, M^+)$], should reflect mainly contributions arising from nonelectrostatic interactions [$\Delta G^{\circ}(\text{non-el})$] and the ionic interaction between the positively charged groups on the ligand and the negatively charged phosphates [$\Delta G^{\circ}(\text{ionic})$]. This relationship is expressed in Eq. (12):

$$\Delta G^{\circ}_{obs} (1 \, M \, M^+) = \Delta G^{\circ}(\text{non-el}) + z\Delta G^{\circ}(\text{ionic}) \tag{12}$$

$\Delta G^{\circ}(\text{ionic})$ is the free energy change per ionic interaction in the absence of counterion release; $\Delta G^{\circ}(\text{ionic})$ is assumed to be independent of salt concentration.

  Based on analyses of the monovalent salt dependence of K_{obs} for a series of oligolysines, $\Delta G^{\circ}(\text{ionic})$ has been estimated to be very nearly zero, with a slightly unfavorable value of $+0.2 \pm 0.1$ kcal/mol (per ionic interaction) for interactions with duplex nucleic acids[18,78] and -0.1 ± 0.1 kcal/mol (per ionic interaction) for poly(U).[68] These values of $\Delta G^{\circ}(\text{ionic})$ are presumably close to zero, since the formation of an ionic interaction within the ligand–nucleic acid complex is partially compensated by the loss of a counterion (M^+)–phosphate interaction; in other words, the exchange reaction has a net free energy change near zero (at the $1 \, M \, M^+$ standard state). Therefore, the value of $\Delta G^{\circ}_{obs} (1 \, M \, M^+)$ provides an estimate of the value of $\Delta G^{\circ}(\text{non-el})$, with a possible small contribution for $\Delta G^{\circ}(\text{ionic})$.[18] However, we emphasize that this approach will not yield meaningful estimates of $\Delta G^{\circ}(\text{non-el})$ if significant preferential interactions with the ligand (protein) exist, since $\log K_{obs}$ will not be a linear function of $\log[M^+]$.

  3. Competitive Effects of Monovalent and Divalent Cations for Binding to Nucleic Acids. As yet, we have discussed ligand–nucleic acid equilibria only in the presence of excess monovalent salt, MX. The effects on K_{obs} caused by divalent cations such as Mg^{2+} are qualitatively similar; however, the coefficient ψ in Eq. (9) is replaced by ϕ, which represents the number of divalent counterions thermodynamically associated per phosphate.[26]

Therefore, in the absence of preferential anion interactions, the relationship in Eq. (13) should hold:

$$\frac{\partial \log K_{obs}}{\partial \log[M^{2+}]} = (\phi/\psi)\frac{\partial \log K_{obs}}{\partial \log[M^+]} \tag{13}$$

For duplex B-form DNA, $\phi = 0.47$, $\phi/\psi = \sim 0.53$,[26] whereas for single-stranded poly(U), $\phi = 0.425$, $\phi/\psi = \sim 0.60$. If Eq. (13) does not hold for a particular ligand–nucleic acid interaction, then this suggests that salt effects other than those due to counterion release from the nucleic acid should be considered. These could result from preferential anion or cation interactions with the protein (see below).

In buffers containing a mixture of monovalent, M^+, and divalent, M^{2+}, cations, the dependence of K_{obs} on $[M^+]$ is complicated owing to the competition between monovalent and divalent cations for binding to the nucleic acid.[20,78] In this case, Eqs. (9) and (10) no longer apply, and Eq. (9) is replaced by Eq. (14):

$$L^{z+} + D \rightleftharpoons LD + x\,M^+ + y\,M^{2+} \tag{14}$$

In the absence of preferential ion interactions with the ligand, the result of the M^+/M^{2+} competition for the nucleic acid will be 2-fold: (1) the value of K_{obs}, as well as the absolute value of the monovalent salt dependence of K_{obs}, $|\partial \log K_{obs}/\partial \log[M^+]|$, will be lower than the values in the absence of divalent cations; and (2) $\partial \log K_{obs}/\partial \log[M^+]$ will no longer be independent of M^+ concentration, that is, a plot of $\log K_{obs}$ versus $\log[M^+]$ will display curvature. Both of these effects result from the fact that $\partial \log K_{obs}/\partial \log[M^+]$ measures only the release of M^+ [the stoichiometric coefficient x in Eq. (14)]. Therefore, when both M^+ and M^{2+} are bound to the nucleic acid, less M^+ will be released on formation of the ligand–nucleic acid complex. Furthermore, the binding of M^{2+} itself will be a function of $[M^+]$, since it also is a positively charged ligand; therefore, the coefficients x and y in Eq. (14) will change with $[M^+]$, resulting in a nonlinear plot of $\log K_{obs}$ versus $\log[M^+]$.[20,78]

B. Cation and Anion Effects on Protein–Nucleic Acid Equilibria

1. General Ion Effects. In general, the equilibrium binding constants for protein–nucleic acid interactions will display a more complicated range of effects owing to changes in salt concentration than is observed for simple positively charged oligocations. This is due to the fact that cations, anions, protons, and water interact with most proteins; hence, preferential interactions of these ions with the protein must be considered in addition to the preferential interactions of cations and water with the nucleic acid. These additional preferential interactions are indicated in Eq. (15),

$$P + D \rightleftharpoons PD + \Delta c\, M^+ + \Delta a\, X^- + \Delta h\, H^+ + \Delta w\, H_2O \qquad (15)$$

where the coefficients Δc, Δa, Δh, and Δw represent the net preferential interactions of cations, anions, protons, and water. The preferential interaction parameter for cation binding is defined as $\Delta c = (c_{PD} - c_P - c_D)$, where c_i represents the moles of cations bound thermodynamically per mole of species i; Δa, Δh, and Δw are defined similarly. As defined, the preferential interaction parameters include contributions from both direct binding and nonideality (activity coefficient) effects. The dependence of K_{obs} on monovalent salt concentration at constant temperature, pressure, and pH for this case is then given by[19]

$$\left(\frac{\partial \log K_{obs}}{\partial \log[MX]} \right) = -(\Delta c + \Delta a) + \frac{2[MX]}{[H_2O]} \Delta w \qquad (16)$$

The justification for the use of monovalent salt concentration, rather than mean ionic activities, in Eq. (16) has been discussed.[18,19] We note that the individual terms Δc, Δa, and Δw can be either positive or negative, indicating a net release or uptake of that species, respectively; however, for positively charged ligands, cation release from the nucleic acid will always be a major component of Δc. The terms Δc and Δw can have contributions from both the protein and the nucleic acid, whereas any anion effects will be due to preferential interactions with the protein. We also note that each of the terms Δc, Δa, and Δw represent net preferential interactions. Therefore, one can imagine the case in which a net release of cations on formation of PD can have separate contributions owing to release of cations from the nucleic acid, which is partially offset owing to an uptake of cations by the protein. This appears to be the case for *E. coli* SSB binding to single-stranded nucleic acids in its $(SSB)_{65}$ binding mode.[87] We also note that Δa and Δc are not constrained by any electroneutrality relationship and can assume any values, depending on the particular interaction.

In general, each of the coefficients in Eq. (15) can vary with the bulk salt concentration, hence the derivative $(\partial \log K_{obs}/\partial \log[MX])$ can also be a function of salt concentration. The one exception to this is the contribution owing to cation release from the nucleic acid, which seems to be constant over a wide range of salt concentration[18,22,71,73] as discussed above. However, for a particular protein–nucleic acid interaction, Eq. (16) can usually

[87] L. B. Overman, W. Bujalowski, and T. M. Lohman, *Biochemistry* **27**, 456 (1988).

be simplified, since not all of the species are likely to interact preferentially. In particular, except at high salt concentrations ($\geq 0.5\ M$), the term arising from preferential hydration is expected to be negligible when compared to the ion release terms.

For simple oligocations,[73] such as polyamines and oligolysines, the net positive charge on the oligocation determines the salt dependence of K_{obs}; however, this is definitely not the case for proteins. This is obvious from the fact that many nucleic acid-binding proteins have a net negative charge at neutral pH ($pI < 7$) and yet still bind tightly to DNA with a net release of cations. An illustration of this point is shown in Fig. 4B for the interaction of the *E. coli* SSB tetramer with poly(U) to form the $(SSB)_{65}$ binding mode. The salt dependence of K_{obs} in NaF is essentially independent of pH over the range from 5.5 to 9.0, even though the net charge of the SSB tetramer should vary drastically over this range, perhaps even changing sign since the protein has a pI of approximately 6.0.[88]

The dependences of K_{obs} on salt concentration for the various ligand–nucleic acid interactions discussed in the previous sections are not due to simple ionic strength effects. The sole use of ionic strength as a parameter to describe the effects of salt concentration on ligand–nucleic acid interactions, or any macromolecular interaction for that matter, implies that the effects of salt are due only to long-range screening effects, without contributions from the direct binding of ions. Although the ionic strength of a solution is the important variable for the description of interactions among ions in simple salt solutions as indicated in De-bye–Hückel theory,[89–91] its use to describe salt effects on ligand–nucleic acid interactions is incorrect. This is most clear from the fact that dramatically different values of K_{obs} are obtained in buffers containing monovalent cations versus divalent cations, even though the solutions are at the same "ionic strength." Therefore, theoretical approaches that attempt to explain salt effects on ligand–nucleic acid interactions, or other macromolecular interactions, solely in terms of ionic strength effects cannot be correct, quantitatively.

Decomposition of Salt Dependence into Contributions from Anions versus Cations. To distinguish the relative contributions to the salt dependence of K_{obs} for a ligand–nucleic acid interaction, arising from preferential cation versus anion interactions, a number of experiments can be performed. First, the effects on K_{obs} and ($\partial \log K_{obs}/\partial \log[MX]$) of a series of

[88] J. Weiner, L. Bertsch, and A. Kornberg, *J. Biol. Chem.* **250,** 1972 (1975).
[89] P. Debye and E. Huckel, *Physik. Z.* **24,** 185 (1923).
[90] P. Debye and E. Huckel, *Physik. Z.* **24,** 384 (1923).
[91] P. Debye and E. Huckel, *Physik. Z.* **25,** 97 (1924).

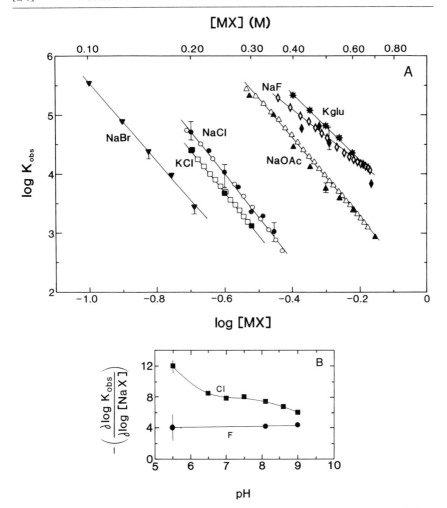

FIG. 4. (A) Dependence of log K_{obs} on total monovalent cation concentration, M^+, for a series of salts differing in anion type, for the equilibrium binding of the *E. coli* SSB tetramer to poly(U) to form the $(SSB)_{65}$ binding mode (25.0°, pH 8.1). Linear least-squares lines are drawn through each set of data. (Data from Overman *et al.*[87]) (B) Dependence of $(\partial \log K_{obs}/\partial \log[NaX])$ on pH for the equilibrium binding of the *E. coli* SSB tetramer to poly(U) to form the $(SSB)_{65}$ binding mode (25.0°) in NaF and NaCl. [Data from L. B. Overman, Ph.D. Thesis, Texas A & M University, College Station, Texas (1989).]

monovalent salts which differ only in the anion type (e.g., KBr, KCl, KCH_3CO_2, KF) can be examined. The interactions of anions with proteins generally follows the Hofmeister series ($F^- < CH_3CO_2^- < Cl^- < Br^-$),[19,92-94] with fluoride showing the weakest preferential interaction.[87] Figure 4A illustrates the dramatic effect of different anions on the interaction of the *E. coli* SSB tetramer with poly(U). In contrast, no effects of anions have been observed for the interaction of simple, short oligolysines ($z \leq 10$) with either duplex or single-stranded nucleic acids.[73,78] This approach should be coupled with changing the pH, since anion binding sites on the protein are generally titratable. Fig. 4B shows data from such an approach for the interaction of the *E. coli* SSB tetramer with poly(U) to form the $(SSB)_{65}$ binding mode.[95] At 25°, the salt dependence in NaCl (∂ log K_{obs}/∂ log[NaCl]) is observed to change dramatically with pH, from a value of -6.7 ± 0.2 at pH 8.6 to -12.0 ± 0.8 at pH 5.5. However, (∂ log K_{obs}/∂ log[NaF]) equals -4.4 ± 0.7, independent of pH. This indicates that the preferential interaction of Cl^- increases with decreasing pH. Furthermore, F^- has a low preferential interaction, and the salt dependence in NaF appears to reflect only the net Na^+ release on complex formation.

Second, the dependence of K_{obs} on monovalent cation concentration, ∂ log K_{obs}/∂ log[M^+], can be compared to its dependence on divalent cation concentration, ∂ log K_{obs}/∂ log[M^{2+}]. If Eq. (13) does not hold, then preferential interactions of either cations or anions with the protein are indicated.

Third, mixtures of salts with a common anion but different monovalent and divalent cations (e.g., KCl and $MgCl_2$) can be used to vary the anion concentration over a wider range than is accessible with a single salt type. The use of this approach generally requires that preferential cation interactions with the protein are negligible, so that the effect of the divalent cation, M^{2+}, is only due to its competition with monovalent cation, M^+, for binding to the nucleic acid.[20,78]

V. Obtaining Thermodynamic Parameters by van't Hoff Analysis

Thermodynamic parameters for a ligand–nucleic acid interaction can be obtained from studies of the temperature dependence of the equilibrium

[92] F. Hofmeister, *Arch. Exp. Pathol. Pharmakol.* **24**, 247 (1888).
[93] P. H. von Hippel and T. Schleich, *in* "Biological Macromolecules" (G. Fasman and S. Timasheff, eds.), p. 417. Dekker, New York, 1969.
[94] K. D. Collins and M. W. Washabaugh, *Q. Rev. Biophys.* **18**, 323 (1985).
[95] L. B. Overman, Ph.D. Thesis, Texas A&M University, College Station, Texas (1989).

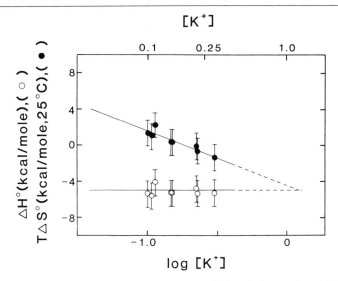

FIG. 5. Dependence of ΔH°_{obs} and $T\Delta S^\circ_{obs}$ on $\log[K^+]$ for the interaction of KWK$_4$-NH$_2$ with poly(U) (25.0°, pH 6.0, KCH$_3$CO$_2$). The values of ΔH°_{obs} at each salt concentration were determined from van't Hoff analysis.[68] Values of $T\Delta S^\circ_{obs}$ were calculated from $T\Delta S^\circ_{obs} = \Delta H^\circ_{obs} + RT \ln K_{obs}$.

constant, K_{obs}, as indicated by

$$[\partial \ln K_{obs}/\partial(1/T)]_P = -\Delta H^\circ_{obs}/R \qquad (17)$$

Therefore, the slope of a plot of $\ln K_{obs}$ versus $1/T$ (van't Hoff plot) yields the standard enthalpy change, ΔH°_{obs}, for the reaction at that temperature. If ΔH°_{obs} is independent of temperature, then the van't Hoff plot will be linear.

For simple oligocations, such as polyamines and oligolysines, which bind exclusively electrostatically, ΔH°_{obs} is nearly zero and independent of salt concentration. This reflects a near balance between the enthalpy change associated with neutralization of a phosphate by a bound counterion (M^+) versus a positive charge on the oligocation.[18,78,79] Figure 5 shows that independence of ΔH° on KCH$_3$CO$_2$ concentration for the equilibrium binding of a positively charged oligopeptide, L-Lys-L-Trp-(L-Lys)$_4$-NH$_2$ (KWK$_4$-NH$_2$), to poly(U).[68] In this case, the salt dependence of ΔG° is due solely to the salt dependence of ΔS°. The slightly negative value of ΔH°, -5 ± 1 kcal/mol, is due to the interaction of the Trp residue.

In general, for a protein–nucleic acid interaction, it is difficult to dissect ΔH°_{obs} into its separate contributions from hydrogen bonds, ionic, hydrophobic interactions, etc. This is primarily because ΔH°_{obs}, as with all

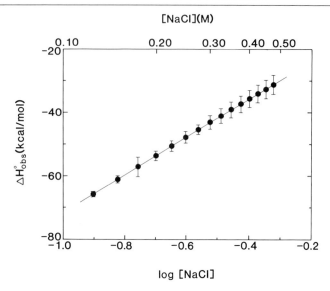

FIG. 6. Dependence of the van't Hoff enthalpy, ΔH°_{obs}, on log[NaCl] for the equilibrium binding of the *E. coli* SSB tetramer to poly(U) to form the (SSB)$_{65}$ binding mode (pH 8.1). The linear least-squares line through the data is $\Delta H^{\circ}_{obs} = 60 \log[\text{NaCl}] - 12$ (kcal/mol). (Data from Overman.[95])

thermodynamic quantities, reflects all changes that occur within the system, not simply those occurring locally at the protein–nucleic acid interface. Furthermore, ΔH°_{obs} (as well as ΔS°_{obs}) is not generally constant, but rather is a function of solution conditions (e.g., salt and pH). This can result if protonation or deprotonation events, which can also be linked to ion binding, are coupled to the formation of the ligand (protein)–nucleic acid complex. Therefore, just as for K_{obs} (ΔG°_{obs}), important information about the thermodynamics of the interaction will be obtained from studies of the dependence of ΔH°_{obs} on solution variables, whereas a determination of ΔH°_{obs} under only one set of solution conditions provides limited information. One must examine how ΔH°_{obs} varies with solution conditions (pH, salt concentration and type, temperature) in order to begin to determine the various contributions to complex stability. For example, the ΔH°_{obs} for the interaction of the *E. coli* SSB tetramer with poly(U) to form the (SSB)$_{65}$ binding mode is quite dependent on the NaCl concentration, ranging from -61 kcal/mol (SSB tetramer) at 0.15 M NaCl to -31 kcal/mol at 0.48 M NaCl as shown in Fig. 6.[95] This seems to reflect the fact that formation of the (SSB)$_{65}$ complex with poly(U) is coupled to preferential ion binding to the protein, and the SSB ion binding sites are titratable.[95] Clearly, an understanding of the origins of the ΔH°_{obs} for this interaction would be impossible, based on measurements at only a single NaCl concentration. The temperature dependence of ΔH°_{obs} is related to the change in heat capacity for the binding reaction, ΔC°_P, as in Eq. (18):

$$(\partial \Delta H^\circ_{obs}/\partial T)_P = \Delta C^\circ_{P,obs} \qquad (18)$$

Therefore, information about the heat capacity change for a reaction can be obtained, in principle, from a study of the temperature dependence of K_{obs} (ΔG°_{obs}). In practice, it is often difficult to study a binding equilibrium over a wide enough range of temperature to obtain accurate values of $\Delta C^\circ_{P,obs}$. In these cases, $\Delta C^\circ_{P,obs}$ is better determined by calorimetric means, although there have yet been no studies of protein–nucleic acid interactions by calorimetry. In any event, a binding equilibrium that yields a linear van't Hoff plot indicates that $\Delta C^\circ_{P,obs}$ equals 0 for the reaction. In general, ligands that bind to nucleic acids in an entirely electrostatic mode will have values of ΔH°_{obs} and ΔC°_P close to zero, as has been observed for the interaction of simple oligocations with nucleic acids.[68,78] Significant values of $\Delta H^\circ_{obs} < 0$ have been measured for the nonspecific binding of E. coli lac repressor to duplex DNA,[26] as well as the nonspecific binding of the E. coli SSB protein to poly(U)[95]; however, in these cases, it appears that the nonzero ΔH°_{obs} results from a linkage of binding to protonation of residues on the protein. On the other hand, significant curvature has been observed in van't Hoff plots for a number of sequence-specific protein–DNA interactions (lac repressor, EcoRI endonuclease, RNA polymerase), indicating $\Delta C^\circ_P \ll 0$ on formation of these complexes. It has been suggested that this large $\Delta C^\circ_P \ll 0$ reflects the hydrophobic effect,[96,97] that is, the removal of large regions of the protein from contact with water, resulting in the release of bound water, which contributes a large favorable entropy change ($\Delta S^\circ_{obs} \gg 0$). The extent to which these large heat capacity effects are general is not yet clear.

VI. Summary

Ligand– and protein–DNA equilibria are extremely sensitive to solution conditions (e.g., salt, temperature, and pH), and, in general, the effects of different solution variables are interdependent (i.e., linked). As a result, an assessment of the basis for the stability and specificity of ligand– or protein–DNA interactions requires quantitative studies of these

[96] J.-H. Ha, R. S. Spolar, and M. T. Record, Jr., J. Mol. Biol. 209, 801 (1989).
[97] R. S. Spolar, J.-H. Ha, and M. T. Record, Jr., Proc. Natl. Acad. Sci. U.S.A. 86, 8382 (1989).

interactions as a function of a range of solution variables. Many of the most dramatic effects on the stability of these interactions result from changes in the entropy of the system, caused by the preferential interaction of small molecules, principally ions which are released into solution on complex formation. A determination of the contributions of these entropy changes to the stability and specificity of protein– and ligand–DNA interactions requires thermodynamic approaches and cannot be assessed from structural studies alone.

Acknowledgments

We thank Tom Record for sharing unpublished data from his laboratory and for many discussions of aspects of this material and Lisa Lohman for help in preparing the figures. Research from the author's laboratory was supported by grants from the National Institutes of Health (GM39062, GM30498) and the American Cancer Society (NP-756). T.M.L. is a recipient of American Cancer Society Faculty Research Award FRA-303.

[25] Nonspecific Ligand–DNA Equilibrium Binding Parameters Determined by Fluorescence Methods

By TIMOTHY M. LOHMAN and DAVID P. MASCOTTI

I. Introduction

Spectroscopy provides sensitive and convenient methods to monitor protein (ligand)–nucleic acid interactions. In this chapter we discuss the use of spectroscopic probes to monitor nonspecific ligand–nucleic acid interactions as well as rigorous methods of analysis to obtain model-independent equilibrium binding isotherms from spectroscopic titrations.[1-3] Such isotherms can then be analyzed, using appropriate statistical thermodynamic models, to yield equilibrium binding parameters for the interaction. Studies of the dependence of the binding parameters on temperature and other solution conditions can then provide thermodynamic information, which is necessary to understand the basis for stability and specificity of the interactions.[4] See also [24] in this volume for further discussion.[5]

---

[1] C. J. Halfman and T. Nashida, *Biochemistry* **11**, 3493 (1972).
[2] W. Bujalowski and T. M. Lohman, *Biochemistry* **26**, 3099 (1987).
[3] T. M. Lohman and W. Bujalowski, this series, Vol. 208, p. 258.
[4] M. T. Record, Jr., J.-H. Ha, and M. A. Fisher, this series, Vol. 208, p. 291.
[5] T. M. Lohman and D. P. Mascotti, this volume [24].

This chapter focuses on the use of steady-state fluorescence techniques to monitor changes in the ligand (protein) that accompany binding, with examples drawn from studies of ligands and proteins that bind nonspecifically to nucleic acids. Spectroscopic methods, such as fluorescence, UV absorbance, and circular dichroism, although indirect, offer many advantages for the study of ligand–nucleic acid as well as other macromolecular interactions. In the case of steady-state fluorescence, one has access to a rapid, nonradioactive method, which can be used at fairly low concentrations of ligand and nucleic acid. For example, micromolar concentrations of tryptophan can be detected easily, hence binding studies with proteins or peptides containing multiple tryptophans can often be performed at concentrations in the 10 nM range. A further general advantage of spectroscopic methods is that titrations of a single solution can be monitored continuously, thus increasing the precision and ease of data collection. These are essential considerations if one is to attempt to investigate systematically the thermodynamics of a protein– or ligand–nucleic acid interaction over a wide range of solution variables. Finally, although assumptions regarding the relationship between the spectroscopic signal and the degree of binding are often made when using spectroscopic techniques, these assumptions are not necessary, since thermodynamic, model-independent methods of analysis are available which enable the determination of absolute binding isotherms.[1-3] Therefore, rigorous investigations of the equilibrium binding and thermodynamic properties of ligand (protein)–nucleic acid interactions that rely on convenient spectroscopic probes are possible.

II. Model-Independent Determination of Ligand–Nucleic Acid Binding Isotherms Using Spectroscopic Approaches to Monitor Binding

A variety of experimental approaches have been described for the study of protein–nucleic acid interactions[6]; however, we shall focus on the use of spectroscopic approaches, specifically changes in steady-state fluorescence, that accompany formation of the complex. To use a change in a spectroscopic signal that is induced on formation of a ligand–nucleic acid complex to obtain a true equilibrium binding isotherm, either the relationship between the signal change and the degree of binding must be known, or a method of analysis must be used that does not require knowledge of this relationship. If some relationship between the signal change and the degree of binding is assumed (e.g., linear), then the resulting

[6] A. Revzin, in "Nonspecific DNA–Protein Interactions" (A. Revzin, ed.), p. 5. CRC Press, Boca Raton, Florida, 1990.

binding isotherm and thermodynamic parameters determined from that binding isotherm are only as valid as the assumed relationship. This major caveat should be kept in mind, since it severely limits the certainty with which quantitative conclusions can be drawn from such data.

The relationship between the spectroscopic signal change and the degree of binding can be obtained through comparisons of isotherms determined by spectroscopic approaches with those determined using thermodynamically rigorous techniques that measure binding directly (e.g., equilibrium dialysis). However, model-independent methods of analysis of spectroscopic titrations have also been described that do not require prior knowledge of the relationship between the signal change and the degree of ligand binding and, in fact, can be used to determine these relationships. We refer to these approaches as binding density function (BDF) analyses. The analysis differs depending on whether the signal change is from the macromolecule (MBDF analysis)[1] or from the ligand (LBDF analysis),[2] and both methods have been reviewed recently.[3] In the following sections we discuss these methods briefly, although we focus on the LBDF analysis for ligand–nucleic acid interactions that are monitored by changes in the fluorescence of the ligand, since the experimental examples that we refer to are of this type. However, it should be noted that these methods of analysis can be used with any signal that reflects binding, independent of its origin (ligand or nucleic acid).

A. Equilibrium Titrations

We mainly discuss the nonspecific binding of ligands to long linear nucleic acids. For this case, the nucleic acid can bind multiple ligands (proteins) and therefore is defined as the macromolecule. However, there are cases involving the interaction of oligonucleotides with multisubunit proteins, in which case the protein is defined as the macromolecule.[7–9] This designation is not simply semantic, since the type of BDF analysis that is applied depends on this identification.[2,3] Based on this nomenclature, there are two types of titrations; each has its own utility, depending on the circumstances and whether the spectroscopic signal that is used to monitor binding originates from the ligand or the nucleic acid.

1. Addition of Ligand to Constant Concentration of Nucleic Acid ("Normal" Titration). Varying the ligand concentration while keeping the nucleic acid concentration constant is the most common type of titration and the easiest to consider, conceptually, since the binding density

[7] T. M. Lohman and W. Bujalowski, *Biochemistry* **27,** 2260 (1988).
[8] W. Bujalowski and T. M. Lohman, *J. Mol. Biol.* **207,** 249 (1989).
[9] W. Bujalowski and T. M. Lohman, *J. Mol. Biol.* **207,** 269 (1989).

(ligands bound per nucleic acid) increases during the course of the titration. We refer to it as a normal titration for this reason. When spectroscopic probes are used to monitor binding, a normal titration is generally used when the spectroscopic signal is from the nucleic acid, since in this case, any change in the spectroscopic signal is a direct reflection of binding.

2. Addition of Nucleic Acid to Constant Concentration of Ligand ("Reverse" Titration). In a reverse titration, the binding density decreases during the course of the titration. It is generally used when the spectroscopic signal is from the ligand, since, in this case, any change in the spectroscopic signal is a direct reflection of binding. However, the use of a reverse titration does not generally enable the entire range of binding densities to be spanned with a single titration, thus placing limitations on the utility of the approach.[2,3] A normal titration can also be used when the spectroscopic signal is from the ligand,[10] thus allowing one to span a wider range of binding densities; however, the data analysis for this case is more complex, and the binding isotherms obtained are generally less accurate than when normal titrations are performed with the signal originating from the nucleic acid.

B. Binding Density Function Analysis

1. Ligand Binding Density Function (LBDF) Analysis. Here we review briefly the general case in which binding is accompanied by a change in the signal (fluorescence) of the ligand, which follows the treatment of Bujalowski and Lohman.[2] Consider the equilibrium binding of a ligand, L, to a nucleic acid, D, such that there can be r states of bound ligand, with each state possessing a different molar fluorescence signal, F_i ($i = 1$ to r). The experimentally observed signal, F_{obs}, from a ligand solution [at total concentration, L_T (moles per liter)] in the presence of nucleic acid [at total nucleotide concentration, D_T (moles per liter)] can then be expressed:

$$F_{obs} = F_F L_F + \sum_{i=1}^{r} F_i L_i \qquad (1)$$

where F_F and L_F are the molar fluorescence signal and concentration of free ligand, respectively, and F_i and L_i are the molar fluorescence signal and concentration, respectively, of the ligand bound in state i. The concentrations of free and bound ligand are related to the total ligand concentration by conservation of mass:

$$L_T = L_F + D_T \sum_{i=1}^{r} \nu_i \qquad (2)$$

[10] N. C. M. Alma, B. J. M. Harmsen, E. A. M. deJong, J. van der Ven, and C. W. Hilbers, *J. Mol. Biol.* **163**, 47 (1983).

where ν_i ($= L_i D_T$) is the ligand binding density (moles of ligand bound per mole of nucleotide) for the ith state.

Based on these definitions, it can be shown[2,3] that the relationship in Eq. (3) holds,

$$Q_{obs}(L_T D_T) = \sum_{i=1}^{r} Q_i \nu_i \tag{3}$$

where $Q_{obs} = (F_{obs} - F_F L_T)/F_F L_T$ is the experimentally observed quenching of the ligand fluorescence at total ligand and total nucleic acid concentrations, L_T and D_T, respectively, and $Q_i = (F_i - F_F)/F_F$ is the quenching of the intrinsic fluorescence of the bound ligand in state i. (Note that the quantity, $F_F L_T \equiv F_0$, is simply the initial fluorescence signal from the free ligand before addition of nucleic acid.)

Equation (3) states that the ligand binding density function (LBDF), $Q_{obs}(L_T/D_T)$, is equal to $\Sigma Q_i \nu_i$, the sum of the binding densities for all i states of ligand binding, weighted by the intrinsic fluorescence quenching for each bound state. Because the Q_i terms are molecular quantities that are constant for each binding state i, under a constant set of solution conditions, then the quantity, $\Sigma Q_i \nu_i$, and hence the LBDF, will also be constant for a given equilibrium binding density distribution, $\Sigma \nu_i$. Therefore, at equilibrium, the values of L_F and $\Sigma \nu_i$ (and each separate value of ν_i) are constant for a given value of $Q_{obs}(L_T/D_T)$, independent of the nucleic acid concentration, D_T. As a result, one can obtain model-independent estimates of the average ligand binding density, $\Sigma \nu_i$, and L_F from an analysis of a plot of $Q_{obs}(L_T/D_T)$ versus D_T for two or more reverse titrations performed at different total ligand concentrations, L_T, under identical solution conditions (see Section IV,A,1 for details of the analysis). As discussed previously,[2] when reverse titrations are used to obtain the binding data, then multiple titrations (6 to 8) are needed to span the entire binding density range, since each individual reverse titration spans only a part of the binding density range (see Fig. 3). This is a major difference between the use of reverse versus normal titrations.[2,3] We also note that the use of the binding density function method of analysis is valid only in the absence of ligand or nucleic acid aggregation (see Section IV,A,1,b for further discussion).

2. *Macromolecule Binding Density Function (MBDF) Analysis.* When the signal is from the macromolecule, that is, the species that binds multiple ligands, the appropriate binding density function is simpler than when the signal is from the ligand. If we again consider the signal to be a fluorescence change (increase or decrease), then the BDF for this case is given in Eq. (4),[1,3]

$$Q_{obs} = \sum_{i=1}^{r} Q_i \nu_i \tag{4}$$

where $Q_{obs} = (F_{obs} - F_F M_T)/F_F M_T$ is the experimentally observed quenching of the macromolecule fluorescence at total ligand and total macromolecule concentrations, L_T and M_T, and $Q_i = (F_i - F_F)/F_F$ is the quenching of the intrinsic fluorescence of the bound macromolecule in state i. This approach involves the analysis of a series of titrations at constant macromolecule concentration, M_{Tx}, in a manner similar to that discussed above for the LBDF analysis (see Lohman and Bujalowski[3,7] and Bujalowski and Lohman[8] for details).

Generation of Binding Isotherms from a Single Titration when $Q_{obs}/Q_{max} = L_B/L_T$. The LBDF analysis allows one to determine rigorously a model-independent binding isotherm and the relationship between Q_{obs} and the fraction of bound ligand, L_B/L_T. However, the LBDF method requires 6 to 8 titrations to construct a single binding isotherm with good precision over a wide range of binding densities (see Fig. 3 and Section IV,A,1), although this is necessary if the relationship between Q_{obs} and L_B/L_T is not known *a priori*. However, if it can be determined from the LBDF analysis, or by comparisons with isotherms determined from more direct methods, that a linear relationship exists between Q_{obs} and L_B/L_T over a wide range of binding densities, then one can use this linear relationship to determine the average binding density, ν, and L_F from a single titration curve.[2] If a linear relationship exists, then Eq. (3) reduces to Eq. (5), which leads to Eqs. (6) and (7),

$$\frac{Q_{obs}}{Q_{max}} = \frac{L_B}{L_T} \tag{5}$$

$$L_F = \left(1 - \frac{Q_{obs}}{Q_{max}}\right)L_T \tag{6}$$

$$\nu_Q = \left(\frac{Q_{obs}}{Q_{max}}\right)\left(\frac{L_T}{D_T}\right) \tag{7}$$

where Q_{max} is the fluorescence quenching when all of the ligand is bound ($L_B/L_T = 1$). Q_{max} can be obtained from the plateau value of Q_{obs} if $L_B/L_T = 1$ can be reached experimentally, or it can be obtained from the LBDF analysis as described in Sections II,B,1 and IV,A,2. In Eq. (7) we have designated the binding density as ν_Q to emphasize that it is calculated based on the relationship in Eq. (5). We emphasize that one should not assume *a priori* that Eqs. (5)–(7) are valid for a particular ligand–nucleic acid interaction in the absence of direct evidence, since serious errors in the calculated isotherms and binding parameters can result. On the other hand, if a direct proportionality does not exist between the signal change and the fraction of bound ligand over a wide range of binding densities, the true binding isotherm can still be constructed without any assumptions through use of the LBDF analysis.

III. Experimental Models

In this section we describe the experimental methods used to obtain equilibrium reverse titrations for a ligand–nucleic acid interaction, in which a fluorescence signal from the ligand is used to monitor binding. The examples used to demonstrate these methods are taken from studies in our laboratory on the nonspecific binding of the *Escherichia coli* single strand binding (SSB) Protein[2,3,8,9,11–13] and synthetic oligolysines to synthetic homopolynucleotides.[14] In these cases, the quenching of the tryptophan fluorescence of the peptide or protein is used to monitor binding. However, the methods of analysis are independent of the signal used to monitor binding, and methods are also available when the signal is from the nucleic acid.[3] Specific ligand–nucleic acid interactions as well as other ligand–macromolecule interactions can also be examined using these approaches if an appropriate signal is available.

A. Equipment

Fluorescence experiments were performed with an SLM-Aminco 8000C spectrofluorometer (Urbana, IL) in its ratiometric mode. The sample temperature is controlled by circulating water from a refrigerated circulating water bath (Lauda RMS-6, Fisher Scientific, Pittsburgh, PA) through the jacketed cuvette holder. Typically, 2-ml samples are used in 4-ml quartz cuvettes [10 × 10 × 4.5 cm; Starna Cell Spectrasil (Atascadero, CA)]. The sample within the cuvette is stirred continuously during the experiment using a cylindrical Teflon-coated stir bar [8 × 8 mm; Bel-Art F37150 (Thomas Scientific, Philadelphia, PA)]; stirring speed is controlled by the magnetic stir plate within the SLM fluorometer. Although this cuvette requires larger sample volumes (with stir bar in place, the cuvette has an effective volume of 3 ml), it enables us to use the cylindrical stir bar, which provides very efficient and continuous but gentle stirring without the need to remove the cuvette during a titration. We have not found a stir bar that provides adequate stirring within a smaller volume cuvette. If smaller cuvettes must be used to conserve precious samples, less efficient mixing methods must be used (e.g., inversion or mixing with

[11] W. Bujalowski and T. M. Lohman, *J. Mol. Biol.* **195,** 897 (1987).
[12] L. B. Overman, W. Bujalowski, and T. M. Lohman, *Biochemistry* **27,** 456 (1988).
[13] L. B. Overman, Ph.D. Thesis, Texas A&M University, College Station, Texas (1989).
[14] D. P. Mascotti and T. M. Lohman, *Proc. Natl. Acad. Sci. U.S.A.* **87,** 3142 (1990).

a Teflon stirring rod), which can introduce additional problems arising from sample denaturation, sticking of the sample to the stir rod, and difficulty in repositioning the cuvette identically within the sample chamber.

Titrations are performed using a glass capillary pipettor (Drummond Microdispenser, Model 105; Thomas Scientific) or Pipetteman pipettor (Rainin, Woburn, MA) for low viscosity solutions. The Drummond is used for additions of 1–5 μl, whereas the Pipetteman is used for larger volumes. A Microman positive displacement pipettor (Rainin) is used when dispensing solutions with viscosities significantly higher than water (e.g., $\geq 10\%$ glycerol).

B. Preliminary Experiments and Considerations

The following preliminary information should generally be obtained in advance of attempting to generate an equilibrium binding isotherm, using fluorescence techniques.

1. Extinction Coefficients. Accurate determination of the ligand and nucleic acid concentrations is essential. Routine measurements of protein or nucleic acid concentrations are best made by absorbance spectroscopy if sufficient material is available; however, this requires knowledge of the extinction coefficient for the protein and nucleic acid at a particular wavelength. These extinction coefficients are also needed in order to apply corrections for "inner filter" effects, as described in Section III,C,2. Extinction coefficients for amino acids, homopolynucleotides, and duplex DNA have been determined,[15-17] and methods for the calculation of extinction coefficients for oligonucleotides, which depend on the nucleotide sequence, have been described.[18] The latter are based on knowledge of the extinction coefficients for a series of dinucleotides.[19] However, extinction coefficients for proteins or peptides, especially nucleic acid-binding proteins that have not been previously investigated, must be determined individually.

An accurate and relatively simple means has been described for determining the extinction coefficients for proteins containing tryptophan

[15] G. Fasman (ed.), "CRC Handbook of Biochemistry and Molecular Biology, Proteins," Vol. 1, 3rd Ed. CRC Press, Boca Raton, Florida, 1976.

[16] G. Fasman (ed.), "CRC Handbook of Biochemistry and Molecular Biology, Nucleic Acids," Vol. 1, 3rd ed. CRC Press, Boca Raton, Florida, 1976.

[17] G. Felsenfeld and H. T. Miles, *Annu. Rev. Biochem.* **236,** 407 (1967).

[18] M. M. Senior, R. A. Jones, and K. J. Breslauer, *Proc. Natl. Acad. Sci. U.S.A.* **85,** 6242 (1988).

[19] C. R. Cantor, M. M. Warshaw, and H. Shapiro, *Biopolymers* **9,** 1059 (1970).

and/or tyrosine and whose amino acid content is known or can be predicted from the DNA sequence of its gene.[20,21] The extinction coefficient, at wavelength λ, of a protein in its native form, $\varepsilon_{P,\text{Native}}$ (M^{-1} cm^{-1}), can be obtained by comparing the absorbance in its native form, A_{Native}, to the absorbance in its denatured form, which can be obtained in the presence of 6 M guanidine hydrochloride (Gu-HCl), $A_{\text{Gu-HCl}}$, as described in Eq. (8):

$$\varepsilon_{P,\text{Native}} = (A_{\text{Native}})(\varepsilon_{P,\text{Gu-HCl}})/(A_{\text{Gu-HCl}}) \tag{8}$$

where $\varepsilon_{P,\text{Gu-HCl}}$ is the extinction coefficient of the denatured protein in 6 M Gu-HCl. $\varepsilon_{P,\text{Gu-HCl}}$ can be calculated from the known extinction coefficients of tryptophan, tyrosine, cystine, and phenylalanine in 6 M Gu-HCl[20–22] using Eq. (9):

$$\varepsilon_{P,\text{Gu-HCl}} \, (M^{-1} \text{ cm}^{-1}) = a\varepsilon_{\text{Tyr}} + b\varepsilon_{\text{Trp}} + c\varepsilon_{\text{Cys}} + d\varepsilon_{\text{Phe}} \tag{9}$$

where a, b, c, and d are the number of each type of residue per protein molecule. This method has been shown to be accurate to $\pm5\%$ for a wide variety of proteins.[21]

2. Stability of Fluorescence Signal. The stability of the ligand fluorescence signal (F_{obs}) should be examined as a function of time ($\pm$ continuous excitation) over the time period and the range of ligand concentrations to be used in the titrations. The effect of the rate of sample stirring should also be examined and optimized to maximize signal stability. If the fluorescence signal changes with time, this may reflect any of the following.

Photobleaching. If a decrease in the fluorescence signal is observed only on continuous excitation, then photobleaching is implicated. Corrections for photobleaching effects can be made when the effect is small ($<5\%$ of the signal) (see Section III,C,2). Alternatively, the effect can often be diminished or eliminated by reducing the excitation slit widths, which reduces the volume of the sample that is irradiated.

Ligand adsorption to cuvette walls. The effect of ligand adsorption to the cuvette walls should be independent of sample irridiation. It can sometimes be eliminated by altering solution conditions or by siliconizing the cuvette walls [5% dimethyldichlorosilane in anhydrous benzene (Sigma, St. Louis, MO)]. However, if this cannot be eliminated, quantitative analysis of the data is not possible.

Change in conformational or assembly state of ligand (protein). Changes in conformational or assembly state can occur if a protein is

---

[20] T. M. Lohman, K. Chao, J. M. Green, S. Sage, and G. T. Runyon, J. Biol. Chem. **264**, 10139 (1989).

[21] S. C. Gill and P. H. von Hippel, Anal. Biochem. **182**, 319 (1989).

[22] H. Edelhoch, Biochemistry **6**,. 1948 (1967).

stored under conditions (buffer, temperature) that are different than those used in the fluorescence experiment and which favor a different conformational or assembly state of the ligand. In this case, the signal should stabilize when the system has reached equilibrium at the new solution conditions. These effects can be minimized by dialyzing the protein against the identical buffer to be used in the experiment. However, if the conformation or assembly state of the ligand is affected by solution conditions, this should be examined in detail. Knowledge of the quaternary structure of the ligand (protein), both free in solution and in complex with the nucleic acid, is essential for determining quantitative equilibrium binding and thermodynamic parameters. Use of a different range of ligand concentrations and different solution conditions should be examined. In some cases, the use of small amounts of detergents can limit aggregation phenomena; however, caution should be exercised with this approach since this clearly perturbs the system under study.

3. Relationship between F_{obs} and Free Ligand Concentration. The relationship between the fluorescence intensity (after applying the appropriate correction factors; see Section III,C,2) and the free ligand concentration should be determined over the range of salt concentrations, pH, and temperature that are planned for investigation. A linear response is necessary in order for the fluorescence signal to be useful as a quantitative measure of ligand concentration. A nonlinear response could result for the following reasons.

Ligand adsorption to cuvette walls. If the ligand adsorbs to the cuvette walls, a plot of the total fluorescence signal versus total ligand concentration will usually display a lag phase at low ligand concentrations followed by a linear signal response above a critical ligand concentration. The lower signal at low ligand concentrations results from adsorption of ligand to "sites" on the cuvette walls, and the linear signal response occurs after the cuvette walls have been saturated with ligand. As stated above, this can sometimes be eliminated by altering solution conditions ([MX], pH, temperature). Raising the salt concentration is also effective in the case of highly charged ligands.

Ligand aggregation. Another potential cause for a nonlinear response, in either direction, is ligand aggregation or self-assembly, if the change in assembly state affects the fluorescence properties. Such aggregation equilibria must be characterized before attempting to interpret binding equilibria quantitatively, since the aggregation will generally affect binding and neglect of this will result in incorrect equilibrium binding parameters. If, the aggregation or self-assembly equilibria are well characterized, then it may be possible to perform nucleic acid binding studies under conditions such that only one aggregation state is favored. However, even in this case, one must determine whether the aggregation state is affected by nucleic acid binding. In general, any potential ligand aggregation phenomena should be examined in advance of undertaking quantitative binding studies. Once the ligand aggregation is characterized quantitatively, it can be incorporated into any model used for analysis of the ligand–nucleic acid titration.

4. *Dependence of Free Ligand Fluorescence on Salt Concentration.* As a prelude to studies of the dependence of K_{obs} on salt concentration, [MX], it is important to determine the salt concentration dependence of the free ligand fluorescence, F_{obs}. If F_{obs} is independent of [MX], then it may be possible to determine the quantitative dependence of K_{obs} on [MX] with much greater ease using an approach commonly referred to as a "salt back-titration" as described in Section III,B,7. However, if F_{obs} is dependent on [MX], then the salt dependence of K_{obs} can only be examined by determining individual isotherms at each [MX]. A salt-dependent F_{obs} could result from salt-dependent changes in the conformational or aggregation state of the ligand. Some salts, containing anions such as bromide or iodide or cations such as cesium, will quench the fluorescence of accessible groups on both the free and bound ligand; hence, fluorescence studies with salts containing these ions usually must be performed at constant salt concentration.[12] On the other hand, one may still be able to obtain important qualitative informtion from performing salt back-titrations, even when the above complications exist. For example, one can determine the general range of salt concentrations over which the ligand–nucleic acid complex dissociates, which is useful for determining the appropriate salt concentration to be used in an individual titration. Changes in F_{obs} as a function of [MX] can also have more trivial origins (although not from an experimental viewpoint), such as ligand adsorption to the cuvette walls at low salt concentration with subsequent release on raising the salt concentration.

5. *Light Scattering.* Another potential complication that must be considered in performing titrations of DNA with some ligands is possible interference with the fluorescence signal due to light scattering that can result from collapse or aggregation of the complexes. This can be a problem for highly positively charged ligands (e.g., polyamines, oligolysines, histones), when the ligand binding density increases beyond a critical value (~30% saturation for oligolysine–single-stranded polynucleotides).[14] Similar observations have been reported for other highly positively charged ligands binding to nucleic acids.[23] When this occurs, such interactions can

[23] D. Pörschke, *Biophys. Chem.* **10**, 1 (1979).

only be investigated at low binding densities if meaningful quantitative binding parameters are to be obtained. Light scattering can be detected by performing the titration as usual in the fluorometer; however, an excitation wavelength is used that is not absorbed by the sample, and emission is monitored at the same wavelength (e.g., 350 nm for proteins or peptides).

6. *Ligand Site Size Determination.* An independent determination of the site size, n, of the ligand when bound to the nucleic acid is important for the analysis of nonspecific ligand–nucleic acid interactions. In principle, the site size can be determined from an analysis of the binding isotherm; however, it is useful to obtain an independent estimate, since this will eliminate one parameter from the model-dependent analysis of a nonspecific ligand–nucleic acid binding isotherm.[11,24,25] Elimination of the site size from these expressions is especially important when investigating a ligand that binds cooperatively, since three parameters (n, K, and ω) are required to describe these isotherms. The equations and definitions of the binding parameters n, K, and ω for two models that describe the nonspecific binding of large ligands to infinitely long linear nucleic acids are given in [24] in this volume.

A site size determination should be made under high affinity conditions (generally $K_{obs} > 10^8\ M^{-1}$), so that a distinct stoichiometric point is observed. Of course, for this very reason, these conditions cannot be used to determine accurately a value for K_{obs}. For ligand (protein)–nucleic acid interactions, a sufficiently high binding constant is most easily achieved by lowering the salt concentration, since the value of K_{obs} generally increases as the salt concentration decreases.[5] To be certain that one is measuring a true site size, uninfluenced by the ligand affinity, at least two titrations should be performed at very different ligand or nucleic acid concentrations. If the apparent stoichiometric points do not agree, then the affinity may not be high enough under these conditions. In this case, if the affinity cannot be raised further by changing solution conditions, then a series of measurements can still be made at different ligand concentrations, and an estimate of the site size can be obtained from an extrapolation of the apparent size to infinite ligand concentration. Figure 1 shows an example of a site size determination for the binding of the *E. coli* SSB protein to poly(dT) under conditions such that the single-stranded DNA interacts with all four subunits of the SSB homotetramer.[26] The linear, nonhyperbolic nature of the titration is generally indicative of the tight

[24] J. D. McGhee and P. H. Von Hippel, *J. Mol. Biol.* **86**, 469 (1974).
[25] W. Bujalowski, T. M. Lohman, and C. F. Anderson, *Biopolymers* **28**, 1637 (1989).
[26] T. M. Lohman and L. B. Overman, *J. Biol. Chem.* **260**, 3594 (1985).

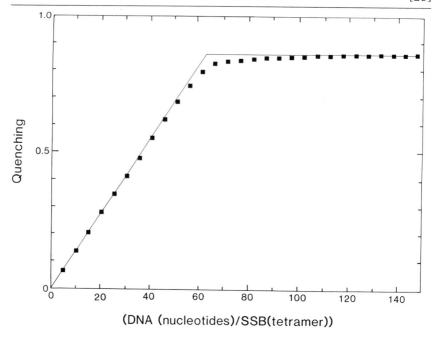

(DNA (nucleotides)/SSB(tetramer))

FIG. 1. Reverse titration of the *E. coli* SSB tetramer with poly(dT) under high-affinity conditions to determine the site size, n, in the $(SSB)_{65}$ binding mode (25.0°, pH 8.1, 0.5 M NaCl). The SSB tetramer concentration was 8.5×10^{-8} M. The excitation and emission wavelengths were 280 and 347 nm, respectively. The site size ($n = 63$ nucleotides/tetramer) was determined from the intersection of the two lines as indicated. (Data from Lohman and Overman.[26])

binding conditions that are necessary for the accurate determination of a site size.

In certain cases, a site size that is determined under one set of conditions (e.g., low salt concentrations) may not reflect the size size under other conditions; hence, the following caveats should be considered. (1) The mode of binding of the ligand (protein) in the high affinity condition (e.g., low salt concentration) used to determine the site size may differ from the mode of binding that exists under the conditions that will be used to determine the equilibrium binding parameters (e.g., higher salt concentration). This has been shown to be the case with the *E. coli* SSB protein, which can bind to single-stranded nucleic acids in at least four different binding modes, depending on the solution conditions.[26,27] (2) Differences may exist in the ligand (protein) conformation or quaternary

[27] W. Bujalowski and T. M. Lohman, *Biochemistry* **25,** 7799 (1986).

structure (assembly state) under the different solution conditions used to measure site size versus equilibrium binding parameters, thus making the site size determined under one set of conditions inapplicable for use under the other conditions. The existence of such complications can sometimes be detected by measuring the site size over a range of solution conditions in order to determine whether the apparent site size varies as a function of solution conditions. One should also check whether the site size determined at low salt conditions is consistent with the equilibrium binding isotherm determined under weaker binding conditions, as well as with any other independent information.

Finally, in the absence of independent structural information, the molecular meaning of an apparent site size determined from a titration can be ambiguous, since it will depend on the mode of binding. This is particularly true for ligands that bind to duplex nucleic acids. For example, a site size of 3 base pairs (bp) may actually represent the interaction of the ligand with 6 nucleotides on each strand of the duplex.[28]

7. *Salt Back-Titrations.* Most methods used to determine equilibrium binding constants are applicable only over a limited range of binding constants. For the nonspecific binding of ligands to nucleic acids using fluorescence, one is typically limited to the range of K_{obs} between 10^3 and $10^6 \, M^{-1}$. A major variable that can be used to bring the observed binding constant into this range is the salt concentration, since most protein–nucleic acid interactions have a significant electrostatic component.[5,29–31] A salt back-titration is a useful preliminary experiment to determine the salt concentration range within which one can measure equlibrium binding constants accurately (at constant temperature and pH). In this experiment, a ligand–nucleic acid complex is preformed under high-affinity conditions, usually low salt concentration, followed by titration with a concentrated salt solution, which will generally cause dissociation of the complex over some salt concentration range, resulting in recovery of the fluorescence signal (e.g., see Fig. 9). The salt concentration range over which one observes approximately 80% of the signal change roughly defines the useful range of salt concentrations that can be used to measure an equilibrium binding isotherm accurately. Of course, this range can also be extended somewhat by changing the ligand or DNA concentration.

Such salt back-titrations can also provide a test of whether the ligand–nucleic acid interaction is fully reversible. For example, if the initial

[28] A. Zlotnick, R. S. Mitchell, and S. L. Brenner, *J. Biol. Chem.* **265**, 17050 (1990).
[29] T. M. Lohman, *Crit. Rev. Biochem.* **19**, 191 (1986).
[30] M. T. Record, Jr., T. M. Lohman, and P. H. deHaseth, *J. Mol. Biol.* **107**, 145 (1976).
[31] M. T. Record, Jr., C. F. Anderson, and T. M. Lohman, *Q. Rev. Biophys.* **11**, 103 (1978).

fluorescence signal of the free ligand is not recovered at high salt concentration (after applying the appropriate corrections for dilution, photobleaching, etc., see Section III,C,2), then this may indicate the presence of some irreversible binding phenomenon (e.g., owing to aggregation of the ligand–DNA complex or the presence of a subset of high-affinity binding modes). This is useful preliminary information before one attempts to examine any system quantitatively. It is possible, in some cases, that the ligand–nucleic acid complex will not dissociate on raising the salt concentration, even into the molar range. This could indicate that the binding constant of the complex, although decreasing with salt, is still high enough under these conditions ($K_{obs} > 10^8 \, M^{-1}$) so that the complex is stable or that K_{obs} is insensitive to changes in salt concentration. Another possibility is that an irreversible complex was formed at low salt concentration and has become trapped kinetically so that dissociation of the complex is very slow (e.g., an aggregate has formed). This latter case can be differentiated from the former two by performing a ligand–DNA titration directly at the high salt concentration; if a complex does not form, then irreversible complex formation at low salt is indicated.

However, if a high-affinity complex does form at the high salt concentration, then one needs to differentiate between the following possibilities: (1) the interaction is salt sensitive, but K_{obs} is greater than $10^8 \, M^{-1}$ even at the high salt concentration, or (2) the binding interaction is not salt sensitive (i.e., mainly nonelectrostatic in nature). These two possibilities can sometimes be differentiated by performing salt back-titrations with divalent cations, since these will often dissociate the complex over a lower range of salt concentrations owing to the high affinity of divalent cations for nucleic acids. The interaction of the E. coli Rho protein with poly(C) appears to be insensitive to the bulk salt concentration,[32] whereas the E. coli SSB protein–poly(dT) interaction, although stable in 5 M NaCl, can be dissociated at high concentrations of NaBr or $MgCl_2$.[12] If the fluorescence intensity observed on dissociating the complex at high salt concentrations is higher than that of the free ligand at low salt concentration, this may indicate that either (1) some ligand had adsorbed to the cuvette walls at the low salt concentration and was subsequently released on raising the salt concentration or (2) the fluorescence of the free ligand increases with increasing salt concentration. However, these effects should have been noted in preliminary experiments that examined the free ligand fluorescence as a function of salt concentration (see Section III,B,2).

If it can be demonstrated that the observed ligand fluorescence quenching (Q_{obs}), the ligand site size (n), and other binding parameters such as

[32] J. A. McSwiggen, D. G. Bear, and P. H. von Hippel, J. Mol. Biol. **199**, 609 (1988).

cooperativity (ω) are independent of [MX], then salt back-titrations such as these can be analyzed quantitatively to determine the salt dependence of the equilibrium binding constant (see Section IV,B).

C. Titration Protocols

1. Excitation and Emission Wavelengths. The choice of the fluorescence excitation and emission wavelengths to be used will depend on the fluorophore that is being monitored, the number of fluorophores per ligand, and the extent to which "inner filter" absorbance effects occur (see below). The majority of our studies have involved proteins or ligands which contain the fluorescent amino acids tryptophan and tyrosine. The intrinsic tryptophan fluorescence of the *E. coli* SSB protein (16 Trp and 16 Tyr residues per SSB tetramer) undergoes a substantial quenching on complexation with single-stranded nucleic acids.[26] A substantial quenching of the tryptophan fluorescence also occurs on binding the oligopeptides (L-Lys)-(L-Trp)-(L-Lys)$_p$-NH$_2$ (KWK$_p$-NH$_2$) (p = 1 to 8) to either single- or double-stranded nucleic acids.[14,33]

The tryptophan fluorescence excitation spectrum has its maximum near 280 nm; however, excitation can still be achieved near 300 nm.[34] If maximum sensitivity is required, an excitation wavelength near 280 nm is used; however, this wavelength also overlaps the nucleic acid absorption spectrum, resulting in the need for inner filter corrections (see below). Therefore, if sensitivity is not an issue, then an excitation wavelength near 295–300 nm should be used to minimize inner filter corrections. The emission wavelength for tryptophan is generally in the range from 340 to 350 nm, sufficiently removed from the nucleic acid absorption spectrum. Unfortunately, the excitation spectrum for tyrosine occurs at relatively lower wavelengths ($\lambda_{ex,max}$ 274 nm), hence one is always faced with significant inner filter corrections in studies of ligands that contain only tyrosine. Generally, when proteins contain both tyrosine and tryptophan, the fluorescence emission spectrum that is observed is that due to tryptophan as a result of energy transfer from tyrosine to tryptophan.

2. Correction Factors

a. Ligand and nucleic acid concentrations. The concentrations of total ligand and nucleic acid at each point in a titration are determined after accounting for dilution. The total ligand concentration, $L_{T,i}$, at each point i in a titration is calculated using Eq. (10),

[33] C. Helene and J. L. Dimicoli, *FEBS Lett.* **26,** 6 (1972).
[34] J. R. Lakowicz, "Principles of Fluorescence Spectroscopy." Plenum, New York, 1983.

$$L_{T,i} = \frac{L_0 V_0}{(V_0 + N_i)} \tag{10}$$

where L_0 is the initial total ligand concentration in the cuvette, V_0 is the initial sample volume, and N_i is the total volume of nucleic acid titrant added up to point i in the titration. The total nucleic acid concentration, $D_{T,i}$, at each point i in the titration is calculated using Eq. (11),

$$D_{T,i} = \frac{D_0 N_i}{(V_0 + N_i)} \tag{11}$$

where D_0 is the stock concentration of the nucleic acid titrant.

b. Fluorescence signal corrections. The fluorescence intensity at each point i in a titration is obtained after corrections for dilution, photobleaching, and inner filter effects. Photobleaching is an excitation-dependent loss of fluorescence intensity. Inner filter effects can result when the absorbance spectrum of the sample overlaps either the excitation or emission wavelength, such that some of the excitation or emission intensity is filtered by the sample, resulting in attenuation of the fluorescence signal.[34,35] These three correction factors are multiplicative, and the relationship between the corrected fluorescence intensity, $F_{i,cor}$, and the observed fluorescence intensity, $F_{i,obs}$ is given by

$$F_{i,cor} = F_{i,obs} \left(\frac{V_0 + N_i}{V_0}\right)\left(\frac{f_0}{f_i}\right)\left(\frac{1}{C_i}\right) \tag{12}$$

The dilution correction is $(V_0 + N_i)/V_0$, where V_0 is the initial volume of the sample and $(V_0 + N_i)$ is the volume at point i in the titration. The photobleaching correction is (f_0/f_i), where f_0 is the ligand fluorescence intensity before the titration is started and f_i is the fluorescence of a ligand sample that has not been titrated but has been irradiated for the same time period as the actual sample.

The inner filter correction, C_i, requires knowledge of the molar extinction coefficient for both the ligand, ε_l, and nucleic acid, ε_d, at both the excitation and emission wavelengths. These are defined in Eq. (13a) and (13b),

$$\varepsilon_l = \varepsilon_{l,ex} + \varepsilon_{l,em} \tag{13a}$$
$$\varepsilon_d = \varepsilon_{d,ex} + \varepsilon_{d,em} \tag{13b}$$

where $\varepsilon_{l,ex}$ and $\varepsilon_{l,em}$ are the extinction coefficients of the ligand at the

[35] B. Birdsall, R. W. King, M. R. Wheeler, C. R. Lewis, Jr., S. R. Goode, R. B. Dunlap, and G. C. K. Roberts, *Anal. Biochem.* **132**, 353 (1983).

excitation (ex) and emission (em) wavelengths, respectively, and $\varepsilon_{d,ex}$ and $\varepsilon_{d,em}$ are those for the nucleic acid. The values of $\varepsilon_{l,ex}$, $\varepsilon_{l,em}$, $\varepsilon_{d,ex}$, and $\varepsilon_{d,em}$ are most readily obtained from the absorbance spectra of the ligand and nucleic acid, respectively, if the extinction coefficient at one wavelength is known (see Section III,B,1).

Given ε_l and ε_d, and the total concentrations of ligand and nucleic acid at each point in the titration, the total absorbance at each point in the titration, A_i, is calculated using Eq. (14):

$$A_i = (\varepsilon_l L_{T,i}) + (\varepsilon_d D_{T,i}) \tag{14}$$

The inner filter correction factor, C_i, at each point i in the titration can then be calculated as follows. If the efficiency of detection is uniform across the observation window, then C_i is given by[35]

$$C_i = \frac{10^{-A_i W_1} - 10^{-A_i W_2}}{2.303 A_i (W_2 - W_1)} \tag{15}$$

here W_1 and W_2 are geometric factors that depend on the geometry of the fluorescence cell compartment and the path length through the sample. The method for determining W_1 and W_2 for an individual spectrofluorometer is described by Birdsall et al.[35] For low absorbances ($A < 0.3$), Eq. (15) simplifies to

$$C_i = \frac{1 - 10^{-A_i}}{2.303 A_i} \tag{16}$$

For absorbances in the range $A \le 0.1$, Eq. (15) can be further simplified to[34]

$$C_i = 10^{-A_i/2} \tag{17}$$

Other empirical inner filter corrections have also been described.[35]

In many cases, neither the ligand nor the nucleic acid absorb at the emission wavelength; hence only inner filter corrections are necessary due to absorbance at the excitation wavelength. If there is an absorbance change on complex formation, then this needs to be considered in calculating the inner filter corrections, unless an isosbestic wavelength can be used for the excitation or emission wavelength. The inner filter correction should also be applied to the initial free ligand fluorescence, F_{init}, to yield the corrected initial fluorescence, F_0. Owing to the obvious complications that inner filter corrections can introduce, experiments should be designed to use wavelengths and concentrations that will minimize the inner filter effects.

3. Stock Solutions

a. Buffers. All concentrated buffers and solutions (e.g., 5 M NaCl, 1 M cacodylic acid) are filtered through 0.22-μm Nalgene filters [115 ml Nalgene (Nalge, Rochester, NY) type S, Cat. NO. 120-0020]. A small aliquot (~50–100 ml) is first filtered and then discarded in order to avoid any contaminants that might be leached from the filter. The solutions to be used in the fluorescence experiments are then prepared by diluting the concentrated stocks with distilled water that has also been passed through a Milli-Q purification system (Continental, Bedford, MA), and filtered through a Millipore (Bedford, MA) 0.22 μm filter. Filtering the buffers eliminates dust, which can cause light scattering, thus increasing the background signal.

To discuss some of the details of the titration methods, we use our studies of the interaction of the positively charged oligopeptide L-Lys-L-Trp-(L-Lys)$_2$-NH$_2$ (KWK$_2$-NH$_2$) with poly(U) as an example.[14] The final conditions are 37 mM Na$^+$, pH 6.0, 25°, and the ligand and nucleic acid stocks used correspond to curve L_{T3} in Fig. 3 (2.6 μM KWK$_2$-NH$_2$).

> Low-salt buffer stock (LS buffer): 1.0 mM NaCl, 0.2 mM Na$_3$EDTA, 10 mM cacodylic acid, titrated to pH 6.0 with 5 N NaOH; the total [Na$^+$] in this buffer is 5.3 mM
>
> High-salt buffer stock (HS buffer): 1.00 M NaCl, 0.2 mM Na$_3$EDTA, 10 mM cacodylic acid, titrated to pH 6.0 with 5 N HCl; the total [Na$^+$] in this buffer is 1.0 M

We note that the pH of buffers containing high concentrations of salts of weak acids (e.g., 2 M KCH$_3$CO$_2$) will decrease on dilution; therefore, the pH of the HS buffer should be adjusted appropriately. (For example, a 2 M KCH$_3$CO$_2$ stock is titrated to pH 6.4 so that on dilution to within the range from 10 mM to 0.5 M it will have a final pH of 6.00 ± 0.03 over the entire salt range.) Note also that the final concentrations of all species that affect the equilibrium should be stated explicitly, since the final concentrations of Na$^+$ and Cl$^-$ may not be equal owing to the presence of buffer and chelating agents such as ethylenediaminetetraacetic acid (EDTA), which contribute Na$^+$.

b. Ligand and nucleic acid stock solutions. The following stock concentrations are typical of those used for a reverse titration of the oligopeptide KWK$_2$-NH$_2$ with the synthetic homopolynucleotide poly(U) (37 mM Na$^+$, 10 mM cacodylic acid, pH 6.0, 25°).

> Ligand master stock: Concentrated stock of 0.1 mM peptide, KWK$_2$-NH$_2$, dialyzed against LS buffer
>
> Nucleic acid master stock: Concentrated stock of poly(U) (7 mM in

nucleotide) ($s_{20,w}$ 9.5 S; ~1100 nucleotides), dialyzed against LS buffer

Nucleic acid titrant stock: Concentrated stock of poly(U) that has been adjusted to 37 mM Na$^+$ by diluting the nucleic acid master stock with HS buffer. In our example, 3.3 μl HS buffer is added to 100.0 ml of nucleic acid master stock. Note that the resulting poly(U) concentration of the nucleic acid titrant stock is now 6.776 mM (nucleotide).

4. *Reverse Titrations.* In this section we describe the experimental procedures for performing reverse titrations (addition of polynucleotide to ligand) to be used to estimate ligand site sizes under high-affinity conditions, or to obtain equilibrium binding isotherms under lower affinity conditions. The information available from and interpretation of these titrations vary depending on whether K_{obs} is in the low- or high-affinity range. Titrations performed under high-affinity conditions ($K_{obs} > 10^8 M^{-1}$) are used to determine binding stoichiometries (site sizes); however, owing to the extremely low free ligand concentration under these conditions, accurate values of K_{obs} cannot be measured. Titrations designed to obtain accurate binding affinities must be performed under lower affinity conditions ($K_{obs} \sim 10^3$ to $10^6 M^{-1}$), so that $L_T \leq (K_{obs})^{-1}$. However, the protocol for the reverse titration is nearly the same for each case, differing only in the solution conditions and the number of data points collected (fewer points are needed for a site size determination).

To illustrate the mechanics of a reverse titration, we describe an experiment in which we titrate a fluorescent oligopeptide, KWK$_2$-NH$_2$, with the synthetic homopolynucleotide, poly(U), and monitor the quenching of the oligopeptide tryptophan fluorescence that accompanies binding. For this case, we use the LS and HS buffer stocks described in Section III,C,3,a and the ligand master stock and nucleic acid titrant stock described in Section III,C,3,b. The nucleic acid titrant stock is brought to room temperature, independent of the temperature of the solution in the cuvette, since this reduces inaccuracies in pipetting that can result from temperature differences between the pipette tip and the titrant stock.

Titration protocol. The solution conditions for the titration are 10 mM cacodylate (pH 6.0), 37 mM Na$^+$, and 2.6 μM KWK$_2$-NH$_2$.

1. The following components are added, in order, to a 4-ml fluorescence cuvette, containing a Teflon-coated magnetic stir bar (3 ml effective volume): LS buffer (2.0 ml), ligand master stock (56.0 μl), and HS buffer (68.0 μl). The mixture is placed in the thermostatted cell holder and stirred using the magnetic stirring mechanism of the SLM-Aminco fluorometer. A reference solution is also made in an identical cuvette by substituting

56.0 μl of LS buffer for the ligand master stock, and this is placed in the reference position of the thermostatted cell holder.

2. The solution is allowed to equilibrate at the designated temperature, which is controlled by a refrigerated circulating water bath. The time required for the sample to equilibrate can be determined by measuring the observed fluorescence intensity at various time intervals until a stable value is reached. If longer than 20 min is required to obtain a stable signal or if a constant fluorescence intensity cannot be achieved, problems such as ligand adsorption to the cuvette or ligand instability, F_{obs}, should be considered (see Section III,B). The initial voltage settings that are used for the photomultiplier tubes (PMT) will depend on whether the fluorescence intensity will decrease or increase on addition of the nucleic acid. Once equilibrium has been reached, initial readings of both the sample, $F_{samp,0}$, and reference cuvettes, $F_{ref,0}$, are taken, and the difference is defined as F_0 ($= F_{samp,0} - F_{ref,0}$).

3. The titration is started by adding a small aliquot of the nucleic acid titrant stock to both the sample and reference cuvettes with the shutters closed, and time is allowed for the system to reach equilibrium. The fluorescence intensity is measured for each cuvette, and the difference between these readings is defined as the observed fluorescence for titration point i, $F_{obs,i}$:

$$F_{obs,i} = F_{samp,i} - F_{ref,i} \tag{18}$$

This definition of $F_{obs,i}$ corrects for fluorescence or light scattering contributions from the nucleic acid or the buffer. Normally, the value of F_{ref} at the end of a titration is less than 5% of the highest value of the sample fluorescence, F_{samp}; however, if it is over 10%, this may indicate that the nucleic stock is contaminated, and such data should be treated with caution or discarded.

The titration is continued in this manner until saturation of the ligand is achieved, although this is not always possible under conditions such that the ligand–nucleic acid affinity is weak. However, even under such weak affinity conditions where a titration end point cannot be achieved directly, model-independent binding isotherms can still be determined through application of the binding density function method, although these methods require multiple titrations at different ligand concentrations[2,3] (see Section II,B,1).

On completion of the titration, corrections for dilution, photobleaching, and inner filter effects should be applied to F_0 and each $F_{obs,i}$ as described in Section III,C,2 in order to obtain the corrected fluorescence intensities, $F_{cor,0}$ and $F_{cor,i}$. Because there is no absolute scale of fluorescence, the

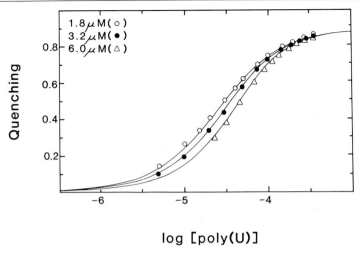

$$\log \; [poly(U)]$$

FIG. 2. Examples of reverse titrations used to construct the binding density plot shown in Fig. 3 for the poly(U)–KWK$_2$-NH$_2$ equilibrium (25.0°, pH 6.0, 37 mM Na$^+$). The excitation and emission wavelengths were 292 and 350 nm, respectively. The smooth curves through the data are based on the McGhee–von Hippel model[24] for large ligands binding noncooperatively to an infinite, homogeneous lattice (see [24] in this volume[5]) using the binding parameters $n = 4$ nucleotides, $Q_{max} = 0.885$, and $K_{obs} = 5.4 \times 10^4 \; M^{-1}$, which were obtained from the analysis of the binding density function plot in Fig. 3.

intensities are normalized by calculating the observed ligand fluorescence quenching at each point i in the titration as in Eq. (19):

$$Q_{obs,i} = (F_{cor,i} - F_{cor,0})/F_{cor,0} \qquad (19)$$

The amount of nucleic acid titrant added in each aliquot will depend on the total ligand concentration, the ligand–nucleic acid affinity, and whether a site size or an equilibrium isotherm is to be determined from the titration. A titration to determine a site size should be designed so that approximately 10–15 points are obtained (7–10 before the saturation plateau is reached). Titrations to be used for the determination of an equilibrium binding isotherm should be designed so that about 20–30 points are obtained before saturation is reached. For these latter titrations, attempts should be made to obtain data points over the greatest binding density range possible. For purposes of analysis of binding isotherms, it is best to plot the signal, Q_{obs}, versus the logarithm of the titrant concentration (D_T), a Bjerrum plot as in Fig. 2, rather than on a linear scale. This allows one to determine easily whether saturation of the ligand has been reached. Therefore, the amount of nucleic acid added to the ligand at each point in the titration should not be constant, but rather should increase throughout the titration in order to obtain roughly evenly spaced points after the data are plotted on a logarithmic scale.

5. *Salt Back-Titrations.* Once a ligand solution has been titrated with nucleic acid as in the previous step, a salt back-titration can be performed. The general aspects of this type of titration are discussed in Section III,B,7. If a number of constraints hold (e.g., $Q_{obs}/Q_{max} = L_B/L_T$; see Section IV,A,2), then the value of Q_{obs} at each point in the salt back-titration can be used to calculate the bound and free ligand concentration, thus enabling a value of K_{obs} to be determined at each salt concentration. The primary advantage of salt back-titrations over repetitive reverse titrations at several salt concentrations is that the dependence of K_{obs} on salt concentration can be obtained with a single stock of reagents in one experiment. Combined with the fact that many more points can be obtained through a continuous titration with salt, the use of a salt back-titration increases the precision with which the salt dependence of K_{obs} can be determined. However, this approach can only be used to calculate K_{obs}, at each salt concentration if Q_{obs} is directly proportional to the fraction of bound ligand, L_B/L_T, and if the free and bound ligand fluorescences are independent of salt concentration.[12]

Salt back-titration protocol. Small aliquots of the HS buffer are added to the pre-formed ligand–nucleic acid complex until the original F_{obs} is approached (see Fig. 9). We note that dilution effects may give the appearance that the value of F_{obs} reaches a premature plateau. An assessment of whether the fluorescence intensity corresponding to fully dissociated ligand has been recovered can only be made accurately after the appropriate corrections have been applied to the values of $F_{obs,i}$ (see Section III,C,2). The results of a salt back-titration of an oligopeptide–poly(U) complex are given in Fig. 9, where both the corrected and uncorrected data are shown to demonstrate the magnitude of the dilution corrections. If full recovery of the initial fluorescence intensity, $F_{cor,0}$ is obtained after the salt back-titration, this indicates that the ligand–nucleic acid interaction is fully reversible. If a discrepancy exists (i.e., a lower or higher fluorescence intensity is recovered), then this may indicate some irreversibility in the system or possibly that the correction factors are in error. In this case, the correction factors (Section III,C,2) should be redetermined before concluding that irreversibility is a problem; however, any such discrepancies should not be ignored.

IV. Analysis of Titration Curves

A. Obtaining Equilibrium Binding Isotherms

1. Ligand Binding Density Function Analysis. For LBDF analysis, a series of reverse titrations are performed at different ligand concentrations, L_{Tx} (x = 1 to i), and the binding density function plot formed from these titrations is analyzed to obtain model-independent equilibrium binding isotherms.[2,3] Because the binding density function analysis requires a comparison of the fluorescence quenching from a number of different titrations, the analysis can be complicated by the fact that each ligand concentration within a given titration does not remain constant throughout a titration owing to dilution. Therefore, to determine the set of concentrations, (L_{Tx}, D_{Tx}), corresponding to a constant value of the binding density function, $Q_{obs}(L_T/D_T)$, one must interpolate between data points of known diluted ligand concentration to obtain the correct value of L_T (D_T is plotted as its final concentration).

A simple modification of the experimental design of the reverse titration can facilitate the BDF analysis. By including ligand in the nucleic acid titrant stock at the same total concentration, L_{Tx}, as in the cuvette, the total ligand concentration will remain constant within an individual titration. Furthermore, if the nucleic acid titrant is "doped" with ligand, it is no longer necessary to apply dilution correction factors to either the total ligand concentration or $F_{obs,i}$ (see Section III,C,2). Note that the reference cuvette should still be titrated with "undoped" nucleic acid titrant stock (replacing the ligand with LS buffer). Therefore, if reverse titrations with ligand-doped nucleic acid are performed, one can determine each pair of L_{Tx} and D_{Tx} values directly from the binding density function plot without the need to interpolate.

a. Data analysis. Data obtained for the equilibrium binding of the fluorescent oligopeptide $KWK_2\text{-}NH_2$ to poly(U) will be used to describe the ligand binding density function (LBDF) analysis.

1. The series of reverse titrations performed at different total ligand concentrations, L_{Tx} (e.g., see Fig. 2), is replotted in the form of an LBDF plot, according to Eq. (3), as in Fig. 3. $Q_{obs}(L_T/D_T)$ should be plotted as a function of the logarithm of D_T so that all of the curves can be compared on a single graph.

2. Smooth curves are drawn empirically through the data points for each titration in order to allow interpolation between points. This can be done either by hand as for the data in Fig. 3 or by using a least-squares polynomial to describe each titration.

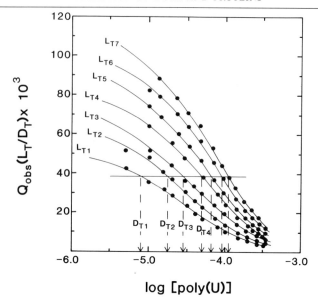

log [poly(U)]

FIG. 3. Binding density function plot constructed from seven reverse titrations of KWK_2-NH_2 with poly(U) (25.0°, pH 6.0, 37 mM Na^+). Each titration is designated as L_{Tx} ($x = 1$ to 7). The total peptide concentrations for each curve are as follows, from top (L_{T7}) to bottom (L_{T1}): 6.0, 5.0, 4.0, 3.2, 2.6, 1.8, and 1.4 μM. The smooth lines were drawn empirically through each set of data in order to facilitate interpolation between data points. A horizontal line has been drawn that intersects each of the curves in order to demonstrate how the pairs of total ligand and total nucleic acid concentrations, L_{Tx} and D_{Tx}, respectively, are obtained from these plots, which result in the estimation of one set of values of $\Sigma \nu_i$ and L_F. The value of the binding density function, $Q_{obs}(L_T/D_T)$, and hence the values of $\Sigma \nu_i$ and L_F, are constant along the horizontal line. Therefore, each point of intersection yields a pair of L_{Tx} (constant for each curve) and D_{Tx} values as indicated (seven total), which can then be plotted as in Fig. 4 to determine one set of values of $\Sigma \nu_i$ and L_F. This is then repeated for a number of horizontal lines to obtain $\Sigma \nu_i$ as a function of L_F, from which a model-independent binding isotherm can be constructed. In actuality, a larger version of this plot was used in order to increase the accuracy with which values of L_{Tx} and D_{Tx} could be obtained from such a plot.

3. A horizontal line is drawn to intersect each of the smooth curves. Because the binding density function, $Q_{obs}(L_T/D_T)$, is constant for a given horizontal line, it follows from Eq. (3) that the average binding density, $\Sigma \nu_i$, and the free ligand concentration, L_F, are also constant at the point where each titration curve intersects that horizontal line [constant $Q_{obs}(L_T/D_T)$.] Therefore, each horizontal line can be analyzed to obtain one set of values of $\Sigma \nu_i$, and L_F, as described in step 4 (see Fig. 3).

4. Each point of intersection of a single horizontal line with each binding density function curve yields a pair of values of L_{Tx} and D_{Tx}. An

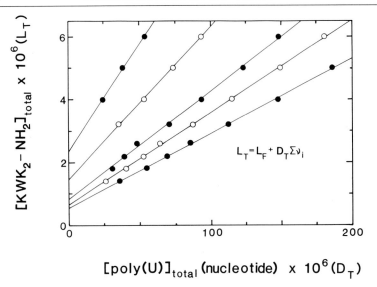

$$L_T = L_F + D_T \Sigma \nu_i$$

[poly(U)]$_{total}$ (nucleotide) × 10^6 (D$_T$)

FIG. 4. Plot of L_T versus D_T for KWK$_2$-NH$_2$ binding to poly(U) (25.0°, pH 6.0, 37 mM Na$^+$). The paired sets (L_T, D_T) for each line were determined from the binding density function plot shown in Fig. 3, based on a single horizontal line similar to the one shown in Fig. 3. The ordinate intercept of each line equals the free peptide concentration, L_F, and the slope of each line equals the average ligand binding density, $\Sigma \nu_i$.

example of this procedure is shown for one horizontal line in Fig. 3. D_{Tx} is determined from the value of the abscissa at the intersection point, and L_{Tx} corresponds to the value of L_T used for that titration. If the reverse titrations were performed with nucleic acid titrant stocks that were "doped" with ligand at concentration L_{Tx}, then L_{Tx} is constant throughout each titration curve. However, if the nucleic acid stocks were not doped with ligand, the value of L_{Tx} at the intersection will vary slightly along each curve owing to dilution effects and therefore must be obtained by interpolation.

5. If a horizontal line intersects r titration curves, then r pairs of values of L_{Tx}, D_{Tx} will be obtained and used to obtain the values of $\Sigma \nu_i$, and L_F for that particular value of the binding density function (one horizontal line). A plot of L_T versus D_T is made using the r pairs of L_{tx}, D_{Tx}, obtained from each horizontal line drawn through the binding density function plot in Fig. 3. The resulting plot should be linear, with a slope equal to $\Sigma \nu_i$ and an ordinate–intercept equal to L_F, as indicated by conservation of mass [Eq. (2)]. A series of such plots to determine $\Sigma \nu_i$ and L_F are shown in Fig. 4.

6. Repetition of this process for a range of values of $Q_{obs}(L_T/D_T)$ yields

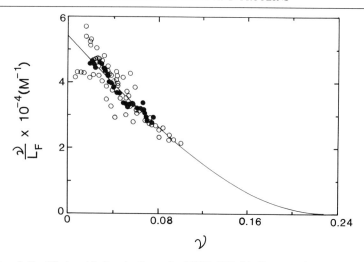

FIG. 5. Equilibrium binding isotherm for KWK_2-NH_2 binding to poly(U) (25.0°, pH 6.0, 37 mM Na$^+$) plotted according to Scatchard.[36] The filled circles represent data obtained from the binding density function analysis of the data shown in Fig. 3. The open circles represent the data calculated independently from each of the seven reverse titrations shown in the binding density function plot in Fig. 3 based on the assumption that $Q_{obs}/Q_{max} = L_B/L_T$, with $Q_{max} = 0.885$ (as determined from Fig. 7). The smooth curve is the simulated binding isotherm based on the best fit binding parameters, $n = 4$ nucleotides, $K = 5.4 \times 10^4 \ M^{-1}$, and a noncooperative model[24] (see [24] in this volume[5]). (Modified from Mascotti and Lohman.[14])

model-independent pairs of $\Sigma \ \nu_i$, and L_F that can be used to construct the binding isotherm as shown in Fig. 5. This isotherm can then be analyzed to extract equilibrium binding parameters using some appropriate model. The filled circles in Fig. 5[36] represent data points obtained from the binding density function analysis, whereas the open symbols represent points calculated from each individual titration curve by assuming that $Q_{obs}/Q_{max} = L_B/L_T$, that is, assuming a linear relationship between quenching and fraction of peptide bound. The points determined from the individual titrations scatter around the points determined from the binding density function analysis.

b. Cautions. The regions of the binding density function plots that are most appropriate for analysis are those in which $Q_{obs}(L_T/D_T)$ displays a nearly linear dependence on $\log D_T$ for each curve that is intersected. The regions at low nucleic acid concentration, where significant curvature occurs, should be avoided since L_T is approximately equal to L_F in this region; hence, the determinations of $\Sigma \ \nu_i$, and L_F are not so accurate.

[36] G. Scatchard, *Ann. N.Y. Acad. Sci.* **51**, 660 (1949).

The most accurate values of $\Sigma \, \nu_i$ and L_F will be those obtained from horizontal lines that intersect the maximum number of curves in the binding density function plot (see Fig. 3), and hence these values should be weighted more in the analysis of any binding isotherm determined by a BDF analysis. A reflection of this is that slight discontinuities in the values of $\Sigma \, \nu_i$, and L_F are sometimes obtained from the analysis of two adjacent horizontal lines that differ in the number of titration curves intersected. The magnitude of these discontinuities will be reduced as more titration curves are included in the BDF plot.

The behavior of the BDF plot when ligand or ligand–nucleic acid aggregation occurs should be mentioned. As pointed out above, Eqs. (3) and (4) are valid only when the molar signal of each species is independent of concentration (i.e., in the absence of aggregation of the ligand or the ligand–nucleic acid complex). To emphasize this, we simulate the behavior of the BDF plots for a ligand–macromolecule equilibria in which ligand self-assembly occurs within the concentration range under examination. In this case, one observes significant deviations in the binding density function plots, which are diagnostic of aggregating systems. Of course, in applying any quantitative analysis of ligand–nucleic acid equilibria it is essential to know if ligand aggregation occurs within the concentration range under study, because, if this is neglected, the analysis will yield only apparent binding parameters that cannot be related directly to free energy changes. However, we show these simulations as examples of the possible behavior when such competitive aggregation equilibria occur. In fact, such deviations can be used to indicate the presence of competing equilibria of this type.

We consider the case in which a protein monomer, P, is in equilibrium with a dimer form, P_2, and all subunits can bind an oligonucleotide, D, but with different equilibrium binding constants as described in Eqs.(20)–(23):

$$P + P \overset{K_\mathrm{D}}{\rightleftharpoons} P_2 \tag{20}$$

$$P + D \overset{K_\mathrm{M}}{\rightleftharpoons} PD \tag{21}$$

$$P_2 + D \overset{K}{\rightleftharpoons} P_2 D \tag{22}$$

$$P_2 D + D \overset{K}{\rightleftharpoons} P_2 D_2 \tag{23}$$

The protein dimer, P_2, binds the oligonucleotides with 10-fold higher affinity than the monomers (i.e., $K = 10 K_\mathrm{M}$). For this hypothetical case, the protein fluorescence is fully quenched on binding oligonucleotide, for both P and P_2; in other words, the dimerization does not influence the fluorescence quenching. To treat this case, it is easiest to define the protein

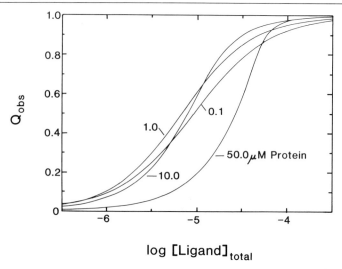

log [Ligand]$_{total}$

FIG. 6. Simulated binding density function plots for the hypothetical case in which a protein exists in a monomer–dimer equilibrium and an oligonucleotide can bind to either form, but with different affinities [see Eqs. (20)–(23) in the text]. Equilibrium binding parameters are $K = 10^6 \, M^{-1}$, $K_M = 10^5 \, M^{-1}$, and $K_D = 10^4 \, M^{-1}$, and it is assumed that $Q_{max} = 1$ for the bound protein monomer, independent of its actual state of assembly. The total protein (monomer) concentration (μM) is indicated for each curve.

as the macromolecule and the oligonucleotide as the ligand; therefore, the macromolecular binding density function (MBDF) given in Eq. (4) is used to analyze this system. Figure 6 shows a set of simulated titrations plotted in the form of a binding density function plot (Q_{obs} versus log[L_T]) for the case in which $K_D = 10^4 \, M^{-1}$, $K = 10^6 \, M^{-1}$, and $K_M = 10^5 \, M^{-1}$. For the protein concentrations used in the simulations, the BDF curves are not well behaved. On increasing the protein concentration in the low protein concentration range (0.1–1 μM), the curves are shifted to lower total ligand concentrations, which is opposite to the behavior expected and observed when no ligand aggregation occurs. At higher protein concentrations (10–50 μM), the curves become shifted to higher ligand concentrations, causing some of the curves to intersect. This type of behavior would clearly suggest the presence of an aggregation phenomenon, which would need to be characterized further if quantitative studies of this interaction were to be pursued.

2. Correlation between Signal Change and Average Binding Density. Once the average binding density, $\Sigma \, \nu_i$, is obtained as a function of L_F, from the BDF analysis, this information can be used to determine the relationship between the average signal change (Q_{obs} in this case) and the

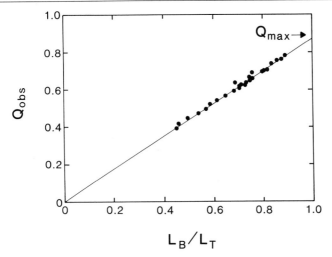

FIG. 7. Determination of the value of Q_{max} for KWK$_2$-NH$_2$ binding to poly(U) (25.0°, pH 6.0, 37 mM Na$^+$) from a plot of the observed quenching, Q_{obs}, versus the fraction of bound ligand (L_B/L_T). The values of L_B/L_T were determined from the binding density function analysis shown in Fig. 3. Linear extrapolation to $L_B/L_T = 1$ yields the maximal quenching of the peptide. (From Mascotti and Lohman.[14])

fraction of bound ligand, L_B/L_T. This is not necessary in order to obtain a binding isotherm, as discussed above; however, if Q_{obs} is found to be directly proportional to L_B/L_T, then binding isotherms can be constructed with much greater ease from a single titration (see Section II,B,3). If it can be demonstrated rigorously that Q_{obs} is directly proportional to L_B/L_T, one can determine the maximum extent of protein fluorescence quenching, Q_{max}, from a linear extrapolation of a plot of Q_{obs} versus L_B/L_T to $L_B/L_T = 1$. An example of such an analysis is shown in Fig. 7 for the KWK$_2$-NH$_2$–poly(U) equilibrium.

For a ligand–nucleic acid interaction, it is possible that the relationship, $Q_{obs}/Q_{max} = L_B/L_T$ is valid only over a limited range of binding densities. A useful check of this relationship can be made over a wide range of binding densities by plotting $[(Q_{obs}/Q_{max})(L_B/L_T) \equiv \nu_Q]$ versus the true average binding density, $\Sigma \nu_i$, as obtained from the LBDF analysis. $[(Q_{obs}/Q_{max})(L_B/L_T)]$ is an apparent binding density, based on the assumption that $Q_{obs}/Q_{max} = L_B/L_T$. This type of plot is shown in Fig. 8 for the KWK$_2$-NH$_2$–poly(U) data presented in Fig. 7. In this case, there is a direct correspondence between these two values over the range from 5 to about 30% coverage of the nucleic acid; hence, it can be concluded that $Q_{obs}/Q_{max} = L_B/L_T$ in this range. On the other hand, one should be cautious

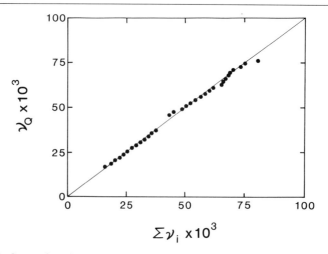

FIG. 8. Comparison between the average binding density, ν_Q, as determined from the assumption that $Q_{obs}/Q_{max} = L_B/L_T$, with $Q_{max} = 0.885$ [see Eq. (7) in the text], with that obtained from a rigorous binding density function analysis, $\Sigma \nu_i$. Data are for KWK$_2$-NH$_2$ binding to poly(U) (25.0°, pH 6.0, 37 mM Na$^+$). The observed relationship demonstrates that for this interaction under these conditions, Q_{obs} is directly proportional to the fraction of ligand bound, L_B/L_T.

about concluding that $Q_{obs}/Q_{max} = L_B/L_T$ for all solution conditions, based on an analysis under only one set of solution conditions.

In Fig. 5 we compare the binding isotherms obtained from application of the BDF analysis with those determined separately from each individual titration curve using Eqs. (5)–(7), which follow from the assumption that $Q_{obs}/Q_{max} = L_B/L_T$. Note that all of the data in Fig. 5 are well described by the noncooperative overlap binding isotherm[24,25] with a single set of binding parameters, n and K, with ω equal to unity (see [24] in this volume[5]). Furthermore, the use of the BDF analysis leads to significantly less scatter since this method serves to average the different titrations.

B. Obtaining Model-Independent Intrinsic Binding Parameters

1. Reverse Titrations. Once a binding isotherm has been determined using the methods described above, the next step is to analyze the isotherm to obtain equilibrium binding parameters. This is then repeated as a function of solution variables. For a nonspecific ligand–nucleic acid interaction, this requires the use of a statistical thermodynamic model, which provides the most realistic physical description of the binding interaction. For example, the unlimited[24,25] and limited[11] nearest-neighbor cooperativ-

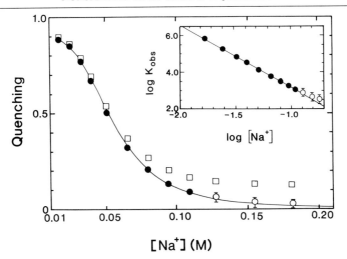

FIG. 9. Salt back-titration of a pre-formed poly(U)–KWK$_2$-NH$_2$ complex (25.0°, pH 6.0) by addition of NaCl. The excitation and emission wavelengths were 292 and 350 nm, respectively. Initial peptide and poly(U) concentrations were 1.48 and 100.5 μM, respectively, in 2096 μl solution volume. ($\square$), Raw data based on observed fluorescence quenching. Circles ($\bigcirc$ and $\bullet$) are the same data after correction for dilution and inner filter effects as described in the text. The inset shows the values of log K_{obs} as a function of log[Na$^+$], as calculated from the corrected salt back-titration data, using the noncooperative binding model^{24} (see [24] in this volume^5). The data represented by the open circles ($\bigcirc$) are not used in the least-squares determination of the salt dependence of K_{obs} (log K_{obs} = -3.42 log[Na$^+$] $-$ 0.26).

ity models have been used to analyze the nonspecific binding of ligands to a linear nucleic acid (see [24] in this volume^5). For these models, the binding parameters (K, n, and ω) are obtained by comparing the experimental binding isotherm to theoretical isotherms generated using the appropriate model. This is best done by using nonlinear regression techniques.$^{37-39}$ For a ligand that displays cooperative binding ($\omega \neq 1$), this requires the determination of three binding parameters; however, as discussed above, if an independent estimate of the site size, n, has been obtained, then the accuracy with which K and ω can be determined from the equilibrium binding isotherm is improved significantly. We note that for the purposes of estimating binding parameters by nonlinear regression techniques, it is preferable to represent the isotherm directly in terms of the observable parameters (e.g., L_B versus log[D_T]), since this enables a

^{37} M. L. Johnson and S. G. Frasier, this series, Vol. 117, p. 301.
^{38} D. F. Senear and D. W. Bolen, this series, Vol. 210, p. 463.
^{39} E. DiCera, this series, Vol. 210, p. 68.

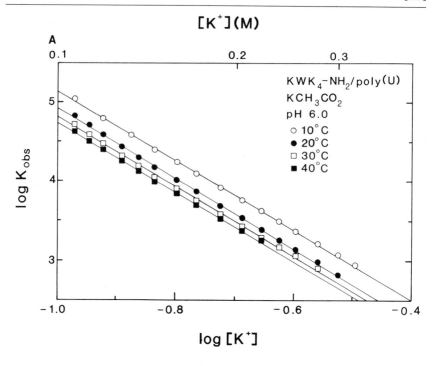

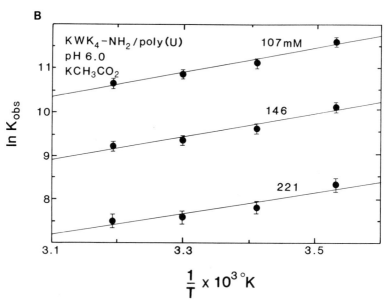

more straightforward error analysis.[32,39] Although the Scatchard[36] representation of binding isotherms as well as other replotting methods provide certain advantages for visualizing the binding isotherm, these are not generally recommended for use in the determination of binding parameters by nonlinear regression methods, since the error analysis is more difficult.[40]

A final point is that such analyses yield model-dependent binding parameters. Therefore, the choice of a particular model should be based on all available data. If a particular model is used for the purposes of analysis, without independent support, then the binding parameters may not correspond to true interaction parameters and hence may not be directly related to thermodynamic quantities.

2. Salt Back-Titrations. If and only if the following conditions are fulfilled can salt back-titrations be used to obtain accurate estimates of K_{obs} as a function of salt concentration: (i) $Q_{obs}/Q_{max} = L_B/L_T$; (ii) the fluorescence of the free and bound ligand are independent of salt concentration; and (iii) the cooperativity parameter, ω, is known and is independent of salt concentration. In addition to comparing the values of K_{obs} determined by the salt back-titration method versus a full titration at constant salt concentration, another useful check is to compare the values of K_{obs} determined from two salt back-titrations that differ only in the salt concentration that is used to form the complex initially. The power of the salt back-titration is that a rapid and very accurate determination of the salt dependence can be obtained for a wide variety of conditions. An example of data obtained from a salt back-titration of a preformed complex between the oligopeptide KWK_2-NH_2 and poly(U) is shown in Fig. 9.

If the above conditions hold, then the free ligand concentration, L_F, and the average binding density, ν, can be calculated at each salt concentration using Eqs. (6) and (7), respectively. To calculate the equilibrium binding constant, K_{obs}, at each salt concentration, it is necessary to know the relationship between ν, L_F, and K_{obs}, which requires a model. We have used the noncooperative overlap binding model of McGhee and von Hippel[24] (see [24] in this volume[5]) to analyze the salt back-titrations of the

[40] C. J. Thompson and I. R. Klotz, *Arch. Biochem. Biophys.* **147**, 178 (1971).

FIG. 10. Dependence of log K_{obs} on log[K^+] at a series of different temperatures as determined from salt back-titrations for the equilibrium binding of KWK_4-NH_2 to poly(U) (25.0°, pH 6.0); KCH_3CO_2 was used to vary the [K^+]. Linear least-squares lines are drawn through each data set. (B) van't Hoff plots constructed at three KCH_3CO_2 concentrations (indicated for each line) from the salt dependences shown in (A). The slopes for each line are identical, indicating that ΔH°_{obs} is independent of [KCH_3CO_2] at pH 6.0 for this interaction.

equilibrium binding of oligolysines to poly(U). The dependence of K_{obs} on [NaCl] for the interaction of the oligopeptide $KWK_2\text{-}NH_2$ with poly(U) as calculated from a salt back-titration is shown in the inset to Fig. 9. Values of K_{obs} determined from individual reverse titrations at constant salt concentration compare quite well to those determined from salt back-titrations for the interaction of poly(U) with both the *E. coli* SSB protein[12] and a series of oligopeptides[14] (see [24] in this volume[5]).

When applicable, salt back-titrations can be used as a rapid means to obtain other thermodynamic parameters. For example, a series of salt back-titrations performed at different temperatures (see Fig. 10A) can be used to obtain not only the salt dependence at each temperature, but also the van't Hoff enthalpy change, ΔH°_{obs}, as a function of salt concentration, as shown in Fig. 10B. Using this approach, a significant amount of data can be obtained in the same day, thus improving the precision of the data so that even small trends in the temperature or salt dependences can be detected.

Acknowledgments

We thank the current and past members of the laboratory, especially Dr. W. Bujalowski, for their contributions to our thinking about ligand-binding equilibria. We also thank Lisa Lohman for help in preparing the figures. Research from the authors' laboratory was supported by grants from the National Institutes of Health (GM39062, GM30498) and the American Cancer Society (NP-756). T.M.L. is a recipient of American Cancer Society Faculty Research Award FRA-303.

Author Index

Numbers in parentheses are footnote reference numbers and indicate that an author's work is referred to although the name is not cited in the text.

Subject Index